Handbook of Radio Engineering and Wireless Technology

Handbook of Radio Engineering and Wireless Technology

Edited by Kevin Merriman

Clanrye International
www.clanryeinternational.com

Clanrye International,
750 Third Avenue, 9th Floor,
New York, NY 10017, USA

ISBN: 978-1-63240-802-0

Cataloging-in-Publication Data

Handbook of radio engineering and wireless technology / edited by Kevin Merriman.
p. cm.
Includes bibliographical references and index.
ISBN 978-1-63240-802-0
1. Radio. 2. Wireless communication systems. 3. Telecommunication. I. Merriman, Kevin.
TK6550 .H36 2019
621.38--dc23

For information on all Clanrye International publications visit our website at www.clanryeinternational.com

CLANRYE INTERNATIONAL

Printed in China.

Contents

Preface

Radio engineering refers to the study and development of devices which have been designed to operate in the radio frequency spectrum between 3 kHz to 300 GHz. This includes all devices that transmit or receive a radio wave, like two-way radios, mobile phones and Wi-Fi. Wireless technology is associated with the transmission of information between two remote points without the presence of any electrical conductor. Radio waves are the primarily used wireless technology in GPS units, satellite television, wireless headphones, etc. The application of radio engineering and wireless technology are present in diverse industries. This book aims to provide detailed information about the modern concepts and theories related to this field. It elucidates the innovative models around prospective developments with respect to radio engineering and wireless technology. The various studies that are constantly contributing towards advancing technologies and evolution of these fields have also been included herein. This book is appropriate for students seeking detailed information in this area as well as for experts.

Significant researches are present in this book. Intensive efforts have been employed by authors to make this book an outstanding discourse. This book contains the enlightening chapters which have been written on the basis of significant researches done by the experts.

Finally, I would also like to thank all the members involved in this book for being a team and meeting all the deadlines for the submission of their respective works. I would also like to thank my friends and family for being supportive in my efforts.

Editor

Simple μ-Negative Half Mode CRLH Antenna Configuration for MIMO Applications

Mahmoud A. ABDALLA[1], *Ahmed A. IBRAHIM*[2,3]

[1] Dept. of Electronic Engineering, MTC College Cairo, Egypt,
[2] Faculty of Engineering, Minia University, Egypt
[3]Inst. of Micro- and Sensor Systems, University of Magdeburg, Germany

maaabdalla@ieee.org, ahmedabdel_monem@mu.edu.eg

Abstract. *A design of a compact size mu-negative half mode composite right left handed metamaterial antenna, using only series capacitive loading, for MIMO application is presented. The proposed configuration is simple as it is realized using via free configuration. The MIMO antenna is formed using two antenna elements designed to operate at 5.8 GHz for wireless applications. The overall MIMO antenna size is only 2.6 × 2.6 cm^2 with in-between separation 1.8 mm (0.034 λ_0). Moreover, the ports mutual coupling reduction between the two antenna elements, achieved without using extra structure, is lower than –20 dB. Compared to conventional two microstrip patch MIMO antennas, our proposed configuration has more than 50% size reduction and 9 dB enhancement in the mutual coupling for the same separation. The antenna design principles, full wave simulations, experimental measurements are introduced with good agreement. Finally, MIMO parameters are extracted and discussed.*

Keywords

Metamaterial, MIMO antenna, composite right left handed transmission line, isolation enhancement

1. Introduction

Left handed metamaterials (LHMs) have gained great interest in past few years in RF/microwave circuit applications. They have been realized in planar transmission line approach, which is basically based on loading a transmission line (TL) with series capacitors and shunt inductors. This transmission line is used to be called a composite right left handed (CRLH) TL [1] or negative refractive index (NRI) TL [2]. Mainly these TLs are characterized by backward waves whose propagation consonant is non-linear, frequency dependent and engineered characteristic impedance. Based on these properties, many compact and multi-band novel antennas have been proposed in the literature by several groups [3–13]. The need for different properties has encouraged the researchers to introduce different configurations of CRLH cell by increasing its order such as extended CRLH TL [14], generalized NRI TL [15], [16]. However, these approaches increase the opportunity for more multi band at the expense of increasing the design complexity. On the other hand, it has been pointed out that Epsilon-negative and Mu-negative structures have imaginary propagation constants such that they have a band stop resonance properties [17], [18]. Consequently, they have been applied in many antennas realizations employing Epsilon-negative [19], [20] and Mu-negative [21], [22]. These structures are simpler than conventional CRLH cells and easier in design.

In the MIMO system, two or more antennas are used on both the transmitter and receiver sides to enhance the channel capacity [23], [24]. However, closely-spaced antenna elements suffer from severe degradation in their diversity performance; due to mutual coupling among the antenna elements. Therefore, reducing the mutual coupling between antenna elements is essential for better performance of MIMO systems. Several works have been carried out to reduce the mutual coupling effect between antenna elements in MIMO systems such as using electromagnetic band gap structures [25]. Also, using defected ground structures (DGSs) etched in different shapes in the ground plane has been employed [26], [27]. However, these structures suffer from complexity and large size. Therefore, different attempts to enhance the isolation have been suggested [28], [29]. Recently, it has been suggested that the different antenna orientations for current reversal can reduce the mutual coupling problem [30–32].

In this paper, we introduce a new two elements metamaterial MIMO antenna. The antenna element is constructed as a Mu-negative resonant structure. The realization has been suggested as half mode realization of CRLH cell by employing series capacitive load only. The proposed MIMO antenna has a compact size and very low coupling isolation without using extra structure. The performance of the designed MIMO antenna has been validated using the electromagnetic full wave simulations. Finally, the simulated results are confirmed by experimental measurements.

2. Half Mode CRLH MIMO Antennas

2.1 Half Mode CRLH Antenna Concept

The CRLH transmission line resonating antenna is based on open circuit termination of one or more CRLH cells as shown in Fig. 1(a). In the figure, L_L and C_L are corresponding to the loading shunt inductor and series capacitor whereas L_R and C_R are corresponding to the patristic element of the loaded transmission line. For N CRLH cells, the design condition for the antenna is to achieve integer multiple of half wave progressive phase. This can be expressed in terms of a constant integer n, the total antenna physical length d, and the guided propagation constant as

$$phase = N\beta d = n\pi . \quad (1)$$

where β is the propagation constant along the CRLH TL, defined as

$$\cos(\beta d) = 1 - \frac{\omega^2}{2}\left(\left(L_R - \frac{1}{\omega^2 C_L}\right)\left(C_R - \frac{1}{\omega^2 L_L}\right)\right). \quad (2)$$

Balanced CRLH TL has a continuous passband with no stopband between the right and left handed pass bands. The transition frequency between the two bands is associated with zero propagation constant β. It has been shown in [33], [34] that the resonance frequency of the zeroth order antenna is either the shunt branch resonance frequency f_{sh} or the series branch resonance frequency f_{se} which are identical. However, it has been explained that for the open circuit resonance mode of the zeroth order mode CRLH antenna, the energy storage is within the shunt branch. The resonant frequencies for the shunt branch f_{0sh} and the series branches f_{0se} can be written as

$$f_{0sh} = \frac{1}{2\pi\sqrt{L_L C_R}}, \quad (3)$$

$$f_{0se} = \frac{1}{2\pi\sqrt{L_R C_L}}. \quad (4)$$

Based on this fact, in case of employing such antenna in MIMO system to reduce the mutual coupling between antenna elements, it is suggested in this paper to design the open circuit antenna element as shown in Fig. 1(b). In other words, the half mode cell is suggested to be realized using the series combination only (C_L and L_R). Thanks to this configuration, the energy storage will be minimized and eliminated and hence the mutual coupling will be reduced. Also, this structure is simpler than full CRLH case since it does not require a shunt inductance.

2.2 Microstrip Half Mode CRLH MIMO Antenna Structure

The 2D layout of the half mode single cell CRLH metamaterial antenna is illustrated in Fig. 2. The antenna is formed from series interdigital capacitors C_L and two parasitic elements which are corresponding to the RH parameters (the series inductor L_R and the shunt capacitor C_R). The proposed configuration is simple since it needs no via connection to the ground. The antenna is printed on FR4 substrate with relative permittivity of 4.4, a dielectric loss tangent of 0.025 and thickness of 1.6 mm.

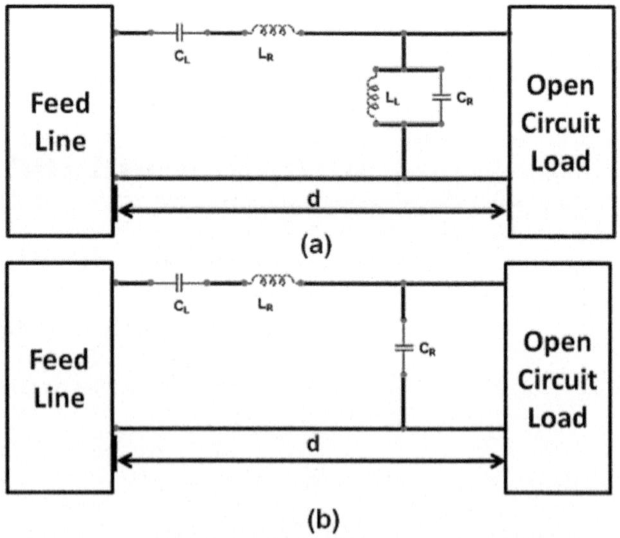

Fig. 1. The equivalent circuit model of the CRLH TL antenna: (a) The full mode case. (b) The practical mu-negative half mode.

The single antenna element is arranged in a two-elements MIMO configuration whose 2D layout is shown in Fig. 3(a). A comparison between the suggested half mode CRLH MIMO antenna and the conventional patch antenna have been established. The 2D layout of two elements patch MIMO antenna is shown in Fig. 3(b). As shown in the figure, the distance between the edges of two antenna elements is only 1.8 mm which equals 0.034 λ_0 at 5.8 GHz in both cases. From Fig. 3, it can be observed that the proposed antenna has compact size (26 × 26 mm^2) but the two-elements patch antenna have size of (54 × 30 mm^2) which is more than double size of the proposed one in Fig. 3(a).

The geometrical design of the antenna layout was done by calculating the loading element C_L in (4) so that the antenna has a resonance at 5.8 GHz. Then C_L was

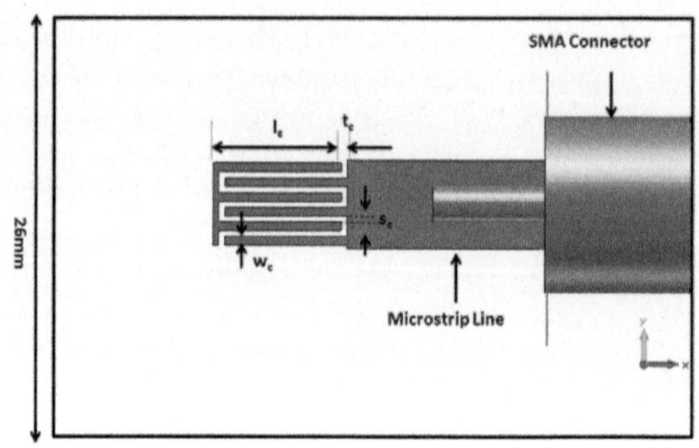

Fig. 2. The 2D layout of single cell half mode CRLH antenna, finger width (W_c = 0.2 mm) and separation (S_c = 0.2 mm), (t_c = 0.2 mm).

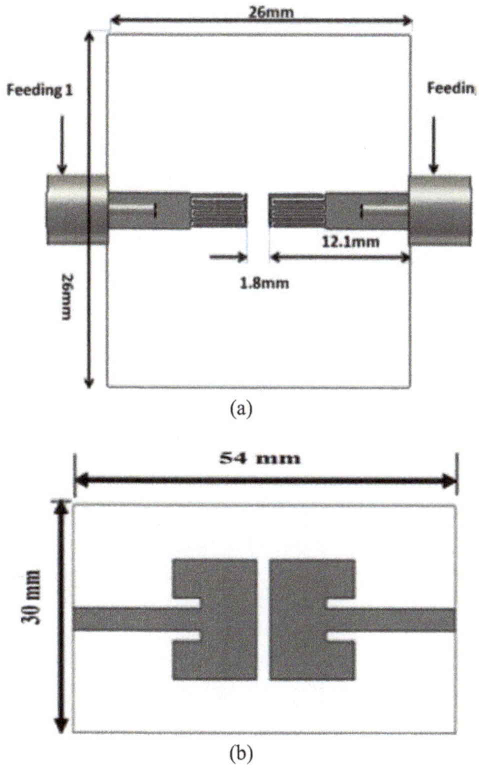

Fig. 3. The 2D layout of (a) half mode CRLH MIMO antenna, $s = 1.8$ mm, (b) the two elements patch MIMO antenna with separation equal to 1.8 mm.

realized as interdigital capacitor. This can be expressed mathematically as [1]

$$C = 3.937\times10^{-5} l_f\left(\varepsilon_r+1\right)\left(0.11\left(n-3\right)+0.252\right)\left[\text{pF}\right] \quad (5)$$

where l_f is the capacitor finger length, n is the capacitor number of fingers. The capacitor width and spacing were selected to match the 50 Ω feeding microstrip transmission line width. The right handed inductor was specified by the patristic inductance of the whole cell length. The design of the proposed MIMO antenna was validated by studying the scattering parameter magnitudes. As shown in Fig. 4(a), the antenna operates at 5.8 GHz at which the return loss is lower than –10 dB approximately; similar results for patch MIMO antenna are observed. On the other hand, the mutual coupling of the half mode MIMO antenna in terms of the transmission coefficient S_{21} magnitude equals –21 dB as shown in Fig. 4(b) whereas it is –12 dB in patch MIMO antenna. Compared to the proposed half mode CRLH MIMO antenna, we can claim that the proposed configuration has 9 dB enhancements in the isolation between the two output ports.

3. Half Mode CRLH MIMO Antenna Performance

The evaluation performance of the half mode CRLH MIMO antenna is done by investigating the envelope correlation, the diversity gain, and total active reflection coefficient (TARC). When the correlation between two antenna elements is low, the diversity gain is high and vice versa. Therefore, the correlation between two antenna elements has to be as low as possible to achieve high performance in MIMO system. The envelope correlation of stub free MIMO antenna is shown in Fig. 5. It is noticed that the envelope correlation coefficient equals 0.02 (–34 dB) at 5.8 GHz. The relationship between the envelope correlation and the diversity gain has been calculated as [35]

$$\rho_e = \frac{\left|S_{11}^{*}S_{12}+S_{21}^{*}S_{22}\right|^2}{\left(1-\left(\left|S_{11}\right|^2+\left|S_{21}\right|^2\right)\right)\left(1-\left(\left|S_{22}\right|^2+\left|S_{12}\right|^2\right)\right)}, \quad (6)$$

$$Gapp = 10\sqrt{1-\left|\rho\right|} \text{ where } \rho_e = \left|\rho\right|. \quad (7)$$

The diversity gain of the stub free MIMO antenna is illustrated in Fig. 6. It is clear that the diversity gain of the antenna equals 9.89.

The last parameter used to measure the performance of the MIMO antenna is TARC. TARC is defined as the ratio of the square root of the total reflected power divided by the square root of the total incident power. The TARC meaning can be defined as the return loss of the whole MIMO antenna elements. The TARC has been calculated from the scattering parameters of a two antenna elements as [35]

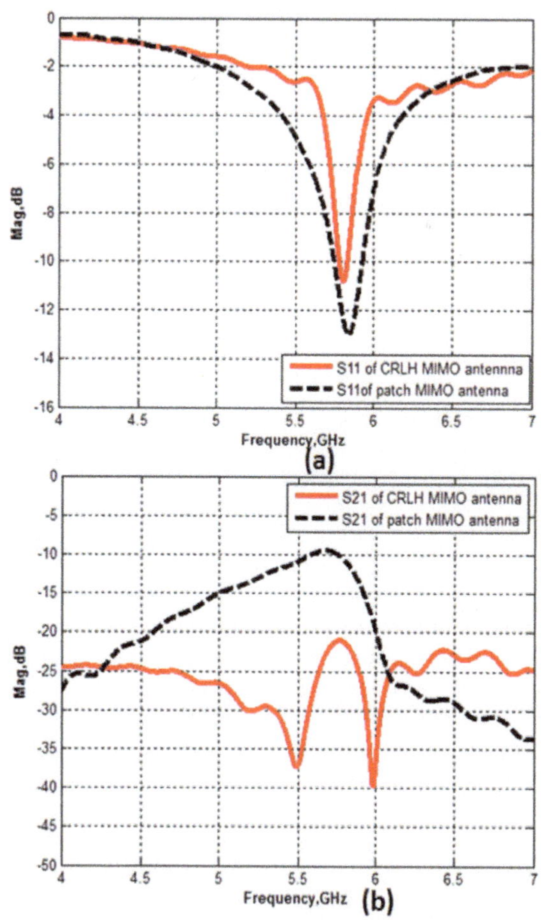

Fig. 4. The scattering parameter magnitudes of simulated results of half mode CRLH MIMO antenna and patch MIMO antenna: (a) S_{11}, (b) S_{21}.

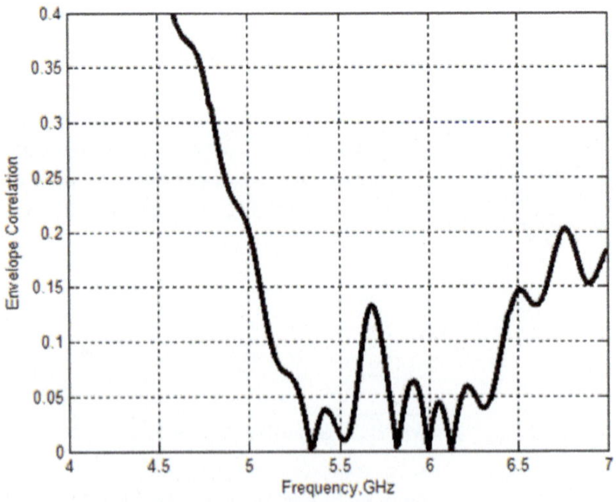

Fig. 5. The envelope correlation coefficient of the half mode CRLH MIMO.

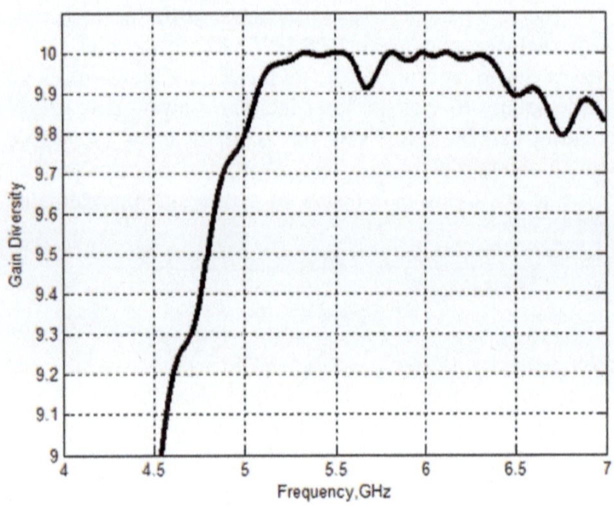

Fig. 6. The diversity gain of the half mode CRLH MIMO antenna.

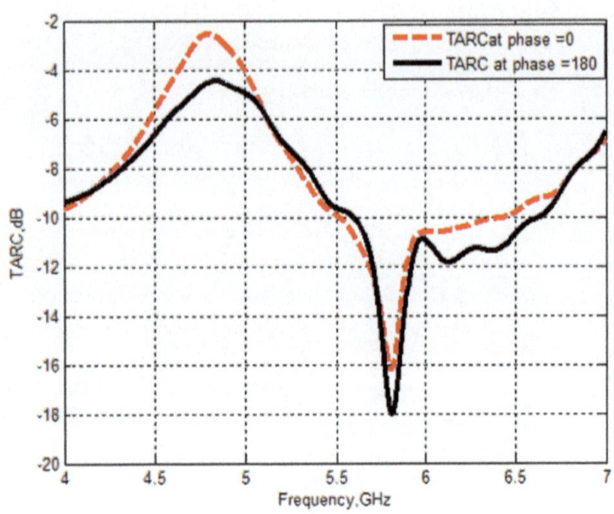

Fig. 7. The TARC of the half mode CRLH MIMO antenna.

$$TARC = \sqrt{\frac{\left|S_{11} + S_{12}\exp(j\theta)\right|^2 + \left|S_{21} + S_{22}\exp(j\theta)\right|^2}{2}} \quad (8)$$

where θ is considered variable angle which is related to the phase of excitation signal. A random phased excitation vectors have been used to calculate average *TARC* and the MIMO channel is assumed to be Gaussian and multipath spread in the propagation channel. *TARC* is calculated at each random phase between 0° and 180°. However, only two phases 0° and 180° are demonstrated as shown in Fig. 7. It is obvious that from Fig. 7 the two MIMO antennas demonstrate the same TARC with two used phases 0° and 180° which equals –18 dB.

4. Experimental Results

The photograph of the fabricated MIMO antenna printed on FR4 substrate is shown in Fig. 8. The simulated and measured scattering parameter magnitudes are shown in Fig. 9. It is clear from the simulated results that the return loss is more than –10 dB and the mutual almost –21 dB. The measured results illustrate that there are small frequency shifts between the measured and simulated results. The return loss is 13 dB at 5.9 GHz and the mutual coupling equals –15 dB. The small shift in frequency beween the simulated and measurement results is due to the

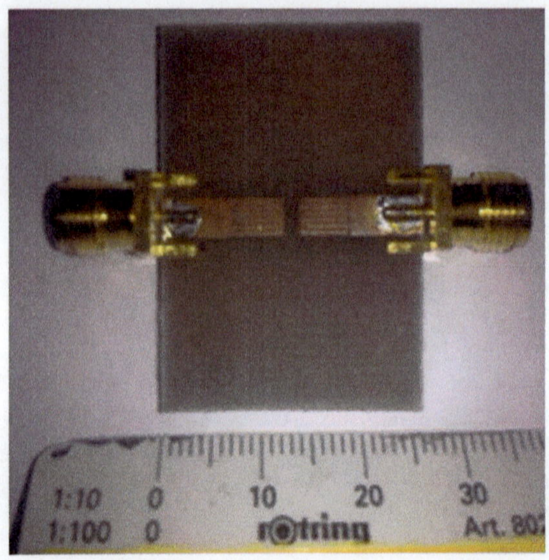

Fig. 8. Photograph of the fabricated half mode CRLH MIMO antenna.

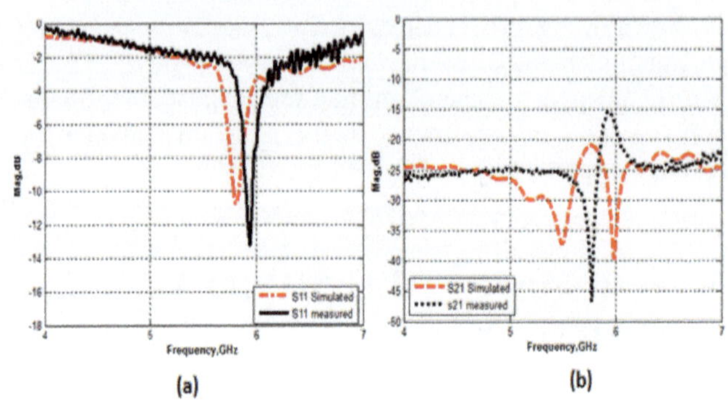

Fig. 9. The simulated and measured scattering parameter magnitudes of the half mode CRLH MIMO antenna: (a) S_{11} results, (b) S_{21} results.

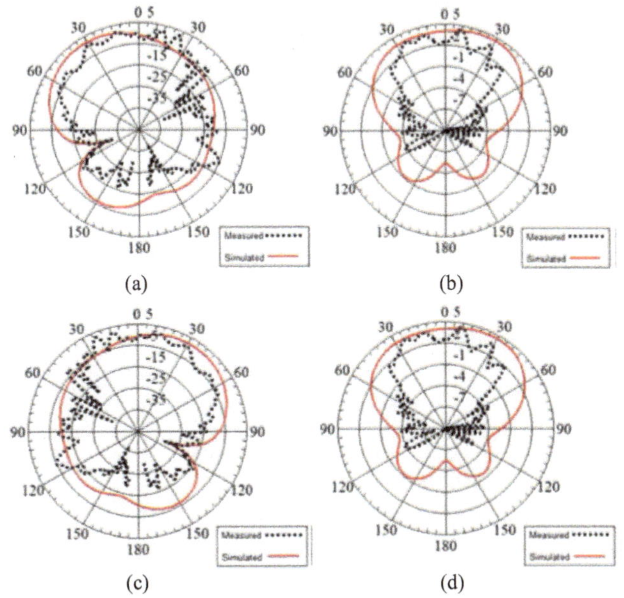

Fig. 10. The simulated and measured directive gain of half mode CRLH MIMO antenna at 5.8 GHz: (a) E plane at port 1, (b) H plane at port 1, (c) E plane at port 2, (d) H plane at port 2.

imperfection of full wave simulation conditions and the fabrication accuracy which cannot be totally avoided. However, it can be still considered a good agreement between the simulated and measured results is achieved.

The measurements of antenna radiation pattern have been done inside an anechoic chamber. The simulated and measured directive gain pattern in both E (*XZ*) and H (*YZ*) planes of the half mode MIMO antenna are shown in Fig. 10. This was obtained assuming the excitation at port 1 and port 2, respectively, while other port is matched. It is obvious that a good agreement is achieved between the measured and simulated results. Finally, we can conclude that the introduced antenna patterns confirm that the proposed antenna can be used for various wireless communication applications. The simulated efficiency of the proposed antenna equals 45%. This relatively small efficiency is due to the compact microstrip size and losses introduced by the used FR4 substrate.

5. Conclusion

A design of a mu-negative half mode CRLH MIMO based antenna has been introduced. The MIMO antenna consists of two antenna elements which operate at 5.8 GHz for wireless applications. The distance between the edges of two antenna elements has been achieved to be only 0.034 λ_0. Also, less than –20 dB mutual coupling between two antenna elements has been achieved with current reversal and without using any extra complicated structure. The proposed MIMO antenna has the advantage of compactness (its size is only 2.6 × 2.6 cm^2). A good agreement has been achieved between simulated and measured results. The retrieved MIMO performance parameters proved low correlation coefficient, high diversity gain and small TARC.

References

[1] CALOZ, C., ITOH, T. *Electromagnetic Metamaterials Transmission Line Theory and Microwave Applications.* New Jersey: John Wiley, 2006. DOI: 10.1002/0471754323

[2] ELEFTHERIADES, G. V., BALMAIN, K. G. *Negative Refraction Metamaterials.* New Jersey: John Wiley, 2005. ISBN: 978-0-471-60146-3

[3] CHENG-JUNG LEE, WEI HUANG, GUMMALLA, A., et al. Small antennas based on CRLH structures: Concept, design, and applications. *IEEE Antennas and Propagation Magazine*, 2011, vol. 53, no. 2, p. 10–25. DOI: 10.1109/MAP.2011.5949321

[4] ZIOLKOWSKI, R. W., JIN, P., LIN, C. Metamaterial-inspired engineering of antennas. *Proceedings of the IEEE*, 2011, vol. 99, no. 10, p. 1720–1731. DOI: 10.1109/JPROC.2010.2091610

[5] DONG, Y., ITOH, T. Metamaterial-based antennas. *Proceedings of the IEEE*, 2012, vol. 100, no. 7, p. 2271–2285. DOI: 10.1109/JPROC.2012.2187631

[6] ZHU, J., ANTONIADES, M. A., ELEFTHERIADES, G. V. A compact tri-band monopole antenna with single-cell metamaterial loading. *IEEE Transactions on Antennas and Propagation,* 2010, vol. 58, no. 4, p. 1031–1038. DOI: 10.1109/TAP.2010.2041317

[7] BARBUTO, M., BILOTTI, F., TOSCANO, A. Design of a multifunctional SRR-loaded printed monopole antenna. *International Journal of RF and Microwave Computer-Aided Engineering,* 2012, vol. 22, no. 4, p. 552–557. DOI: 10.1002/mmce.20645

[8] JIN, P., LIN, C. C., ZIOLKOWSKI, R. W. Multifunctional, electrically small, planar near-field resonant parasitic antennas. *IEEE Antennas and Wireless Propagation Letters*, 2012, vol. 11, p. 200–204. DOI: 10.1109/LAWP.2012.2187322

[9] BARBUTO, M., TROTTA, F., BILOTTI, F., TOSCANO, A. Design of a low-profile antenna by using orthogonal parasitic meandered monopoles. *Progress In Electromagnetics Research Letters*, 2015, vol. 55, p. 23–29. DOI: 10.2528/PIERL15061903

[10] LI XUE, JINWEN TIAN. Low-profile fully-printed multifrequency monopoles loaded with complementary metamaterial transmission line. *Radioengineering*, 2015, vol. 24, no. 1, p. 64–69. DOI: 10.13164/re.2015.0064

[11] VRBA, D., POLIVKA, M. Radiation efficiency improvement of zeroth-order resonator antenna. *Radioengineering*, 2009, vol. 18, no. 1, p. 1–8.

[12] UGARTE-MUNOZ, E., HERRAIZ-MARTÍNEZ, F. J., GONZALEZ-POSADAS, V., et al. Patch antenna based on metamaterials for a RFID transponder. *Radioengineering,* 2008, vol. 17, no. 2, p. 66–71.

[13] WAHBA, W., ABDALLA, M. A., ALLAM, A. M. Experimental verification of a compact zeroth order metamaterial substrate integrated waveguide antenna. *Progress in Electromagnetic Research C*, 2016, vol. 67, p. 193–201. DOI: 10.2528/PIERC16052906

[14] RENNINGS, A., OTTO, S., MOSIG, J., et al. Extended composite right/left-handed (E-CRLH) metamaterial and its application as quadband quarter-wavelength transmission line. In *Asia-Pacific Microwave Conference (APMC).* Yokohama (Japan), 2006, p. 1405–1408. DOI: 10.1109/APMC.2006.4429669

[15] RYAN, C. G. M., ELEFTHERIADES, G. V. Design of a printed dual-band coupled-line coupler with generalised negative-refractive index transmission lines. *IET Microwaves, Antennas and Propagation*, 2012, vol. 6, p. 705–712. DOI: 10.1049/iet-map.2011.0508

[16] FOUAD HAGAG, M. A., ABDALLA, M. A. Ultra compact CPW dual band filter based on Π-generalized metamaterial NRI transmission line. *Journal of Electromagnetic Waves and Applications*, 2015, vol. 29, no. 8, p. 1093–1103. DOI: 10.1080/09205071.2015.1044123

[17] ALÙ, A., ENGHETA, N. Pairing an epsilon-negative slab with a mu-negative slab: resonance, tunneling and transparency. *IEEE Transactions on Antennas and Propagation*, 2003, vol. 51, no. 10, p. 2558–2571. DOI: 10.1109/TAP.2003.817553

[18] ABDALLA, M. A., HU, Z. Nonreciprocal left handed coplanar waveguide over ferrite substrate with only shunt inductive load. *Microwave and Optical Technology Letters,* 2007, vol. 49, p. 2810–2814. DOI: 10.1002/mop.22848

[19] PARK, J.-H., RYU, Y.-H., LEE, J.-G., LEE, J.-H. Epsilon negative zeroth-order resonator antenna. *IEEE Transactions on Antennas and Propagation,* 2007, vol. 55, no. 12, p. 3710–3712. DOI: 10.1109/TAP.2007.910505

[20] ABDALLA, M. A., FOUAD, M., A., AHMED, A., ZHIRUN HU. A new compact microstrip triple band antenna using half mode CRLH transmission line. In *2013 IEEE International Antennas and Propagation Society International Symposium Digest (APSURSI).* Orlando (USA), Jul. 2013, p. 634–635. DOI: 10.1109/APS.2013.6710977

[21] KUNPENG WEI, ZHANG, Z., FENG, Z., et al. A wideband MNG-TL dipole antenna with stable radiation patterns. *IEEE Transactions on Antennas and Propagation,* 2013, vol. 61, no. 5, p. 2418–2424. DOI: 10.1109/TAP.2013.2241717

[22] SIMORANGKIR, R., LEE, Y. A planar dual-band periodic leaky-wave antenna based on a mu-negative (MNG) transmission line. *IEEE Transactions on Antennas and Propagation*, 2015, vol. 63, no. 5, p. 2370–2374. DOI: 10.1109/TAP.2015.2410802

[23] PAULRAJ, A. J., GORE, D. A., NABAR, R. U., BOLCSKEI, H. An overview of MIMO communications - A key to gigabit wireless. *Proceedings of the IEEE*, 2004, vol. 92, no. 2, p. 198 to 217. DOI: 10.1109/JPROC.2003.821915

[24] DIOUM, I., CLEMENTE, M., DIALLO, A., LUXEY, C., et al. Meandered monopoles for 700MHz LTE handsets and improved MIMO channel capacity performance. *Radioengineering*, 2011, vol. 20, no. 4, 2011, p. 726–732.

[25] KIM, S.-H., LEE, J.-Y., NGUYEN, T. T., JANG, J.-H. High-performance MIMO antenna with 1-D EBG ground structures for handset application. *IEEE Antennas and Wireless Propagation Letters*, 2013, vol. 12, p. 1468–1471. DOI: 10.1109/LAWP.2013.2288797

[26] IBRAHIM, A. A., ABDALLA, M. A., ABDEL-RAHMAN, A. B., et al. Compact MIMO antenna with optimized mutual coupling reduction using DGS. *International Journal of Microwave and Wireless Technologies*, 2014, vol. 6, no. 2, p 173–180. DOI: 10.1017/S1759078713001013

[27] ADDACI, R., DIALLO, A., LUXEY, C., et al. Design of multi-antenna system for UMTS clamshell mobile phones with ground plane effects considerations. *Radioengineering*, 2014, vol. 23, no. 2, p. 724–732.

[28] SHARAWI, M. S., NUMAN, A. B., KHAN, M. U., et al. A dual-element dual-band MIMO antenna system with enhanced isolation for mobile terminals. *IEEE Antennas and Wireless Propagation Letters*, 2012, vol. 11, p. 1006–1009. DOI: 10.1109/LAWP.2012.2214433

[29] ABDALLA, M. A., ABDELRAHEEM, A. Compact transmit receive hybrid electromagnetic isolation in antenna array transceiver system for full duplex applications. *IET Microwaves, Antennas and Propagation*, 2016. DOI: 10.1049/iet-map.2016.0215

[30] ABDALLA, M. A., IBRAHIM, A. A. Design of close, compact, and high isolation meta-material MIMO antennas. In *2013 IEEE International Antennas and Propagation Society International Symposium Digest (APSURSI).* Orlando (USA), Jul. 2013, p. 186 to 187. DOI: 10.1109/APS.2013.6710754

[31] IBRAHIM, A. A., ABDALLA, M. A. CRLH MIMO antenna with reversal configuration. *AEU-International Journal of Electronics and Communications*, 2016, vol. 70, p. 1134–1141. DOI: 10.1016/j.aeue.2016.05.012

[32] ABDALLA, M. A., IBRAHIM, A. A. Design and performance evaluation of metamaterial inspired MIMO antennas for wireless applications. *Wireless Personal Communications*, 2016, p. 1–17. DOI: 10.1007/s11277-016-3809-4.

[33] RENNINGS, A., LIEBIG, T., CALOZ, C., WALDOW, P. MIM CRLH series mode zeroth order resonant antenna (ZORA) implemented in LTCC technology. In *Asia-Pacific Microwave Conference (APMC 2007).* 2007, p. 1–4. DOI: 10.1109/APMC.2007.4554577

[34] ABDALLA, M. A., ZHIRUN HU, MUVIANTO, C. Analysis and design of triple band metamaterial simplified CRLH cells loaded monopole antenna. *International Journal of Microwave and Wireless Technologies*, June 2016, p. 1–11. DOI: 10.1017/S1759078716000738

[35] TOKTAS, A., AKDAGLI, A. Compact multiple-input multiple-output antenna with low correlation for ultra-wide-band applications. *IET Microwaves, Antennas and Propagation*, 2015, vol. 9, no. 8, p. 822–829. DOI: 10.1049/iet-map.2014.0086

About the Authors ...

Mahmoud ABDALLA was born in 1973. He received the M.Sc. degree in Electrical Engineering from the Military Technical College in 2000, and the Ph.D. degree from the Microwave and Communication Group, School of Electrical Engineering, Manchester University, UK, in 2009. His research includes microwave/ millimeter components, miniaturized multiband antennas/wideband antennas, electromagnetic energy harvesting systems, smart antenna systems for DOA estimation and interference cancellations, metamaterial/EBG structures. Also, his work is oriented into MIMO antenna with better performance. Under these directions, Dr. Abdalla has published more than 100 journal / conference papers.

Ahmed A. IBRAHIM was born in 1986. He received the B.Sc. degree, with grade of very good, in Electrical Engineering from the Electronic and Communication Engineering Department, Elminia University, Elminia, Egypt in 2007. He was awarded the M.Sc. degree in Electronic and Communication Engineering from Elminia University in 2011, and the Ph.D. degree in Electronic and Communication Engineering from Elminia University in 2014. He is now post-doctoral fellow in the Otto-von-Guericke-Universität Magdeburg,Germany. Also he is a lecturer in Electronic and Communication Engineering Department. His research focused on the design and analysis of microstrip antennas, microstrip filters, and their application in wireless communications, also on metamaterial MIMO antennas and different metamaterial applications in microwave bands.

Directly Printable Frequency Signatured Chipless RFID Tag for IoT Applications

Hafsa ANAM[1], Ayesha HABIB[1], Syeda Irum JAFRI[1], Yasar AMIN[1, 2], Hannu TENHUNEN[2, 3]

[1] ACTSENA Research Group, University of Engineering and Technology, 47050- Taxila, Pakistan
[2] iPack VINN Excellence Center, Royal Institute of Technology, SE-16440, Stockholm, Sweden
[3]TUCS, University of Turku, Turku-20520, Finland

ayesha.habib@uettaxila.edu.pk

Abstract. *This paper proposes a low-cost, compact, flexible passive chipless RFID tag that has been designed and analyzed. The tag is a bowtie-shaped resonator based structure with 36 slots; where each patch is loaded with 18 slots. The tag is set in a way that each slot in a patch corresponds to a metal gap in the other patch. Hence there is no mutual interference, and high data capacity of 36 bits is achieved in such compact size. Each slot corresponds to a resonance frequency in the RCS curve, and each resonance corresponds to a bit. The tag has been realized for Taconic TLX-0, PET, and Kapton®HN (DuPont™) substrates with copper, aluminum, and silver nanoparticle-based ink (Cabot CCI-300) as conducting materials. The tag exhibits flexibility and well optimized while remaining in a compact size. The proposed tag yields 36 bits in a tag dimension of 24.5 × 25.5 mm². These 36 bits can tag 2^{36} number of objects/items. The ultimate high capacity, compact size, flexible passive chipless RFID tag can be arrayed in various industrial and IoT-based applications.*

Keywords

Chipless, Radio Frequency Identification (RFID), Radar Cross-Section (RCS), backscattering

1. Introduction

Internet of things (IoT) is a combination of number of smart objects, connected via wired/wireless networks to the internet [1–5]. RFID and wireless sensor networks (WSN) are major entities of IoT system. Latest developments in RFID have enabled IoT [6]. RFID is an emerging contactless data capturing technology which is widely used for tracking purposes, theft control, health monitoring, food monitoring, luggage tracking, clothing, electronic cards and pollution control, etc., [7]. An estimated 75 billion products equipped with RFID tags will be sold yearly till the year 2019 [8]. RFID has to upswing for the latest requirement of IoT development and emergence to meet the demands of modern era [9]. Limitations of RFID technology are cost, reliability and recycling aspects [10]. The main hindrance in RF identification deployment depends on its cost per tag [11]. The emerging aspect of RFID technology and such limitations have motivated the researchers to move towards chipless tags that outperform compared to conventional chip-based tags hence tremendously reducing the cost compared to chip-based tags [10], [12]. Chipless RFID involves information coding in the form of electromagnetic signature (EMS) [13]. Chipless RFID does not need any communication protocol for identification process [13]. The most promising benefit of chipless RFID tag is that they can be printed directly on the products [8]. The reliability and versatility of chipless RFID tags can be depicted from the fact that they can replace ten trillion barcodes yearly [8].

Chipless RFID has been an interesting field for researchers because of some challenging features like enhancing the coding capacity, miniaturization, within a suitable frequency band and an enhanced read range [12]. A number of papers have appeared addressing various such aspects [12], [14–16]. One of the major aspects is to enhance data density while maintaining a suitable tag size in a reasonable frequency band [12]. Various researchers have addressed such aspect [12], [17–28]. In [29], a 3.8 bits/cm² compact polarization independent, discrete slot ring resonator based chipless RFID tag has been proposed. It gives back-scattered frequency signature in a compact size with enhanced coding capacity. Another spurline resonator based chipless RFID tag in a size of 40 × 27 mm² yielding 8-bit data capacity has been proposed [30]. In [27], a low-profile data encoding chipless RFID tag is designed. In this design, the data is encoded as complex natural resonances (CNRs) on the structure within an area of 24 × 24 mm² yielding 24-bit data. Similarly, a compact, flexible 24-bit dual polarized chipless RFID tag in a size of 20.6 × 19.9 mm² is designed [7]. It discusses flexibility in very compact size.

In this paper, a novel 36-bit chipless RFID tag is presented. The novelty of the tag relies on its flexibility, compact size and high data capacity of 36 bits, which has not been done so far in such a compact size. Also, the tag is

Parameters	Tag-1	Tag-2	Tag-3	Tag-4	Tag-5
Substrate	Taconic TLX-0	PET	PET	Kapton®HN	Kapton®HN
Permittivity	2.45	2.9	2.9	3.5	3.5
Radiator	Copper	Aluminum	Silver nano ink	Aluminum	Silver nano ink
Thickness [mm]	0.035	0.007	0.015	0.007	0.015
Trans. bits	36	36	36	36	36
No. of tagged objects	68,719,476,736	68,719,476,736	68,719,476,736	68,719,476,736	68,719,476,736
Freq. band [GHz]	5–15.5	5.3–18.2	4–18.2	5–17	4–17
Size [mm^2]	27 × 26	24.5 × 25.5	24.5 × 25.5	24.5 × 25.5	24.5 × 25.5
Flexibility	x	✓	✓	✓	✓

Tab. 1. Characteristic comparison table.

equipped with flexibility which is the recent requirement of RFID industry applications especially green electronics. PET and Kapton®HN are considered in the tag as flexible substrates, whereas, silver nanoparticle-based ink is used for cheap printability of the design. The conductivity of nanoparticle-based ink is 9×10^6 S/m.

The tag is a resonator based structure that is excited by a linear polarized incident plane wave in a tag dimension of 24.5 × 25.5 mm^2. Firstly, copper is used as a radiator for the Taconic TLX-0 substrate to achieve desired RCS response. Then, aluminum and silver nanoparticle-based ink are deployed as a radiator for PET and Kapton®HN substrates to achieve printability along with flexibility in a reduced tag design. The entire tag yields a data capacity of 36-bit, hence 2^{36} number of objects can be tagged. The frequency ranges for Taconic TLX-0 along with copper as the radiator is 5–15.5 GHz, for PET along with aluminum is 5.3–18.2 GHz and 4–18.2 GHz with silver nanoparticle-based ink. The frequency range for Kapton®HN along with aluminum as the radiator is 5–17 GHz and 4–17 GHz for silver nanoparticle-based ink as a radiator.

2. Theory and Fundamental Principle

RFID involves electromagnetic waves to identify a tagged object [13] remotely. The tag is a resonator structure, where each slot corresponds to one dip and each dip corresponds to one bit. Hence 36 slots yield 36 bits data density. 36-bit tag is designed using CST STUDIO SUITE®. The tag is excited by using linearly incident plane wave. The E-field plane wave equation is given as

$$\mathbf{E}(x,y,z,t) = E_0 \exp\left[\mathrm{j}(wt-kz)\right]\hat{\mathbf{x}} + E_0 \exp\left[\mathrm{j}\left(wt-kz+\frac{\pi}{2}\right)\right]\hat{\mathbf{y}} \quad (1)$$

where **E** is an electric field, w is the angular frequency, t is time, k is wave vector, (x,y,z) is the position vector.

Chipless RFID tags are classified as retransmission based tags and backscattering based tags [31]. The main working principle of chipless RFID tags is backscattering. Identification is based on unique frequency signature generation in a desired frequency range that is measured as radar cross-section (RCS) of the tag [12], [32]. The radar cross-section (RCS) versus frequency shows the electromagnetic behavior of tags. The RCS is analyzed at far-field distance given by Fraunhoffer distance formula

$$R = \frac{2D^2}{\lambda} \quad (2)$$

where D is the radiator's largest dimension and λ is the wavelength of radio wave [33].

To measure RCS response of chipless RFID tag, we need two antennas: one for transmitting and the other for receiving. Reader antenna sends an electromagnetic wave (EW) also known as '*interrogator signal*' towards the tag [34]. The tag then encodes the data information in that signal and sends the '*backscattered signal*' containing encoded information towards the reader [35]. The back-

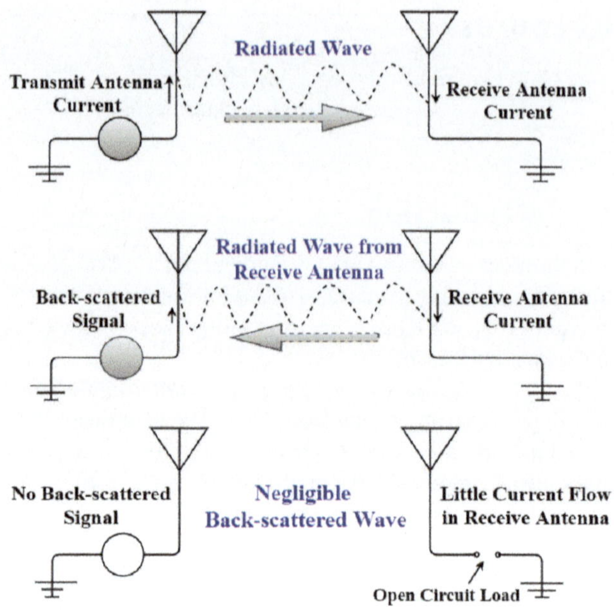

Fig. 1. Simplified physics of backscattering phenomena.

Parameters		**Tag-1**		**Tag-2**		**Tag-3**		**Tag-4**		**Tag-5**	
Substrate		**Taconic TLX-0**		**PET**		**PET**		**Kapton®HN**		**Kapton®HN**	
Radiator		**Copper** [mm]		**Aluminum** [mm]		**Silver nano-particle-based ink** [mm]		**Aluminum** [mm]		**Silver nano-particle-based ink** [mm]	
Slot width	**Metal width**										
*S*1	*M*1	0.3	0.2	0.4	0.5	0.5	0.4	0.6	0.3	0.6	0.3
*S*2	*M*2	0.3	0.2	0.2	0.2	0.2	0.2	0.2	0.2	0.2	0.2
*S*3	*M*3	0.3	0.2	0.2	0.2	0.2	0.2	0.2	0.2	0.2	0.2
*S*4-*S*32	*M*4-*M*32	0.3	0.2	0.3	0.2	0.3	0.2	0.3	0.2	0.3	0.2
*S*33	*M*33	0.3	0.3	0.3	0.6	0.3	0.6	0.3	0.6	0.3	0.6
*S*34	*M*34	0.3	0.3	0.3	0.5	0.3	0.5	0.3	0.5	0.3	0.5
*S*35	*M*35	0.6	1	0.6	0.7	0.6	0.7	0.6	0.7	0.6	0.7
*S*36	*M*36	0.4	0.9	0.4	0.7	0.4	0.7	0.4	0.7	0.4	0.7

Tab. 2. Proposed tags dimensions.

scattered signal contains unique frequency signature for identification. So, there is no need for any integrated circuit to encode the data. Figure 1 shows the backscattering phenomenon.

The power received from a transmitting antenna by a receiving antenna is given by Friis transmission equation [36]

$$\frac{P_{\mathrm{RX}}}{P_{\mathrm{TX}}} = \left(\frac{\lambda}{4\pi r}\right)^2 G_{\mathrm{TX}} G_{\mathrm{RX}} \tag{3}$$

where r is the distance between transmitter and receiver, G_{TX} and G_{RX} are the gain of the transmitter and receiver, P_{TX} is power transmitted and P_{RX} is power received.

All the tagged items/objects should lie in the read range/working space of system for proper RFID operation [37]. The maximum theoretical read range of chipless RFID system can be calculated from (4) [23], [38]

$$R^{\max} = \sqrt[4]{P_{\mathrm{TX}} G_{\mathrm{TX}} G_{\mathrm{RX}} \frac{\lambda^2}{(4\pi)^3 P_{\mathrm{RX}}} \sigma_{\min}} \tag{4}$$

where P_{TX} is transmitting power, G_{TX} is transmitting antenna gain, G_{RX} is receiving antenna gain, λ is the wavelength, P_{RX} is the sensitivity of the receiver and $\sigma_{\min}$ is the most minimum RCS level possible to be detected by the reader.

3. Proposed Tag Design

The proposed tag is loaded with 36 slots, each of varying length in a tag dimension of 24.5 × 25.5 mm^2. Each slot is numbered according to its length. Each slot of different length corresponds to a dip that resonates at a particular frequency. So there are 36 dips corresponding to 36 bits yielding 2^{36} number of possible tag ID combinations. The tag is designed in a way that each slot in upper patch corresponds to metal in the lower patch and vice versa. Therefore, the slots are at alternate positions with metal gaps for adjacent patches. Hence, each slot will be of different length, resonating at a different frequency. Ultimately, there will be no mutual coupling and high dense data in a compact size is achieved while fully utilizing the frequency band. The proposed tag design is shown in Fig. 2.

There are five tags that have been designed using Taconic TLX-0, PET, and Kapton®HN substrates along with copper, aluminum and silver nanoparticle-based ink as radiators. We can analyze that along with changing substrate and radiator; there is variation in tag electrical properties. The tag has been designed and optimized for different substrates, so there is a slight variation in dimensions while optimizing the tag for flexible substrates. The detailed characteristic comparison of all the tags is shown in Tab. 1. It can be observed from Tab. 1 that the tag has been initially designed for the Taconic TLX-0 substrate. Then the tag has further been optimized for PET substrate to achieve flexibility. To meet the modern application requirements there is a trade-off between the bandwidth and

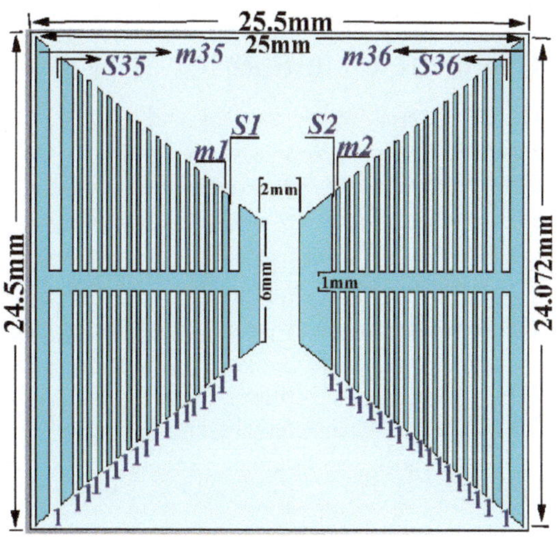

Fig. 2. The layout of designed tag.

tag size/flexibility. Moreover, for efficient band utilization while using flexible substrate, the tag design is optimized for Kapton®HN.

4. Results and Discussion

The proposed tag design encodes 36 bits data. The tag has been designed for different substrates of varying electrical properties using different conducting materials. The detailed analysis of tag dimensions for all the designed tags is shown in Tab. 2.

4.1 Taconic TLX-0 Substrate

The tag designed using Taconic TLX-0 substrate and copper radiator is referred as 'Tag-1'. The RCS vs. frequency response for Tag-1 is shown in Fig. 3. The electrical permittivity of Taconic TLX-0 is 2.45, deployed using copper as a radiator with a thickness of 0.035 mm. 36-bit tag response has been analyzed in the frequency range between 5 GHz and15.5 GHz.

4.2 PET Substrate

Tags referred as 'Tag-2,' and 'Tag-3' are designed using PET substrate along with aluminum and silver nanoparticle-based ink as conducting materials, respectively. The electrical permittivity of PET is 2.9. The tag designed using PET as substrate and aluminum as radiator is referred as 'Tag-2'. The RCS vs. frequency response for Tag-2 is shown in Fig. 4. Aluminum used as radiator has thickness of 0.007 mm. The tag yields 36 bits in the frequency range of 5.3–18.2 GHz.

The tag represented as 'Tag-3' is designed using PET substrate along with silver nanoparticle-based ink as conducting material. The RCS response for Tag-3 is shown in Fig. 5. The thickness of the silver nanoparticle-based ink is 0.015 mm. The RCS curve for Tag-3 lies in the frequency range of 4–18.2 GHz.

4.3 Kapton®HN Substrate

The tags referred as 'Tag-4,' and 'Tag-5' are designed on Kapton®HN substrate using aluminum and silver nanoparticle-based ink as radiators.

The electrical permittivity of Kapton®HN is 3.5. Kapton®HN is deployed for its easy availability and low-cost along with flexibility. By changing the radiator, there is variation in resonances of the tag. 'Tag-4' has been designed using aluminum radiator on Kapton®HN substrate. The RCS vs. frequency response for Tag-4 is shown in Fig. 6. Aluminum in Tag-4 has thickness of 0.007 mm and yields 36-bit data in the frequency range between 5 GHz and 17 GHz.

The tag designed using silver nano ink radiator on Kapton®HN substrate is referred as 'Tag-5'. The RCS response for Tag-5 along with fabricated design is shown in Fig. 7.

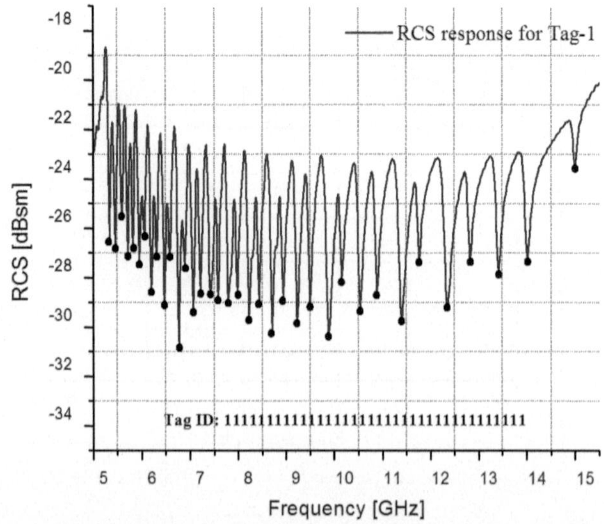

Fig. 3. RCS response for Tag-1.

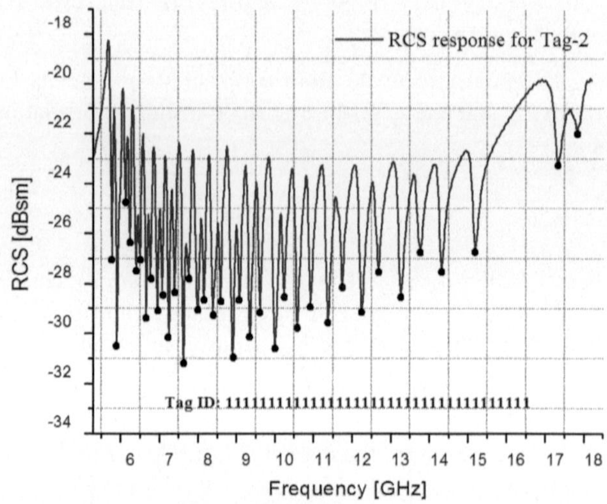

Fig. 4. RCS response for Tag-2.

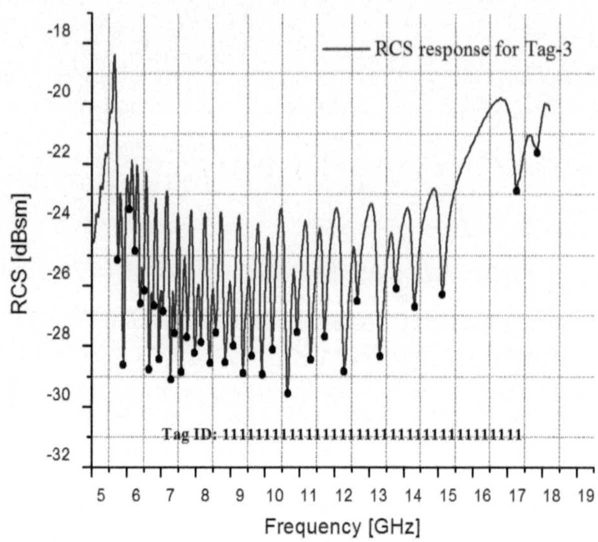

Fig. 5. RCS response for Tag-3.

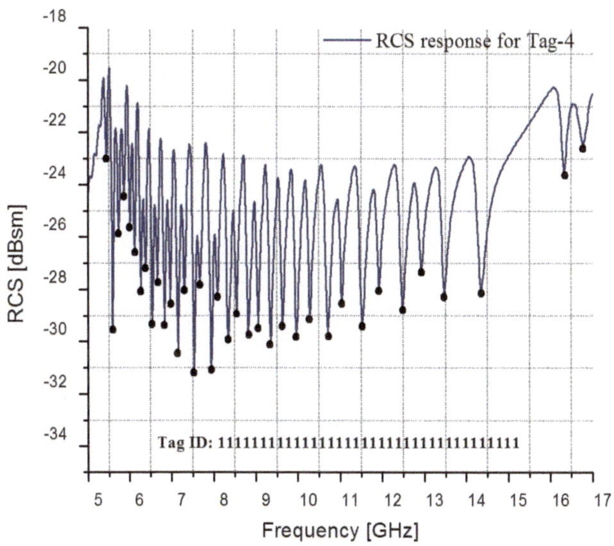

Fig. 6. RCS response for Tag-4.

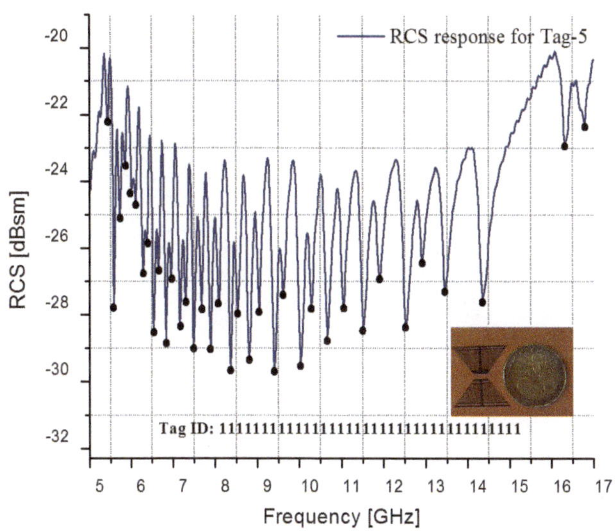

Fig. 7. RCS response for Tag-5.

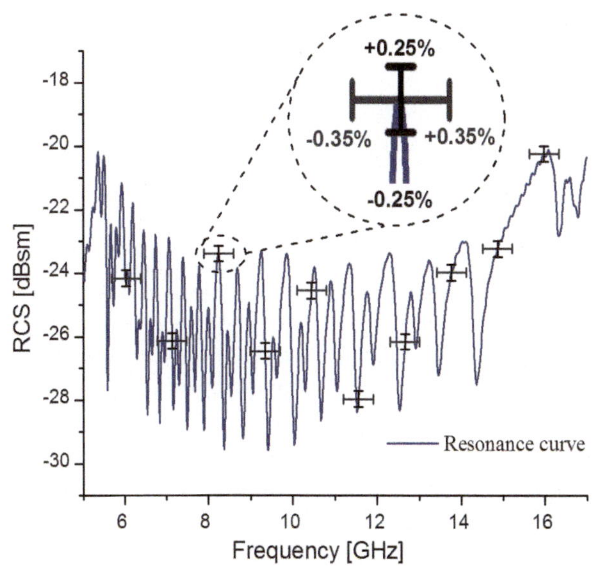

Fig. 8. Reliability curve of Tag.

The tag RCS response for 36 bits using silver nano ink radiator of 0.015 mm thickness lies in the frequency range of 4–17 GHz. We analyze that by the change of conducting material, there is variation in resonances of the tags.

To analyze the reliability parameters, five prototypes of each substrate are manufactured and tested. The corresponding reliable performance is shown in Fig. 8. It has been observed that the tag shows 0.25 % tolerance in RCS values and exhibits 0.35 % tolerance in the frequency band of interest.

By changing the substrate, there is variation in electrical properties which consequently results in frequency shift in RCS curve. With decreasing permittivity, there is a right shift in resonances which can be analyzed from '*LSB*' (Least significant bit) of comparison graph shown in Fig. 9. A comparative analysis of previous research works and the proposed research work is shown below in Tab. 3.

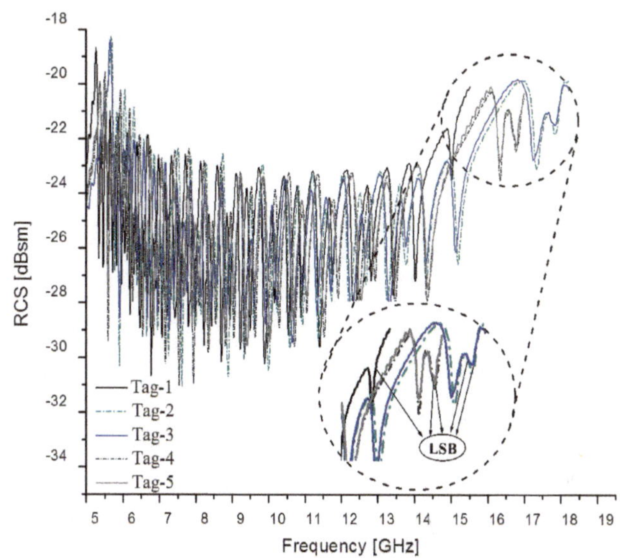

Fig. 9. Comparison graph of all presented tags.

Parameters	Paper 1 [36]	Paper 2 [7]	Proposed paper
Trans. Bits	12	24	36
No. of tagged items	4096	16777216	68,719,476,736
Size [mm^2]	35 × 33	20.6 × 19.9	24.5 × 25.5
Flexibility	x	✓	✓

Tab. 3. Comparison table.

To analyze different coded combinations, various tag ID's have been generated, simulated and tested. A comparative analysis of presented full tag with two different tag ID's along with their prototypes is shown in Fig. 10. Tag-A corresponds to all 1's with a tag ID

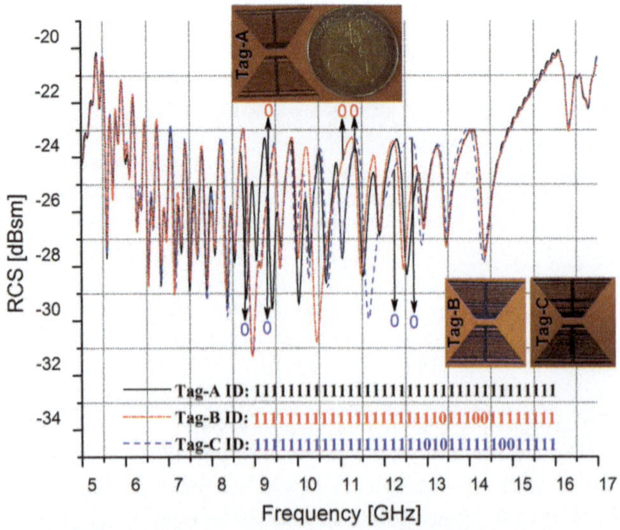

Fig. 10. Comparison of different ID combinations.

111111111111111111111111111111111111. For Tag-B, *S*9, *S*10 and *S*14 slots are shorted leading towards a coded combination of ID having '0' bits: 111111111111111111111101110011111111. Again, another data word having '0' bits representing tag ID: 111111111111111111110101111110011111 is presented as Tag-C by shorting *S*6, *S*7, *S*14, and *S*16. It has been analyzed that the occurrence of 0-bit has a slight effect on the amplitude and resonance frequency of neighboring peaks.

5. Measured Results

After testing, the results are measured and analyzed. The measured and computed results are shown in Fig. 11. RCS evaluation for each bit at a particular resonance can be calculated theoretically by using (5) [39]

$$f = \frac{c}{2L}\sqrt{\frac{2}{\varepsilon_r + 1}} \tag{5}$$

where ε_r is the relative permittivity of substrate, L is the slot length and c is the speed of light.

The experimental setup of chipless RFID for RCS evaluation includes two horn antennas; one transmitting and the other receiving as shown in Fig. 12. The tag deployed on the item is set at a far-field distance from the antennas. The transmitting antenna bombards an interrogator signal on the tag and receiver antenna then reads its response for identification using Vector Network Analyzer (VNA) R&S®ZVL13. The tag is printed using a table-top printer available from Dimatix "DMP2800" inkjet printer.

6. Conclusion

The research has proposed a novel, 36-bit passive chipless RFID resonator based tag. The tag has been designed, printed and tested. The tag consists of slots of different widths and lengths etched on the radiating patch. The proposed tag is of 24.5 × 25.5 mm^2 dimensions, designed on Taconic TLX-0, PET, and Kapton®HN substrates with copper, aluminum, and silver nanoparticle-based ink as radiators. The novelty of the tag relies on its compact size, high-density data and flexibility; deployed all together for green electronics, economic, environment-friendly and IoT-based applications.

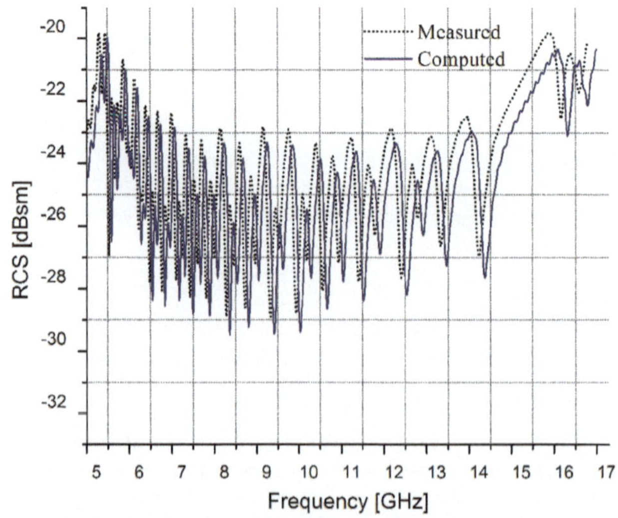

Fig. 11. Measured and computed RCS response.

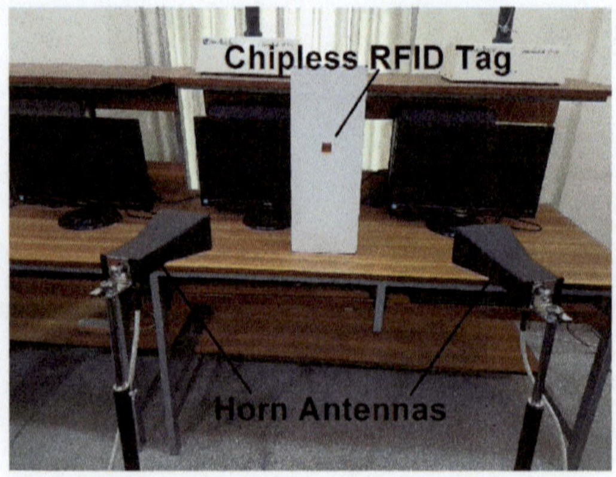

Fig. 12. Experimental set-up.

Acknowledgments

This work was financially supported by Vinnova (The Swedish Governmental Agency for Innovation Systems) and University of Engineering and Technology Taxila, Pakistan through the Vinn Excellence Centers program and ACTSENA research group funding, respectively.

References

[1] KHAN, U. H., ASLAM, B., AZAM, M. A., et al. Compact RFID enabled moisture sensor. *Radioengineering,* Sept. 2016, vol. 25, no. 3, p. 449–456. DOI: 10.13164/re.2016.0449

[2] GUBBI, J., BUYYA, R., MARUSIC, S., et al. Internet of Things (IoT): A vision, architectural elements, and future directions. *Future Generation Computer Systems*, 2013, vol. 29, no. 7, p. 1645–1660. DOI: 10.1016/j.future.2013.01.010

[3] WANT, R., SCHILIT, B. N., JENSON, S. Enabling the internet of things. *Computer*, 2015, vol. 48, no. 1, p. 28–35. DOI: 10.1109/MC.2015.12

[4] TRUONG, H. L., DUSTDAR, S. Principles for engineering IoT cloud systems. *IEEE Cloud Computing*, 2015, vol. 2, no. 2, p. 68 to 76. DOI: 10.1109/MCC.2015.23

[5] KHAN, M. S., ISLAM, M. S., DENG, H. Design of a reconfigurable RFID sensing tag as a generic sensing platform towards the future internet of things. *IEEE Internet of Things Journal*, 2014, vol. 1, no. 4, p. 300–310. DOI: 10.1109/JIOT.2014.2329189

[6] FUQAHA, A. A., GUIZANI, M., MOHAMMADI, M., et al. Internet of Things: A survey on enabling technologies, protocols, and applications. *IEEE Communications Surveys & Tutorials*, 2015, vol. 17, no. 4, p. 2347–2376. DOI: 10.1109/COMST.2015.2444095

[7] JAVED, N., HABIB, A., AMIN, Y., et al. Directly printable moisture sensor tag for intelligent packaging. *IEEE Sensors Journal,* 2016, vol. 16, no. 16, p. 6147–6148. DOI: 10.1109/JSEN.2016.2582847

[8] IDTechEx, *Printed and Chipless RFID Forecasts, Technologies & Players* 2009–2019. [Online] Available at: www.IdtechEx.com

[9] RAZZAQUE, M. A., JEVRIC, M. M., PALADE, A. Middleware for Internet of Things: A survey. *IEEE Internet of Things Journal*, 2016, vol. 3, no. 1, p. 70–95. DOI: 10.1109/JIOT.2015.2498900

[10] RODRIGUES, R. A. A., GURJAO, E. C., de ASSIS, F. M. Radar cross-section and electric field analysis of backscattering elements of chipless RFID tag. In *IEEE RFID Technology and Applications Conference (RFID-TA).* Tampere (Finland), 2014, p. 103–108. DOI: 10.1109/RFID-TA.2014.6934209

[11] KARMAKAR, N. C. Tag, You're it radar cross section of chipless RFID tags. *IEEE Microwave Magazine,* 2016, vol. 17, no. 7, p. 64 to 74. DOI: 10.1109/MMM.2016.2549160

[12] KHAN, M. M., TAHIR, F. A., CHEEMA, H. M. High capacity polarization sensitive chipless RFID tag. In *IEEE International Symposium on Antennas and Propagation & USNC/URSI National Radio Science Meeting*. Vancouver (Canada), 2015, p. 1770–1771. DOI: 10.1109/APS.2015.7305274

[13] VENA, A., PERRET, E., TEDJINI, S. High-capacity chipless RFID tag insensitive to the polarization. *IEEE Transactions on Antennas and Propagation*, 2012, vol. 60, no. 10, p. 4509–4515. DOI: 10.1109/TAP.2012.2207347

[14] SHAO, B., AMIN, Y., CHEN, Q., et al. Directly printed packaging-paper-based chipless RFID tag with coplanar LC resonator. *IEEE Antennas and Wireless Propagation Letters,* Feb. 2013, vol. 12, p. 325–328. DOI: 10.1109/LAWP.2013.2247556

[15] VENA, A., PERRET, E., TEDJINI, S. Design of compact and auto-compensated single-layer chipless RFID tag. *IEEE Transactions on Microwave Theory and Techniques*, 2012, vol. 60, no. 9, p. 2913–2924. DOI: 10.1109/TMTT.2012.2203927

[16] AMIN, Y., CHEN, Q., ZHENG, L. R., et al. Development and analysis of flexible UHF RFID antennas for 'green' electronics. *Progress In Electromagnetics Research*, 2012, vol. 130, p. 1–15. DOI: 10.2528/PIER12060609

[17] FENG, Y., XIE, L., CHEN, Q., et al. Low-cost printed chipless RFID humidity sensor tag for intelligent packaging. *IEEE Sensors Journal,* June 2015, vol. 15, no. 6, p. 3201–3208. DOI: 10.1109/JSEN.2014.2385154

[18] BHUIYAN, M. S., AZAD, A., KARMAKAR, N. Dual-band modified complementary split ring resonator (MCSRR) based multi-resonator circuit for chipless RFID tag. In *2013 IEEE Eighth International Conference on Intelligent Sensors, Sensor Networks and Information Processing.* Melbourne (VIC), 2013, p. 277–281. DOI: 10.1109/ISSNIP.2013.6529802

[19] AMIN, E. M., BHUIYAN, M. S., KARMAKAR, N. C., et al. Development of a low cost printable chipless RFID humidity sensor. *IEEE Sensors Journal*, 2014, vol. 14, no. 1, p. 140–149. DOI: 10.1109/JSEN.2013.2278560

[20] VENA, A., PERRET, E., TEDJINI, S., et al. Design of chipless RFID tags printed on paper by flexography. *IEEE Transactions on Antennas and Propagation*, 2013, vol. 61, no. 12, p. 5868–5877. DOI: 10.1109/TAP.2013.2281742

[21] AMIN, E. M., KARMAKAR, N., PRERADOVIC, S. Towards an intelligent EM barcode. In *7th International Conference on Electrical and Computer Engineering (ICECE).* Dhaka (Bangladesh), 2012, p. 826–829. DOI: 10.1109/ICECE.2012.6471678

[22] MARTINEZ-IRANZO, U., MORADI, B., GARCIA-GARCIA, J. Open ring resonator structure for compact chipless RFID tags. In *IEEE MTT-S International Microwave Symposium*. Phoenix (AZ, USA), 2015, p. 1–3. DOI: 10.1109/MWSYM.2015.7166790

[23] PRERADOVIC, S., KARMAKAR, N. Design of fully printable planar chipless RFID transponder with 35-bit data capacity. In *The European Microwave Conference EuMC.* Rome (Italy), 2009, p. 013-016. ISBN: 978-1-4244-4748-0

[24] ISLAM, M. A., KARMAKAR, N. C. Real-world implementation challenges of a novel dual-polarized compact printable chipless RFID tag. *IEEE Transactions on Microwave Theory and Techniques*, 2015, vol. 63, no. 12, p. 4581–4591. DOI: 10.1109/TMTT.2015.2495285

[25] HERROJO, C., NAQUI, J., PAREDES, F., et al. Spectral signature barcodes implemented by multi-state multi-resonator circuits for chipless RFID tags. In *IEEE MTT-S International Microwave Symposium (IMS).* San Francisco (CA, USA), 2016, p. 1–4. DOI: 10.1109/MWSYM.2016.7539981

[26] HARAZ, O. M., ASHRAF, M., ALSHEBILI, S., et al. Design of UWB chipless RFID tags using 8-bit open circuit stub resonators. In *17th International Symposium on Antenna Technology and Applied Electromagnetics (ANTEM).* Montreal (QC, Canada), 2016, p. 1–2. DOI: 10.1109/ANTEM.2016.7550145

[27] REZAIESARLAK, R., MANTEGHI, M. Complex-natural-resonance-based design of chipless RFID tag for high-density data. *IEEE Transactions on Antennas and Propagation*, 2014, vol. 62, no. 2, p. 898–904. DOI: 10.1109/TAP.2013.2290998

[28] HABIB, A., AZAM, M. A., AMIN, Y., et al. Chipless slot resonators for IoT system identification. In *IEEE International Conference on Electro Information Technology (EIT).* Grand Forks (ND), 2016, p. 0341–0344. DOI: 10.1109/EIT.2016.7535262

[29] VENA, A., PERRET, E., TEDJINI, S. Chipless RFID tag using hybrid coding technique. *IEEE Transactions on Microwave Theory and Techniques*, 2011, vol. 59, no. 12, p. 3356–3364. DOI: 10.1109/TMTT.2011.2171001

[30] SUMI, M., DINESH, R., NIJAS, C. M., et al. Frequency coded chipless RFID tag using spurline resonators. *Radioengineering*, 2014, vol. 23, no. 1, p. 203–208. ISSN: 1805-9600

[31] ZUO, Y. Survivable RFID systems: Issues, challenges, and techniques. *IEEE Transactions on Systems, Man, and Cybernetics, Part C (Applications and Reviews)*, 2010, vol. 40, no. 4, p. 406 to 418. DOI: 10.1109/TSMCC.2010.2043949

[32] YANG, C., SHEN, W., WANG, X. Applications of internet of things in manufacturing. In *IEEE 20th International Conference on Computer Supported Cooperative Work in Design (CSCWD).*

Nanchang (China), 2016, p. 670–675. DOI: 10.1109/CSCWD.2016.7566069

[33] Wikimedia Foundation, Inc. *Near and Far Field.* 6 pages. [Online] Cited 2016-09-25. Available at: https://en.wikipedia.org /wiki/Near_and_far_field

[34] MARTINEZ, M., WEIDE, D. V. D. Circular polarization on depolarizing chipless RFID tags. In *IEEE Radio and Wireless Symposium (RWS)*. Austin (TX, USA), 2016, p. 145–147. DOI: 10.1109/RWS.2016.7444388

[35] DOBKIN, D. M. *The RF in RFID: Passive UHF RFID in Practice.* Elsevier Inc, 2008. ISBN: 978-0-7506-8209-1

[36] BALANIS, C. A. *Antenna Theory, Analysis and Design.* Wiley, 2005. ISBN: 0-471-66782-X

[37] JANKOWSKI-MIHUŁOWICZ, P., WĘGLARSKI, M. Determination of passive and semi-passive chip parameters required for synthesis of interrogation zone in UHF RFID systems. *Elektronika Ir Elektrotechnika,* 2014, vol. 20, no. 9, p. 65–73. DOI: 10.5755/j01.eee.20.9.5007

[38] XU, L., HUANG, K. Design of compact trapezoidal bow-tie chipless RFID tag. *International Journal of Antennas and Propagation*, 2015, 7 p. DOI: 10.1155/2015/502938

[39] JAVED, N., HABIB, A., AKRAM, A., et al. 16-bit frequency signatured directly printable tag for organic electronics. *IEICE Electronics Express,* 2016, vol. 13, no. 11, p. 1–6. DOI: 10.1587/elex.13.20160406

About the Authors ...

Hafsa ANAM is working as a research scholar at the Telecommunication Engineering Department, University of Engineering and Technology, Taxila, Pakistan. She has received her BSc degree in Telecommunication Engineering from the University of Engineering and Technology, Taxila, Pakistan in 2015. The same year, she joined the University of Engineering and Technology, Taxila as full-time MS researcher under the supervision of Dr.Yasar Amin, where she is pursuing her master's degree focused on chipless RFID tags.

Ayesha HABIB is working as a research scholar at the Telecommunication Engineering Department, University of Engineering and Technology Taxila, Pakistan. She did her BSc in Telecommunication Engineering from the Army Public College of Management and Sciences, in 2012. She has completed her MSc. in Telecommunication Engineering from the University of Engineering and Technology, Taxila, Pakistan in 2014. In 2014, she has joined ACTSENA research group as a research scholar and has started her Ph.D. under the supervision of Dr. Yasar Amin. Her main research focus is on chipless RFID tags with integrated sensor.

Syeda Irum JAFRI is working as a lecturer at the Telecommunication Engineering Department, University of Engineering and Technology, Taxila, Pakistan. She has received her MSc. degree in Telecommunication in 2014, from UET, Taxila, Pakistan. Her area of specialization is antenna system design.

Yasar AMIN is a Chairman and Associate Professor of the Telecommunication Engineering Department, University of Engineering and Technology, Taxila, Pakistan. He is the founder of ACTSENA Research Group at UET Taxila, Pakistan. He did his BSc. in Electrical Engineering in 2001 with specialization in Telecommunication, and M.Sc. in Electrical Engineering in 2003 with specialization in System-on-Chip Design from the Royal Institute of Technology (KTH), Sweden. His Ph.D. is in Electronic and Computer Systems from the Royal Institute of Technology (KTH), Sweden, with the research focus on printable green RFID antennas for embedded sensors, while he has MBA in Innovation and Growth from Turku School of Economics, University of Turku, Finland. He has done several specialized courses from Stanford University, California, USA and Massachusetts Institute of Technology (MIT), USA. He has supervised over 15 M.Sc. thesis, and presently supervising 8 doctoral thesis. He is presently serving as a leading Guest Editor at two international journals and an active reviewer of more than a dozen well reputed international journals. He has contributed to over 20 journal papers, over 30 reviewed international conference papers. Dr. Yasar is a member of IEEE, IET, ACM and ACES.

Hannu TENHUNEN is a chair professor of Electronic Systems at Royal Institute of Technology (KTH), Stockholm, Sweden. Prof. Tenhunen has held professor position as full professor, invited professor or visiting honorary professor in Finland (TUT, UTU), Sweden (KTH), USA (Cornel U), France (INPG), China (Fudan and Beijing Jiaotong Universities), and Hong Kong (Chinese University of Hong Kong), and has an honorary doctorate from Tallinn Technical University. He has been the director of multiple national large-scale research programs or being an initiator and director of national or Knowledge and Innovation Community EIT ICT Labs.European graduate schools. He has actively contributed to VLSI and SoC design in Finland and Sweden via creating new educational programs and research directions, most lately at European level as being the EU-level Education Director of the new European flagship initiative European Institute of Technology and Innovations (EIT), and its Knowledge and Innovation Community EIT ICT Labs.

A Cascaded Reduced-Dimension STAP Method for Airborne MIMO Radar in the Presence of Jammers

Ting WANG, Yongjun ZHAO, Shiwen CHEN, Kunfan ZHANG

Zhengzhou Institute of Information Science and Technology, Science Road 62, Zhengzhou, People's Republic of China

wangtingsp@163.com, zhaoyjzz@163.com, ndsccsw@126.com, my_chocolate@163.com

Abstract. *A cascaded reduced-dimension (RD) space-time adaptive processing (STAP) method for airborne multiple-input multiple-output (MIMO) radar in the presence of jammers is proposed in this paper. The proposed MIMO-STAP method for clutter plus jamming suppression proceeds in two steps. Firstly, the jamming is suppressed by its orthogonal complementary subspace obtained in the passive radar mode, while the receive dimension is reduced. Secondly, the tri-iterative algorithm (TRIA) is utilized to suppress the clutter combining the remaining receive degree of freedom (DOF) with the transmit DOF and the Doppler DOF, and further dimension reduction is implemented. The proposed method can effectively realize the separate jamming and clutter elimination. Moreover, the training sample number and the computational complexity are significantly decreased. Simulation results verify the validity of the proposed cascaded RD MIMO-STAP method under jamming condition.*

Keywords

Multiple-input multiple-output (MIMO) radar, space-time adaptive processing (STAP), dimension reduction, jammer, tri-iterative algorithm (TRIA)

1. Introduction

Multiple-input multiple-output (MIMO) radar has drawn considerable attention in recent years [1–5]. According to the system configuration, two classes of MIMO radar are mainly investigated, i.e., distributed MIMO radar [2], [3] and collocated MIMO radar [4]. Because of the waveform diversity property, the collocated MIMO radar can produce extended virtual array aperture and enlarge the system degree of freedom (DOF) [4], [5]. Thus the collocated MIMO radar is applicable for the airborne platform which severely limits the aperture length and the weight of the airborne radar [5], [6], [7]. Meanwhile, space-time adaptive processing (STAP) [8], [9], [10] has become a crucial technique for the airborne radar to suppress the clutter and realize the ground moving target indication (GMTI). Nowadays, the airborne MIMO radar is combined with STAP technique to improve the clutter suppression and GMTI performance [11–15].

Due to the extra DOF provided by the waveform diversity, MIMO-STAP system enlarges the requirement of the independent and identically distributed (IID) training samples according to the well-known Reed-Mallett-Brennan (RMB) rule [16], and suffers a tremendous increase in computational cost. These practical problems have severely limited the development and application of MIMO-STAP. Up to now, various reduced-dimension (RD) STAP algorithms [12–15] have been developed for airborne MIMO radar to decrease the training sample requirement and the computational burden, such as the classical tri-iterative algorithm (TRIA) [14], [15].

Moreover, the practical operation environment of airborne radar becomes progressively complicated, and the target detection capability of airborne radar is usually affected by jamming and other interferences [10], [17], [18]. The existence of jammers will further bring additional processing difficulty to the conventional STAP system [10]. Therefore, we should investigate effective STAP approaches in the presence of jammers to mitigate the clutter plus jamming. To the best of our knowledge, the state-of-art research focusing on the MIMO-STAP method under jamming condition is still relatively less.

Aiming at the above problem and the research status, a cascaded RD MIMO-STAP method is proposed in this paper. Two stages are involved in this method. In the first stage, the orthogonal complementary subspace of jamming is obtained in the passive radar mode, and partial receive DOF is utilized to cancel the jamming. In the second stage, we conduct the TRIA to suppress the clutter after the matched-filtering and time-delaying, and to further reduce the dimension. Therefore, the separate jammer and clutter rejection is performed, and the computational cost is efficiently reduced as well as the required IID sample number. Simulation results are provided to verify the effectiveness of the proposed MIMO-STAP method for clutter plus jamming suppression.

The rest of the paper is organized as follows. The signal model of airborne MIMO radar STAP is established in Sec. 2, including the jamming and clutter signals. Then

Section 3 illustrates the principle of the proposed cascaded RD MIMO-STAP method according to two steps, and analyzes the sample requirement and the computational complexity. The simulation results are presented in Sec. 4 to demonstrate the performance of the proposed method. Finally, we draw the conclusion in Sec. 5.

Notations. The operators $(\cdot)^{\mathrm{T}}, (\cdot)^{\mathrm{H}}, (\cdot)^{-1}, E[\cdot]$ and $\otimes$ denote the transpose, conjugate transpose, inverse, expectation and Kronecker product, respectively. $\delta(\cdot)$ stands for the unit impulse function. The symbol $\mathrm{diag}\{r_1, r_2\}$ represents a diagonal matrix with diagonal entries r_1, r_2. $\mathbf{I}_M$ denotes the $M \times M$ identity matrix.

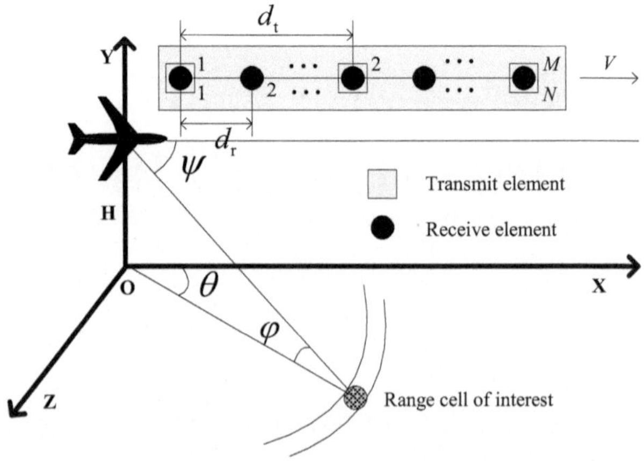

Fig. 1. Configuration of airborne MIMO radar.

2. Signal Model of Airborne MIMO Radar STAP

Figure 1 presents a side-looking monostatic MIMO radar system equipped with collocated transmit and receive linear arrays, in which there are M transmit elements with uniform spacing d_t and N receive elements with uniform spacing d_r, and sparse coefficient $\alpha = d_t/d_r$, where $d_r = \lambda/2$ and λ is the wavelength. A coherent processing interval (CPI) consists of K pulses with a constant radar pulse repetition interval (PRI) T. The radar platform travels at height H and at velocity V. The cone angle between the line-of-sight (LOS) and the velocity vector is ψ, while θ and φ are the azimuth angle and the elevation angle, respectively. The number of clutter patches which are uniformly distributed in a range cell is N_c. The clutter data vector of MIMO-STAP can be expressed as

$$\begin{aligned}\mathbf{y}_c &= \sum_{i=1}^{N_c} \rho_i \mathbf{v}(f_{sc,i}, f_{dc,i}) \\ &= \sum_{i=1}^{N_c} \rho_i \mathbf{a}_d(f_{dc,i}) \otimes \mathbf{a}_t(f_{sc,i}) \otimes \mathbf{a}_r(f_{sc,i})\end{aligned} \quad . \tag{1}$$

where ρ_i is the reflect coefficient of the clutter patch at θ_i, $\mathbf{v}(f_{sc,i}, f_{dc,i}) = \mathbf{a}_d(f_{dc,i}) \otimes \mathbf{a}_t(f_{sc,i}) \otimes \mathbf{a}_r(f_{sc,i})$ is the whole space-time steering vector of MIMO-STAP for the clutter patch at θ_i, $\mathbf{a}_d(f_{dc,i}) = [1 \;\; \exp(j2\pi f_{dc,i}) \;\; \cdots \;\; \exp(j2\pi(K-1)f_{dc,i})]^{\mathrm{T}}$ $= [1 \;\; \exp(j2\pi\beta f_{sc,i}) \;\; \cdots \;\; \exp(j2\pi(K-1)\beta f_{sc,i})]^{\mathrm{T}}$, $\mathbf{a}_t(f_{sc,i}) = [1 \;\; \exp(j2\pi\alpha f_{sc,i}) \;\; \cdots \;\; \exp(j2\pi(M-1)\alpha f_{sc,i})]^{\mathrm{T}}$ and $\mathbf{a}_r(f_{sc,i}) = [1 \;\; \exp(j2\pi f_{sc,i}) \;\; \cdots \;\; \exp(j2\pi(N-1)f_{sc,i})]^{\mathrm{T}}$ are steering vectors of temporal Doppler, spatial transmit array, and spatial receive array, respectively, $f_{dc,i} = 2VT\cos\psi_i / \lambda$ $= 2VT\cos\theta_i \cos\varphi / \lambda$ and $f_{sc,i} = d_r \cos\psi_i / \lambda = d_r \cos\theta_i \cos\varphi / \lambda$ are the normalized Doppler and spatial frequencies of the clutter patch at θ_i, respectively, $\beta = f_{dc,i} / f_{sc,i} = 2VT / d_r$ and $i = 1, 2, \cdots, N_c$.

The clutter covariance matrix can be expressed as

$$\mathbf{R}_C = E\left[\mathbf{y}_c \mathbf{y}_c^{\mathrm{H}}\right] = \sum_{i=1}^{N_c} \sigma_{c,i}^2 \mathbf{v}(f_{sc,i}, f_{dc,i}) \mathbf{v}^{\mathrm{H}}(f_{sc,i}, f_{dc,i}) \tag{2}$$

where $\sigma_{c,i}^2$ is the variance of the i th clutter patch, which is in direct proportion to the radar cross section (RCS) and satisfies $E\left[\left|\rho_i \rho_k^*\right|\right] = \sigma_{c,i}^2 \delta(i-k)$, $i, k = 1, 2, \cdots, N_c$.

Assume there exist J independent jammers in practical circumstance. And each jammer is independent at different times in temporal domain, i.e., it is Doppler white. Then the jamming covariance matrix can be written as

$$\mathbf{R}_J = \mathbf{I}_{KM} \otimes \mathbf{R}_{rJ} = \mathbf{I}_{KM} \otimes [\sum_{i=1}^{J} \sigma_{j,i}^2 \mathbf{a}_r(f_{sj,i}) \mathbf{a}_r^{\mathrm{H}}(f_{sj,i})] \tag{3}$$

where $\sigma_{j,i}^2$ is the variance of the i th jammer, and $f_{sj,i}$ is the normalized spatial frequency of the i th jammer.

Assume the noise obeys white Gaussian distribution in both the spatial and temporal domains, thus the noise covariance matrix can be written as

$$\mathbf{R}_W = \sigma_w^2 \mathbf{I}_{KMN} \tag{4}$$

where σ_w^2 is the variance of the additive white noise.

Considering the clutter, the jammer and the noise exist at the same time, the fully-adaptive weight vector $\mathbf{w} \in \mathbb{C}^{KMN \times 1}$ for MIMO radar STAP can be calculated by

$$\mathbf{w} = \frac{\mathbf{R}^{-1}\mathbf{v}(f_{s,0}, f_{d,0})}{\mathbf{v}^{\mathrm{H}}(f_{s,0}, f_{d,0})\mathbf{R}^{-1}\mathbf{v}(f_{s,0}, f_{d,0})} \tag{5}$$

where $\mathbf{R} = \mathbf{R}_C + \mathbf{R}_J + \mathbf{R}_W$ represents the clutter-plus-jamming-plus-noise covariance matrix, and $\mathbf{v}(f_{s,0}, f_{d,0}) = \mathbf{a}_d(f_{d,0}) \otimes \mathbf{a}_t(f_{s,0}) \otimes \mathbf{a}_r(f_{s,0})$ is the space-time steering vector of target, $f_{s,0}$ and $f_{d,0}$ are the normalized spatial and Doppler frequencies of the target.

The full-dimension optimum STAP can suppress the clutter and jamming simultaneously. With the extension of processing dimension in MIMO radar, the training sample number L of the full-dimension MIMO-STAP should meet the requirement $L \geq 2KMN$ to estimate the matrix $\mathbf{R}$ efficiently [16]. However, it is very difficult to acquire enough IID samples in the practical non-homogeneous clutter scenarios. Moreover, the complexity to compute the matrix estimation and inversion is $O(LK^2M^2N^2)+O(K^3M^3N^3)$. The problem of computational cost is so challenging that the real-time processing can hardly be realized [8], [9], [10]. Therefore, the effective RD techniques should be investigated for MIMO-STAP under jamming condition.

3. Cascaded RD MIMO-STAP Method for Clutter plus Jamming Suppression

Exploiting the different signal model characteristics between jamming and clutter, we put forward a cascaded RD MIMO-STAP method in this section. The proposed method is conducted in two steps: (1) suppressing the jamming by its orthogonal complementary subspace; (2) suppressing the clutter by utilizing TRIA. The processing diagram is illustrated in Fig. 2.

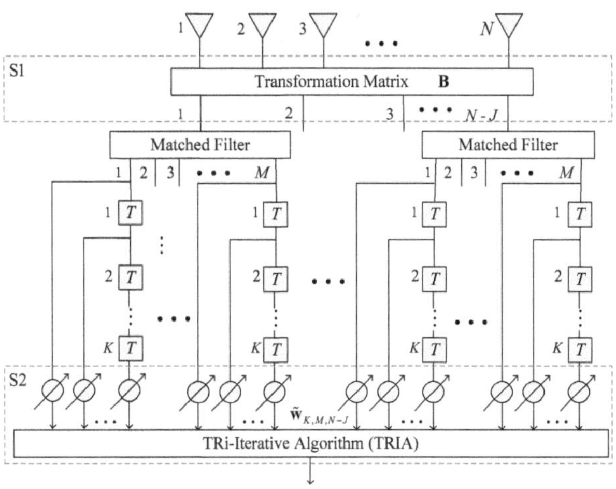

Fig. 2. Processing diagram of the proposed method.

3.1 Jamming Cancellation Based on Subspace Orthogonality

In the first stage, we utilize a transformation matrix $\mathbf{B}$ to suppress the jamming in the spatial receive domain, which is determined by the $N\times 1$ receive signals of the first pulse in passive radar mode. When the airborne MIMO radar operates in the passive mode, only the jamming and the noise are acquired at the receive site, while the target signal and the clutter are not contained.

Then the receive jamming plus noise covariance matrix in passive mode can be described as

$$\mathbf{R}_{\mathrm{r}}=\sum_{i=1}^{J}\sigma_{\mathrm{j},i}^2\mathbf{a}_{\mathrm{r}}(f_{\mathrm{sj},i})\mathbf{a}_{\mathrm{r}}^{\mathrm{H}}(f_{\mathrm{sj},i})+\mathbf{I}_N\,. \quad (6)$$

Through performing eigenvalue decomposition (EVD) on $\mathbf{R}_{\mathrm{r}}$, we can obtain

$$\mathbf{R}_{\mathrm{r}}=\mathbf{E}_{\mathrm{J}}\boldsymbol{\Lambda}_1\mathbf{E}_{\mathrm{J}}^{\mathrm{H}}+\mathbf{E}_{\mathrm{J}}^{\perp}\boldsymbol{\Lambda}_2\mathbf{E}_{\mathrm{J}}^{\perp\mathrm{H}} \quad (7)$$

where $\boldsymbol{\Lambda}_1=\mathrm{diag}(\lambda_1,\lambda_2,\cdots\lambda_J)$, $\boldsymbol{\Lambda}_2=\mathrm{diag}(\lambda_{J+1},\lambda_{J+2},\cdots\lambda_N)$, $\lambda_1\ge\lambda_2\ge\cdots\ge\lambda_J\ge\lambda_{J+1}=\lambda_{J+2}=\cdots=\lambda_N$, $\mathbf{E}_{\mathrm{J}}\in\mathbb{C}^{N\times J}$ and $\mathbf{E}_{\mathrm{J}}^{\perp}\in\mathbb{C}^{N\times(N-J)}$ are the matrix of eigenvectors corresponding to $\boldsymbol{\Lambda}_1$ and $\boldsymbol{\Lambda}_2$, respectively. Based on the subspace property in matrix theory [19], $\mathbf{E}_{\mathrm{J}}$ constitutes the jamming subspace, while $\mathbf{E}_{\mathrm{J}}^{\perp}$ constitutes the orthogonal complementary subspace of jamming. Then we select $\mathbf{B}=\mathbf{E}_{\mathrm{J}}^{\perp}$ as the transformation matrix to function on the radar echo, which is orthogonal to the jamming subspace. Thus, the jamming signal is suppressed when MIMO radar operates in active mode, and the dimension reduction is implemented in the spatial receive domain at the same time (from N to $N-J$).

3.2 Clutter Rejection Based on TRIA

After the jamming suppression stage in active radar mode, the k th echo pulse of the clutter signal received by $(N-J)$ channels can be written as

$$\tilde{\mathbf{x}}_{\mathrm{c}}(t,k)=\sum_{i=1}^{N_{\mathrm{c}}}\rho_i\exp\left(\mathrm{j}2\pi(k-1)f_{\mathrm{dc},i}\right)\mathbf{B}^{\mathrm{H}}\mathbf{a}_{\mathrm{r}}(f_{\mathrm{sc},i})\mathbf{a}_{\mathrm{t}}^{\mathrm{T}}(f_{\mathrm{sc},i})\mathbf{s}(t) \quad (8)$$

where $\mathbf{s}(t)$ represents the transmit waveform vector of MIMO radar which satisfies the orthogonality condition

$$\int_{T_0}\mathbf{s}(t)\mathbf{s}^{\mathrm{H}}(t)=\mathbf{I}_M\,. \quad (9)$$

where T_0 is the pulse width.

Next, the clutter output of the matched filters with M transmit waveforms behind each channel is expressed as

$$\begin{aligned}\tilde{\mathbf{X}}_{\mathrm{c}}(k)&=\int_{T_0}\tilde{\mathbf{x}}_{\mathrm{c}}(t,k)\mathbf{s}^{\mathrm{H}}(t)\mathrm{d}t\\&=\sum_{i=1}^{N_{\mathrm{c}}}\rho_i\exp\left(\mathrm{j}2\pi(k-1)f_{\mathrm{dc},i}\right)\mathbf{B}^{\mathrm{H}}\mathbf{a}_{\mathrm{r}}(f_{\mathrm{sc},i})\mathbf{a}_{\mathrm{t}}^{\mathrm{T}}(f_{\mathrm{sc},i}).\end{aligned} \quad (10)$$

Further, by stacking the K pulse data $\tilde{\mathbf{Y}}_{\mathrm{c}}=[\tilde{\mathbf{X}}_{\mathrm{c}}(1)\ \ \tilde{\mathbf{X}}_{\mathrm{c}}(2)\ \cdots\ \tilde{\mathbf{X}}_{\mathrm{c}}(K)]$ of the delay lines according to the column order, we can obtain the space-time data vector of the clutter signal after jamming suppression

$$\tilde{\mathbf{y}}_{\mathrm{c}}=\sum_{i=1}^{N_{\mathrm{c}}}\rho_i\mathbf{a}_{\mathrm{d}}(f_{\mathrm{dc},i})\otimes\mathbf{a}_{\mathrm{t}}(f_{\mathrm{sc},i})\otimes[\mathbf{B}^{\mathrm{H}}\mathbf{a}_{\mathrm{r}}(f_{\mathrm{sc},i})]\,. \quad (11)$$

Comparing (11) to (1), we can find that the spatial receive dimension of the MIMO-STAP signal model has been reduced from N to $N-J$, as a consequence of the transformation matrix $\mathbf{B}$ in the first stage.

Meanwhile, the clutter covariance matrix can be written as

$$\begin{aligned}\tilde{\mathbf{R}}_{\mathrm{C}}&=E\left[\tilde{\mathbf{y}}_{\mathrm{c}}\tilde{\mathbf{y}}_{\mathrm{c}}^{\mathrm{H}}\right]\\&=\sum_{i=1}^{N_{\mathrm{c}}}\sigma_{\mathrm{c},i}^2[\mathbf{a}_{\mathrm{d}}(f_{\mathrm{dc},i})\mathbf{a}_{\mathrm{d}}^{\mathrm{H}}(f_{\mathrm{dc},i})]\otimes[\mathbf{a}_{\mathrm{t}}(f_{\mathrm{sc},i})\mathbf{a}_{\mathrm{t}}^{\mathrm{H}}(f_{\mathrm{sc},i})]\\&\quad\otimes[\mathbf{B}^{\mathrm{H}}\mathbf{a}_{\mathrm{r}}(f_{\mathrm{sc},i})\mathbf{a}_{\mathrm{r}}^{\mathrm{H}}(f_{\mathrm{sc},i})\mathbf{B}].\end{aligned} \quad (12)$$

Then after the similar corresponding processing, the jamming covariance matrix and the noise covariance matrix can be respectively expressed as

$$\tilde{\mathbf{R}}_{\mathrm{J}} = \sum_{i=1}^{J} \sigma_{\mathrm{j},i}^2 \mathbf{I}_{KM} \otimes [\mathbf{B}^{\mathrm{H}} \mathbf{a}_{\mathrm{r}}(f_{\mathrm{sj},i}) \mathbf{a}_{\mathrm{r}}^{\mathrm{H}}(f_{\mathrm{sj},i}) \mathbf{B}], \quad (13)$$

$$\tilde{\mathbf{R}}_{\mathrm{W}} = \sigma_{\mathrm{w}}^2 \mathbf{I}_{KM(N-J)} \otimes [\mathbf{B}^{\mathrm{H}} \mathbf{B}] = \sigma_{\mathrm{w}}^2 \mathbf{I}_{KM(N-J)}. \quad (14)$$

Due to the orthogonality between $\mathbf{B}$ and the jamming subspace [19], we can come to the conclusion that the jamming covariance matrix $\tilde{\mathbf{R}}_{\mathrm{J}} = \mathbf{0}$ in theory. We define $\tilde{\mathbf{R}} = \tilde{\mathbf{R}}_{\mathrm{C}} + \tilde{\mathbf{R}}_{\mathrm{W}}$ and $\tilde{\mathbf{v}}(f_{\mathrm{s},0}, f_{\mathrm{d},0}) = \mathbf{a}_{\mathrm{d}}(f_{\mathrm{d},0}) \otimes \mathbf{a}_{\mathrm{t}}(f_{\mathrm{s},0}) \otimes [\mathbf{B}^{\mathrm{H}} \mathbf{a}_{\mathrm{r}}(f_{\mathrm{s},0})]$, then the clutter suppression can be expressed as the following optimization problem

$$\min_{\tilde{\mathbf{w}}} \ \tilde{\mathbf{w}}^{\mathrm{H}} \tilde{\mathbf{R}} \tilde{\mathbf{w}} \quad \text{subject to } \tilde{\mathbf{w}}^{\mathrm{H}} \tilde{\mathbf{v}}(f_{\mathrm{s},0}, f_{\mathrm{d},0}) = 1. \quad (15)$$

Then based on the idea of TRIA [14], [15] and the structure of the novel clutter model, the whole weight vector $\tilde{\mathbf{w}} \in \mathbb{C}^{KM(N-J)\times 1}$ can be described as the following separable form

$$\tilde{\mathbf{w}} = \tilde{\mathbf{w}}_{\mathrm{d}} \otimes \tilde{\mathbf{w}}_{\mathrm{t}} \otimes \tilde{\mathbf{w}}_{\mathrm{r}} \quad (16)$$

where $\tilde{\mathbf{w}}_{\mathrm{d}} \in \mathbb{C}^{K\times 1}$, $\tilde{\mathbf{w}}_{\mathrm{t}} \in \mathbb{C}^{M\times 1}$ and $\tilde{\mathbf{w}}_{\mathrm{r}} \in \mathbb{C}^{(N-J)\times 1}$ are the reduced-dimension weight vectors corresponding to the Doppler, the transmit and the receive parts, respectively.

Thus (15) can be further rewritten as

$$\min_{\tilde{\mathbf{w}}_{\mathrm{d}}, \tilde{\mathbf{w}}_{\mathrm{t}}, \tilde{\mathbf{w}}_{\mathrm{r}}} \ (\tilde{\mathbf{w}}_{\mathrm{d}} \otimes \tilde{\mathbf{w}}_{\mathrm{t}} \otimes \tilde{\mathbf{w}}_{\mathrm{r}})^{\mathrm{H}} \tilde{\mathbf{R}} (\tilde{\mathbf{w}}_{\mathrm{d}} \otimes \tilde{\mathbf{w}}_{\mathrm{t}} \otimes \tilde{\mathbf{w}}_{\mathrm{r}}) \quad \text{subject to } (\tilde{\mathbf{w}}_{\mathrm{d}} \otimes \tilde{\mathbf{w}}_{\mathrm{t}} \otimes \tilde{\mathbf{w}}_{\mathrm{r}})^{\mathrm{H}} \tilde{\mathbf{v}}(f_{\mathrm{s},0}, f_{\mathrm{d},0}) = 1. \quad (17)$$

According to the above formulation, the implementation steps to determine the low-dimension adaptive weight vectors $\tilde{\mathbf{w}}_{\mathrm{d}}$, $\tilde{\mathbf{w}}_{\mathrm{t}}$ and $\tilde{\mathbf{w}}_{\mathrm{r}}$ through solving (17) are summarized as follows:

(a) Given the normalized initial values $\tilde{\mathbf{w}}_{\mathrm{d}}(0) = \mathbf{a}_{\mathrm{d}}(f_{\mathrm{d},0}) / \|\mathbf{a}_{\mathrm{d}}(f_{\mathrm{d},0})\|$ and $\tilde{\mathbf{w}}_{\mathrm{t}}(0) = \mathbf{a}_{\mathrm{t}}(f_{\mathrm{s},0}) / \|\mathbf{a}_{\mathrm{t}}(f_{\mathrm{s},0})\|$;

(b) Construct the receive reduced-dimension transformation matrix $\mathbf{T}_{\mathrm{r}} = \tilde{\mathbf{w}}_{\mathrm{d}}(i-1) \otimes \tilde{\mathbf{w}}_{\mathrm{t}}(i-1) \otimes \mathbf{I}_{N-J} \in \mathbb{C}^{KM(N-J)\times(N-J)}$, and derive the receive reduced-dimension covariance matrix $\tilde{\mathbf{R}}_{\mathrm{r}} = \mathbf{T}_{\mathrm{r}}^{\mathrm{H}} \tilde{\mathbf{R}} \mathbf{T}_{\mathrm{r}}$ and the corresponding receive steering vector $\tilde{\mathbf{v}}_{\mathrm{r}} = \mathbf{T}_{\mathrm{r}}^{\mathrm{H}} \tilde{\mathbf{v}}(f_{\mathrm{s},0}, f_{\mathrm{d},0})$. Then calculate the receive weight vector $\tilde{\mathbf{w}}_{\mathrm{r}}(i) = \tilde{\mathbf{R}}_{\mathrm{r}}^{-1} \tilde{\mathbf{v}}_{\mathrm{r}} / \tilde{\mathbf{v}}_{\mathrm{r}}^{\mathrm{H}} \tilde{\mathbf{R}}_{\mathrm{r}}^{-1} \tilde{\mathbf{v}}_{\mathrm{r}}$, where $i = 1, 2, \cdots$ is the iteration number;

(c) Construct the transmit reduced-dimension transformation matrix $\mathbf{T}_{\mathrm{t}} = \tilde{\mathbf{w}}_{\mathrm{d}}(i-1) \otimes \mathbf{I}_M \otimes \tilde{\mathbf{w}}_{\mathrm{r}}(i) \in \mathbb{C}^{KM(N-J)\times M}$, and derive the transmit reduced-dimension covariance matrix $\tilde{\mathbf{R}}_{\mathrm{t}} = \mathbf{T}_{\mathrm{t}}^{\mathrm{H}} \tilde{\mathbf{R}} \mathbf{T}_{\mathrm{t}}$ and the corresponding transmit steering vector $\tilde{\mathbf{v}}_{\mathrm{t}} = \mathbf{T}_{\mathrm{t}}^{\mathrm{H}} \tilde{\mathbf{v}}(f_{\mathrm{s},0}, f_{\mathrm{d},0})$. Then calculate the transmit weight vector $\tilde{\mathbf{w}}_{\mathrm{t}}(i) = \tilde{\mathbf{R}}_{\mathrm{t}}^{-1} \tilde{\mathbf{v}}_{\mathrm{t}} / \tilde{\mathbf{v}}_{\mathrm{t}}^{\mathrm{H}} \tilde{\mathbf{R}}_{\mathrm{t}}^{-1} \tilde{\mathbf{v}}_{\mathrm{t}}$;

(d) Construct the Doppler reduced-dimension transformation matrix $\mathbf{T}_{\mathrm{d}} = \mathbf{I}_K \otimes \tilde{\mathbf{w}}_{\mathrm{t}}(i) \otimes \tilde{\mathbf{w}}_{\mathrm{r}}(i) \in \mathbb{C}^{KM(N-J)\times K}$, and derive the Doppler reduced-dimension covariance matrix $\tilde{\mathbf{R}}_{\mathrm{d}} = \mathbf{T}_{\mathrm{d}}^{\mathrm{H}} \tilde{\mathbf{R}} \mathbf{T}_{\mathrm{d}}$ and the corresponding Doppler steering vector $\tilde{\mathbf{v}}_{\mathrm{d}} = \mathbf{T}_{\mathrm{d}}^{\mathrm{H}} \tilde{\mathbf{v}}(f_{\mathrm{s},0}, f_{\mathrm{d},0})$. Then calculate the Doppler weight vector $\tilde{\mathbf{w}}_{\mathrm{d}}(i) = \tilde{\mathbf{R}}_{\mathrm{d}}^{-1} \tilde{\mathbf{v}}_{\mathrm{d}} / \tilde{\mathbf{v}}_{\mathrm{d}}^{\mathrm{H}} \tilde{\mathbf{R}}_{\mathrm{d}}^{-1} \tilde{\mathbf{v}}_{\mathrm{d}}$;

(e) Repeat performing (b), (c) and (d) to calculate $\tilde{\mathbf{w}}_{\mathrm{d}}(i)$, $\tilde{\mathbf{w}}_{\mathrm{t}}(i)$ and $\tilde{\mathbf{w}}_{\mathrm{r}}(i)$, respectively, until $\|\tilde{\mathbf{w}}_{\mathrm{d}}(i) - \tilde{\mathbf{w}}_{\mathrm{d}}(i-1)\| / \|\tilde{\mathbf{w}}_{\mathrm{d}}(i)\| < \varepsilon_1$ ($0 < \varepsilon_1 \ll 1$) and $\|\tilde{\mathbf{w}}_{\mathrm{t}}(i) - \tilde{\mathbf{w}}_{\mathrm{t}}(i-1)\| / \|\tilde{\mathbf{w}}_{\mathrm{t}}(i)\| < \varepsilon_2$ ($0 < \varepsilon_2 \ll 1$) (the tolerance thresholds ε_1 and ε_2);

(f) In the end, the three low-dimension weight vectors $\tilde{\mathbf{w}}_{\mathrm{d}}$, $\tilde{\mathbf{w}}_{\mathrm{t}}$ and $\tilde{\mathbf{w}}_{\mathrm{r}}$ are resolved respectively based on tri-iterative optimization to synthesize the whole weight vector $\tilde{\mathbf{w}} = \tilde{\mathbf{w}}_{\mathrm{d}} \otimes \tilde{\mathbf{w}}_{\mathrm{t}} \otimes \tilde{\mathbf{w}}_{\mathrm{r}}$.

3.3 Analysis of the Required Sample Number and the Computational Burden

For the efficient estimation of $\tilde{\mathbf{R}}_{\mathrm{d}}$, $\tilde{\mathbf{R}}_{\mathrm{t}}$ and $\tilde{\mathbf{R}}_{\mathrm{r}}$ in our proposed MIMO-STAP method, the required sample number $\tilde{L}$ should be larger than or equal to $2\max\{K, M, (N-J)\}$. Moreover, the computational complexity of the proposed method is merely $\mathrm{O}[I(\tilde{L}K^2 + \tilde{L}M^2 + \tilde{L}(N-J)^2)] + \mathrm{O}[I(K^3 + M^3 + (N-J)^3)]$ which involves the matrix estimation and inversion, where I is the iteration number, and $I = 4 \sim 5$ in the simulation when the iteration thresholds are set as $\varepsilon_1 = \varepsilon_2 = 0.001$. Therefore, the sample requirement and computational load of the proposed MIMO-STAP method are far less than those of the full-dimension MIMO-STAP introduced in Sec. 2, which are $L \geq 2KMN$ and $\mathrm{O}(LK^2M^2N^2) + \mathrm{O}(K^3M^3N^3)$.

Meanwhile, the required sample number $\hat{L}$ of the direct TRIA for clutter plus jamming suppression should be larger than or equal to $2\max\{K, M, N\}$, and the computational complexity under the same iteration number is $\mathrm{O}[I(\hat{L}K^2 + \hat{L}M^2 + \hat{L}N^2)] + \mathrm{O}[I(K^3 + M^3 + N^3)]$. Since $N < K$ usually holds in general, the sample requirement of the proposed method approximates to that of the direct TRIA. However, the computational burden of the proposed method is further reduced, while the near clutter plus jamming suppression performance of MIMO-STAP can be achieved compared to the direct TRIA.

Under the typical parameters: temporal pulse number $K = 16$, transmit array element number $M = 10$, receive array element number $N = 10$, jammer number $J = 2$, and iteration number $I = 5$, the corresponding quantitative comparisons based on the above analysis are listed in Tab. 1 to illustrate the advantage of our method in terms of the sample requirement and computational complexity.

	Full-dimension	Direct TRIA	Proposed method
Sample requirement	≥ 3200	≥ 32	≥ 32
Computational complexity	1.4653×10^{6}	1.0344×10^{5}	9.524×10^{4}

Tab. 1. Quantitative comparisons of the sample requirement and computational complexity.

4. Simulation Results

In the simulation, the basic parameters for MIMO-STAP are $K = 16$, $M = 10$, $N = 10$, $\lambda = 0.23$ m, $d_t = d_r = 0.115$ m, $\beta = 1$, $T = 500$ μs, $V = 115$ m/s, $H = 8000$ m, $N_c = 180$. Besides, the cone angle of the target is assumed to be located at the side-looking direction of the airborne radar, which means the normalized spatial frequency of the target is $f_{s,0} = 0$. The tolerance thresholds in TRIA are set as $\varepsilon_1 = \varepsilon_2 = 0.001$. The clutter-to-noise ratio (CNR) is 40 dB.

Assume there exist two independent jammers at the jamming directions $\psi_1 = 60°$ and $\psi_2 = 120°$ in the scenario, i.e., $f_{sj,1} = 0.25$ and $f_{sj,2} = -0.25$, respectively, and the jamming-to-noise ratios (JNRs) are both 40 dB. Figure 3 presents the corresponding clutter plus jamming spectrum and its projection. From Fig. 3, we can observe that the jamming spectrum is distributed along the Doppler frequencies and is discrete at spatial frequencies, which meets the analysis of the jamming signal model shown in Sec. 2.

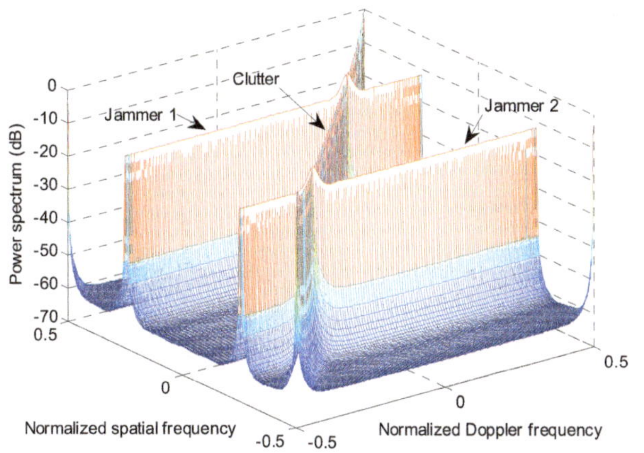

a)

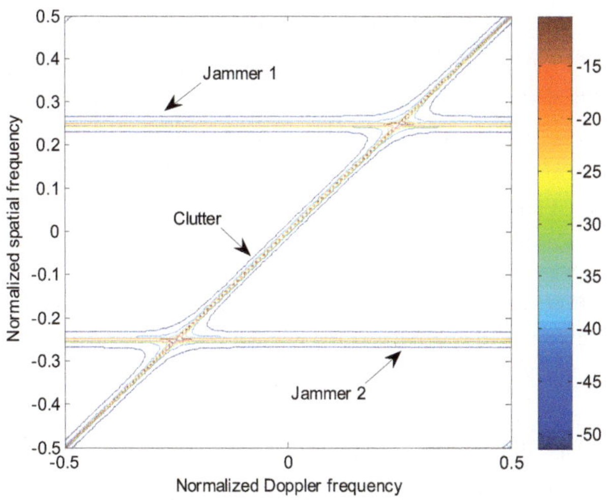

b)

Fig. 3. a) Clutter plus jamming spectrum. b) Two-dimensional spectrum projection.

The STAP performance is evaluated by improvement factor (IF), which is defined as the ratio of output signal-to-clutter-plus-jamming-plus-noise ratio (SCJNR) to the input SCJNR. Then in the following simulations, the receive jamming plus noise sample number applied to estimate $\mathbf{R}_r$ in the passive radar mode is fixed to 40. As comparison, the optimum IF value is provided in each simulation figure indicated by "Optimum", which can be expressed as

$$\begin{aligned}\mathrm{IF}_{\mathrm{opt}} &= \frac{\mathrm{SCJNR}_{\mathrm{opt}}}{\mathrm{SCJNR}_{\mathrm{in}}} \\ &= \frac{\left|\mathbf{w}^{\mathrm{H}}\mathbf{v}(f_{s,0}, f_{d,0})\right|^2 (\mathrm{CNR}+\mathrm{JNR}+1)\sigma_w^2}{\mathbf{w}^{\mathrm{H}}\mathbf{R}\mathbf{w}} \\ &= \mathbf{v}^{\mathrm{H}}(f_{s,0}, f_{d,0})\mathbf{R}^{-1}\mathbf{v}(f_{s,0}, f_{d,0})(\mathrm{CNR}+\mathrm{JNR}+1)\sigma_w^2\end{aligned} \tag{18}$$

where the covariance matrix $\mathbf{R} = \mathbf{R}_C + \mathbf{R}_J + \mathbf{R}_W$ is ideal and accurately known, not the estimated one produced by the training samples.

Figure 4 plots the IF against the iteration number in TRIA at the assumed normalized Doppler frequency of the target $f_{d,0} = 0.25$. It is notable that $\tilde{L} = 200$ training sample number is adopted in TRIA, which is obviously lower than the MIMO-STAP dimension KMN and accords with the real clutter scenario. It can be seen from Fig. 4 that the proposed method converges to stability very well after 4~5 iterations, and the convergence value is approximate to that of the direct TRIA, which is about 3 dB lower than the ideal optimum IF value of the full-dimension MIMO-STAP. Moreover, the computational load is lower than that of the direct TRIA because of the receive dimension reduction in the jamming suppression stage.

Additionally, we consider the influence of different JNRs on the proposed method. Under the new JNR condi-

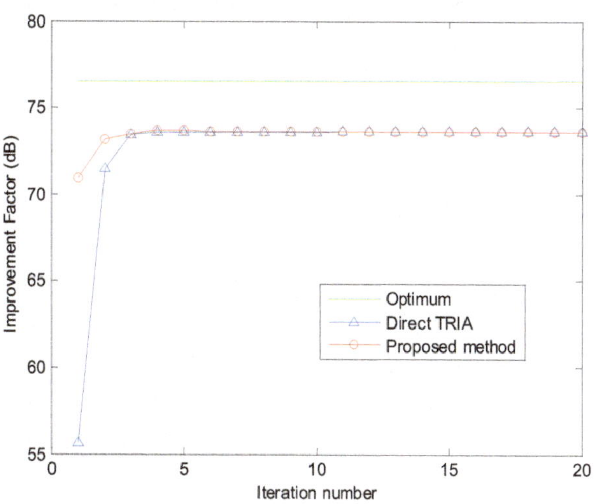

Fig. 4. IF versus the iteration number.

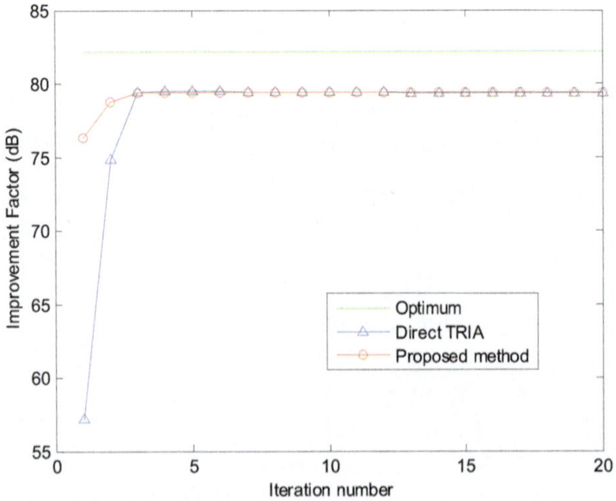

Fig. 5. IF versus the iteration number ($JNR_1 = 30$ dB, $JNR_2 = 50$ dB).

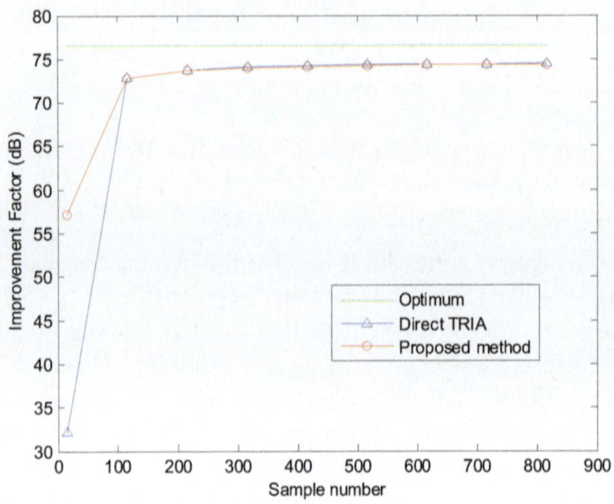

Fig. 6. IF versus the training sample number.

tion that $JNR_1 = 30$ dB and $JNR_2 = 50$ dB, which is different from the previous simulation, the IF against the iteration number in TRIA at $f_{d,0} = 0.25$ is depicted in Fig. 5. From Fig. 5, it can be found that the proposed method still owns stable convergence property in the changed JNR case, and only the convergence value is different from that in Fig. 4. Therefore, the theoretical correctness of our method is further validated. Afterwards, we return to the original JNR setting ($JNR_1 = JNR_2 = 40$ dB) to implement the following simulations.

Figure 6 presents the IF versus the number of training samples at $f_{d,0} = 0.25$. Monte Carlo experiments are carried out 100 times to obtain the average values. The minimum sample number is taken as $\tilde{L}_{\min} = \max\{K, M, N\} = 16$, for the feasibility of each low-dimension covariance matrix inversion in the direct TRIA. It can be observed that the proposed method converges fast with the training sample number, and exhibits superior IF performance than the direct TRIA under small sample number condition. In the practical non-homogeneous circumstance with insufficient IID samples, the proposed method owns advantage for anti-clutter and anti-jamming.

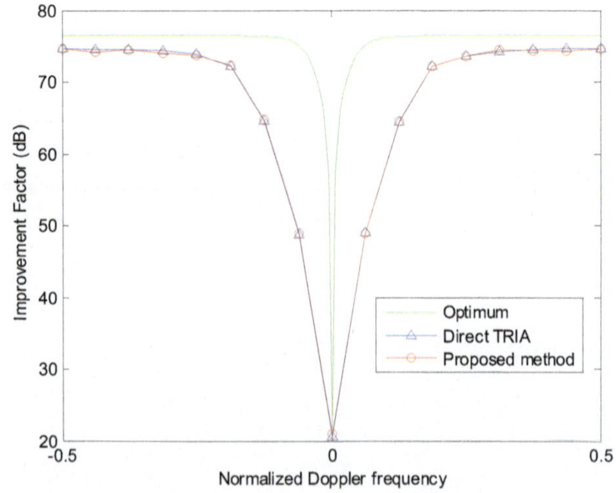

Fig. 7. IF versus normalized Doppler frequency.

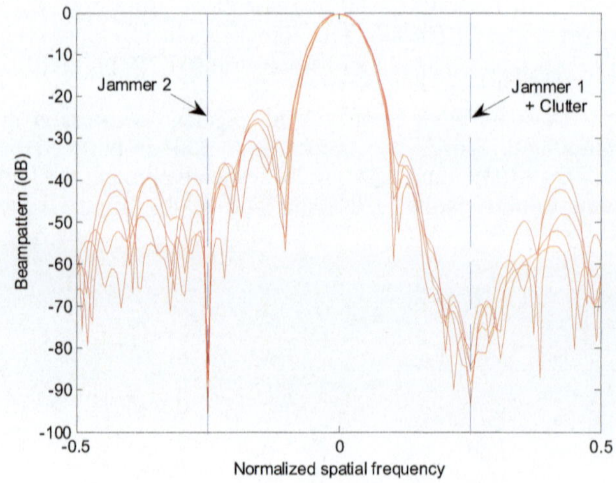

Fig. 8. Beampattern in the presence of jammers and clutter.

Figure 7 shows the IF performance against the Doppler frequency for the fixed $\tilde{L} = 200$ samples. From Fig. 7, we can see that the IF curve of the proposed method resembles that of the direct TRIA versus different Doppler frequencies. This implies that the excellent minimum detectable velocity (MDV) performance can be achieved by the proposed cascaded MIMO-STAP method with lower computational cost, which is beneficial for the detection of the low-speed target in the clutter plus jamming background.

Figure 8 shows the spatial beampattern at $f_{d,0} = 0.25$ in the presence of jammers and clutter of 5 independent trials by employing $\tilde{L} = 200$ samples. We can find that the nullings are formed accurately in the jamming directions $f_{sj,1} = 0.25$ and $f_{sj,2} = -0.25$, as well as the corresponding clutter notch, which further demonstrates the effectiveness of the proposed method.

5. Conclusion

Aiming at the typical jamming scenario, we propose a cascaded RD STAP method for airborne MIMO radar in this paper. Based on the different characteristics between jamming and clutter, we first cancel the jamming through utilizing the orthogonal property of its complementary subspace, with the reduction of the receive dimension. Next, the clutter is eliminated by TRIA based on the separation of the Doppler, transmit and receive weight vectors for MIMO-STAP. Simulation results demonstrate that the proposed method can provide excellent capability of clutter plus jamming suppression. Besides, the proposed cascaded MIMO-STAP method can effectively reduce the sample requirement and the computational burden, which is valuable for practical engineering application.

In our future research, the real scenario will be taken into consideration, including the clutter and the jammer. Meanwhile, more application tools can be utilized to execute our proposed method and produce practical results, such as field programmable gate array (FPGA) and C program, etc. Under the novel FPGA framework, the real-time requirement is a key factor when conducting the proposed cascaded MIMO-STAP algorithm. Thus, reasonable allocation and parallel implementation of the whole task are essential. On the other hand, efficient C program for realizing the conventional and fixed operations can be packaged into digital signal processor (DSP) and assembled with the new application system to further accelerate the processing speed.

Acknowledgments

This work was supported by the National Natural Science Foundation of China (61501513).

References

[1] FISHLER, E., HAIMOVICH, A., BLUM, R., et al. MIMO radar: an idea whose time has come. In *Proceedings of the IEEE Radar Conference*. Philadelphia (USA), 2004, p. 71–78. DOI: 10.1109/NRC.2004.1316398

[2] FISHLER, E., HAIMOVICH, A. M., BLUM, R. S., et al. Spatial diversity in radars—models and detection performance. *IEEE Transactions on Signal Processing*, 2006, vol. 54, no. 3, p. 823 to 838. DOI: 10.1109/TSP.2005.862813

[3] HAIMOVICH, A. M., BLUM, R. S., CIMINI, L. J. MIMO radar with widely separated antennas. *IEEE Signal Processing Magazine*, 2008, vol. 25, no. 1, p. 116–129. DOI: 10.1109/MSP.2008.4408448

[4] LI, J., STOICA, P. MIMO radar with collocated antennas. *IEEE Signal Processing Magazine*, 2007, vol. 24, no. 5, p. 106–114. DOI: 10.1109/MSP.2007.904812

[5] BLISS, D.W., FORSYTHE, K.W. Multiple-input multiple-output (MIMO) radar and imaging: degrees of freedom and resolution. In *Proceedings of the 37th Asilomar Conference on Signals, Systems and Computers*. Pacific Grove (USA), 2003, p. 54–59. DOI: 10.1109/ACSSC.2003.1291865

[6] BLISS, D.W., FORSYTHE, K.W., DAVIS, S.K., et al. GMTI MIMO radar. In *Proceedings of International Waveform Diversity and Design Conference*. Kissimmee (USA), 2009, p. 118–122. DOI: 10.1109/WDDC.2009.4800327

[7] ZATMAN, M. The applicability of GMTI MIMO radar. In *Proceedings of Asilomar Conference on Signals, System, and Computers*. Pacific Grove (USA), 2010, p. 2138–2142. DOI: 10.1109/ACSSC.2010.5757928

[8] WARD, J. *Space-Time Adaptive Processing for Airborne Radar*. Technical Report 1015, Lexington (USA): MIT Lincoln Laboratory, 1994

[9] GUERCI, J. R. *Space-Time Adaptive Processing for Radar*. Norwood (USA): Artech House, 2003. ISBN: 1580533779

[10] KLEMM, R. *Principles of Space-Time Adaptive Processing*. London (UK): IEE, 2002. ISBN: 0863415660

[11] WU, Y., TANG, J., PENG, Y.N. Models and performance evaluation for multiple-input multiple-output space-time adaptive processing radar. *IET Radar, Sonar and Navigation*, 2009, vol. 3, no. 6, p. 569–582. DOI: 10.1049/iet-rsn.2008.0025

[12] CHEN, C. Y., VAIDYANATHAN, P.P. MIMO radar space-time adaptive processing using prolate spheroidal wave functions. *IEEE Transactions on Signal Processing*, 2008, vol. 56, no. 2, p. 623 to 635. DOI: 10.1109/TSP.2007.907917

[13] ZHANG, W., HE, Z.S., LI, J., et al. Multistage multiple-beam beamspace reduced-dimension space-time adaptive processing for multiple-input-multiple-output radar based on maximum cross-correlation energy. *IET Radar, Sonar and Navigation*, 2015, vol. 9, no. 7, p. 772–777. DOI: 10.1049/iet-rsn.2014.0226

[14] XIANG, C., FENG, D.Z., LV, H. Three-dimensional reduced-dimension transformation for MIMO radar space-time adaptive processing. *Signal Processing*, 2011, vol. 91, no. 8, p. 2121–2126. DOI: 10.1016/j.sigpro.2011.01.017

[15] HE, J., FENG, D.Z., MA, L. Reduced-dimension clutter suppression method for airborne multiple-input multiple-output radar based on three iterations. *IET Radar, Sonar & Navigation*, 2015, vol. 9, no. 3, p. 249–254. DOI: 10.1049/iet-rsn.2014.0149

[16] REED, I., MALLETT, J., BRENNAN, L. Rapid convergence rate in adaptive arrays. *IEEE Transactions on Aerospace and Electronic Systems*, 1974, vol. 10, no. 6, p. 853–863. DOI: 10.1109/TAES.1974.307893

[17] KLEMM, R. Adaptive air- and spaceborne MTI under jamming conditions. In *Proceedings of IEEE National Radar Conference*. Boston (USA), 1993, p. 167–172. DOI: 10.1109/NRC.1993.270472

[18] RICHARDSON, P.G. STAP covariance matrix structure and its impact on clutter plus jamming suppression solutions. *Electronics Letters*, 2001, vol. 37, no. 2, p. 118–119. DOI: 10.1049/el:20010090

[19] HORN, R. A., JOHNSON, C. R. *Matrix Analysis*. New York (USA): Cambridge University Press, 1990. ISBN: 0-521-30586-1

About the Authors ...

Ting WANG was born in 1988. He is a Ph.D. candidate at Zhengzhou Institute of Information Science and Technology. He received his M.Sc. degree from Hefei Electronic Engineering Institute in 2013. His research interests include space-time adaptive processing and MIMO radar signal processing.

Yongjun ZHAO was born in 1964. He is a professor and Ph.D. advisor at Zhengzhou Institute of Information

Science and Technology. He received his M.Sc. degree from Shanghai Jiaotong University in 1992 and his Ph.D. degree from Beijing Institute of Technology in 2008. His research interests include radar signal processing and array signal processing.

Shiwen CHEN was born in 1974. He is an associate professor at Zhengzhou Institute of Information Science and Technology. He received his M.Sc. degree and Ph.D. degree from Zhengzhou Institute of Information Science and Technology in 2004 and 2013, respectively. His research interests include radar signal analyzing and processing.

Kunfan ZHANG was born in 1979. He is a lecturer at Zhengzhou Institute of Information Science and Technology. He received his M.Sc. degree and Ph.D. degree from Zhengzhou Institute of Information Science and Technology in 2003 and 2014, respectively. His research interests include radar system design and real-time signal processing.

Knowledge-Aided Non-Homogeneity Detector for Airborne MIMO Radar STAP

Ting WANG, Yongjun ZHAO, Jiantao WANG

Zhengzhou Institute of Information Science and Technology, Science Road 62, Zhengzhou, People's Republic of China

wangtingsp@163.com, zhaoyjzz@163.com, 18538324298@126.com

Abstract. *The target detection performance decreases in airborne multiple-input multiple-output (MIMO) radar space-time adaptive processing (STAP) when the training samples contaminated by interference-targets (outliers) signals are used to estimate the covariance matrix. To address this problem, a knowledge-aided (KA) generalized inner product non-homogeneity detector (GIP NHD) is proposed for MIMO-STAP. Firstly, the clutter subspace knowledge is constructed by the system parameters of MIMO radar STAP. Secondly, the clutter basis vectors are utilized to compose the clutter covariance matrix offline. Then, the GIP NHD is integrated to realize the effective training samples selection, which eliminates the effect of the outliers in training samples on target detection. Simulation results demonstrate that in non-homogeneous clutter environment, the proposed KA-GIP NHD can eliminate the outliers more effectively and improve the target detection performance of MIMO radar STAP compared with the conventional GIP NHD, which is more valuable for practical engineering application.*

Keywords

Airborne multiple-input multiple-output (MIMO) radar, space-time adaptive processing (STAP), knowledge-aid (KA), general inner product non-homogeneity detector (GIP NHD), outliers

1. Introduction

Multiple-input multiple-output (MIMO) radar has been an attractive research field in recent years [1–12]. Compared with traditional radar, relevant researches have demonstrated that MIMO radars own a lot of potential advantages, such as superior angle resolution, increased Doppler resolution, better clutter mitigation capability, etc. By utilizing airborne MIMO radar for the application of ground moving target indication (GMTI) [6–8], the characteristics of both the aperture size and the pulse integration interval can be improved to achieve lower minimum detectable velocity (MDV). Therefore, the specialty of MIMO radars will lead to better GMTI performance, and bring opportunity to the development of space-time adaptive processing (STAP) technique [8–12].

To realize GMTI in strong clutter environment based on STAP technique [13–15], the clutter covariance matrix has to be estimated from the training samples set. Besides, all the training samples should obey the independent and identical distribution (IID), and the required sample number should be at least as double as the system degrees of freedom (DOFs) according to the well-known Reed-Mallett-Brennan (RMB) rule [16]. In addition, STAP in airborne MIMO radar becomes even more challenging in complexity and convergence, because of the extra DOFs created by the orthogonal waveforms [8–12]. In practice, it is difficult to acquire enough IID samples to ensure the performance of MIMO radar STAP. Moreover, the training samples non-homogeneity will lead to the estimation error of the clutter covariance matrix and severely degrade the performance of MIMO-STAP [11], [17], [18]. Especially, as a classic factor for forming the non-homogeneous clutter environment, when there exist interference-targets (outliers) signals in the training samples set, the specific phenomenon named target self-nulling will be generated [17–22].

Non-homogeneity detector (NHD) [18–22] is well-known for its ability to improve the target detection performance of STAP in non-homogeneous environment. The generalized inner products (GIP) method [18–22] is a typical NHD for identifying the outliers with non-homogeneity, under the condition on the accurate estimation of the clutter covariance matrix. However, strong outliers may exist in the training samples set, which will result in severe performance degradation of the GIP NHD. Furthermore, if there are more than one strong outlier in the training samples set, the performance of the GIP NHD to determine and eliminate the weaker outliers will as well be deteriorated significantly [21], [22]. Recently, researchers have proposed the knowledge-aided (KA) STAP methods [22–28] to improve the STAP performance and its robustness in the practical application. Thus the prior knowledge, such as the system parameters, can be taken into consideration in the training sample selection, to guarantee that all the chosen samples satisfy the IID property. Reference [22] derived a robust NHD method to eliminate the outliers from the training samples set by utilizing the constructed covariance matrix based on the prolate spheroidal wave functions (PSWF). The method can be seen as a KA technique, which is also applicable to MIMO-STAP case with a slight modification. However, the computation and the application of PSWF are somewhat complicated, since it involves time-band-limited sampling theory and the computation of PSWF eigenvector [8], [22].

In this paper, we propose a KA-GIP NHD for airborne MIMO radar STAP, which is very simple and robust in practice. The clutter subspace knowledge is first calculated by the MIMO-STAP system parameters, and then the clutter basis vectors are utilized to construct the corresponding clutter covariance matrix conveniently. Next, the GIP NHD is integrated with KA-clutter covariance matrix to obtain the GIP statistics. Therefore, by comparing the statistics with the setting threshold, the outliers that contaminate the training samples are eliminated. The effectiveness of the proposed KA-GIP NHD for MIMO-STAP is verified by simulation results.

The remainder of the paper is organized as follows. In Sec. 2, we establish the STAP signal model for side-looking airborne MIMO radar. In Sec. 3, a brief review of the conventional GIP NHD is given and the effect of outliers on the conventional GIP NHD is analyzed to formulate the problem. Then in Sec. 4, we present the construction of the knowledge-aided clutter subspace of MIMO radar, and propose the novel GIP NHD for MIMO-STAP based on the KA-clutter subspace. The simulation results are provided in Sec. 5 to show the performance advantage of the KA-GIP NHD for MIMO-STAP over the conventional GIP NHD. In the end, Section 6 summarizes the conclusion of the paper.

Notations. The operations $(\cdot)^{\mathrm{T}}$, $(\cdot)^{\mathrm{H}}$, $(\cdot)^{-1}$, $\otimes$ denote the transpose, conjugate transpose, inverse, and Kronecker product, respectively. $E[\cdot]$ and $trace(\cdot)$ denote the expectation and trace of a matrix. $\mathbf{I}_M$ denotes the $M \times M$ identity matrix. $\delta(\cdot)$ stands for the unit impulse function, and $\lceil a \rceil$ stands for the smallest integer larger than a.

2. MIMO Radar STAP Signal Model

Figure 1 presents a sidelooking monostatic MIMO radar system equipped with collocated transmit and receive linear arrays, in which there are M transmit elements with uniform spacing d_{T} and N receive elements with uniform spacing d_{R}, and sparse coefficient $\alpha = d_{\mathrm{T}}/d_{\mathrm{R}}$, where $d_{\mathrm{R}} = \lambda/2$ and λ is the wavelength. A coherent processing interval (CPI) consists of K pulses with a constant radar pulse repetition interval (PRI) T. The radar platform travels at height H and at velocity V. The cone angle between the line-of-sight (LOS) and the velocity vector is ψ, while θ and φ are the azimuth angle and the elevation angle, respectively. The number of clutter patches which are uniformly distributed in a range cell is N_{c}. The whole space-

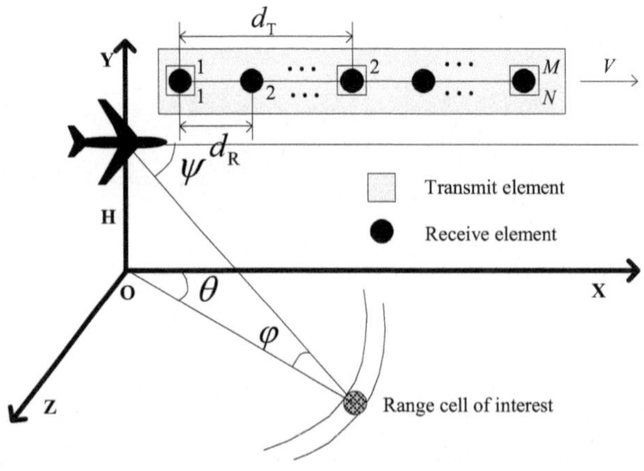

Fig. 1. Configuration of airborne MIMO radar.

time steering vector of MIMO-STAP for a certain clutter patch is written as

$$\mathbf{v}(f_{\mathrm{sc},i}, f_{\mathrm{dc},i}) = \mathbf{a}_{\mathrm{D}}(f_{\mathrm{dc},i}) \otimes \mathbf{a}_{\mathrm{T}}(f_{\mathrm{sc},i}) \otimes \mathbf{a}_{\mathrm{R}}(f_{\mathrm{sc},i}). \tag{1}$$

where

$\mathbf{a}_{\mathrm{D}}(f_{\mathrm{dc},i}) = [1 \quad e^{j2\pi f_{\mathrm{dc},i}} \quad \cdots \quad e^{j2\pi(K-1)f_{\mathrm{dc},i}}]^{\mathrm{T}}$
$= [1 \quad e^{j2\pi\beta f_{\mathrm{sc},i}} \quad \cdots \quad e^{j2\pi(K-1)\beta f_{\mathrm{sc},i}}]^{\mathrm{T}}$,
$\mathbf{a}_{\mathrm{T}}(f_{\mathrm{sc},i}) = [1 \quad e^{j2\pi\alpha f_{\mathrm{sc},i}} \quad \cdots \quad e^{j2\pi(M-1)\alpha f_{\mathrm{sc},i}}]^{\mathrm{T}}$ and
$\mathbf{a}_{\mathrm{R}}(f_{\mathrm{sc},i}) = [1 \quad e^{j2\pi f_{\mathrm{sc},i}} \quad \cdots \quad e^{j2\pi(N-1)f_{\mathrm{sc},i}}]^{\mathrm{T}}$ are steering vectors of temporal Doppler, spatial transmit array, and spatial receive array, respectively, $f_{\mathrm{dc},i} = 2VT\cos\theta_i\cos\varphi/\lambda$ and $f_{\mathrm{sc},i} = d_{\mathrm{R}}\cos\theta_i\cos\varphi/\lambda$ are the normalized Doppler and spatial frequencies of the clutter patch at θ_i, respectively, $\beta = f_{\mathrm{dc},i}/f_{\mathrm{sc},i} = 2VT/d_{\mathrm{R}}$ and $i = 1, 2, \cdots, N_{\mathrm{c}}$.

The space-time clutter-plus-noise data vector from the l th range cell is denoted as

$$\begin{aligned}\mathbf{x}(l) &= \mathbf{x}_{\mathrm{c}}(l) + \mathbf{x}_{\mathrm{n}}(l) \\ &= \sum_{i=1}^{N_{\mathrm{c}}} \rho_i \mathbf{v}(f_{\mathrm{sc},i}, f_{\mathrm{dc},i}) + \mathbf{x}_{\mathrm{n}}(l)\end{aligned} \tag{2}$$

where ρ_i is the reflect coefficient of the i th clutter patch, and $\mathbf{x}_{\mathrm{n}}(l)$ represents the additive Gaussian white noise.

The clutter covariance matrix can be expressed as [13], [27]

$$\mathbf{R}_{\mathrm{c}} = \sum_{i=1}^{N_{\mathrm{c}}} \sigma_{\mathrm{c},i}^2 \mathbf{v}(f_{\mathrm{sc},i}, f_{\mathrm{dc},i}) \mathbf{v}^{\mathrm{H}}(f_{\mathrm{sc},i}, f_{\mathrm{dc},i}) \tag{3}$$

where $\sigma_{\mathrm{c},i}^2$ is the variance of the i th clutter patch, which is in direct proportion to the radar cross section (RCS) and satisfies $E\left[\left|\rho_i \rho_k^*\right|\right] = \sigma_{\mathrm{c},i}^2 \delta(i-k)$, $i, k = 1, 2, \cdots, N_{\mathrm{c}}$.

Assume the noise obeys the Gaussian distribution and is white in both the spatial and temporal domains, thus the noise covariance matrix can be written as

$$\mathbf{R}_{\mathrm{n}} = \sigma_{\mathrm{n}}^2 \mathbf{I}_{KMN} \tag{4}$$

where σ_{n}^2 is the variance of the additive white noise.

The fully-adaptive weight vector for MIMO radar STAP can be calculated by

$$\mathbf{w} = \frac{\mathbf{R}_{\mathrm{c+n}}^{-1}\mathbf{v}(f_{\mathrm{s},0}, f_{\mathrm{d},0})}{\mathbf{v}^{\mathrm{H}}(f_{\mathrm{s},0}, f_{\mathrm{d},0})\mathbf{R}_{\mathrm{c+n}}^{-1}\mathbf{v}(f_{\mathrm{s},0}, f_{\mathrm{d},0})} \quad (5)$$

where $\mathbf{R}_{\mathrm{c+n}} = \mathbf{R}_{\mathrm{c}} + \mathbf{R}_{\mathrm{n}}$ is the clutter-plus-noise covariance matrix, and $\mathbf{v}(f_{\mathrm{s},0}, f_{\mathrm{d},0})$ is the space-time steering vector of target.

Usually, $\mathbf{R}_{\mathrm{c+n}}$ is unknown a priori in practice, thus it should be estimated from the adjacent training sample set beside the range cell under test

$$\hat{\mathbf{R}} = \frac{1}{L}\sum_{l=1}^{L}\mathbf{x}(l)\mathbf{x}^{\mathrm{H}}(l) \quad (6)$$

where $\mathbf{x}(l)$ denotes the training sample of the lth range cell. Through replacing $\mathbf{R}_{\mathrm{c+n}}$ by $\hat{\mathbf{R}}$, (5) can be used to calculate the weight vector of the practical MIMO-STAP system. However, with the dimension expansion of the MIMO-STAP system, it is tough to obtain sufficient IID samples in practice, since the practical clutter circumstances are generally non-homogeneous. In particular cases when there exist outliers in the training samples set [11], [17], [18], they will lead to the estimation error of the clutter covariance matrix and seriously degrade the performance of MIMO radar STAP. To solve the non-homogeneity problem, NHDs are utilized to remove outliers from the training samples set before implementing the STAP algorithms to airborne MIMO radar system.

3. Conventional GIP NHD

The GIP method [18–22] is a representative criterion of NHDs. We assume the original training samples set $\mathbf{\Omega}$ consists of L range cell samples in the adjacent clutter region of the range cell under test, i.e., $\mathbf{\Omega} = \{\mathbf{x}(l), l = 1,2,\ldots,L\}$, then we can define the GIP statistics as

$$\eta_{\mathrm{GIP}} = \mathbf{x}^{\mathrm{H}}(l)\mathbf{R}_{\mathrm{E}}^{-1}\mathbf{x}(l) = \left|\mathbf{R}_{\mathrm{E}}^{-1/2}\mathbf{x}(l)\right|^2 \quad (7)$$

where $\mathbf{R}_{\mathrm{E}}$ is the test covariance matrix estimated by using the original training samples in set $\mathbf{\Omega}$. The GIP statistics can be explained as the inner product of the whitening filter output vector by using $\mathbf{R}_{\mathrm{E}}^{-1/2}$, and the expectation of which can be obtained by

$$E\left[\eta_{\mathrm{GIP}}\right] = trace\left\{\mathbf{R}_{\mathrm{E}}^{-1}E\left[\mathbf{x}(l)\mathbf{x}^{\mathrm{H}}(l)\right]\right\} = D \quad (8)$$

where $D = KMN$ denotes the DOFs of MIMO-STAP system. Therefore it can be found from (8) that when the non-homogeneous training sample cannot be effectively whitened by the whitening matrix $\mathbf{R}_{\mathrm{E}}^{-1}$, the GIP statistics in (7) will deviate from expectation D. Then the non-homogeneous sample can be easily determined by the deviation level between the GIP statistics and the expectation D. Thus, we can eliminate the non-homogeneous training samples when estimating the clutter covariance matrix for MIMO radar STAP.

From the above analysis, we can see that the GIP statistics rely on the accurate estimation of the test covariance matrix $\mathbf{R}_{\mathrm{E}}$. However, different from the homogeneous data shown in (2), the non-homogeneous training sample can be expressed as

$$\mathbf{x}(l) = \mathbf{x}_{\mathrm{c}}(l) + \mathbf{x}_{\mathrm{n}}(l) + \Delta\mathbf{x}(l) \quad (9)$$

where $\Delta\mathbf{x}(l)$ is the additional item introduced by the non-homogeneous environment. Assume the training samples set $\mathbf{\Omega}$ contains P non-homogeneous training samples, and then the covariance matrix estimated by using the training samples in $\mathbf{\Omega}$ can be written as

$$\begin{aligned}
\hat{\mathbf{R}} &= \frac{1}{L}\sum_{l=1}^{L}\mathbf{x}(l)\mathbf{x}^{\mathrm{H}}(l) \\
&= \frac{1}{L}\sum_{l=1}^{L-P}[\mathbf{x}_{\mathrm{c}}(l)+\mathbf{x}_{\mathrm{n}}(l)][\mathbf{x}_{\mathrm{c}}(l)+\mathbf{x}_{\mathrm{n}}(l)]^{\mathrm{H}} \\
&\quad + \frac{1}{L}\sum_{l=L-P+1}^{L}[\mathbf{x}_{\mathrm{c}}(l)+\mathbf{x}_{\mathrm{n}}(l)+\Delta\mathbf{x}(l)][\mathbf{x}_{\mathrm{c}}(l)+\mathbf{x}_{\mathrm{n}}(l)+\Delta\mathbf{x}(l)]^{\mathrm{H}} \\
&= \frac{1}{L}\sum_{l=1}^{L}[\mathbf{x}_{\mathrm{c}}(l)+\mathbf{x}_{\mathrm{n}}(l)][\mathbf{x}_{\mathrm{c}}(l)+\mathbf{x}_{\mathrm{n}}(l)]^{\mathrm{H}} \\
&\quad + \frac{1}{L}\sum_{l=L-P+1}^{L}[\Delta\mathbf{x}(l)\Delta\mathbf{x}^{\mathrm{H}}(l)+(\mathbf{x}_{\mathrm{c}}(l)+\mathbf{x}_{\mathrm{n}}(l))\Delta\mathbf{x}^{\mathrm{H}}(l)+ \\
&\quad \Delta\mathbf{x}(l)(\mathbf{x}_{\mathrm{c}}(l)+\mathbf{x}_{\mathrm{n}}(l))^{\mathrm{H}}] \\
&= \mathbf{R}_{\mathrm{E}} + \Delta\mathbf{R}_{\mathrm{E}}.
\end{aligned} \quad (10)$$

From (10), we can see that the additional term $\Delta\mathbf{R}_{\mathrm{E}} = \frac{1}{L}\sum_{l=L-P+1}^{L}[\Delta\mathbf{x}(l)\Delta\mathbf{x}^{\mathrm{H}}(l)+(\mathbf{x}_{\mathrm{c}}(l)+\mathbf{x}_{\mathrm{n}}(l))\Delta\mathbf{x}^{\mathrm{H}}(l)+\Delta\mathbf{x}(l)(\mathbf{x}_{\mathrm{c}}(l)+\mathbf{x}_{\mathrm{n}}(l))^{\mathrm{H}}]$ will be produced if the training sample set $\mathbf{\Omega}$ is contaminated by the outliers, which has a negative effect on the GIP statistics. This will make the conventional GIP NHD difficult to distinguish the non-homogeneous training sample and cause the undetected error. Therefore, the non-homogeneity of the training samples will also degrade the performance of the conventional GIP NHD in identification of the outliers in the range cell under test [21], [22].

Hence, in order to enhance the robustness of the test covariance and eliminate the negative effects of the outliers on the conventional GIP NHD, we consider utilizing the KA technique in this paper. The MIMO-STAP system parameters, such as platform height and velocity, transmit array element number and spacing, receive array element number and spacing, temporal pulse number and interval, etc., are taken into account as priori knowledge to construct the clutter covariance matrix offline. Then the clutter covariance matrix in the GIP statistics is data-independent and only the homogenous clutter information of the range cell under test is included. Accordingly, the estimation of $\mathbf{R}_{\mathrm{E}}$ is not a requisite, and the knowledge-aided GIP NHD can realize the outliers elimination efficiently.

4. Knowledge-Aided GIP NHD for MIMO Radar STAP

4.1 KA Clutter Subspace Construction

Let $\mathbf{C}$ be clutter subspace of MIMO radar STAP. From (3), $\mathbf{C}$ can be spanned by all the space-time vectors of the N_c clutter patches in a range cell, that is

$$\mathbf{C}=span\left\{\mathbf{v}(f_{sc,i},f_{dc,i}):f_{dc,i}=\beta f_{sc,i},\ i=1,2,\cdots,N_c\right\}. \quad (11)$$

The clutter DOFs of MIMO radar, or the rank of the clutter covariance matrix $\mathbf{R}_c$, is denoted as [8]

$$r_C=rank(\mathbf{R}_c)=N+\alpha(M-1)+\beta(K-1). \quad (12)$$

Then the space-time vector of the ith clutter patch can be expressed as

$$\mathbf{v}(f_{sc,i},f_{dc,i})=$$

$$=\begin{bmatrix}1\\ e^{j2\pi f_{sc,i}}\\ \vdots\\ e^{j2\pi(N-1)f_{sc,i}}\\ e^{j2\pi\alpha f_{sc,i}}\\ \vdots\\ e^{j2\pi[\alpha(M-1)+N-1]f_{sc,i}}\\ e^{j2\pi\beta f_{sc,i}}\\ \vdots\\ e^{j2\pi[\beta(K-1)+\alpha(M-1)+N-1]f_{sc,i}}\end{bmatrix}=\mathbf{B}\begin{bmatrix}1\\ e^{j2\pi f_{sc,i}}\\ e^{j2\pi 2f_{sc,i}}\\ \vdots\\ e^{j2\pi[\beta(K-1)+\alpha(M-1)+N-1]f_{sc,i}}\end{bmatrix} \quad (13)$$

where $\mathbf{B}$ is a matrix of dimension $KMN\times r_C$, and the ith column and pth row element of $\mathbf{B}$ is denoted as

$$b_{i,p}=\begin{cases}1 & i=(k-1)MN+(m-1)N+n \\ & \text{and } p=\beta(k-1)+\alpha(m-1)+n \\ 0 & \text{others}\end{cases} \quad (14)$$

where $k = 1, 2,\ldots, K$, $m = 1, 2,\ldots, M$, and $n = 1, 2,\ldots, N$. The matrix $\mathbf{B}$ consists of r_C row vectors, that is $\mathbf{B} = [\mathbf{b}_1\,\mathbf{b}_2\ldots\mathbf{b}_{r_C}]$. From (13), the space-time vector of the ith clutter patch can be expressed as the linear combination of r_C row vectors of $\mathbf{B}$, which means these row vectors can span the clutter subspace of MIMO radar, that is, $\mathbf{C} = span\{\mathbf{b}_1\,\mathbf{b}_2\ldots\mathbf{b}_{r_C}\}$. And $\{\mathbf{b}_i\}$ is orthogonal to each other, $i = 1, 2,\ldots, N_c$. Therefore $\{\mathbf{b}_i\}$ can be seen as the clutter basis vectors for MIMO radar. The orthonormal basis vectors $\{\mathbf{b}_{c,i}\}$ for clutter can be derived by the normalizing operation on basis vectors $\{\mathbf{b}_{c,i}\}$.

It is should be mentioned that, in the case where α and β are non-integers because of the parameter preferences in reality, the clutter DOFs may not be an integer. Then the clutter rank should be set as the smallest integer larger than the non-integer, that is $r_C=\lceil N+\alpha(M-1)+\beta(K-1)\rceil$. In practice, $\mathbf{B}$ can be obtained through offline sampling and stored in memory. Thus, the training data is not required any more to carry out the clutter subspace estimation. In this section, the given clutter subspace is applicable to all the range cells and it can be calculated by the known parameters of transmit array element number M, receive array element number N, temporal pulse number K, coefficients α and β according to (14).

4.2 GIP NHD Based on the KA Clutter Subspace of MIMO Radar

Based on the theory of the KA clutter subspace presented in the above section, the clutter covariance matrix of MIMO radar can be constructed as

$$\mathbf{R}_{KA}=\sum_{i=1}^{r_C}\sigma_{b,i}^2\mathbf{b}_{c,i}\mathbf{b}_{c,i}^H \quad (15)$$

where $\sigma_{b,i}^2$ is the covariance corresponding to each $\mathbf{b}_{c,i}$ with the average value $\sigma_{b,i}^2=CNR\cdot\sigma_n^2=\mathbf{R}_c(1,1)/r_C=(\sum_{k=1}^{N_c}\sigma_{c,k}^2)/r_C$ to maintain the same clutter-to-noise ratio (CNR) with the clutter covariance matrix $\mathbf{R}_c$ in (3) [13], [14], under the precondition of the same noise variance σ_n^2. Consequently, we can calculate the proposed KA-GIP statistics by substituting (15) into (7)

$$\eta_{KA}=\mathbf{x}^H(l)\mathbf{R}_{KA}^{-1}\mathbf{x}(l). \quad (16)$$

It can be observed that the clutter covariance matrix in the proposed KA-GIP statistics merely contains the homogeneous clutter information in the range cell under test. Furthermore, the estimated clutter covariance matrix $\mathbf{R}_E$ will not have any influence on the KA-GIP statistics because the KA-clutter covariance matrix $\mathbf{R}_{KA}$ has replaced $\mathbf{R}_E$. So the non-homogeneous training samples can be identified and eliminated accurately from estimating the clutter covariance matrix for MIMO-STAP. Thus, the outliers can be eliminated from the training samples set more effectively by the proposed KA-GIP NHD, and the target detection performance of MIMO-STAP will be improved.

It is worth noting that these clutter basis vectors can be calculated offline by the MIMO-STAP system parameters (M, N, K, α, β) and memorized in storage. Compared with the PSWF-based clutter subspace knowledge in [22], the construction of the data-independent clutter covariance matrix proposed in this paper is more convenient, though these two methods can achieve approximately the same knowledge-aided effects. Besides, if we constructed the covariance matrix directly by (3) [13], [27], although the discrete representation can approximate the continuous distribution of the real ground clutter, the number and angles of the clutter patches are unknown definitely. This means that the parameters N_c and θ_i are undetermined, and the clutter patches have to be divided and located in the modeling process. Then the construction of the clutter matrix by the columns of $\mathbf{B}$ has lower computational com-

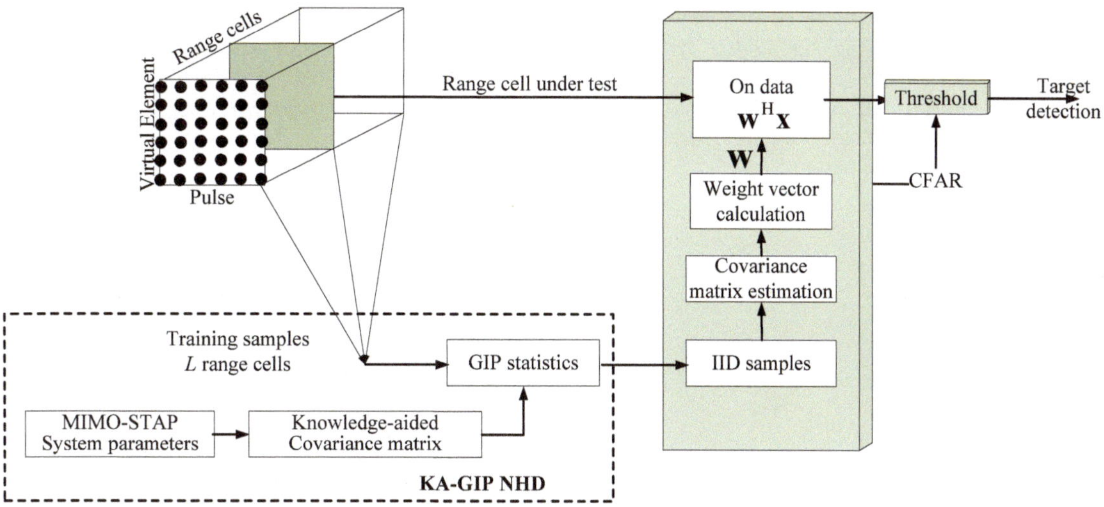

Fig. 2. MIMO-STAP procedure with KA-GIP NHD.

plexity, rather than by (3). Hence for the practical MIMO-STAP system, the proposed KA-GIP NHD is very simple to be executed.

The computational complexity of the proposed KA-GIP NHD is herein analyzed. Firstly, in (15), the construction of the KA clutter covariance matrix $\mathbf{R}_{\mathrm{KA}}$ can be realized through r_{C} multiplications of a $KMN \times 1$ vector $\mathbf{b}_{c,i}$ by its conjugate transpose $\mathbf{b}_{c,i}^{\mathrm{H}}$. In this step, the computational cost is $O(r_{\mathrm{C}}K^2M^2N^2)$. Then, equation (16) involves the inversion of matrix $\mathbf{R}_{\mathrm{KA}}$, and the computational complexity of $\mathbf{R}_{\mathrm{KA}}^{-1}$ is $O(K^3M^3N^3)$. Last, in (16) showing the GIP statistics, the computational burden of $\mathbf{x}^{\mathrm{H}}(l)\,\mathbf{R}_{\mathrm{KA}}^{-1}\,\mathbf{x}(l)$ aiming at L samples is $O(LK^2M^2N^2 + LKMN)$. Taking all these above factors together, we can come to the conclusion that the total computational complexity of the proposed KA-GIP NHD procedure is $O(r_{\mathrm{C}}K^2M^2N^2) + O(K^3M^3N^3) + O(LK^2M^2N^2 + LKMN)$.

Figure 2 illustrates the corresponding procedure of MIMO radar STAP with this newly-developed NHD. It also should be indicated that the spatial data of MIMO-STAP is equivalent to the $MN \times 1$ virtual array output after matched filtering [8]. The spatial data is corresponding to "Virtual Element" in Fig. 2, i.e., the data of virtual array elements. Meanwhile, the temporal data of MIMO-STAP belongs to each "Pulse" shown in Fig. 2. Afterwards, the spatial data and the temporal data in each range cell are integrated as a whole when processed via the KA-GIP NHD procedure, as we illustrated in the previous principle. In the final target detection step of Fig. 2, CFAR refers to "constant false alarm rate". CFAR detection is a common adaptive algorithm used in radar systems to detect target returns against a background of noise, clutter and interference. According to the changing strength level of the noise, clutter, and interference, the threshold level can be adjusted adaptively to maintain a constant probability of false alarm. Thus the target detection can be efficiently realized by comparing the MIMO-STAP output with the adaptive threshold. And this portion belongs to the next processing step after our proposed NHD.

5. Simulations

The basic MIMO-STAP parameters for the simulation are: transmit array element number $M = 5$, receive array element number $N = 8$, temporal pulse number $K = 10$, radar wavelength $\lambda = 0.23$ m, transmit array element spacing $d_{\mathrm{T}} = 0.92$ m, receive array element spacing $d_{\mathrm{R}} = 0.115$ m, $\alpha = N$, $\beta = 1$, pulse repetition interval $T = 500$ µs, platform velocity $V = 115$ m/s, platform height $H = 8000$ m, clutter patch number $N_{\mathrm{c}} = 180$. The CNR is 40 dB. We assume that the cone angle of each outlier is 90° and the normalized Doppler frequency is 0.25, which are the same as those of the target. A non-homogeneous clutter scenario is presented, and there are totally $L = 500$ range cells in the training set. It should be noted that the value of L is less than double of the system DOFs, i.e. $L < 2MNK$. This implies that the sample number requirement of RMB rule could not be satisfied, and the setting of the training sample number in this simulation accords with the real non-homogeneous clutter scenario. In this simulation four outliers are taken into consideration and Table 1 lists the corresponding outlier parameters.

Outlier number	Range cell	Power relative to noise [dB]
1	100	20
2	200	30
3	300	30
4	400	20

Tab. 1. Parameters of outliers.

Experiment 1. Verification of KA clutter subspace basis

Figure 3 shows the clutter power distribution on each KA clutter subspace basis vector of MIMO radar, which can be expressed as $\mathbf{b}_{c,i}^{\mathrm{H}}\mathbf{R}_{c}\mathbf{b}_{c,i}$, $i = 1, 2, \ldots, r_{\mathrm{C}}$. As comparison, Figure 3 illustrates the eigenvalues of $\mathbf{R}_{\mathrm{c}}$. The estimated clutter rank is correctly $N + \alpha(M-1) + \beta(K-1)$. From Fig. 3 we can see that the proposed KA subspace basis vectors possess almost all clutter power, which can be utilized to approximate the actual clutter subspace of MIMO radar. Compared with the eigenvalue decomposition (EVD) of the clutter covariance matrix, no samples are needed in obtaining the KA clutter subspace basis vectors, and these basis vectors can be easily updated with the corresponding MIMO radar system parameters.

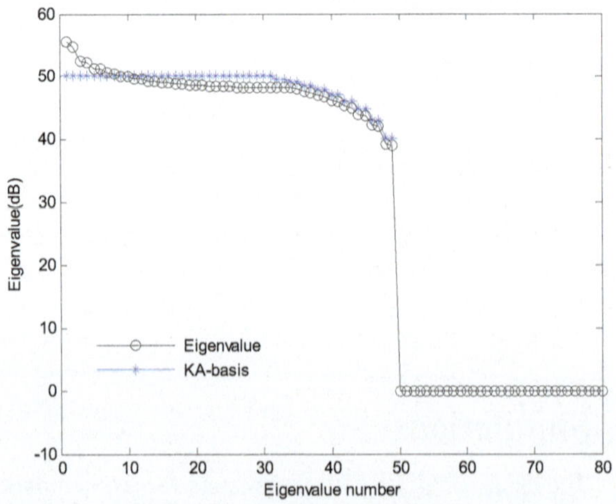

Fig. 3. Clutter power distribution on each KA clutter subspace basis.

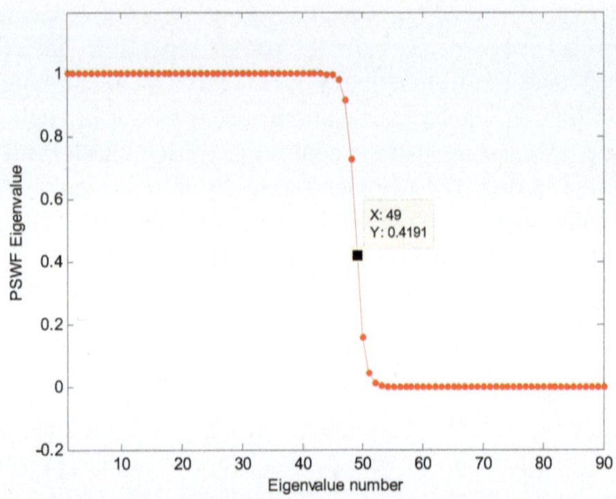

Fig. 4. PSWF eigenvalue.

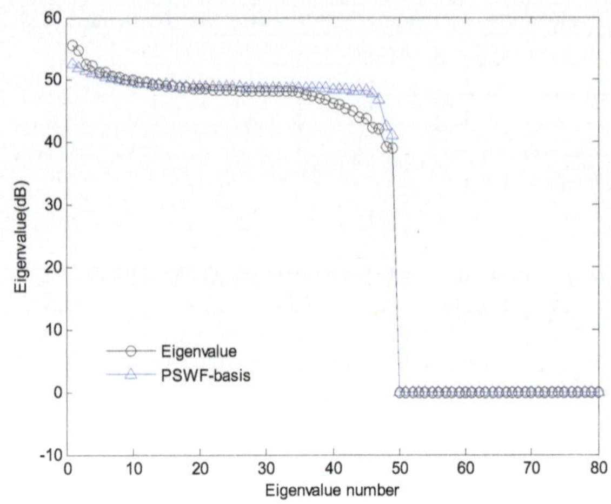

Fig. 5. Clutter power distribution on each PSWF clutter subspace basis.

The eigenvalues of PSWF and the clutter power distribution on each PSWF basis vector are shown in Fig. 4 and Fig. 5, respectively. From the simulation results, it is proved that the PSWF-GIP method in [22] can also be applied to MIMO radar STAP. However, the EVD computation for obtaining PSWF eigenvector has to be implemented when using the PSWF-GIP method with the computational complexity of $O(K^3M^3N^3)$. Thus, the KA-GIP NHD proposed in this paper is more convenient for application compared with the PSWF-GIP method, although the approximate KA consequence can be attained by both methods.

Experiment 2. Effectiveness and robustness of outliers detection

The corresponding calculation results of the conventional GIP statistics and the proposed KA-GIP statistics are shown in Fig. 6a) and b), respectively. For each method, the average values are obtained through 100 Monte Carlo trials. It should be noted that the Monte Carlo simulations, as well as other simulation experiments in this paper, are performed by MATLAB software on a computer with CPU Core i7 2.6 GHz and 8 GB of RAM. Comparing Fig. 6a) to Fig. 6b), it can be seen that the noise level in the background of Fig. 6b) is obviously much lower, and the proposed KA-GIP NHD improves the average statistics significantly. The conventional GIP NHD only detects two strong outliers, while the KA-GIP NHD has detected all the outliers. It indicates that our proposed NHD is much more effective for MIMO radar STAP than the conventional one.

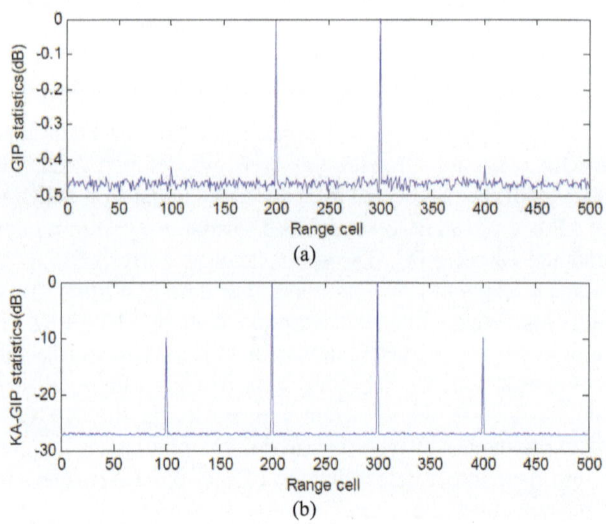

Fig. 6. Comparison of GIP statistics in non-homogeneous scenario. a) Conventional GIP statistics. b) KA-GIP statistics.

Experiment 3. Clutter suppression and target detection performance of MIMO radar STAP

The STAP performance is measured by improvement factor (*IF*), where *IF* is defined as the ratio of output signal-to-clutter-plus-noise-ratio (*SCNR*) to the input *SCNR*. Figure 7 shows the *IF* of the fully-adaptive MIMO-STAP after the outliers in the training samples set are eliminated by the KA-GIP NHD, compared with that for which the conventional GIP NHD is applied to eliminate the outliers. It can be seen that another notch emerges in the outlier Doppler frequency when using the conventional GIP NHD, which is the so-called target self-nulling phenomenon. On the contrary, after using the new NHD proposed in this paper to eliminate the outliers, the MIMO-STAP performance is improved significantly because only IID samples are taken into the STAP weight vector calculation and the effect of outliers has been totally avoided as well as the self-nulling phenomenon.

Suppose there exist two targets which are located in the 130th range cell and the 170th range cell, respectively. The signal-to-noise ratio (*SNR*) of the target in the 130th range cell is assumed to be 0 dB while the *SNR* of the other in the 170th range cell is –10 dB. The corresponding MIMO-STAP outputs are shown in Fig. 8, after applying the conventional GIP and the KA-GIP NHDs to remove the outliers, respectively. It can be observed from Fig. 8 that by STAP with conventional GIP NHD, the remaining clutter powers in all range cells under test are still too strong, and the weaker target signal in the 170th range cell cannot be detected accurately because of the self-nulling effect. At the same time, the peak value of the relatively stronger target in the 130th range cell is also not very distinct. On the other hand, MIMO-STAP with KA-GIP NHD can suppress clutter effectively and detect both targets correctly in the 130th and 170th range cells.

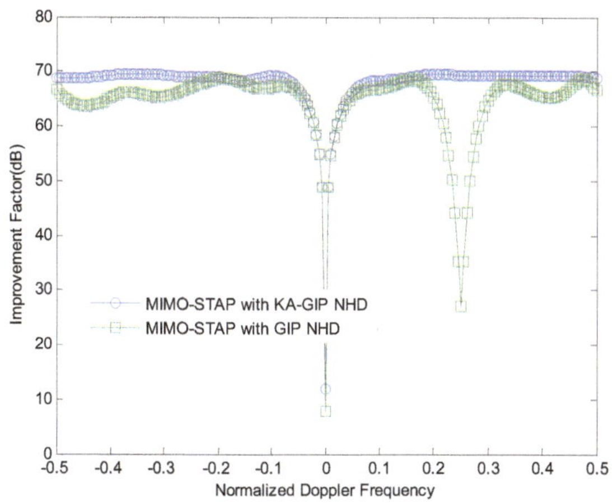

Fig. 7. Comparison of improvement factors.

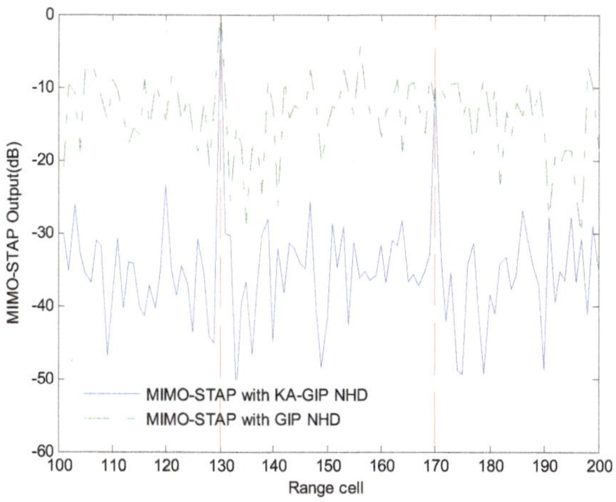

Fig. 8. Comparison of MIMO-STAP filtering outputs.

6. Conclusion

In this paper, we investigate the non-homogeneity detection technique in airborne MIMO radar STAP, and propose a knowledge-aided GIP NHD for MIMO-STAP. The KA-covariance matrix is constructed to replace the estimated covariance matrix, which is only determined by system parameters and not affected by the outliers at all. Due to this, the outliers can be eliminated more effectively by the novel NHD we put forward, and the target detection performance of MIMO radar STAP in non-homogeneous clutter environments will be improved significantly compared with the conventional GIP NHD. Moreover, the proposed KA-GIP NHD is very convenient to be implemented in practical MIMO-STAP system and has great value for engineering application.

The real scenario can be more complicated and severe than the case we considered in this paper, such as the very rough surface. Under that clutter condition, most regular training samples do not obey the property of IID, and it is difficult to sufficiently estimate the clutter characteristic. Thus, we should deal with the more complicated clutter environment in our future research, and promote relative practical experiment to test our method. Moreover, how to integrate priori knowledge with the real scenario and develop a more effective KA-GIP NHD for MIMO-STAP is also a valuable perspective which will be explored in the future work.

Acknowledgments

This work was supported by the National Natural Science Foundation of China (61501513).

References

[1] FISHLER, E., HAIMOVICH, A., BLUM, R., et al. MIMO radar: an idea whose time has come. In *Proceedings of the IEEE Radar Conference*. Philadelphia (USA), 2004, p. 71–78. DOI: 10.1109/NRC.2004.1316398

[2] HAIMOVICH, A. M., BLUM, R. S., CIMINI, L. J. MIMO radar with widely separated antennas. *IEEE Signal Processing Magazine*, 2008, vol. 25, no. 1, p. 116–129. DOI: 10.1109/MSP.2008.4408448

[3] LI, J., STOICA, P. MIMO radar with colocated antennas. *IEEE Signal Processing Magazine*, 2007, vol. 24, no. 5, p. 106–114. DOI: 10.1109/MSP.2007.904812

[4] FRANKFORD, M. T., STEWART, K. B., MAJUREC, N., et al. Numerical and experimental studies of target detection with MIMO radar. *IEEE Transactions on Aerospace and Electronic Systems*, 2014, vol. 50, no. 2, p. 1569–1575. DOI: 10.1109/TAES.2014.120180

[5] LI, N., CUI, G., KONG, L., et al. MIMO radar moving target detection against compound-Gaussian clutter. *Circuits, Systems, and Signal Processing*, 2014, vol. 33, no. 6, p. 1819–1839. DOI: 10.1007/s00034-013-9718-9

[6] BLISS, D. W., FORSYTHE, K. W. Multiple-input multiple-output (MIMO) radar and imaging: degrees of freedom and resolution. In *Proceedings of the 37th Asilomar Conference on Signals, Systems and Computers*. Pacific Grove (USA), 2003, p. 54–59. DOI: 10.1109/ACSSC.2003.1291865

[7] BLISS, D. W., FORSYTHE, K. W., DAVIS, S. K., et al. GMTI MIMO radar. In *Proceedings of International Waveform Diversity and Design Conference*. Kissimmee (USA), 2009, p. 118–122. DOI: 10.1109/WDDC.2009.4800327

[8] CHEN, C. Y., VAIDYANATHAN, P. P. MIMO radar space-time adaptive processing using prolate spheroidal wave functions. *IEEE Transactions on Signal Processing*, 2008, vol. 56, no. 2, p. 623–635. DOI: 10.1109/TSP.2007.907917

[9] WU, Y., TANG, J., PENG, Y. N. Models and performance evaluation for multiple-input multiple-output space-time adaptive processing radar. *IET Radar, Sonar & Navigation*, 2009, vol. 3, no. 6, p. 569–582. DOI: 10.1049/iet-rsn.2008.0025

[10] ZHANG, W., HE, Z.S., LI, J., et al. Beamspace reduced-dimension space-time adaptive processing for multiple-input-multiple-output radar based on maximum cross-correlation energy. *IET Radar, Sonar & Navigation*, 2015, vol. 9, no. 7, p. 772–777. DOI: 10.1049/iet-rsn.2014.0226

[11] AHMADI, M., MOHAMED-POUR, K. Space-time adaptive processing for phased-multiple-input-multiple-output radar in the non-homogeneous clutter environment. *IET Radar, Sonar & Navigation*, 2014, vol. 8, no. 6, p. 585–596. DOI: 10.1049/iet-rsn.2013.0246

[12] XIANG, C., FENG, D. Z., LV, H. Three-dimensional reduced-dimension transformation for MIMO radar space-time adaptive processing. *Signal Processing*, 2011, vol. 91, no. 8, p. 2121–2126. DOI: 10.1016/j.sigpro.2011.01.017

[13] WARD, J. *Space-Time Adaptive Processing for Airborne Radar*. Technical Report 1015, Lexington (USA): MIT Lincoln Laboratory, 1994.

[14] GUERCI, J. R. *Space-Time Adaptive Processing for Radar*. Norwood (USA): Artech House, 2003. ISBN: 1580533779

[15] KLEMM, R. *Principles of Space-Time Adaptive Processing*. London (UK): IEE, 2002. ISBN: 0863415660

[16] REED, I., MALLETT, J., BRENNAN, L. Rapid convergence rate in adaptive arrays. *IEEE Transactions on Aerospace and Electronic Systems*, 1974, vol. 10, no. 6, p. 853–863. DOI: 10.1109/TAES.1974.307893

[17] MELVIN, W. L. Space-time adaptive radar performance in heterogeneous clutter. *IEEE Transactions on Aerospace and Electronic Systems*, 2000, vol. 36, no. 2, p. 621–633. DOI: 10.1109/7.845251

[18] MELVIN, W. L., GUERCI, J. R. Adaptive detection in dense target environments. In *Proceedings of IEEE Radar Conference*, Atlanta (USA), 2001, p. 187–192. DOI: 10.1109/NRC.2001.922975

[19] SCHOENING, G. N., PICCIOLO, M. L., MILI, L. Improved detection of strong non-homogeneities for STAP via projection statisics. In *Proceedings of IEEE International Radar Conference*. Arlington (USA), 2005, p. 720–725. DOI: 10.1109/RADAR.2005.1435920

[20] TANG, B., TANG, J., PENG, Y. N. Detection of heterogeneous samples based on loaded generalized inner product method. *Digital Signal Processing*, 2012, vol. 22, no. 4, p. 605–613. DOI: 10.1016/j.dsp.2012.03.001

[21] RANGASWAMY, M., MICHELS, J. H., HIMED, B. Statistical analysis of the non-homogeneity detector for STAP applications. *Digital Signal Processing*, 2004, vol. 14, no. 3, p. 253–267. DOI: 10.1016/S1051-2004(03)00021-6

[22] YANG, X. P., LIU, Y. X., LONG, T. Robust non-homogeneity detection algorithm based on prolate spheroidal wave functions for space-time adaptive processing. *IET Radar, Sonar & Navigation*, 2013, vol. 7, no. 1, p. 47–54. DOI: 10.1049/iet-rsn.2011.0404

[23] STOICA, P., LI, J., ZHU, X. M., et al. On using a priori knowledge in space-time adaptive processing. *IEEE Transactions on Signal Processing*, 2008, vol. 56, no. 6, p. 437–444. DOI: 10.1109/TSP.2007.914347

[24] ZHU, X. M., LI, J., STOICA, P. Knowledge-aided space-time adaptive processing. *IEEE Transactions on Aerospace and Electronic Systems*, 2011, vol. 47, no. 2, p. 1325–1336. DOI: 10.1109/TAES.2011.5751261

[25] TANG, B., TANG, J., PENG, Y.N. Performance of knowledge aided space time adaptive processing. *IET Radar, Sonar & Navigation*, 2011, vol. 5, no. 3, p. 331–340. DOI: 10.1049/iet-rsn.2010.0131

[26] WU, Y., TANG, J., PENG, Y. N. On the essence of knowledge-aid clutter covariance estimate and its convergence. *IEEE Transactions on Aerospace and Electronic Systems*, 2011, vol. 47, no. 1, p. 569–585. DOI: 10.1109/TAES.2011.5705692

[27] YANG, Z., DE LAMARE, R. C., LI, X., et al. Knowledge-aided STAP using low rank and geometry properties. *International Journal of Antennas and Propagation*, 2014, vol. 2014, no. 2, p. 341–346. DOI: 10.1155/2014/196507

[28] ZHANG, S., HE, Z., LI, J., et al. A robust colored-loading factor optimization approach for knowledge-aided STAP. In *Proceedings of IEEE Radar Conference*. Philadelphia (USA), 2016, p. 1–5. DOI: 10.1109/RADAR.2016.7485266

About the Authors ...

Ting WANG was born in 1988. He is a Ph.D. candidate at Zhengzhou Institute of Information Science and Technology. He received his M.Sc. degree from Hefei Electronic Engineering Institute in 2013. His research interests include space-time adaptive processing and MIMO radar signal processing.

Yongjun ZHAO was born in 1964. He is a professor and Ph.D. advisor at Zhengzhou Institute of Information Science and Technology. He received his M.Sc. degree from Shanghai Jiaotong University in 1992 and his Ph.D. degree from Beijing Institute of Technology in 2008. His research interests include radar signal processing and array signal processing.

Jiantao WANG was born in 1984. He received his M.Sc. degree and Ph.D. degree from the National University of Defense Technology in 2008 and 2014, respectively. He is currently working as a lecturer at Zhengzhou Institute of Information Science and Technology. His research interests include radar signal processing and data processing.

Optimal Energy-Efficient Power Allocation Scheme with Low Complexity for Distributed Antenna System

Hao WANG[1], Xiangbin YU[1,3], Weiye XU[2], Benben WEN[1]*

[1]College of Electronic and Information Engineering, Nanjing University of Aeronautics and Astronautics, Nanjing, China
[2]School of Communication Engineering, Nanjing Institute of Technology, Nanjing, China
[3]National Mobile Communications Research Laboratory, Southeast University, Nanjing, China

*yxbxwy@gmail.com

Abstract. *In this paper, by maximizing the energy efficiency (EE), an optimal power allocation scheme is developed for downlink distributed antenna system (DAS). Different from conventional optimal power allocation schemes that need iterative calculation, the developed scheme can provide closed-form power allocation and no iteration is required. Based on the definition of EE, the optimized objective function is firstly formulated, and then a computationally efficient algorithm is proposed to obtain the optimal number of active remote antennas and the corresponding power allocation. Using the optimal number, the multidimensional solution for the optimized function is transformed into searching one-dimensional solution. As a result, closed-form expression of power allocation coefficients is attained. Numerical results verify the effectiveness of the proposed scheme. The scheme can obtain the same EE as the conventional optimal scheme but with lower complexity, and it has more accuracy than the existing low-complexity scheme.*

Keywords

Energy efficiency, distributed antenna system, power allocation, low complexity, Rayleigh channel

1. Introduction

Recently, green communication, where energy efficiency (EE) is pursued [1–3], has received much attention, since energy demand and prices increase dramatically. As an important technology for green communication, distributed antenna system (DAS) can expand cell coverage and improve the EE [4–6]. For this reason, different energy-efficient power allocation (PA) schemes have been developed for DAS [7–11]. In [7], using iterative search and numerical calculation, an approximate PA scheme is proposed for generalized DAS. An adaptive PA scheme for achieving maximum EE while satisfying spectral efficiency requirement in multiuser DAS is developed in [8]. In [9], by maximizing the upper bound of average EE, a suboptimal energy-efficient PA scheme is presented for DAS over composite fading channel. An optimal PA algorithm to maximize EE is proposed in [10], where the optimization problem is solved by using the Karush-Kuhn-Tucker (KKT) conditions and Lambert function. Considering that the above schemes need iterations, a low-complexity PA scheme aiming at maximizing EE is presented in [11], but there exist some errors in the theoretical derivations on PA. Hence, the accuracy of the optimal PA cannot be guaranteed. Besides, the above schemes basically consider single receive antenna for analysis convenience, and thus the corresponding EE performance will be limited.

Motivated by the reasons above, we propose a low-complexity energy-efficient power allocation scheme for DAS in composite Rayleigh channel, where path-loss, shadowing, fading, and multiple receive antennas are all considered. Also we present a new computationally efficient algorithm to achieve the optimal number of active remote antennas and the corresponding closed-form power allocation coefficients. It is shown that the proposed power allocation scheme and algorithm are both valid, the power allocation scheme can obtain the same energy efficiency as the conventional optimal one [10], and the algorithm has more accuracy than the existing low-complexity algorithm [11]. Furthermore, no iteration is required.

The rest of this paper is organized as follows. Section 2 introduces the system model. In Section 3, a low-complexity power allocation scheme is developed, and the comparison between different PA schemes is presented. Simulation results are provided in Sec. 4, and Section 5 concludes the paper.

The notations throughout this paper are as follows. Bold lower case letters denote column vectors. The superscript $(\cdot)^{\mathrm{T}}$ denotes transposition, $E\{\cdot\}$ denotes statistical expectation, $W(\cdot)$ denotes Lambert W function.

2. System Model

We consider a downlink DAS with N_{t} remote antennas (RAs) and multiple receive antennas operating in a single cell environment as illustrated in Fig. 1, where N_t RAs are distributed in the cell, the i-th RA is denoted as RA_i, and Base Station (BS) is referred as RA_1. All RAs are linked to the central processing unit via dedicated wired connection. The mobile station (MS) is equipped with N_t antennas, and the received signal at MS from RA_i is given by

$$\mathbf{y}_i = \sqrt{P_i}\left[h_i^1, \cdots, h_i^{N_r}\right]^{\mathrm{T}} x_i + \mathbf{z} \tag{1}$$

where P_i, $i = 1, \ldots, N_t$, is the transmit power consumed by the RA_i, x_i stands for the transmitted symbol from RA_i with unit energy, $\mathbf{z}$ is the complex Gaussian noise vector with zero mean and variance σ_n^2, and $h_i^j = g_i^j \Omega_i$, $j = 1, \ldots, N_r$, is the composite channel fading coefficient between RA_i and the j-th antenna of MS, where g_i^j denotes the small-scale Rayleigh fading between RA_i and the j-th receive antenna, $\Omega_i = (S_i d_i^{-\alpha_i})^{1/2}$ denotes the large-scale fading between RA_i and the MS, where α_i is the path loss exponent, d_i is the distance from RA_i to the MS, and S_i denotes the shadowing effect.

In general, EE is defined as the ratio of data transmission rate to the total consumed power. Based on this, the EE of DAS can be written as

$$\eta_{\mathrm{EE}} = \frac{\log_2\left(1 + \sum_{i=1}^{N_t} \gamma_i P_i\right)}{\sum_{i=1}^{N_t} P_i + P_c} \tag{2}$$

where $\gamma_i = \sum_{j=1}^{N_r} |g_i^j|^2 \Omega_i^2 / \sigma_n^2$ is defined as the channel to noise ratio (CNR) after maximal ratio combining at the receiver. Per RA power constraint is $0 \le P_i \le P_{\max,i}$, where $P_{\max,i}$ is the maximum transmit power available at RA_i. P_c denotes the circuit power and is a constant.

3. Low-complexity Power Allocation Scheme

In this section, we will develop an optimal power allocation scheme with low complexity, and present a computationally efficient algorithm to obtain the optimal number of active remote antennas and the corresponding PA coefficients.

Subject to the power constraint, we firstly formulate the objective function of the optimal PA as

$$\max_{\mathbf{P}} \quad \eta_{\mathrm{EE}} = \frac{\log_2(1 + \sum_{i=1}^{N_t} \gamma_i P_i)}{\sum_{i=1}^{N_t} P_i + P_c} \tag{3}$$
$$\text{subject to} \quad 0 \le P_i \le P_{\max,i}, \forall i \in \{1, \ldots, N_t\}$$

where $\mathbf{P} = [P_1, \ldots, P_{N_t}]^{\mathrm{T}}$. Without loss of generality, we assume that $\gamma_1 > \gamma_2 > \ldots > \gamma_{N_t}$. It has been proved in [10] that the optimal PA solution will have the following general form:

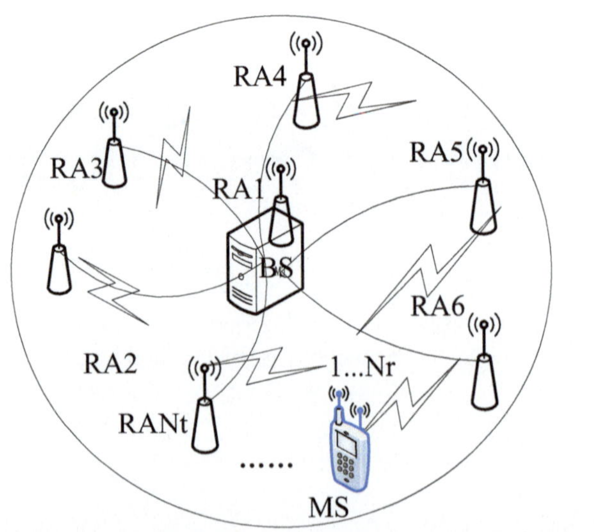

Fig. 1. A circle-cell DAS structure.

$$\mathbf{P}^* = [P_{\max,1}, \ldots, P_{\max,N_0-1}, P_{N_0}^*, \underbrace{0, \ldots, 0}_{N_t - N_0}]^{\mathrm{T}} . \tag{4}$$

According to this, the optimization of PA in (3) will become to search the optimal number of active RAs (i.e., N_0) and the corresponding PA. The optimized problem in (3) can be converted into finding the optimal solution of the following segmented function:

$$\psi(k, \mathbf{P}_k) = V_k(P_k), \quad k=1, \ldots, N_t \tag{5}$$

where $\mathbf{P}_k = [P_{\max,1}, \ldots, P_{\max,k-1}, P_{\max,k}, 0, \ldots, 0]^{\mathrm{T}}$ and $V_k(P_k)$ is shown as

$$V_k(P_k) = \frac{\log_2(1 + \sum_{i=1}^{k-1} \gamma_i P_{\max,i} + \gamma_k P_k)}{\sum_{i=1}^{k-1} P_{\max,i} + P_k + P_c} . \tag{6}$$

Obviously, (5) is a continuous function. Taking the derivative of $V_k(P_k)$ with respect to P_k yields

$$V_k'(P_k) = \frac{\gamma_k \Phi_k - (1 + \Psi_k)\ln(1 + \Psi_k)}{(1 + \Psi_k)\Phi_k^2 \ln 2} = \frac{A_k(P_k)}{B_k(P_k)} \tag{7}$$

where $\Phi_k = \sum_{i=1}^{k-1} P_{\max,i} + P_k + P_c$, $\Psi_k = \sum_{i=1}^{k-1} \gamma_i P_{\max,i} + \gamma_k P_k$. Since $B_k(P_k)$ is positive, the sign of $V'_k(P_k)$ is determined by $A_k(P_k)$ only.

Taking the derivative of $A_k(P_k)$ with respect to P_k gives

$$A_k'(P_k) = -\gamma_k \ln\left(1 + \sum_{i=1}^{k-1} \gamma_i P_{\max,i} + \gamma_k P_k\right) < 0 . \tag{8}$$

It means that $A_k(P_k)$ is decreasing in any segment. With (7), we have:

$$A_k(P_{\max,k}) = \gamma_k\left(\sum_{i=1}^{k} P_{\max,i} + P_c\right) - \left(1 + \sum_{i=1}^{k} \gamma_i P_{\max,i}\right)\ln\left(1 + \sum_{i=1}^{k} \gamma_i P_{\max,i}\right), \tag{9}$$

$$A_{k+1}(0)=\gamma_{k+1}\left(\sum_{i=1}^{k}P_{\max,i}+P_{c}\right)-\left(1+\sum_{i=1}^{k}\gamma_{i}P_{\max,i}\right)\ln\left(1+\sum_{i=1}^{k}\gamma_{i}P_{\max,i}\right). \tag{10}$$

Because $\gamma_k > \gamma_{k+1}$, $A_k(P_{\max,k}) > A_{k+1}(0)$. Based on this, considering that $A_k(P_k)$ is decreasing function in a segment, the $A_k(P_k)$ in different segments has the following relation:

$$\forall P_k, P_{k+1}, A_k(P_k) > A_{k+1}(P_{k+1}). \tag{11}$$

It means that $A_k(P_k)$ is always greater than $A_{k+1}(P_{k+1})$. According to the analysis above, the optimal PA, $P^*_{N_0}$ will be achieved. Specifically:

If $A_{N_0}(P_{\max,N_0})>0$ and $A_{N_0+1}(0)<0$ hold, then we have $V'_{N_0}(P_{\max,N_0})>0$ and $V'_{N_0+1}(0)<0$. From (11), it is easily found that $V_{N_0+1}(P_{N_0+1})$ is decreasing function for $0<P_{N_0+1}<P_{\max,N_0+1}$, and $V_{N_0}(P_{N_0})$ is increasing function for $0<P_{N_0}<P_{\max,N_0}$. Moreover, (5) is a continuous function. Hence, $P^*_{N_0}$ in (4) equals $P_{\max,N_0}$ under this case.

If $\exists 0<P^*_{N_0}<P_{\max,N_0}$, $A_{N_0}(P^*_{N_0})=0$, then with (11), for $k<N_0, A_k(P_k)>A_{N_0}(P^*_{N_0})=0$, so $V'_k(P_k)>0$ for $k<N_0$. Similarly, $V'_k(P_k)<0$ for $k>N_0$. According to (8), and considering $V'_{N_0}(P^*_{N_0})=A_{N_0}(P^*_{N_0})=0$, we have: $V'_{N_0}(P_{N_0})>0$ for $0\le P_{N_0}<P^*_{N_0}$, and $V'_{N_0}(P_{N_0})<0$ for $P^*_{N_0}<P_{N_0}<P_{\max,N_0}$. Hence $A_{N_0}(P_{N_0})=0$ has sole solution, i.e., $P^*_{N_0}$.

From (7), $A_{N_0}(P_{N_0})$ can be written as

$$A_{N_0}(P_{N_0})=\gamma_{N_0}\left(\sum_{i=1}^{N_0-1}P_{\max,i}+P_{N_0}+P_c\right)-\left(1+\sum_{i=1}^{N_0-1}\gamma_iP_{\max,i}+\gamma_{N_0}P_{N_0}\right)\ln\left(1+\sum_{i=1}^{N_0-1}\gamma_iP_{\max,i}+\gamma_{N_0}P_{N_0}\right). \tag{12}$$

Let $\phi_1=\gamma_{N_0}(\sum_{i=1}^{N_0-1}P_{\max,i}+P_c)$, $\phi_2=1+\sum_{i=1}^{N_0-1}\gamma_{N_0}P_{\max,i}$, then (12) is changed to

$$A_{N_0}(P_{N_0})=\phi_1+\gamma_{N_0}P_{N_0}-(\phi_2+\gamma_{N_0}P_{N_0})\ln(\phi_2+\gamma_{N_0}P_{N_0}). \tag{13}$$

Equating $A_{N_0}(P_{N_0})$ to zero gives

$$\phi_1-\phi_2=(\phi_2+\gamma_{N_0}P_{N_0})[\ln(\phi_2+\gamma_{N_0}P_{N_0})-1]. \tag{14}$$

With (14), we have:

$$\frac{\phi_1-\phi_2}{\mathrm{e}}=\frac{\phi_2+\gamma_{N_0}P_{N_0}}{\mathrm{e}}\ln\left(\frac{\phi_2+\gamma_{N_0}P_{N_0}}{\mathrm{e}}\right). \tag{15}$$

Using (15) and the Lambert function [12], we can obtain the solution of P_{N_0}, i.e., $P^*_{N_0}$, as follows:

$$P^*_{N_0}=\frac{1}{\gamma_{N_0}}\left\{\exp\left[W\left(\frac{\phi_1-\phi_2}{e}\right)+1\right]-\phi_2\right\} \tag{16}$$

where $W(\cdot)$ denotes the Lambert function. Equation (16) provides a closed-form expression of $P^*_{N_0}$ in (4). Based on the analysis above, a low-complexity optimal PA algorithm is proposed, and correspondingly, the algorithm procedure is summarized as follows:

Algorithm 1 Efficient PA calculation method

1: Sort the γ_i with descending order.

2: Calculating the $V_k(0)$, for $k = 1,\ldots, N_t$, and finding K that makes $V_K(0)$ be the largest among $\{V_k(0)\}$.

3: If $V'_K(0) > 0$ & $V'_{K-1}(P_{\max,K-1}) > 0$, then

  Return $N_0 = K$, go to 4.

Else

  If $V'_K(0) < 0$ & $V'_{K-1}(P_{\max,K-1}) < 0$ then

    Return $N_0 = K-1$, go to 4.

  Else

    Return $N_0=K-1, P_{N_0}=P_{\max,N_0}$, go to 5.

  End

End

4: Using (16) to obtain the $P^*_{N_0}$

5: Return $\mathbf{P}^*$ in (4), end the algorithm.

Using the above algorithm, the optimal number of active RAs is determined correctly, and the multi-dimensional root finding for the optimized objective function becomes one-dimensional root finding. As a result, closed-form power allocation is computed and no iteration is required. For this algorithm, the number of Lambert W function is calculated by one or zero. Note that in [11], a low-complexity PA scheme is also proposed for maximizing the energy efficiency in DAS, but the scheme cannot always guarantee the optimal values since there are errors in [11, (9)] and [11, (11)], i.e., the derived derivative $V'_k(P_k)$ and optimal PA $P^*_{N_0}$ are not accurate so that the obtained power value may exceed the given maximum transmit power. For example, using the CNRs as $\{\gamma_i\}_{i=1,\ldots,7}=$ {6.3608e+05, 4.8028e+05, 1.6016e+05, 1.3177e+05, 6.6877e+04, 6.4877e+03, 1.2921e+03}, we plot the corresponding numerical results of the energy efficiency in Fig. 2.

As shown in Fig. 2, the proposed scheme can be in good agreement with the conventional optimal scheme [10], but the existing low-complexity scheme [11] has obvious difference because the derived (9) and (11) in [11] have errors, which result in inaccurate power allocation.

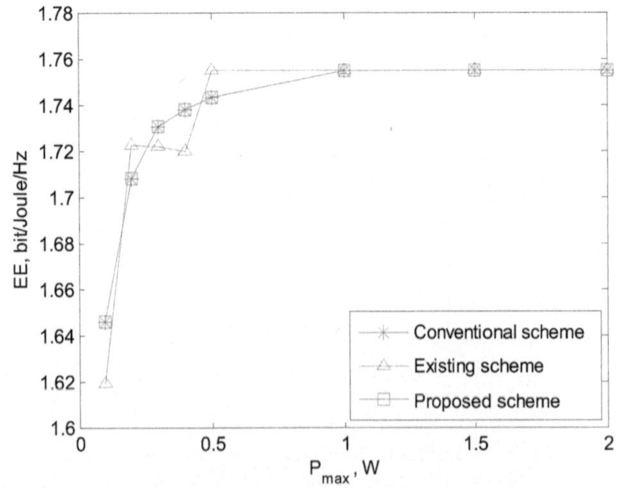

Fig. 2. EE comparison of DAS with different PA schemes.

	Proposed scheme	Scheme [11]	Scheme [10]
Lambert W function calculations	0 or 1	1	N_t
N_0	accurate	inaccurate	-
$P^*_{N_0}$	accurate	inaccurate	accurate

Tab. 1. Comparison of different schemes.

Taking the example of Fig. 2, let $P_{max,i} = P_{max} = 0.5$, then $V'_2(0) = -0.1460 < 0$, so $N_0 = 1$ according to the algorithm in [11] (this algorithm is based on the [(9),11]), but in fact, $V'_2(0) = 0.0417 > 0$, and correspondingly, $N_0 = 2$ according to our algorithm. Besides, using [(11), 11], $p^*_{N_0} = \frac{1}{\gamma_{N_0}}\left[\exp\left\{W\left(\frac{\phi_1-\phi_2}{e}\right)+1\right\}-\left(\sum_{i=1}^{N_0-1} P_{max,i}+P_c\right)\right] = 0.8219 > 0.5$, i.e., $p^*_{N_0} > P_{max}$, but according to our algorithm, $p^*_{N_0} = \frac{1}{\gamma_{N_0}}\left[\exp\left\{W\left(\frac{\phi_1-\phi_2}{e}\right)+1\right\}-\left(1+\sum_{i=1}^{N_0-1} \gamma_i P_{max,i}\right)\right] = 0.1650 < 0.5$, where $\varphi_1 = \gamma_{N_0}(\sum_{i=1}^{N_0-1} P_{max,i} + P_c)$ and $\varphi_2 = 1 + \sum_{i=1}^{N_0-1} \gamma_i P_{max,i}$. Obviously, the result from [(11), 11] is larger than P_{max,N_0}, whereas our result can satisfy the maximum power constraint.

The above results indicate that the proposed scheme is valid. Besides, the complexity analysis measured by the number of Lambert function calculation and comparison among three schemes are listed in Tab. 1. From Tab. 1, it is shown that the proposed scheme has lower calculation complexity than the other two and is more accurate than the existing scheme [11].

4. Simulation Results

In this section, we will assess the validity of the proposed scheme via computer simulation. In simulation, it is assumed that $P_{max,i} = P_{max}$ ($i = 0, \ldots, N_t$) for analysis convenience. The BS (RA_1) is in the center of the cell and the RAs are evenly and symmetrically placed in the cell. The main parameters used in simulations are listed in Tab. 2. The simulation results are shown in Fig. 3 and Fig. 4.

Parameters	value
Number of remote antennas N_t	7
Path loss exponent α_i	3, 3.5, 4
Shadow fading standard deviation σ_i	8 dB
Noise power N_0	–104 dBm
Circuit power P_c	10 W
Cell radius D	1000 m
MS distribution	Uniform
(D_1, τ_1)	(0,0)
(D_i, τ_i), $i = 2, \ldots, N_t$	$(\sqrt{\frac{3}{7}}D, \frac{2\pi(i-1)}{N_t-1})$
Number of receive antennas N_r	1, 2, 3

Tab. 2. Simulation parameters.

Figure 3 illustrates the EE performance of DAS with different receive antennas. Considering that the optimal scheme in [10] is based on single receive antenna, we extend the scheme in [10] to multiple receive antennas case after some modifications. From Fig. 3, it is observed that the EE of the proposed scheme is the same as that of the conventional optimal scheme [10]. Moreover, the EE increases with the growth of the number of receive antennas. Namely, the EE of DAS with three receive antennas ($N_r = 3$) is higher than that with two receive antennas ($N_r = 2$) because of more spatial diversity gain, and due to the same reason, the EE of DAS with two receive antennas ($N_r = 2$) is higher than that with single receive antenna ($N_r = 1$). The results above indicate that the developed scheme is effective and reasonable, and the application of multiple receive antennas does improve the EE performance effectively.

In Fig. 4 we plot the EE performance of DAS with different path loss exponents, where two receive antennas are considered. It is found that the EE of the proposed scheme is still consistent with that of conventional optimal

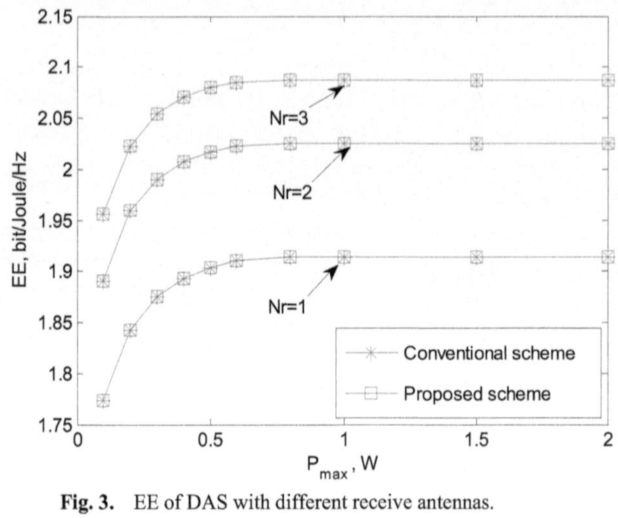

Fig. 3. EE of DAS with different receive antennas.

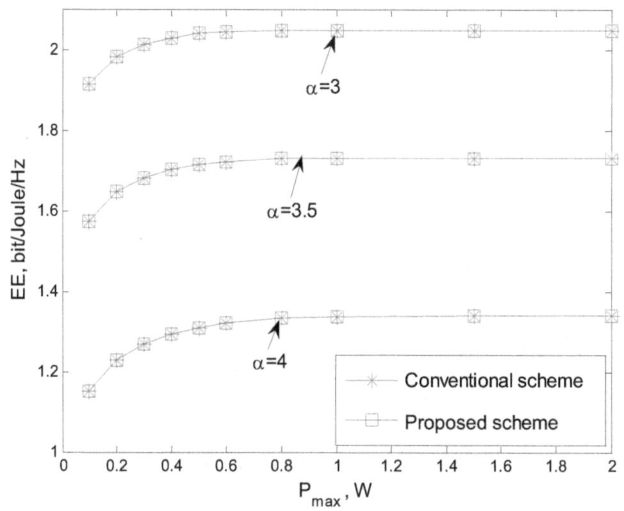

Fig. 4. EE of DAS with different path loss exponents.

scheme [10]. Besides, the EE decreases with the growth of the path loss exponent α as expected. Namely, the EE of DAS with $\alpha = 4$ is lower than that with $\alpha = 3.5$, and the EE of DAS with $\alpha = 3.5$ is lower than that with $\alpha = 3$. This is because the system with higher path loss exponent will experience greater path loss, which leads to a decline in EE performance.

5. Conclusions

We have developed an optimal energy-efficient power allocation scheme for DAS using multiple receive antennas in composite Rayleigh fading channel, and proposed a computationally efficient algorithm to obtain the optimal number of active RAs. Based on this, optimal power allocation is derived, and the corresponding closed-form expression is attained. The results indicate that our scheme has lower complexity than the existing schemes. Moreover, it can obtain the same EE performance as the existing optimal scheme, and provide the optimal active antenna number and power allocation accurately.

Acknowledgements

This work is supported by the National Natural Science Foundation of China (61571225, 61571224), Foundation of Graduate Innovation Center in NUAA (kfjj20160409), National High Technology Research and Development Program of China (2015AA01A709), Open Research Fund of National Mobile Communications Research Laboratory of Southeast University (2017D03), Qing Lan Project of Jiangsu, and Six Talent Peaks Project in Jiangsu (DZXX-007).

References

[1] LI, G. Y., XU, Z., XIONG, C., et al. Energy-efficient wireless communications: tutorial, survey, and open issues. *IEEE Transactions on Wireless Communications*, 2011, vol. 18, no. 6, p. 28–35. DOI: 10.1109/MWC.2011.6108331

[2] FENG, D., JIANG, C., LIM, G., et al. A survey of energy-efficient wireless communications. *IEEE Communication Surveys and Tutorials*, 2013, vol. 15, no. 1, p. 167–178. DOI: 10.1109/SURV.2012.020212.00049

[3] HELIOT, F., IMRAN, M. A., TAFAZOLLI, R. On the energy efficiency-spectral efficiency trade-off over the MIMO Rayleigh fading channel. *IEEE Transactions on Communications*, 2012, vol. 60, no. 5, p. 1345–1356. DOI: 10.1109/TCOMM.2012.031712.110215

[4] HEATH, R., PETERS, S., WANG, Y., ZHANG, J. A current perspective on distributed antenna systems for the downlink of cellular systems. *IEEE Communications Magazine*, 2013, vol. 51, no. 4, p. 161–167. DOI: 10.1109/MCOM.2013.6495775

[5] PARK, E., LEE, S. R., LEE, I. Antenna placement optimization for distributed antenna systems. *IEEE Transactions on Wireless Communications*, 2012, vol. 11, no. 6, p. 2468–2477. DOI: 10.1109/TWC.2012.051712.110670

[6] KIM, H., LEE, S. R., LEE, K. J., LEE, I. Transmission schemes based on sum rate analysis in distributed antenna systems. *IEEE Transactions on Wireless Communications*, 2012, vol. 11, no. 3, p. 1201–1209. DOI: 10.1109/TWC.2012.011812.111008

[7] CHEN, X., XU, X., TAO, X. Energy efficient power allocation in generalized distributed antenna system. *IEEE Communications Letters*, 2012, vol. 16, no. 7, p. 1022–1025. DOI: 10.1109/LCOMM.2012.051512.120241

[8] HE, C., SHENG, B., ZHU, P. C., YOU, X. H. Energy efficiency and spectral efficiency trade-off in downlink distributed antenna systems. *IEEE Wireless Communications Letters*, vol. 1, no. 3, p. 153–156. DOI: 10.1109/WCL.2012.022812.120048

[9] WANG, Y., YU, X. B., LI, Y., WU, B. B. Energy efficient power allocation for distributed antenna system over shadowed Nakagami fading channel. *Radio Engineering*, 2015, vol. 24, no. 4, p. 1077 to 1083. DOI: 10.13164/re.2015.1077

[10] KIM, H., LEE, S. R., SONG, C., LEE, K. J., LEE, I. Optimal power allocation scheme for energy efficiency maximization in distributed antenna systems. *IEEE Transactions on Communications*, 2015, vol. 63, no. 2, p. 431–440. DOI: 10.1109/TCOMM.2014.2385772

[11] WU, J., LIU, W. J., YOU, X. Low-complexity power allocation for energy efficiency maximization in DAS. *IEEE Communications Letters*, 2015, vol. 19, no. 6, p. 925–928. DOI: 10.1109/LCOMM.2015.2415781

[12] CORLESS, R. M., GONNET, G. H., HARE, D. E., JEFFREY, D. J., KNUTH, D. E. On the Lambert W function. *Advances in Computational Mathematics*, 1996, vol. 5, p. 329–359. DOI: 10.1007/BF02124750

About the Authors ...

Hao WANG was born in Jiangsu, China. He is currently working towards the M.Sc. degree at Nanjing University of Aeronautics and Astronautics (NUAA).

Xiangbin YU (corresponding author) was born in Jiangsu, China. He received his Ph.D. in Communication and Information Systems in 2004 from the National Mobile Communications Research Laboratory at Southeast University, China. He is a full Professor of Information and Communication Engineering at NUAA. Currently, he also works as a Visiting Scholar at University of Delaware, USA. He has served as a technical program committee of International Conference on Communications (ICC 2016, 2015), Globecom'2006, International Conference on Communications Systems (ICCS 2008, 2010), International Conference on Communications and Networking in China (Chinacom 2010, 2014) and International Conference on Wireless Communications and Signal Processing 2011. He has been a member of IEEE ComSoc Radio Communications Committee (RCC) since May 2007. His research interests include MIMO, distributed antenna systems, energy efficiency, adaptive modulation and cross-layer design.

Weiye XU was born in Neimengu, China. She received her M.S. degree in Communication and Information Systems from Hohai University. She is an associate professor at the School of Communication Engineering at Nanjing Institute of Technology. Her research interests include digital communication, MIMO technique, and distributed antenna systems.

Benben WEN was born in Henan, China. He is currently working towards the M.Sc. degree at Nanjing University of Aeronautics and Astronautics.

Superstrate-integrated Switchable Beam Rectangular Microstrip Antenna for Gain Enhancement

Supakit KAWDUNGTA [1], *Puwanai JAIBANAUEM* [1], *Rattiya PONGGA* [1], *Chuwong PHONGCHAROENPANICH* [2]

[1] Faculty of Engineering, Rajamangala University of Technology Lanna, Chiang Mai 50300, Thailand
[2] Faculty of Engineering, King Mongkut's Institute of Technology Ladkrabang, Bangkok 10520, Thailand

supakitting@rmutl.ac.th, kpchuwon@kmitl.ac.th, pchuwong@gmail.com

Abstract. *This research has proposed a switchable beam rectangular microstrip antenna with double parasitic patches, in which two PIN diodes were deployed for manipulation of the main beam direction and a superstrate (i.e. either a dielectric slab or metamaterial) for enhancement of the antenna gain. The dielectric slab is a second FR4 substrate while the metamaterial (MTM) is the 7 × 17 periodic structure of planar cycloid dipoles (PCD). Simulations were carried out and three different antenna prototypes (i.e. the proposed switchable beam rectangular microstrip antenna, the proposed antenna either with dielectric slab or MTM) fabricated and experimented. The simulation and experimental results are in good agreement and exhibit good impedance matching ($|S_{11}| < -10$ dB) along the operating frequency. The average measured gain is 7 dBi with the unidirectional radiation pattern along the operating frequency. The proposed switchable antennas with and without the superstrate are operable in the 2.4-2.5 GHz WLAN system and switchable in three directions of 0°, 30° and 330° (xy-plane). Moreover, the findings validate the applicability of either the dielectric slab or three MTM block-layers as the superstrate to improve the antenna gain.*

Keywords

Gain enhancement, metamaterial, microstrip, parasitic patch, PIN diode, switched beam, superstrate

1. Introduction

The exponential growth in smartphone use, together with the greater availability of budget internet-connected devices, has contributed to a phenomenon in which the demand for digital wireless communications outstrips the supply (i.e. capacity). In addition, the digital wireless technology suffers from the multipath fading effect where the waves reflect and scatter upon impacting the obstacles. According to [1], [2], these issues could be efficiently addressed with the utilization of a smart antenna, beam steering antenna or switchable beam antenna.

In [1], the authors applied the path-diversity reception technique of the constant modulus algorithm (CMA) to an adaptive array antenna (AAA) to reduce the co-channel interference (CCI) in the terrestrial digital mobile communications. The aim was to efficiently utilize the frequency spectrum. In [2], a smart antenna based on the beamforming technique with spatial diversity using a fuzzy interference system and a neural network (NN) was proposed to reduce the signal loss associated with low beamforming. A genetic algorithm (GA) was used in the optimization of the fuzzy logic and NN, and the technique could achieve a lower side lobe level (SLL) than did the non-algorithm-integrated methods.

By comparison, the switchable beam antenna is less costly and of lower profile. Nevertheless, its beam steering is of discrete fashion. In [3], the switchable beam disc antenna was proposed for a wireless communications network. The main beam of the antenna could be rotated 360° around the azimuth angle and possesses a wide impedance bandwidth. The four elements of the slot antenna with single feed point were also presented. This antenna controls the main beam by alternating the ON/OFF state of four PIN diodes over the azimuth angle [4].

In [5], the authors proposed a switchable beam antenna using a printed circuit board (PCB) and an FR4 substrate with a square microstrip ring structure. A PIN diode was also utilized to manipulate the radiation pattern. In [6], the authors proposed a single-fed switchable beam antenna using the quasi-Yagi structure and a PIN diode to select an antenna director. Meanwhile, further improvements were made to the switchable beam antenna for the dual-band operation (2.4 GHz and 5.8 GHz) [7] using frequency selective surfaces [8]. In addition, several other components have been utilized to manipulate the antenna main beam, e.g. a transistor, field effect transistor (FET), and radio frequency microelectromechanical system (RF MEMS) switch [9], [10].

Wireless communications require a high-gain antenna, especially for the point-to-point and point-to-multipoint applications. Many gain improvement techniques have been proposed and experimented, e.g. the multilayer substrates or superstrates in the printed antenna [11], the metamaterial (MTM) to tune the impedance matching [12]. In addition, the multiband loaded monopole was integrated into the MTM to achieve the multiband operation for the WLAN/WiMAX applications [13]. In [14], the MTM was deployed in the design of an array antenna and thereby could suppress the mutual coupling while achieving the higher directivity. Moreover, the MTM was utilized to improve the impedance bandwidth [15], [17].

This current research proposes a switchable beam rectangular microstrip antenna with double parasitic patches. Two PIN diodes are deployed for manipulation of the main beam direction and a superstrate (i.e. either a dielectric slab or metamaterial) for enhancement of the antenna gain. The dielectric slab is a second FR4 substrate while the metamaterial (MTM) is the 7 × 17 periodic structure of planar cycloid dipoles (PCD). The proposed switchable antennas with and without the superstrate are operable in the 2.4-2.5 GHz WLAN system and switchable in three directions of 0°, 30° and 330° (*xy*-plane).

The organization of this research is as follows: Section 1 is the introduction. Section 2 deals with the proposed metamaterial (MTM), while Section 3 is concerned with the switchable beam antenna structure and design procedure. Section 4 discusses the deployment of the superstrate (either the dielectric slab or the MTM) to enhance the antenna gain. Three different prototype antennas and their respective experimental results are presented in Sec. 5. The concluding remarks are provided in Sec. 6.

2. The Configuration of Metamaterial (MTM)

Metamaterials (MTM) are synthetic periodic-structure materials with a negative refractive index n or with negative permittivity ε and permeability μ. MTM have been utilized in a variety of applications, including in ultra-wideband (UWB) antennas, monopole antennas, lens antennas and waveguides [15–17]. For double negative (DNG) media, the thin metal wire (TW) structure and the split ring resonator (SRR) structure are collectively employed for negative ε and μ.

Figure 1 illustrates the schematic of a *planar cycloid dipole* (PCD) of the MTM structure functioning as the DNG media. The PCD is fashioned from TW and SRR. In the study, a PCD was simulated under the normal incidence in a waveguide using CST Microwave Studio [18] and, due to its symmetrical characteristic, only S_{11} and S_{21} were investigated [19], [20].

Figure 2(a) illustrates $|S_{11}|$ and $|S_{21}|$ in dB within the frequency range of 1–4 GHz. In Fig. 2(a), $|S_{11}|$ and $|S_{21}|$ are at their respective peak and trough at about the frequency of 2.45 GHz, where the effective constitutive parameters could be extracted from the S parameters using the Nicolson-Ross-Weir (NRW) method [19], [20].

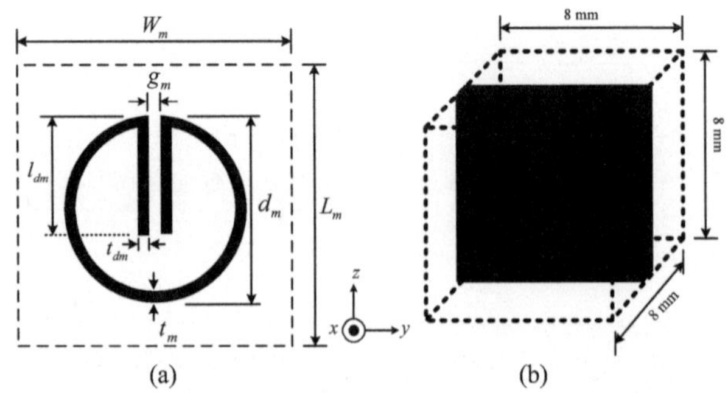

Fig. 1. The geometry of the planar cycloid dipole: (a) 2D view, (b) 3D view.

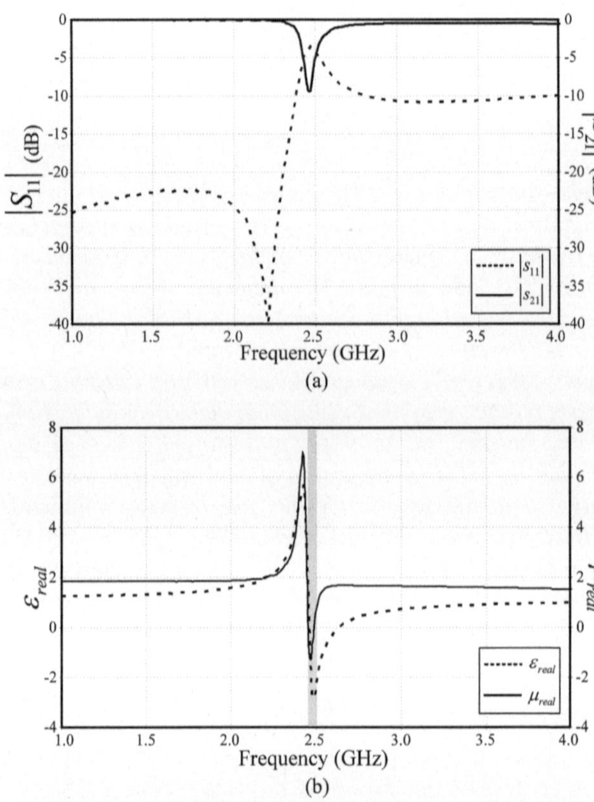

Fig. 2. The simulated results of the planar cycloid dipole: (a) $|S_{11}|$ and $|S_{21}|$ in dB, (b) the *real* effective permittivity and permeability.

Parameter	Description	Physical Size (mm)
W_m	Total width of MTM	8.00
L_m	Total length of MTM	8.00
d_m	Cycloid diameter	6.80
t_m	Cycloid thickness	0.20
g_m	Distance between two dipoles	0.20
t_{dm}	Dipole thickness	0.15
l_{dm}	Dipole length	5.60

Tab. 1. The parameters and physical dimensions of the planar cycloid dipole.

Based on Tab. 1, the simulated $|S_{11}|$ and $|S_{21}|$ were estimated and subsequently extracted for the real effective permittivity and permeability. In its design, the planar cycloid dipole (PCD) is mounted on an FR4 substrate

($\varepsilon_r = 4.3$) with a thickness of 1.8 mm. The simulation results indicated that the real effective permittivity and permeability of the PCD were both negative in the 2.43–2.5 GHz frequency range, as shown in Fig. 2(b).

3. Switchable Beam Antenna Structure

Figure 3 illustrates the schematic of the initial rectangular microstrip antenna whose feeding structure is that of a microstrip line. Initially, the microstrip feed line was simulated at the center frequency f_r of 2.45 GHz on an FR4 substrate ($\varepsilon_r = 4.3$) with a thickness h of 3.2 mm. The design of the feed line was carried out, whereby the patch antenna and the microstrip line match with the characteristic impedance of 50 Ω [21]. The thickness t and length l_2 of the microstrip line were thus 6.16 mm and 17 mm, respectively. The position of the feed line was on one edge of the rectangular patch.

Meanwhile, the design of the rectangular patch was carried out using the transmission line model with the dominant mode (TM_{010}), as expressed in (1). The initial width W and length L of the rectangular patch were 37.6 mm and 61.2 mm (half wavelength) at 2.45 GHz.

$$W = \frac{1}{2 f_r \sqrt{\mu_0 \varepsilon_0}} \sqrt{\frac{2}{\varepsilon_r + 1}}. \tag{1}$$

To simulate, W and L of the radiating patch were varied. It was found that the change in W influenced the resonant frequency. In other words, as the width W increased, the resonant frequency was shifted to the lower frequency. Moreover, simulations were carried out to determine the optimal position of the feed line as it influences the resonant frequency and the polarization of the antenna. The simulation results revealed that, with the feed line positioned in one of the corners, the resonant frequency was subject to W and the polarization was of horizontal direction. On the other hand, with the feed line positioned at the center, the length of the radiating patch L played a significant part in the resonant frequency while the polarization was of vertical direction [21–23]. The chosen dimensions of the radiating patch were thus 27 mm and 48 mm for W and L, respectively, achieving the resonant frequency at 2.43 GHz with an impedance bandwidth of 70 MHz ($|S_{11}| < -10$ dB).

Due to the narrow impedance bandwidth (i.e. 70 MHz) of the rectangular microstrip antenna and its lowered efficiency as a result of the microstrip feed line, a tuning stub was incorporated into the feed line for improvement of the impedance bandwidth (Fig. 3). In the simulation, p_1, l_1 and t were varied and the results revealed that p_1, l_1 and t of 9.75 mm, 9 mm and 3 mm, respectively, could achieve an impedance bandwidth of 88 MHz with the unidirectional radiation pattern.

In the design of the switchable beam rectangular microstrip antenna, two parasitic patches were individually mounted on either side of the radiating patch (Fig. 4(a)). In principle, the introduction of the parasitic patch influences the impedance of the antenna due to the mutual coupling effect between the parasitic and radiating patches. In addition, the distance between the parasitic and radiating patches dictates the radiation pattern. Moreover, two PIN diodes (D_1, D_2) were deployed to short the parasitic patches to the substrate and ground plane. The status of the PIN diodes (i.e. ON or OFF) was manipulated by the DC voltage. Figure 4 illustrates the schematic of the proposed switchable beam rectangular microstrip antenna with double parasitic patches.

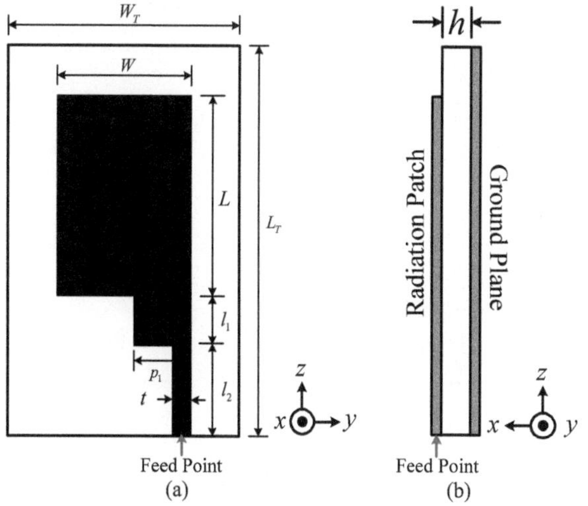

Fig. 3. The schematic of the rectangular microstrip antenna with tuning stub: (a) Top view, (b) side view.

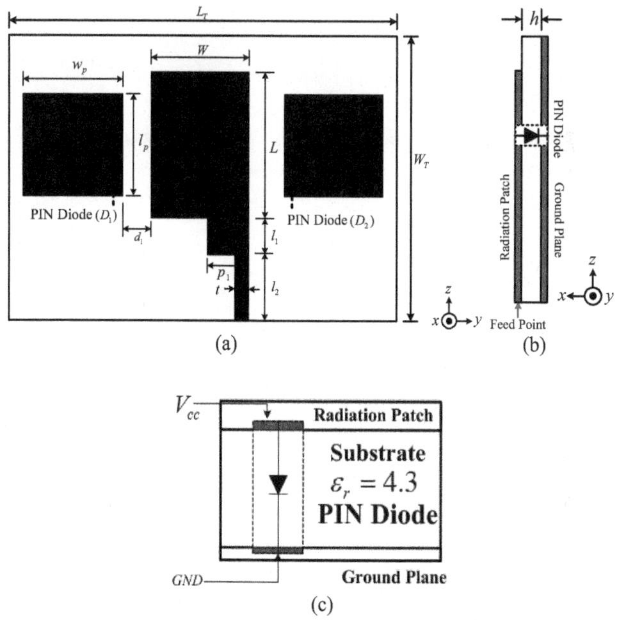

Fig. 4. The schematic of the switchable beam rectangular microstrip antenna: (a) Top view, (b) Side view, (c) Position of PIN diode.

Simulations were then carried out and the suitable antenna parameters are tabulated in Tab. 2. In addition, the proposed switchable beam rectangular microstrip antenna with two parasitic patches is operable in three modes:

mode 1 when both PIN diodes (D_1 and D_2) are OFF, mode 2 when D_1 is ON while D_2 is OFF, and mode 3 when D_1 is OFF and D_2 is ON.

Figure 5(a) compares the simulated $|S_{11}|$ of the switchable beam rectangular microstrip antenna with double parasitic patches under the three operation modes. The simulated $|S_{11}|$ varied with the modes of operation but still covered the operating frequency. The *xy*-plane radiation pattern also varied with the variations in the status of the PIN diodes, where the radiation pattern was directed forward under mode 1 and in the left and right directions under modes 2 and 3, respectively. Figure 5(b) illustrates the *xy*-plane radiation patterns of the switchable beam rectangular microstrip antenna under the three operation modes.

Antenna Parameter	Description	Physical Size (mm)
W_T	Total width	90
L_T	Total length	144
W	Radiation patch width	26.50
L	Radiation patch length	48
p_1	Tuning stub width	12
l_1	Tuning stub length	10
l_2	Microstrip line length	17
t	Microstrip line thickness	3
d_1	Distance between radiation and parasitic patches	8
w_p	Parasitic patch width	32
l_p	Parasitic patch length	30

Tab. 2. The parameters of the proposed switchable beam rectangular microstrip antenna.

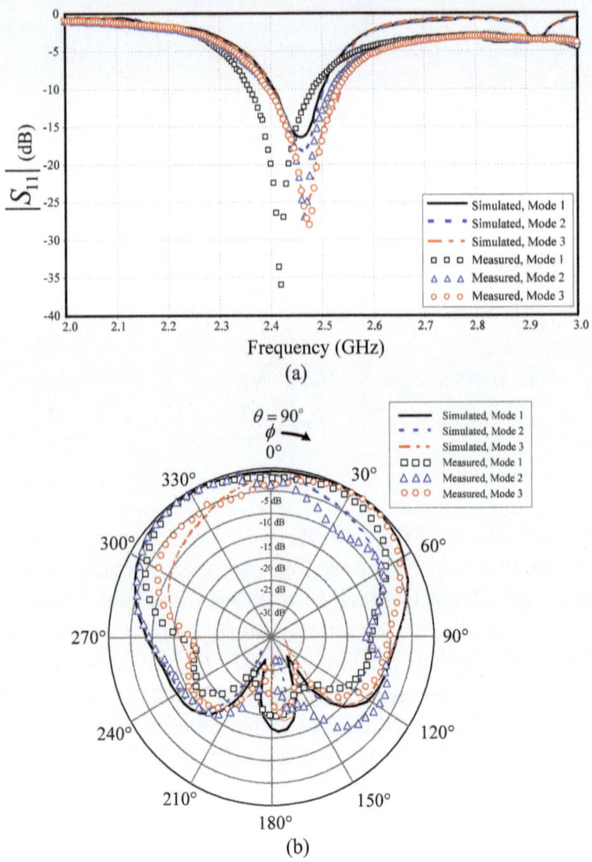

Fig. 5. Simulated and measured characteristics of the switchable beam rectangular microstrip antenna under three operation modes: (a) $|S_{11}|$, (b) the *xy*-plane radiation patterns.

Mode	PIN Diodes		Direction of Main Beam	Maximum Gain (dBi)
	D_1	D_2		
1	OFF	OFF	0° (Center)	5.6
2	ON	OFF	330° (Left)	6.7
3	OFF	ON	30° (Right)	6.5

Tab. 3. The simulated performance of the proposed antenna under three operation modes.

Table 3 summarizes the results of the proposed antenna with regard to the main beam direction and maximum gain under the three operation modes. Under mode 1, the main beam directs at 0° (center) with HPBW of 120°. The main beam directs at 330° (left) with HPBW of 60° under mode 2 and at 30° (right) with HPBW of 60° under mode 3. In addition, $|S_{11}|$ is less than –10 dB along the 2.4–2.5 GHz frequency range under all three operation modes.

4. Gain Enhancement Using Dielectric Slab or MTM

This section examines the deployment of either a dielectric slab or the MTM as the superstrate for enhancement of the antenna gain. In the first scenario, a dielectric slab (i.e. another FR4 substrate) was introduced and mounted onto the switchable beam rectangular microstrip antenna, forming two additional layers of superstrate: free space ($\varepsilon_r = 1$) and the additional FR4 ($\varepsilon_r = 4.3$).

In Fig. 6, the distance between the antenna and the dielectric slab *d* was varied between 1–11 mm, and the FR4 thickness was either 1.8 mm or 3.2 mm. It was found that the bandwidth and gain increased with increase in *d* and that the 3.2 mm-thick FR4 outperformed the 1.8 mm-thick superstrate. Thus, *d* and the FR4 thickness, respectively, of 10 mm and 3.2 mm were adopted for the improved bandwidth (i.e. 132 MHz) with the maximum gain of 6.2 dBi.

Figures 7(a) and (b) respectively illustrate the simulated radiation patterns and $|S_{11}|$ of the proposed switchable beam rectangular microstrip antenna with dielectric slab under the three operation modes. The simulation results indicated that the main beam of the proposed antenna with dielectric slab could be manipulated in three directions. Under mode 1, the main beam was at 0° (center) with HPBW of 100°, a bandwidth of 132 MHz and a gain of 6.2 dBi. Under mode 2, the main beam was at 330° (left) with HPBW of 50°, a bandwidth of 130 MHz and a gain of 7.30 dBi, while it was at 30° (right) with HPBW of 50°, a bandwidth of 135 MHz and a gain of 7.46 dBi under mode 3. Thus, the dielectric slab could be deployed as the superstrate to improve the bandwidth and gain of the switchable beam rectangular microstrip antenna.

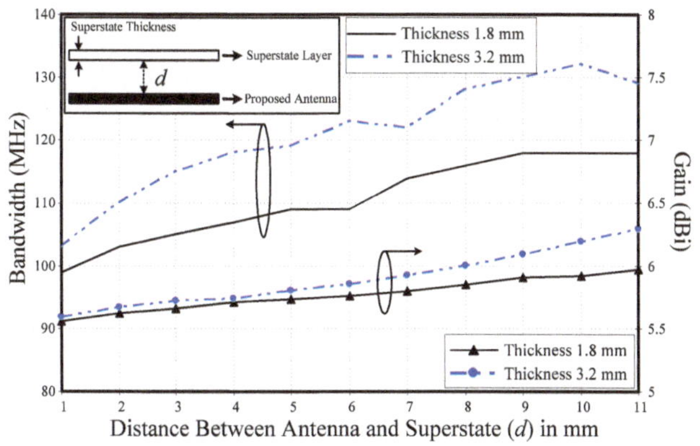

Fig. 6. Simulated bandwidths and gains for various d and two thicknesses of FR4.

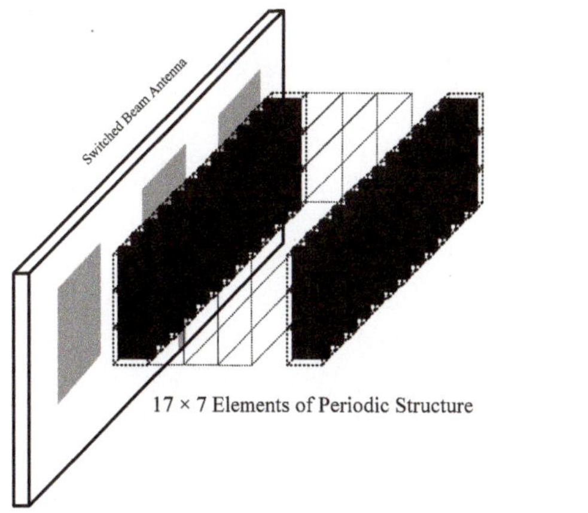

Fig. 8. Depiction of the switchable beam rectangular microstrip antenna with the MTM superstrate.

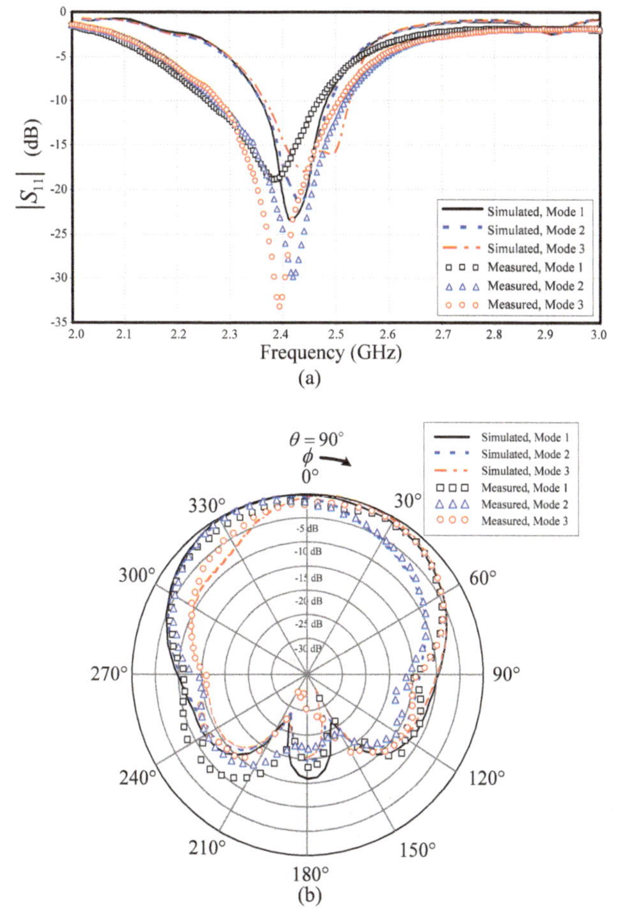

Fig. 7. Simulated and measured characteristics of the switchable beam rectangular microstrip antenna with dielectric slab under three operation modes: (a) $|S_{11}|$, (b) *xy*-plane radiation patterns.

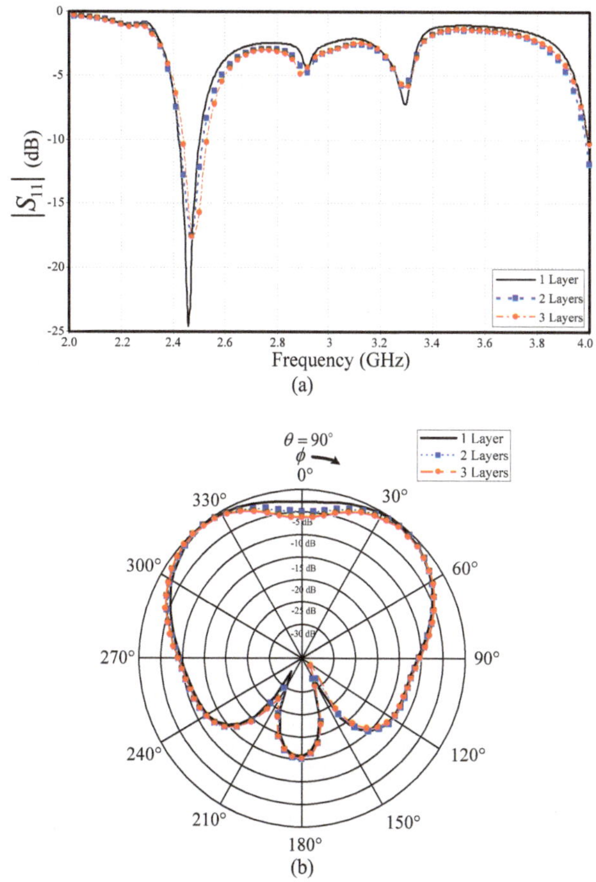

Fig. 9. The simulated characteristics of the switchable beam rectangular microstrip antenna for various MTM block-layers: (a) $|S_{11}|$, (b) *xy*-plane radiation patterns.

In the second scenario, the proposed MTM was utilized as the superstrate of the switchable beam rectangular microstrip antenna to enhance the antenna gain. The MTM was placed in front of the switchable antenna with a distance of 2 mm, as shown in Fig. 8. The arrangement of the planar cycloid diodes (PCD) was such that one MTM block-layer was of 7 × 17 periodic structure. Simulations were then carried out by increasing the number of MTM block-layers while observing $|S_{11}|$ and the radiation pattern (Fig. 9(a)-(b)).

The simulation results revealed that $|S_{11}|$ shifted to the higher resonant frequency with increase in the number of MTM block-layers. In addition, the antenna gain increased with increase in the MTM block-layers, while the beamwidth became narrower. This research thus utilized three MTM block-layers so that the final antenna would not be too oversized while achieving the improved antenna gain in relation to the non-MTM antenna.

Further simulations were carried out with the switchable beam rectangular microstrip antenna with three MTM block-layers under the three operation modes. Figures 10(a) and (b) respectively illustrate the simulated $|S_{11}|$ of the MTM-integrated antenna and the radiation patterns under the three operation modes. Specifically, under mode 1, the main beam directs at 330° and 30° with HPBW of 50°, a bandwidth of 130 MHz and a gain of 6.9 dBi. Under mode 2, the main beam directs at 330° (left) with HPBW of 48°, a bandwidth of 130 MHz and a gain of 7.5 dBi, while the main beam directs at 30° (right) with HPBW of 47°, a bandwidth of 135 MHz and a gain of 7.8 dBi under mode 3. Figure 11 shows the radiation efficiency of the three proposed antennas. The trends of the radiation efficiency for these three antennas are decreased when the frequency is higher. The loss of the FR4 substrate causes the degraded radiation efficiency at the higher frequency. At the frequency of 2.45 GHz, the radiation efficiency of the proposed antenna is 72%. When this structure is added by either the dielectric slab or MTM, the efficiency is increased to 74.5%. It is obvious that the radiation efficiency of the proposed antenna with either the dielectric slab or MTM is better than the proposed antenna without superstrate. The additional superstrate (dielectric slab or MTM) can improve the radiation efficiency over the relevant range of frequencies because the losses of surface waves are reduced. It is an inherent anisotropy of the additional structure and back-ward wave in superstrate.

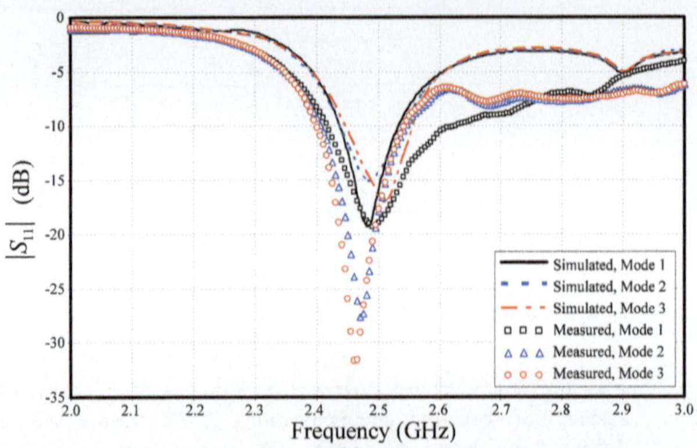

(a)

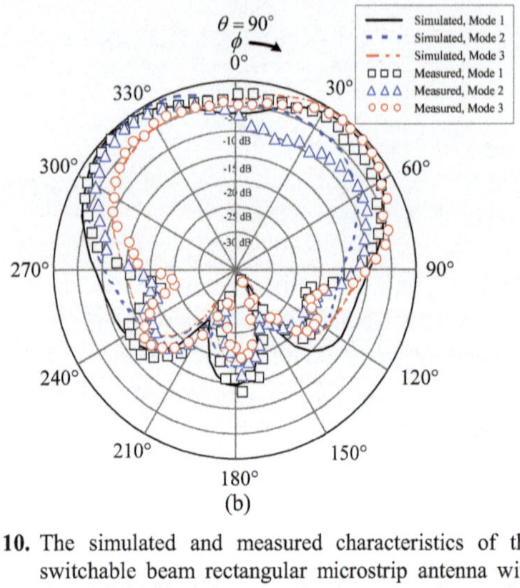

(b)

Fig. 10. The simulated and measured characteristics of the switchable beam rectangular microstrip antenna with three MTM block-layers under three operation modes: (a) $|S_{11}|$, (b) xy-plane radiation patterns.

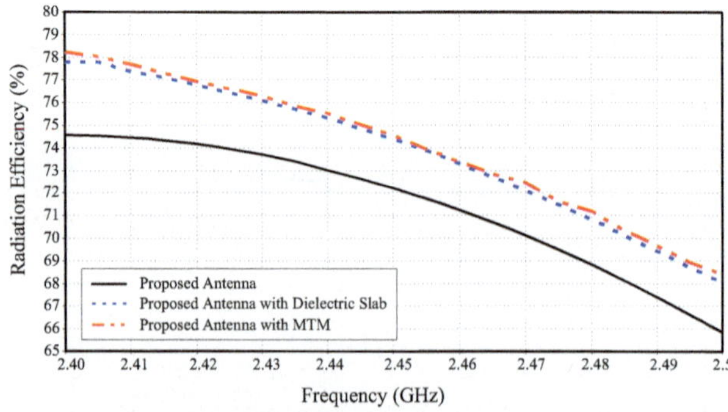

Fig. 11. The simulated radiation efficiency of the three proposed antennas.

Fig. 12. Prototype of the switchable beam rectangular microstrip antenna.

5. Experimental Results

To verify, three prototype antennas were fabricated and experiments undertaken. The three prototype antennas were the proposed switchable beam rectangular microstrip antenna, the proposed antenna with either the dielectric slab or MTM.

Figure 12 depicts the prototype of the switchable beam rectangular microstrip antenna with two parasitic patches on a 3.2 mm-thick FR4 substrate ($\varepsilon_r = 4.3$). Two PIN diodes with R_{on} of 1.8 Ω and C_{on} of 9.3 pF were deployed to affix the parasitic patches to the substrate and the ground plane. The experimental results revealed that the proposed switchable beam rectangular microstrip antenna could achieve a bandwidth of 120 MHz. In addition, the xy-plane radiation patterns under the three operation modes were as follows: under mode 1, the direction of the main beam was of 0° with HPBW of 120° and the maximum gain of 5.6 dBi; under mode 2, the direction was of 330° with HPBW of 60° and the maximum gain of 6.7 dBi; and under mode 3, the direction was of 30° with HPBW of 60° and the maximum gain of 6.5 dBi (Fig. 5(a), (b)).

The second prototype antenna was the switchable beam rectangular microstrip antenna with dielectric slab as the superstrate with the distance between the antenna and the slab of 10 mm. This antenna dimension is 90 mm × 144 mm × 16.4 mm. In Fig. 7(a), the prototype antenna with the dielectric slab could achieve a bandwidth of 200 MHz. In addition, the xy-plane radiation patterns under the 3 operation modes are as follows: under mode 1

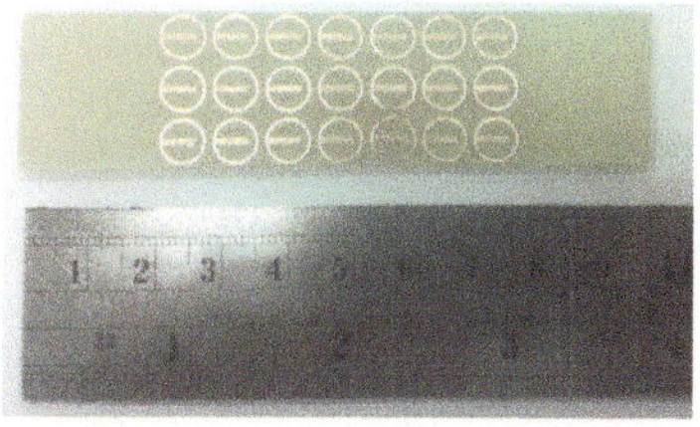

Fig. 13. Periodic structure of planar cycloid dipoles.

the direction of the main beam was of 0° with HPBW of 100° and the maximum gain of 6.2 dBi; under mode 2, the direction was of 330° with HPBW of 50° and the maximum gain of 7.3 dBi; and under mode 3, the direction was of 30° with HPBW of 50° and the maximum gain of 7.46 dBi (Fig. 7(b)).

The third prototype antenna was the switchable beam rectangular microstrip antenna with three MTM block-layers, as shown in Fig. 13. The antenna size is 90 mm × 144 mm × 29.2 mm. The prototype antenna with three MTM block-layers could achieve a bandwidth of 150 MHz. The *xy*-plane radiation patterns under the three operation modes are as follows: under mode 1, the directions of the main beam were of 30° and 330° with HPBW of 50° and the maximum gain of 6.9 dBi; under mode 2, the direction was of 330° with HPBW of 48° and the maximum gain of 7.5 dBi; and under mode 3, the direction was of 30° with HPBW of 47° and the maximum gain of 7.8 dBi (Fig. 10(a), (b)).

The measured results indicated that the switchable beam rectangular microstrip antennas of all three configurations were operable under all three operation modes along the 2.4–2.5 GHz frequency range. Table 4 compares the performances of the three prototype antennas and it is evident that the deployment of either the dielectric slab or MTM as the superstrate could enhance the antenna gain.

6. Conclusions

This research has presented the switchable beam rectangular microstrip antenna in which two PIN diodes were deployed for manipulation of the main beam direction and the superstrate (i.e. either a dielectric slab or the MTM) for enhancement of the antenna gain. In this research, the dielectric slab is a second FR4 substrate while the metamaterial (MTM) is the 7 × 17 periodic structure of planar cycloid dipoles (PCD). The proposed switchable antennas with and without the superstrate are operable in the 2.4–2.5 GHz WLAN system and switchable in three directions of 0°, 30° and 330° (*xy*-plane). The simulation and experimental results validate the applicability of the dielectric slab and three MTM block-layers as the superstrate to improve the antenna gain. In addition, the proposed superstrate-integrated switchable beam rectangular microstrip antenna could be employed in the point-to-point and point-to-multipoint applications.

Mode	Prototype Antenna			Prototype Antenna with Dielectric Slab			Prototype Antenna with MTM		
	Direction of Main Beam	HPBW	Maximum Gain (dBi)	Direction of Main Beam	HPBW	Maximum Gain (dBi)	Direction of Main Beam	HPBW	Maximum Gain (dBi)
1	0°	120°	5.6	0°	100°	6.2	330° and 30°	50°	6.9
2	330°	60°	6.7	330°	50°	7.3	330°	48°	7.5
3	30°	60°	6.5	30°	50°	7.46	30°	47°	7.8

Tab. 4. Comparison of the measured results of the three prototype antennas under three operation modes.

Acknowledgments

The authors would like to express deep gratitude to Rajamangala University of Technology Lanna for the financial sponsorship under the Hands-on Research and Development Project (UR-17).

References

[1] FURUKAWA, H., KAMIO, Y., SASAOKA, H. Cochannel interference reduction and path-diversity reception technique using CMA adaptive array antenna in digital land mobile communications. *IEEE Transactions on Vehicular Technology*, 2001, vol. 50, no. 2, p. 605–616. DOI: 10.1109/25.923072

[2] GHOUSE BASHA, T. S., ALOYSIUS, G., RAJAKUMAR, B. R., GIRI PRASAD, M. N., SRIDEVI, P. V. A constructive smart antenna beam-forming technique with spatial diversity. *IET Microwaves, Antennas and Propagation*, 2012, vol. 6, no. 7, p. 773 – 780. DOI: 10.1049/iet-map.2011.0356

[3] CERRI, G., LEO, R. D., PRIMIANI, V. M., MONTEVERDE, C., RUSSO, P. Design and prototyping of a switching beam disc antenna for wideband communications. *IEEE Transactions on Antennas and Propagation*, 2006, vol. 54, no. 12, p. 3721–3726. DOI: 10.1109/TAP.2006.886555

[4] LAI, M.-I., WU, T.-Y., HSIEH, J.-C., WANG, C.-H., JENG, S.-K. Compact switched-beam antenna employing a four-element slot antenna array for digital home applications. *IEEE Transactions on Antennas and Propagation*, 2008, vol. 56, no. 9, p. 2929–2936. DOI: 10.1109/TAP.2008.928775

[5] SHYNU NAIR, S. V., AMMANN, M. J. Reconfigurable antenna with elevation and azimuth beam switching. *IEEE Antennas and Wireless Propagation Letters*, 2010, vol. 9, p. 367–370. DOI: 10.1109/LAWP.2010.2049332

[6] QIN, P.-Y., GUO, Y. J., DING, C. A beam switching quasi-Yagi dipole antenna. *IEEE Transactions on Antennas and Propagation*, 2013, vol. 61, no. 10, p. 4891–4899. DOI: 10.1109/TAP.2013.2274635

[7] NOR, M. Z. M., RAHIM, S. K. A., SABRAN, M. I., SOH, P. J., VANDENBOSCH, G. A. E. Dual-band, switched-beam, reconfigurable antenna for WLAN applications. *IEEE Antennas and Wireless Propagation Letters*, 2013, vol. 12, p. 1500–1503. DOI: 10.1109/LAWP.2013.2289919

[8] EDALATI, A., DENIDNI, T. A. Frequency selective surfaces for

beam-switching applications. *IEEE Transactions on Antennas and Propagation*, 2013, vol. 61, no. 1, p. 195–200. DOI: 10.1109/TAP.2012.2219842

[9] CHENG, S., RANTAKARI, P., MALMQVIST, R., SAMUELSSON, C., VAHA-HEIKKILA, T., RYDBERG, A., VARIS, J. Switched beam antenna based on RF MEMS SPDT switch on quartz substrate. *IEEE Antennas and Wireless Propagation Letters*, 2009, vol. 8, p. 383–386. DOI: 10.1109/LAWP.2009.2018712

[10] REBEIZ, G. M., *RF MEMS: Theory, Design and Technology*. New York: Wiley, 2003. ISBN: 0471201693

[11] YANG, H. Y., ALEXOPOULOS, N. G. Gain enhancement methods for printed circuit antennas through multiple superstrates. *IEEE Transactions on Antennas and Propagation*, 1987, vol. 35, no. 7, p. 860–864. DOI: 10.1109/TAP.1987.1144186

[12] SELVANAYAGAM, M., ELEFTHERIADES, G. V. A compact printed antenna with an embedded double-tuned metamaterial matching network. *IEEE Transactions on Antennas and Propagation*, 2010, vol. 58, no. 7, p. 2354–2361. DOI: 10.1109/TAP.2010.2048876

[13] HUANG, H., LIU, Y., ZHANG, S., GONG, S. Multiband metamaterial-loaded monopole antenna for WLAN/WiMAX applications. *IEEE Antennas and Wireless Propagation Letters*, 2015, vol. 14, p. 662–665. DOI: 10.1109/LAWP.2014.2376969

[14] BUELL, K., MOSALLAEI, H., SARABANDI, K. Metamaterial insulator enabled superdirective Array. *IEEE Transactions on Antennas and Propagation*, 2007, vol. 55, no. 4, p. 1074–1085. DOI: 10.1109/TAP.2007.893373

[15] PANDEY, G. K., SINGH, H. S., BHARTI, P. K., MESHRAM, M. K. Metamaterial-based UWB antenna. *Electronics Letters*, 2014, vol. 50, no. 18, p. 1266–1268. DOI: 10.1049/el.2014.2366

[16] ANTONIADES, M. A., ELEFTHERIADES, G. V. A broadband dual-mode monopole antenna using NRI-TL metamaterial loading. *IEEE Antennas and Wireless Propagation Letters*, 2009, vol. 8, p. 258–261. DOI: 10.1109/LAWP.2009.2014402

[17] CALOZ, C., ITOH, T. *Electromagnetic Metamaterials: Transmission Line Theory and Microwave Applications*. John Wiley & Sons, 2006. ISBN: 978-0-471-66985-2

[18] CST Microwave Studio, User's Manual.

[19] ZIOLKOWSKI, R. W. Design, fabrication, and testing of double negative metamaterials. *IEEE Transactions on Antennas and Propagation*, 2003, vol. 51, no. 7, p. 1516–1529. DOI: 10.1109/TAP.2003.813622

[20] WU, B.-I., WANG, W., PACHECO, J., CHEN, X., GRZEGORCZYK, T., KONG, J. A. A study of using metamaterials as antenna substrate to enhance gain. *Progress In Electromagnetics Research*, 2005, PIER 51, p. 295–328. DOI: 10.2528/PIER04070701

[21] BALANIS, C. A., *Antenna Theory: Analysis and Design*. 3rd ed., John Wiley & Sons, 2005. ISBN: 978-0-471-66782-7

[22] CHEN, W.-L., WANG, G.-M. Small size edge-fed Sierpinski carpet microstrip patch antennas. *Progress In Electromagnetics Research C*, 2008, vol. 3, p. 195–202. DOI: 10.2528/PIERC08050302

[23] POZAR, D. M., VODA, S. M. A rigorous analysis of a microstripline fed patch antenna. *IEEE Transactions on Antennas and Propagation*, 1987, vol. 35, no. 12, p. 1343–1350. DOI: 10.1109/TAP.1987.1144041

[24] SABAPATHY, T., JAMLOS, M. F., AHMAD, R. B., JUSOH, M., JAIS, M. I., KAMARUDIN, M. R. Electronically reconfigurable beam steering antenna using embedded RF pin based parasitic arrays (ERPPA). *Progress In Electromagnetics Research*, 2013, vol. 140, p. 241–261. DOI: 10.2528/PIER13042906

About the Authors ...

Supakit KAWDUNGTA received the Bachelor of Engineering (2006) in Telecommunications Engineering, Master of Engineering (2007) in Telecommunications Engineering and Doctor of Engineering (2011) in Electrical Engineering from King Mongkut's Institute of Technology Ladkrabang, Bangkok, Thailand. From 2011 to present, he is an Assistant Professor in electronic and telecommunications engineering, Rajamangala University of Technology Lanna, Chiang Mai, Thailand.

Puwanai JAIBANAUEM received the Bachelor of Engineering (2016) in Electronic Engineering from Rajamangala University of Technology Lanna, Chiang Mai, Thailand. From 2016 to present, he is an engineer in Advanced Info Service PLC., Chiang Mai, Thailand.

Rattiya PONGGA received the Bachelor of Engineering (2015) in Electronic Engineering from Rajamangala University of Technology Lanna, Chiang Mai, Thailand. From 2015 to present, she is an engineer in Thai NJR Co., Ltd., Lamphun, Thailand.

Chuwong PHONGCHAROENPANICH received the Bachelor of Engineering (Hons), Master of Engineering, and Doctor of Engineering degrees from the Faculty of Engineering, King Mongkut's Institute of Technology Ladkrabang (KMITL), Bangkok, Thailand, in 1996, 1998, and 2001, respectively. He is currently an Associate Professor at the Department of Telecommunications Engineering, KMITL where he also serves as the Leader of the Wireless Communication Laboratory at the same institute. He was the board committee of ECTI Association in 2008–2011. He serves as the chair of IEEE MTT/AP/ED Thailand chapter. He is the associate editor of IEICE Trans. Com. and ECTI Trans. EEC. He has been the organizing committee of several international conferences including the TPC chair of 2009 International Symposium on Antennas and Propagation (ISAP2009) and TPC member of ISAP2012. He is the reviewer of many journals including IEEE Trans. Antennas and Propagation and Electronics Letters, ECTI Trans. and many international conferences including ISAP and Asia-Pacific Microwave Conference (APMC). His research interests are antenna design for various mobile and wireless communications, conformal antennas and array antenna theory. Dr. Phongcharoenpanich is a Member of IEEE, IEICE and ECTI.

UWB Monopole Antenna with Band Notched Characteristics Mitigating Interference with WiMAX

Ahmed IBRAHIM [1,2], *Wael ALI* [3], *Jan MACHAC* [4]

[1] Dept. of Electrical Engineering, Faculty of Engineering, Minia University, Minia, Egypt
[2] Institute of Micro- and Sensor Systems, University of Magdeburg, Germany
[3] Dept. of Electronics & Comm. Engineering, College of Engineering, Arab Academy for Science, Technology and Maritime Transport (AASTMT), Alexandria, Egypt
[4] Dept. of Electromagnetic Field, Faculty of Electrical Engineering, Czech Technical University in Prague, Czech Republic

wael_abd_ellatif@yahoo.com, ahmedabdel_monem@mu.edu.eg, machac@fel.cvut.cz

Abstract. *Compact asymmetric coplanar strip (ACS)-fed monopole antenna is presented in this paper for operation in ultra wide-band (UWB) applications. The antenna is composed of a semi-elliptical monopole and lateral coplanar ground plane to operate in the UWB range. It occupies a very small size of 28 × 11.5 mm² and is mounted on low cost FR4 substrate of 1.6 mm thickness and dielectric constant of 4.4. Band notched structure is applied to the proposed antenna to assure the rejection of WiMAX frequency band centered at 3.5 GHz to minimize the interference of the communication system with this service. The radiation characteristic of the proposed ACS-fed monopole antenna is nearly omnidirectional with moderate gain over the entire UWB frequency range. The measurements of the fabricated model confirm the designed behavior. The compactness in size, cheap cost and simple feeding technique make it as a good candidate to be integrated within the portable devices for wireless communication.*

Keywords

UWB monopole antenna, band notched characteristics, WiMAX

1. Introduction

After the allocation of the unlicensed frequency band (3.1–10.6 GHz) for ultra wide-band (UWB) communications by the Federal Communications Commission (FCC) in 2002 [1], many researchers focus their interests to design antennas working in this band with attractive specifications. Compact size, low cost, light weight and easy fabrication process are the main promising requirements of the antenna to be integrated in portable devices that can operate with different wireless communication applications in the UWB frequency range [2–5]. One of the serious problems facing any UWB antenna is the electromagnetic interference (EMI) problem since there are other wireless communication systems operating in the same frequency band of UWB communications such as Worldwide Interoperability for Microwave Access (WiMAX) working at 3.3–3.6 GHz [6]. To overcome the presented problem, a band notched structure is urgently needed. Various methods have been reported to achieve a band rejection behavior, such as by etching slots in the radiating patch [7], [8], or in the ground plane [9] or in the microstrip line [10]. Parasitic resonators beside the partial ground plane are loaded as in [11]. Also, resonators beside the feeding line are added in [12]. However, all the aforementioned UWB antennas do not have their structure compact enough in order to be easily integrated in systems without deteriorating the overall performance. Also, some of these antennas do not have uniplanar structure, and require double sided etching and alignment. This increases the fabrication complexity and decreases the accuracy of the fabricated designs.

In order to reduce the complexity of the fabricated antennas, a coplanar waveguide (CPW) structure is used which facilitates the fabrication process and is preferable for the microwave integrated circuit applications [13]. Furthermore, in order to minimize the overall size of an antenna, asymmetric coplanar strip (ACS) structure is introduced. The overall size of this antenna can be reduced to about one half of the common coplanar waveguide CPW-fed monopole antennas as reported in [13–19].

In this paper, a semi-elliptical monopole antenna with ACS structure is proposed for UWB applications. Covering the frequency band of UWB communications (3.1 to 10.6 GHz) is the main goal of the designed antenna. An inherent band rejection structure is utilized to achieve a notch at 3.5 GHz to mitigate the interference with WiMAX applications allocated in 3.3–3.6 GHz frequency band. The proposed antennas were fabricated and then scattering parameters and radiation characteristics were measured to verify the simulations done by the commercial software (CST). Good agreement between simulated and measured results confirms the ability of the proposed band notched ACS-fed antenna for operation in the UWB communication systems without interfering the allocated WiMAX frequency band at 3.5 GHz.

2. Configuration of ACS-Fed Antenna without and with Stub Responsible for Band Notched Characteristic

The geometry of the proposed UWB monopole antenna with the optimized dimensions is shown in Fig. 1(a). First of all, an elliptic CPW feed antenna is designed. Asymmetric coplanar strip feed with half ground plane is used to reduce the antenna size. Also, instead of using complete elliptic radiator, half of the elliptic radiator is used as shown in Fig. 1(a). The impedance matched from 3.1 GHz to more than 10.6 GHz as shown in Fig. 2 has been achieved by properly changing the width of the gap between the microstrip line and ground strip, and by making the ground metallization round. The antenna is mounted on low cost FR4 substrate with h = 1.6 mm, ε_r = 4.4, and tanδ = 0.02. The antenna consists of semi-elliptical radiator with ACS feeding line. The ACS feeding line has a signal strip 3 mm in width, and distance between the signal strip and the coplanar ground plane equal to 0.3 mm to match the 50 Ω SMA connector. The lateral ground plane is presented with bended edge to efficiently enhance the impedance characteristics of the proposed UWB antenna. The overall dimensions of the antenna are $28 \times 11.5 \times 1.6$ mm^3. The antenna was successfully designed and optimized by using the CST Microwave Studio software. To mitigate the interference with WiMAX systems, the band rejection function is desirable in UWB communication systems. The geometry of the proposed band notched UWB monopole antenna with the optimized dimensions is shown in Fig. 1(b). A nearly U-shaped stub is utilized to achieve the band rejection at 3.5 GHz for the proposed structure. The central frequency f_{notch} of the band notched response is controlled by the overall length of the resonator using the following equation:

$$L_{total} \approx \frac{\lambda_g}{2} = \frac{c}{2 f_{notch} \sqrt{\varepsilon_{eff}}} \quad (1)$$

where the rough estimate of the effective permittivity is

$$\varepsilon_{eff} \approx \frac{\varepsilon_r + 1}{2}. \quad (2)$$

c is speed of light, ε_r is substrate permittivity. The notch bandwidth is controlled namely by the stub strip width that determines its wave impedance.

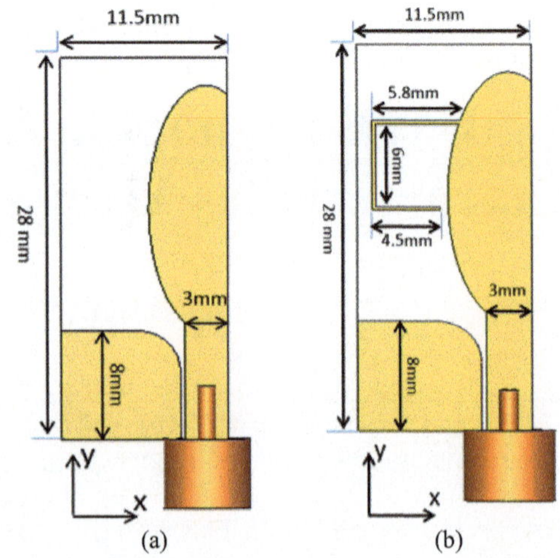

Fig. 1. Geometry of the proposed ACS-fed UWB monopole antenna.

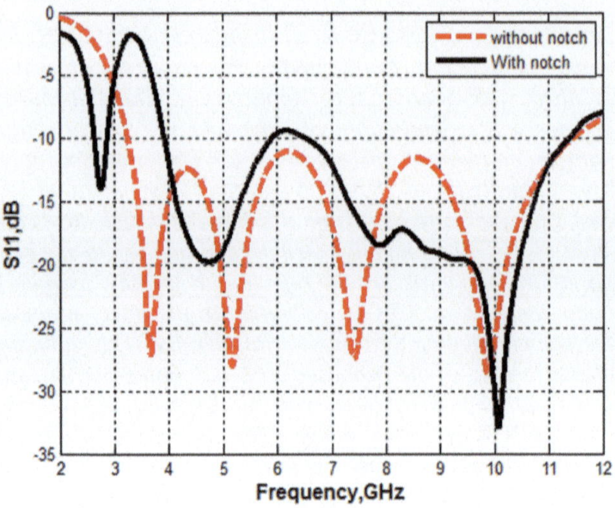

Fig. 2. The simulated reflection coefficient of the proposed structures.

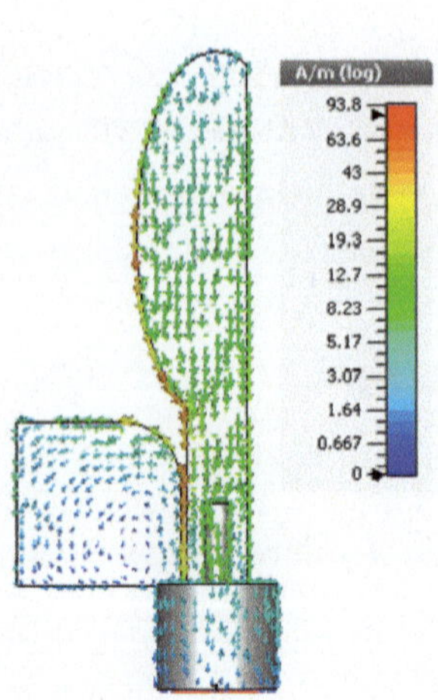

Fig. 3. Simulated surface current distributions of UWB antenna at 3.5 GHz.

3. Results and Discussion

The surface current distribution of the proposed ACS-fed UWB antenna without and with the stub creating the band notched characteristic was simulated by the CST software. Figure 3 shows the current distribution on the surface of UWB antenna at 3.5 GHz and it can be noticed that the current flows uniformly on the antenna surface. This is the standard current distribution of any radiating antenna. This current distribution can be significantly distorted by connecting a stub to the layout edge.

For better demonstration of the band notched characteristics, the magnitude of the current distribution is investigated at two different frequencies, one in the coverage-band and the other in the rejection-band as shown in Fig. 4. It can be observed that at the notched frequency band at

3.5 GHz, the surface current is highly concentrated over the resonating stub as shown in Fig. 4(a), which means that the energy is stored in the resonator rather than radiated in the air. In Fig. 4(b), the current is uniformly distributed over the antenna surface at 7 GHz, which confirms that the antenna radiates.

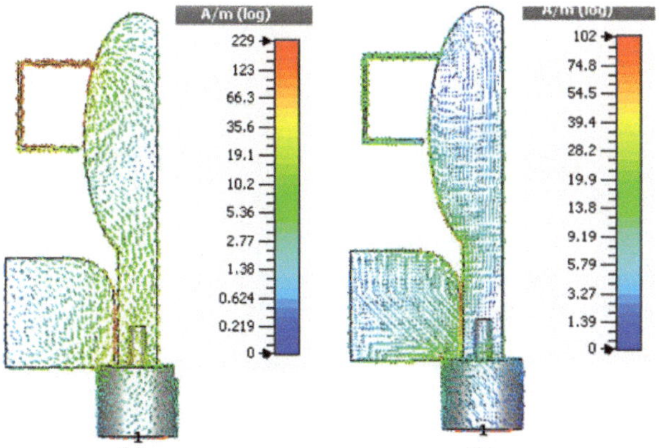

Fig. 4. Simulated surface current distributions of UWB antenna with band notched filter (a) at 3.5 GHz, (b) at 7 GHz.

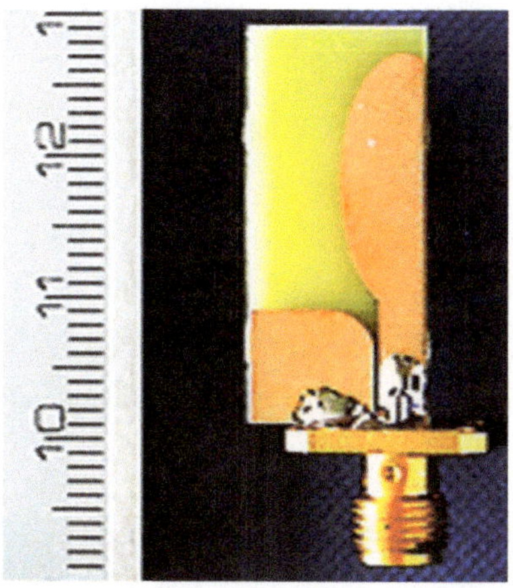

Fig. 5. Photograph of fabricated UWB monopole antenna.

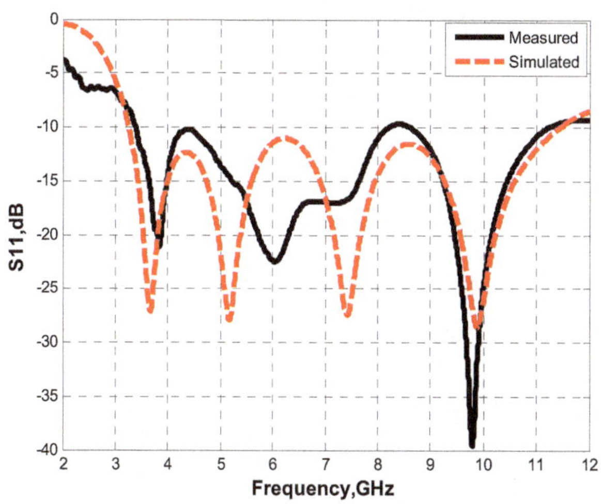

Fig. 6. The simulated and measured reflection coefficient of the UWB monopole antenna.

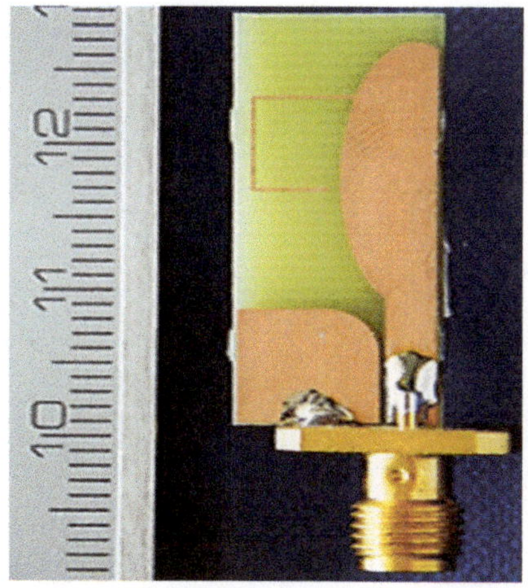

Fig. 7. Photograph of fabricated band notched UWB antenna.

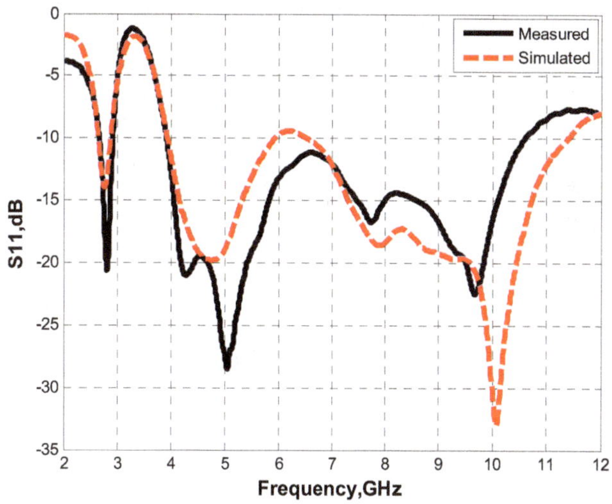

Fig. 8. The simulated and measured reflection coefficient of the band notched UWB antenna.

The semi-elliptical radiator connected with 50 Ω ACS line and the lateral partial ground plane are printed on the top layer of low cost FR4 epoxy substrate defined above as shown in Fig. 5. Figure 6 shows simulated and measured reflection coefficient. They are in a good consistency. The antenna is working in the UWB range from 3.2 GHz till 11.4 GHz with a reflection coefficient lower than –10 dB over the achieved frequency band. Figure 7 shows the fabricated prototype of the band notched UWB antenna with compact size of 28 × 11.5 mm^2. This antenna uses the resonant stub in order to achieve a notch in the operating band of WiMAX. As can be seen from Fig. 8, the measured S_{11} tracks the simulated S_{11} well. Both results cover the operating band of UWB application from 2.7 GHz to 10.7 GHz except rejected band from 2.9 GHz to 3.9 GHz with reflection coefficient greater than –10 dB as illustrated in Fig. 8 confirming that the proposed antenna is capable of mitigating the interference with WiMAX service. The reflection coefficient characteristics were measured using R&S vector network analyzer ZVA40, the radiation patterns were measured in an anechoic chamber using the NSI 800F-30 system.

A comparison between simulated realized gains of the two proposed UWB antennas without and with band notch is demonstrated in Fig. 9. The simulated realized gain of both antennas is almost constant over the achieved frequency band except the case of band notched antenna, which has notched gain between 3 GHz to 3.9 GHz with a minimum value of –5.5 dB at 3.4 GHz. The average measured gain is 3.5 dBi. For the case of UWB antenna without notch, the simulated efficiency has a maximum value of 98 % and a minimum value of 83 % within the frequency range from 3–11 GHz as depicted in Fig. 10. Furthermore, the simulated efficiency for the case of band notched UWB antenna has a maximum value of 93 % and minimum value of 14 % at 3.4 GHz.

The radiation patterns of the proposed antenna with band notched characteristic in the *x-z* plane ($\varphi = 0°$), *y-z* plane ($\varphi = 90°$) and *x-y* plane ($\theta = 90°$) were measured inside an anechoic chamber. The normalized simulated and measured radiation patterns are measured at 3.5 GHz, 6 GHz and 10 GHz in the three aforementioned planes of the proposed antenna as shown in Fig. 12, Fig. 13 and Fig. 14, respectively. It is clearly deduced that the UWB monopole antenna has nearly omni-directional patterns in the three planes at the three different frequencies except the occurrence of bi-directional pattern in *y-z* plane at 3.5 GHz. From comparison of radiation patterns of the proposed notched antenna with those of the original antenna (not presented here), one can conclude that the resonant stub does not affect the shape of radiation patterns. Of course, the radiating field intensities at 3.5 GHz of the notched antenna are much lower, relative to gain behavior shown in Fig. 9.

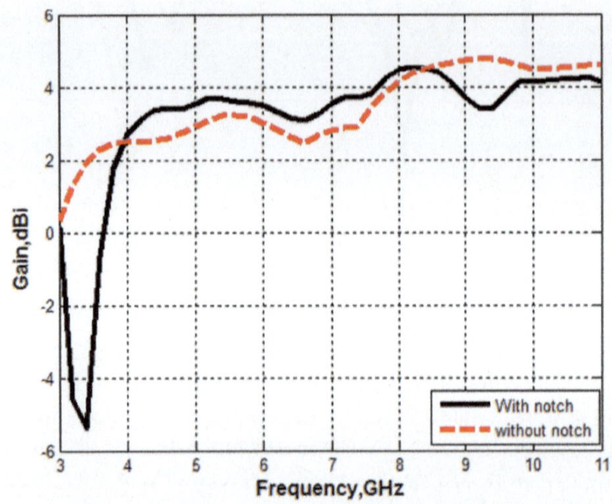

Fig. 9. The simulated realized gain of the fabricated antennas.

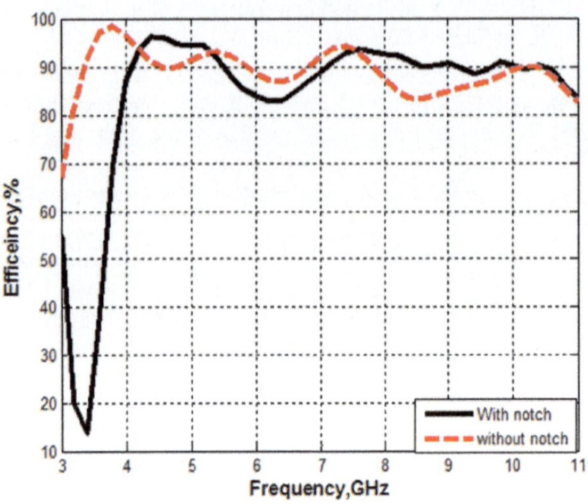

Fig. 10. The simulated efficiency of the fabricated antennas.

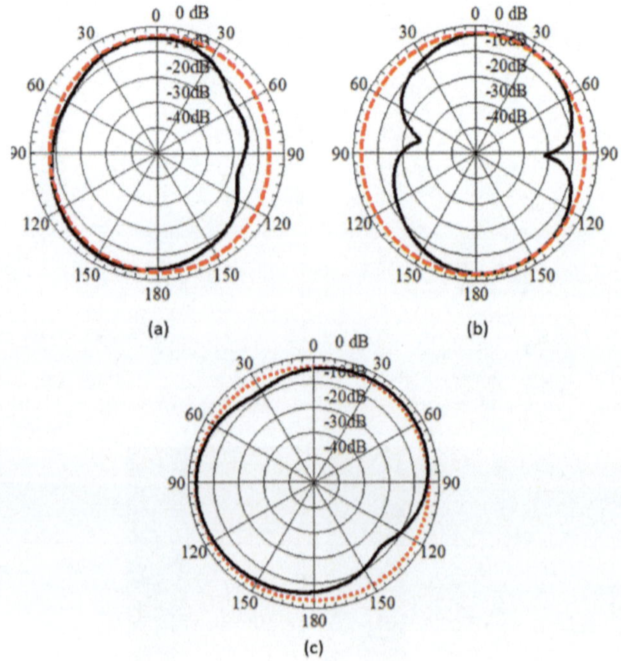

Fig. 12. The simulated red (dashed) and measured black (solid) normalized radiation patterns at f = 3.5 GHz: (a) *x-z* plane, (b) *y-z* plane, (c) *x-y* plane.

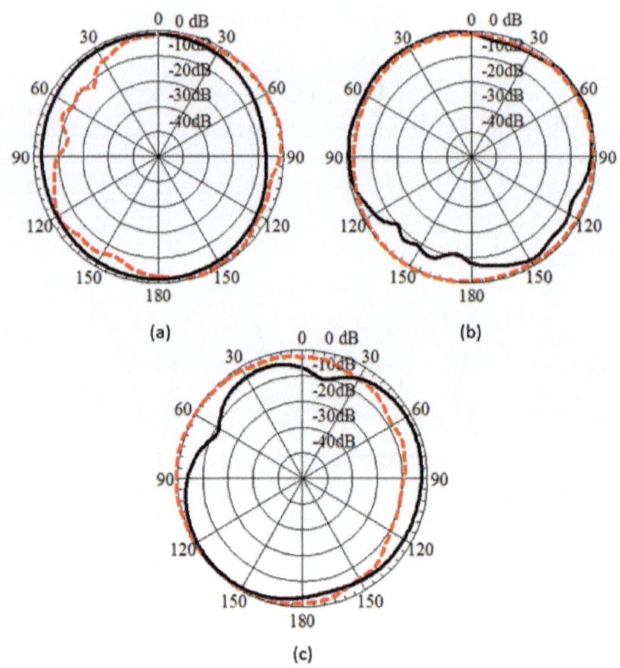

Fig. 13. The simulated red (dashed) and measured black (solid) normalized radiation patterns at f = 6 GHz: (a) *x-z* plane, (b) *y-z* plane, (c) *x-y* plane.

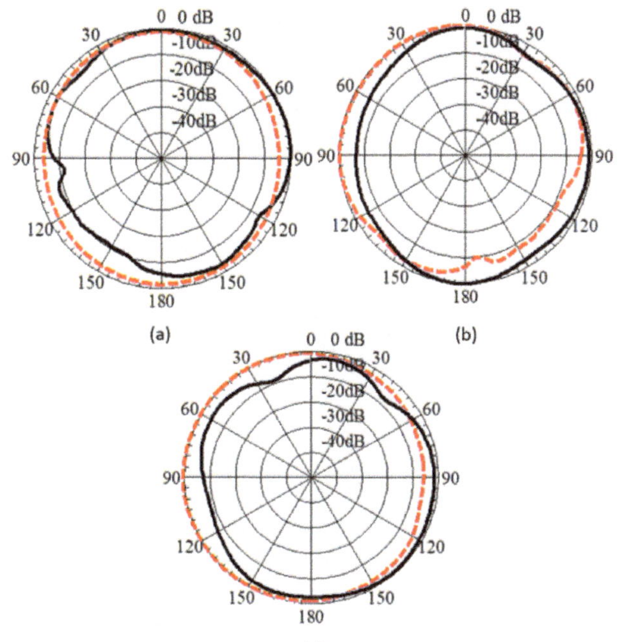

Fig. 14. The simulated red (dashed) and measured black (solid) normalized radiation patterns at $f = 10$ GHz: (a) x-z plane, (b) y-z plane, (c) x-y plane.

Ref.	Size [mm^2]	Frequency range [GHz]	Notch frequencies [GHz]	Realized gain at notch [dBi]
[20]	25 × 32	3-11	5.5	–5.2
			8.7	–3.6
[21]	58 × 65.5	3.1-11	-	-
[22]	24 × 32	2.9-12.5	5.5	–4.9
			7	–3.9
[23]	45 × 55	2-10.5	2.4	-
			3.5	-
			5.2	-
[16]	12.3 × 28	3-12.8	3.5	–3.3
			8.2	–2.5
[24]	30 × 32	3.1-14	5.5	–2
			6.8	–0.9
			11.5	–0.7
[25]	24 × 32	2.7-12	3.5	–4.3
			8	–2.1
This work	11.5 × 28	2.7-11.4	3.4	–5.5

Tab. 1. Comparison of the proposed antenna and other published works.

A comparison between the proposed antenna and other published work is demonstrated in Tab. 1. The presented antenna occupies the smallest area 11.5 × 28 mm^2 and shows the lowest gain at the notch frequency –5.5 dBi.

4. Conclusion

Compact size UWB monopole antennas without/with band notched characteristics have been designed, fabricated and tested. A semi-elliptical radiator connected with 50 Ω ACS line is printed on low cost FR4 substrate together with an asymmetric coplanar ground. An excellent impedance bandwidth has been experimentally achieved from 2.7 GHz to 11.4 GHz with notched band 2.9–3.9 GHz for interference mitigation purpose. The rejected band coincides with the band allocated for WiMAX applications (3.3–3.6 GHz). Furthermore, good omnidirectional radiation patterns with almost constant gain and efficiency over the UWB range have been achieved. The very small size due to ACS-feeding, makes the proposed UWB band notched antenna suitable for size-restricted systems.

Acknowledgments

The experimental work has been supported by the Grant Agency of Czech Republic under project number No. 17-00607S.

References

[1] FEDERAL COMMUNICATIONS COMMISSION. *First Report and order, Revision of part 15 of the commission's Rule Regarding Ultra-Wideband Transmission System FCC 02-48*. 2002.

[2] REDDY, G. S., KAMMA, A., MISHRA, S. K., MUKHERJEE, J. Compact bluetooth/UWB dual-band planar antenna with quadruple band-notch characteristics. *IEEE Antennas and Wireless Propagation Letters*, 2014, vol. 13, p. 872–875. DOI: 10.1109/LAWP.2014.2320892

[3] BOUTEJDAR, A., ABD ELLATIF, W. A novel compact UWB monopole antenna with enhanced bandwidth using triangular defected microstrip structure and stepped cut technique. *Microwave and Optical Technology Letters*, 2016, vol. 58, no. 6, p. 1514–1519. DOI: 10.1002/mop.29820

[4] ALI, W. A., ZAKI, A. I., ABDOU, M. H. Design and fabrication of rectangular ring monopole array with parasitic elements for UWB applications. *Microwave and Optical Technology Letters*, 2016, vol. 58, p. 2268–2273. DOI: 10.1002/mop.30021

[5] ABDALLA, M. A., IBRAHIM, A. A., BOUTEJDAR, A. Resonator switching for UWB antenna in wireless applications. *IET Microwaves, Antennas and Propagation*, 2015, vol. 9, no. 13, p. 1468–1477. DOI: 10.1049/iet-map.2014.0838

[6] LI, T., HUO, F. F., ZHAI, H. Q., LIANG, C. H. Compact ultra-wideband antenna with sharp band-notched characteristics for WiMax and WLAN. *Electronics Letters*, 2012, vol. 48, no. 22, p. 1380–1382. DOI: 10.1049/el.2012.2924

[7] XU, J. SHEN, D., ZHANG, X., WU, K. A compact disc ultrawide band (UWB) antenna with quintuple band rejections. *IEEE Antennas and Wireless Propagation Letters*, 2012, vol. 11, p. 1517-1520. DOI: 10.1109/LAWP.2012.2234075

[8] DAS, S., MITRA, D., CHAUDHURI, S. R. B. Design of UWB planar monopole antennas with etched spiral slot on the patch for multiple band-notched characteristics. *International Journal of Microwave Science and Technology*, 2015, Article ID 303215, 9 p. DOI: 10.1155/2015/303215

[9] OJAROUDI, N., GHADIMI, N., OJAROUDI, Y. Compact multi-resonance monopole antenna with dual band-stop property for UWB wireless communications. *Wireless Personal Communications*, 2015, vol. 81, no. 2, p. 563–579. DOI: 10.1007/s11277-014-2145-9

[10] CAI, Y. Z., YANG, H. C., CAI, L. Y. Wideband monopole antenna with three band-notched characteristics. *IEEE Antennas and Wireless Propagation Letters*, 2014, vol. 13, p. 607–610. DOI: 10.1109/LAWP.2014.2313178

[11] LI, L., ZHOU, Z. L., HONG, J. S. Compact UWB antenna with four band-notches for UWB applications. *Electronics Letters*, 2011, vol. 47, no. 22, p. 1211–1212. DOI: 10.1049/el.2011.2334

[12] SARKAR, D., SRIVASTAVA, K. V., K. SAURAV, K. A compact microstrip-fed triple band-notched UWB monopole antenna. *IEEE Antennas and Wireless Propagation Letters*, 2014, vol. 13, p. 396–399. DOI: 10.1109/LAWP.2014.2306812

[13] FEI, P., JIAO, Y.-C., ZHU, Y., ZHANG, F.-S. Compact CPW-fed monopole antenna and miniaturized ACS-fed half monopole antenna for UWB applications. *Microwave and Optical Technology Letters*, 2012, vol. 54, p. 1605–1609. DOI: 10.1002/mop.26909

[14] LIU, Y.-F., WANG, P., QIN, H. A compact triband ACS-fed monopole antenna employing inverted-L branches for WLAN/WiMAX applications. *Progress In Electromagnetics Research C*, 2014, vol. 47, p. 131–138. DOI: 10.2528/PIERC14012002

[15] LI, X., SHI, X. W., HU, W., FEI, P., YU, J. F. Compact triband ACS-fed monopole antenna employing open-ended slots for wireless communication. *IEEE Antennas and Wireless Propagation Letters*, 2013, vol. 12, p. 104–107. DOI: 10.1109/LAWP.2013.2252414

[16] LI, Y., LI, W., YE, Q. Miniaturization of asymmetric coplanar strip-fed staircase UWB antenna with reconfigurable notch band. *Microwave and Optical Technology Letters*, 2012, vol. 55, no. 7, p. 1467–1470. DOI: 10.1002/mop.27634

[17] DEEPU, V., RAJ, R. K., JOSEPH, M., SUMA, M. N., MOHANAN, P. Compact asymmetric coplanar strip fed monopole antenna for multiband applications. IEEE Transactions on Antennas and Propagation, 2007, vol. 55, no. 8, p. 2351–2357. DOI: 10.1109/TAP.2007.901847

[18] ASHKARALI, P., SREENATH, S., SUJITH, R., et al. A compact asymmetric coplanar strip fed dual-band antenna for DCS/WLAN applications. *Microwave and Optical Technology Letters*, 2012, vol. 54, p. 1087–1089. DOI: 10.1002/mop.26731

[19] SOLTANI, S., AZARMANESH, M. N., LOTFI, P. Design of small ACS fed band-notch UWB monopole antenna using particle swarm optimization, *Microwave and Optical Technology Letters*, 2010, vol. 52, p. 1510–1513. DOI: 10.1002/mop.25269

[20] LI, Y., LI, W., YU, W. A switchable UWB slot antenna using SIS-HSIR and SIS-SIR for multi-mode wireless communications applications. *Applied Computational Electromagnetics Society Journal*, 2012, vol. 27, no. 4, p. 340–351.

[21] TAWK Y., CHRISTODOULOU, C. G. A new reconfigurable antenna design for cognitive radio. *IEEE Antennas and Wireless Propagation Letters*, 2009, vol. 8, p. 1378–1381. DOI: 10.1109/LAWP.2009.2039461

[22] LI, Y., LI, W., YE, Q. A reconfigurable wide slot antenna integrated with sirs for UWB/multiband communication applications. *Microwave and Optical Technology Letters*, 2012, vol. 55, p. 52–55. DOI: 10.1002/mop.27253

[23] *Al-Husseini, M., Ramadan, A., El-Hajj, A.*, et al. Design based on complementary split-ring resonators of an antenna with controllable band notches for UWB cognitive radio applications," In *Proceedings of the IEEE International Symposium on Antennas and Propagation and USNC/URSI National Radio Science Meeting (APSURSI '11)*. Spokane (WA, USA), 2011, p. 1120 to 1122. DOI: 10.1109/APS.2011.5996479

[24] LI, Y., LI, W., YE, Q. A reconfigurable triple-notch-band antenna integrated with defected microstrip structure band-stop filter for ultra-wideband cognitive radio applications. *International Journal of Antennas and Propagation*, 2013, Article ID 472645, 13 p. DOI: 10.1155/2013/472645

[25] LI, Y., LI, W., MITTRA, R. A compact CPW-fed circular slot antenna with reconfigurable dual band-notch characteristics for UWB communication applications. *Microwave and Optical Technology Letters*, 2014, vol. 56, no. 2, p. 465–468. DOI: 10.1002/mop.28087

About the Authors...

Ahmed IBRAHIM was born in 1986. He received the B.Sc. degree, with grade of very good, in Electrical Engineering from the Electronic and Communication Engineering Dept., Elminia University, Elminia, Egypt in 2007. He was awarded the M.Sc. degree from Elminia University in 2011 and the Ph.D. degree from Minia University in 2014, both in Electronic and Communication Engineering. He is now post-doctoral fellow in the Otto-von-Guericke-Universität Magdeburg, Germany. He is also a lecturer in Electronic and Communication Engineering Dept. His research focused on the design and analysis of microstrip antennas, microstrip filters, and its application in wireless communications, and also metamaterial MIMO antenna and different metamaterial applications in microwave bands.

Wael ALI received his B.Sc. and M.Sc. in Electronics and Communications Engineering from Arab Academy for Science, Technology and Maritime Transport (AASTMT), Alexandria, Egypt in 2004 & 2007 respectively. He obtained his Ph.D. in Electronics and Communications Engineering from Alexandria University, Alexandria, Egypt in 2012. He is a lecturer in Arab Academy for Science, Technology and Maritime Transport (AASTMT), Alexandria, Egypt. His research interests include smart antennas, microstrip antennas, microwave filters, and metamaterials.

Jan MACHAC is a professor at the Dept. of Electromagnetic Field, Faculty of Electrical Engineering, Czech Technical University in Prague, Czech Republic. His main scientific interests are: investigation of planar passive elements and subsystems of microwave technology, planar antennas, planar microwave filters, propagation of electromagnetic waves in periodic structures, and metamaterials. Prof. Machac is an author or co-author of more than 250 publications in scientific journals and scientific international and national conferences. He is a Senior Member of IEEE Microwave Theory and Technique Society, member of the IEEE MTT-S Int. Microwave Symposium TPRC, since 2008, member of the TPC of the European Microwave Conference during 1995-1997, and since 2013, reviewer of: IEEE Transactions on Microwave Theory and Technique, IEEE Microwave and Wireless Components Letters, Electronics Letters, and IET Microwave Antennas and Propagation.

Mutual Coupling Reduction for Dual-Band MIMO Antenna with Simple Structure

Chao-ming LUO, Jing-song HONG, Muhammad AMIN

School of Physical Electronics, University of Electronic Science and Technology of China, No. 4, Section 2, Chengdu 610054, China

461339852@163.com, cemlab@uestc.edu.cn, aminssphysics@hotmail.com

Abstract. *In this paper, a novel dual-band MIMO (multiple input, multiple output) antenna for WLAN (wireless local area network) applications is presented. The MIMO antenna contains two dual-band antenna elements, each of which comprises a T-shaped monopole and a special ├-shaped stub resonator. Two operating bands with center frequencies of 5.5 GHz and 2.5 GHz are crested by the monopole of T shape and the stub resonator of ├ shape, accordingly. The ├-shaped stub also works as an isolation structure at the higher band, which can simplify the dual-band isolation design into a single-band problem. Moreover, the isolation is enhanced at the lower band by inserting a metal strip which can cancel out original coupling. The inserted metal strip is the only additional decoupling structure in this design and has a simple texture with a compact size. The measured and simulated results reveal that the designed MIMO antenna can cover all the 2.4/5.2/5.8 GHz WLAN operating bands and within the recommended bands the isolations exceed by 20 dB.*

Keywords

Dual-band, high isolation, multiple-input multiple-output (MIMO) antenna, wireless local area network (WLAN)

1. Introduction

Multiple-input and multiple output (MIMO) technology has attracted more research attention due to the demands for higher transmission rate and more reliable link in wireless communications. MIMO is a complex technology by utilizing multiple antennas to increase the channel capacity and overcome multipath fading propagation problem [1], [2]. However, due to the space limit at the size-shrinking terminal devices, the most critical problem in a MIMO antenna design is how mutual coupling can be reduced between the closely packed antenna elements.

Various studies had been performed to boost the isolation of the antenna ports [3–13]. A tree-like parasitic structure [3], or mushroom like EBG structure [4–6] can minimize the mutual coupling among radiating components though restraining the surface wave propagation. The defected ground structure [7] or a simple ground plane modification [8] performing as a band-stop filter was designed to increase the isolation. Use of slot has been useful to include a notch in the isolation between antennas [9–11]. In [12], a method for isolation enhancement, based on metamaterial was given for the MIMO antenna. However, these methods have the common deficiency of complex structures, which will occupy a large space on the antenna. In [13], a neutralization line is physically linked to the antenna components for improving isolation. In this structure, the exiting mutual currents on the sufferer antenna are counterbalance by implanting 180° out-of-phase currents from an excitation antenna. The neutralization-line technique is extensively used as it occupies small space. However, this method markedly deteriorates the reflection coefficient and therefore, the trial and error process are required to obtain the suitable dimensions which need lots of time.

Moreover, most decoupling methods as mentioned above are only suited for offering narrow band isolation among two antennas. In previous studies on dual-band MIMO antennas [14–17], mutual coupling reduction is usually achieved by two isolation structures. One is used for lower band, while the other one is for higher band. However, combination of the two isolation structures may increase complexity of system, especially when the isolation structures are complex.

In this design, a novel structure of the antenna element is applied, which can avoid this problem. The dual-band antenna element comprises a monopole of T shape and a stub resonator of special ├ shape. At the lower frequencies, the ├-shaped stub can be considered as an antenna because it is coupled-fed by the T-shaped monopole. At the higher frequencies, it can reduce mutual coupling though suppressing surface wave propagation, which works as a reflective component. Thus, the space of the antenna is maximum utilized and the design of dual-band MIMO antenna is simplified to a single-band problem, which only needs to consider enhancing isolation at the lower band. Moreover, instead of adding a neutralization line that is physically joined to the antenna elements [13], we put a metal strip between the antenna elements in order to artificially generate a supplementary coupling route for enhancing the isolation. The metal strip, which is the only additional decoupling structure, has small effect on initial antenna impedance as it is not physically connected to the antenna. Due to the dual function of ├-shaped stub and the small occupied space of the metal strip, this antenna has a small size and simple structure.

2. Antenna Structure

The geometry of the proposed dual-band MIMO antenna can be seen in Fig. 1. The antenna is printed on a $26 \times 50\ mm^2$ FR4 substrate. This substrate has thickness of 0.8 mm with relative dielectric constant of 4.4 and loss tangent of 0.02. The MIMO antenna contains of two T-shaped monopoles, two ├-shaped grounded branches and a metal strip. A T-shaped monopole and a ├-shaped stub compose a dual-band antenna component and the two antenna components are printed symmetrically. The desirable high isolation is obtained by the metal strip and the unique structure of the antenna element. In the following, more details are studied.

3. Antenna Design

3.1 Dual-band Antenna Element

In Fig. 2(a), the composition of an element of the dual-band antenna can be seen. The antenna is made of a monopole of T shape and a special stub resonator of ├ shape, which are engraved on the substrate at the different surfaces. A 50-Ω microstrip line is used to directly excite the T-shaped monopole, whereas the stub resonator having ├ shape extended from ground is coupled-fed from the driven the T-shaped monopole. In Fig. 2(b), the simulated S_{11} of the antenna component without and with the resonator stub having ├ shape is shown. As observed, the T-shaped monopole without ├-shape stub has an operating frequency at around 5.5 GHz. The ├-shaped stub provides an additional resonance frequency of 2.5 GHz and improves the impedance matching characteristic from 5 to 6 GHz. Moreover, the current distributions for the two operating frequencies can be seen in Fig. 3. As inspected, the main currents are on the T-shaped monopole at 5.5 GHz, while the stronger currents are distributed near the ├-shaped stub at 2.5 GHz. Thus, we can conclude that, the T-shaped monopole determines the resonance of the upper frequency band at 5.5 GHz, while at the lower frequency 2.5 GHz, the ├-shaped stub regulates the impedance matching.

For better illustrating, the parametric studies are plotted in Fig. 4. As observed in Fig. 4, increasing lengths of the resonant structures cause the operating band to shift to lower frequencies. The optimal length of the resonant structures is approximately equivalent to a quarter-wavelength at the corresponding frequency. The concept of using a parasitic element has been also useful to design multiband monopole antennas [18]. Figure 5 shows the simulated radiation patterns of the antenna element. As suspected, the cross polarization is relatively high because of the introducing of the horizontal segment of the ├-shaped stub. This is not a serious problem for WLAN applications since the propagation channels are usually multipath, especially in the indoor environment.

The ├-shaped stub extended from ground works as a main radiator of the antenna at the 2.4–2.6 GHz band, while it works as a reflective component at other bands. This indicates that when the proposed dual-band antenna

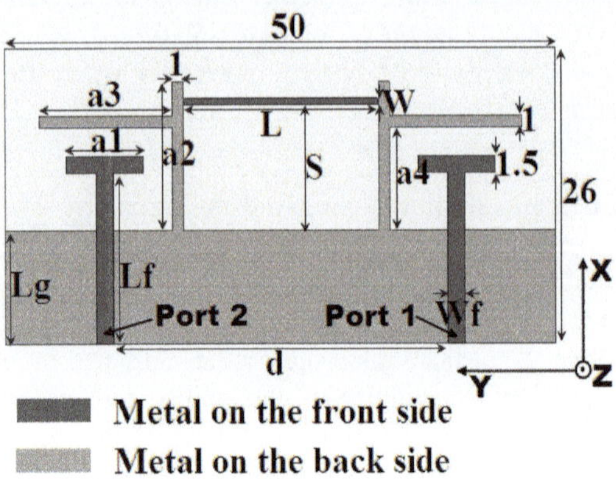

Fig. 1. Geometry of the dual-band MIMO antenna: $L_g = 10$, $L_f = 15$, $W_f = 1.5$, $a_1 = 7$, $a_2 = 13$, $a_3 = 12$, $a_4 = 9$, $d = 30.5$, $L = 17.6$, $S = 11$, and $W = 0.5$. (All dimensions are in millimeters).

Fig. 2. (a) Geometry and (b) S_{11} of dual-band antenna element.

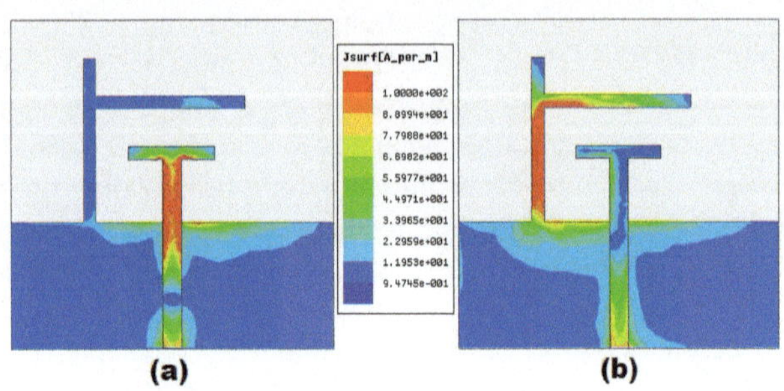

Fig. 3. Simulated, surface current distributions at (a) 2.5 GHz and (b) 5.5 GHz.

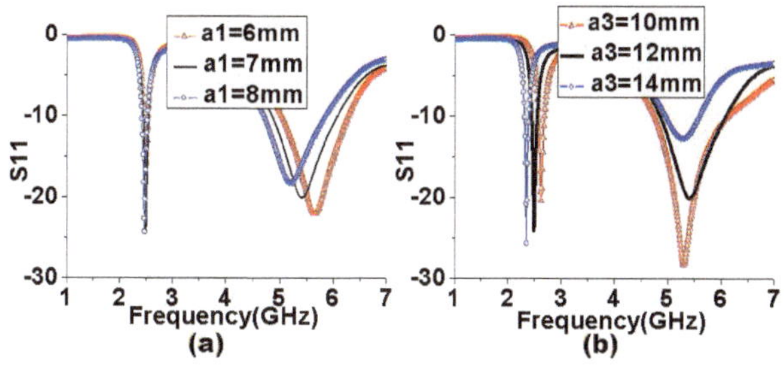

Fig. 4. Simulated S_{11} for different values of (a) $a1$ and (b) $a3$.

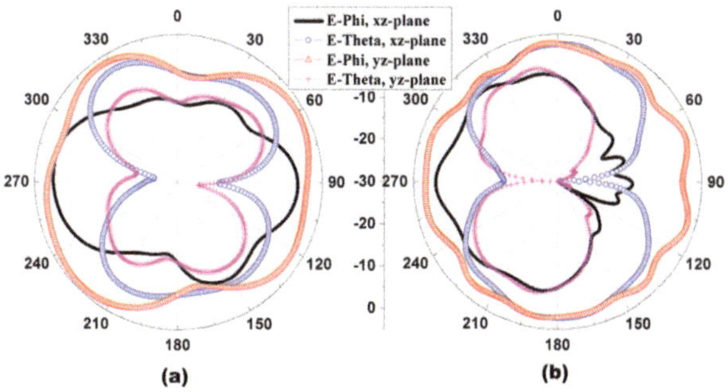

Fig. 5. Simulated radiation patterns of dual-band antenna element at (a) 2.5 GHz and (b) 5.5 GHz.

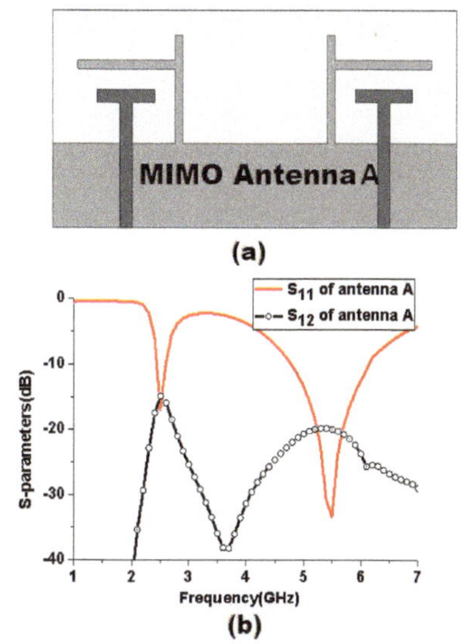

Fig. 6. (a) Geometry, (b) S_{11} and S_{12} of MIMO antenna A.

component is operating to construct a two-component MIMO antenna structure (MIMO Antenna A), a good isolation may be obtained at the higher frequencies 5–6 GHz. It can be proved by the S-parameters of antenna A, as shown in Fig. 6. The S_{12} is lower than –20 dB in the frequency range 5–6 GHz and –15 dB over 2.4–2.6 GHz. Moreover, in both the upper band (5–6 GHz) and lower band (2.4–2.6 GHz), the S_{11} is lower than –10 dB.

3.2 Metal Strip

As mentioned above, high isolation can be accomplished without any additional decoupling structures for the upper band and that is because of the unique structure of the antenna component. In order to lessen the mutual coupling for the lower band, a metal strip of size 17.6×0.5 mm^2 is placed between the elements, which can introduce new coupling to counteract the initial coupling. Different from the traditional neutralization line, which markedly deteriorates the reflection coefficient, the metal strip in this design is not physically connected to the antenna elements and is perpendicular to the microstrip line. To understand the working mechanism deeply, parameter studies are carried out. The simulated S-parameters for different L and S are shown in Figs. 7 and 8, respectively. For clarity, these figures for S_{12} have been enlarged around the frequency of 2.5 GHz. Figures 7(a) and 8(a) show that S_{11} varies less with the parameters, indicating a minor

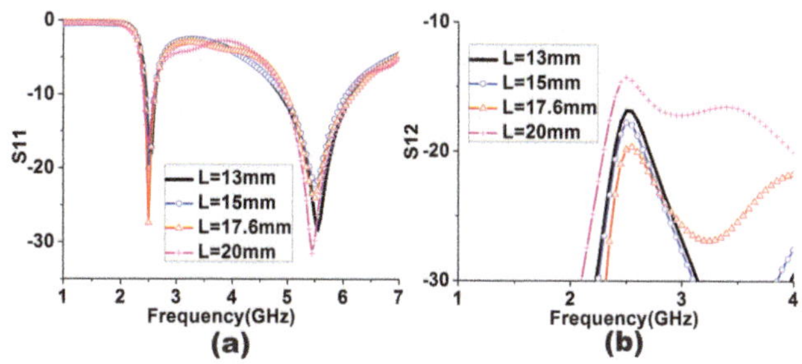

Fig. 7. S-parameters with different lengths of the metal strip: (a) S_{11} (b) S_{12}.

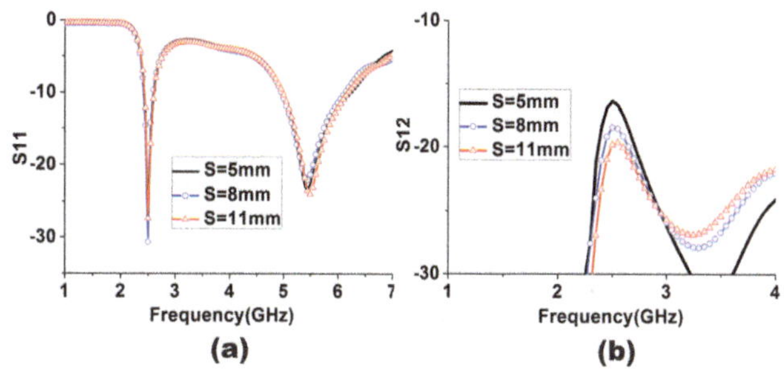

Fig. 8. S-parameters with different positions of the metal strip: (a) S_{11} (b) S_{12}.

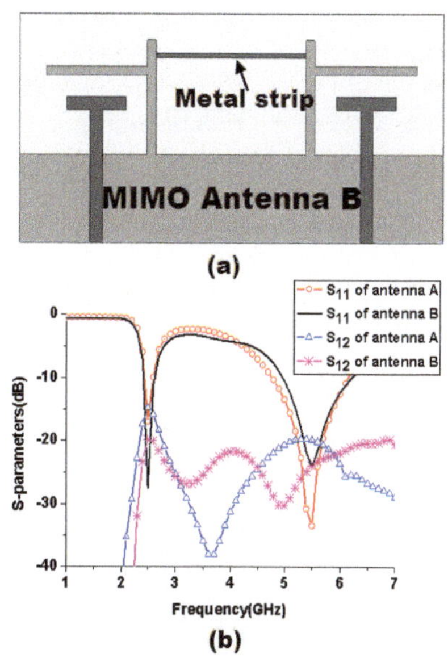

Fig. 9. (a) Structure of MIMO antenna B, (b) S_{11} and S_{12} of antenna A and B.

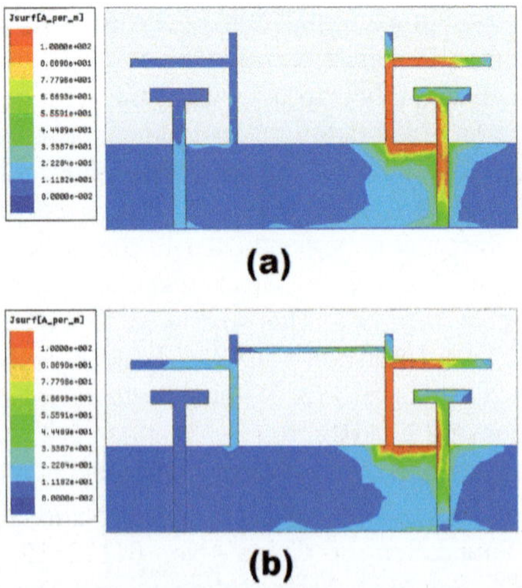

Fig. 10. At 2.5 GHz, current distributions on MIMO (a) A and (b) B antennas.

effect of the metal strip on reflection coefficient. As shown in Figs. 7(b) and 8(b), when L increases from 13 mm to 17.6 mm or S varies from 5 mm to 11 mm, the isolation at 2.5 GHz is improved about 5 dB. The reason for this phenomenon is that the additional coupling produced by the metal strip is insufficient if the metal strip is short or S is small. As expected, when the metal strip is too long ($L = 20$ mm), the isolation gets worse caused by excessive additional coupling. Moreover, at other frequencies, the metal strip also introduces more or less coupling to change the isolation characteristics, but remains within an acceptable range i.e. higher than 20 dB.

In Fig. 9, the return losses of antenna A (without the metal strip) and antenna B (with the metal strip) are given. The mutual coupling is reduced to less than –20 dB over the entire WLAN band by adding the metal strip within the two antenna elements. It can also be observed that the effect of the metal strip on S_{11} is small. The mechanism of decoupling is that the metal strip yields a supplementary route for the coupling currents, and the original coupling should be mostly offset. The currents from the excited antenna component to the non-excited component are significantly reduced, it is shown in Fig. 10.

4. Experimental Results

In order to verify the results obtained from simulations, a prototype antenna was fabricated and tested, which can be seen in Fig. 11.

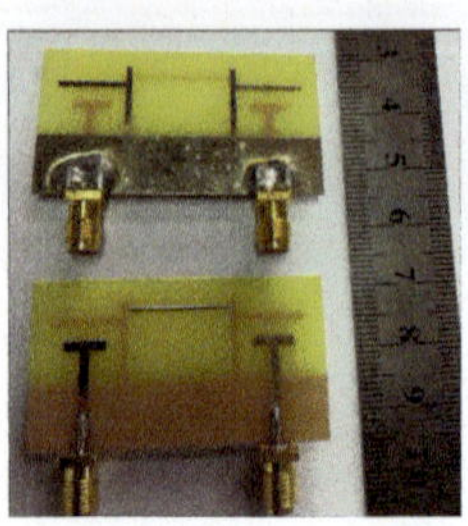

Fig. 11. Photograph of the fabricated antenna.

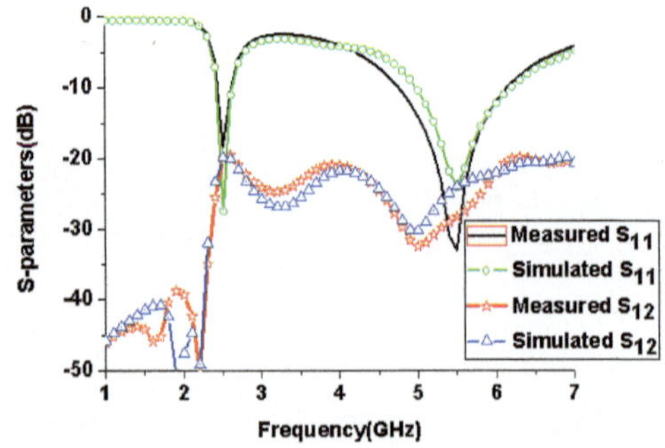

Fig. 12. S_{11} and S_{12} of the antenna (proposed).

Figure 12 shows the measured and simulated S-parameters. The measured –10 dB impedance bandwidth for the 2.4 GHz is 300 MHz (from 2.35 GHz to 2.65 GHz) band and for the 5.2/5.8 GHz bands is 1.3 GHz (from 4.9 to 6.2 GHz). For the higher and lower frequency bands, the measured mutual couplings are lower than –26 dB and –20 dB, respectively. It can be confirmed by the nearness of the measured and simulated frequency responses of the fabricated antenna that the operating band satisfies the design demands.

For evaluating the proposed MIMO antenna functioning, the (ECC) envelope correlation coefficient, a measure to describe how much the communication channels are isolated with each other, is taken as an analytical parameter. It can be calculated from the S-parameters using the following formula [19]:

$$ECC = \frac{\left|S_{11}^{*}S_{12} + S_{21}^{*}S_{22}\right|^{2}}{\left|\left(1-\left|S_{11}\right|^{2}-\left|S_{21}\right|^{2}\right)\left(1-\left|S_{22}\right|^{2}-\left|S_{12}\right|^{2}\right)\right|}. \quad (1)$$

The measured and simulated ECCs across the desired frequency bands are shown in Fig. 13. It can be seen that both the ECCs of the proposed MIMO antenna are below 0.06 over the complete WLAN frequency band, which is much lower than criterion of low ECC (ECC < 0.5)[19]. In addition, equation (1) illustrates that small $|S_{11}|$ and $|S_{12}|$ can cause low ECC. It can be inferred that desired ECC demands good performances in impedance matching and isolation. With the low return loss and high isolation, the ECC of the proposed MIMO antenna is almost 0 at the operating bands (2.35 GHz to 2.65 GHz, and 4.9 GHz to 6.2 GHz). The difference in simulated and measured ECCs

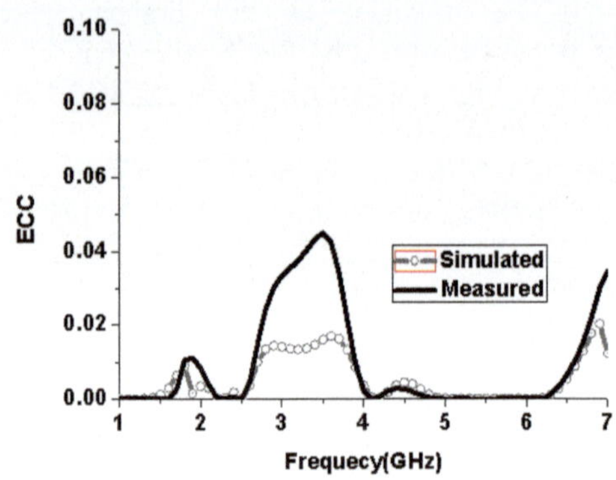

Fig. 13. Simulated and measured ECC.

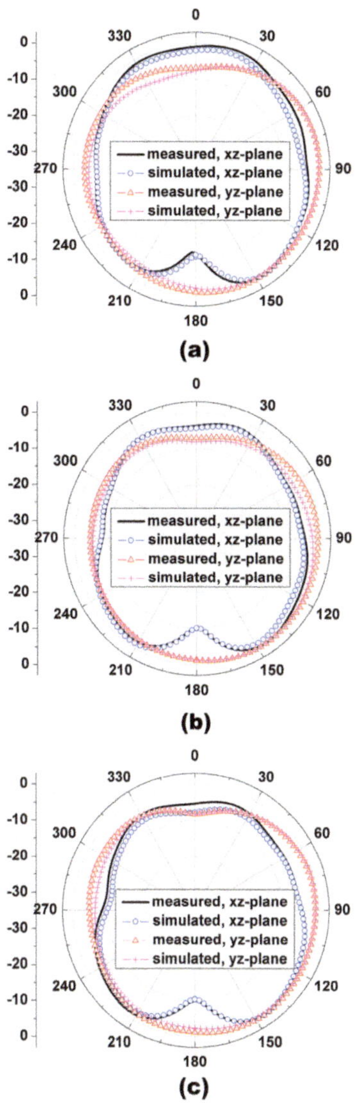

Fig. 14. Radiation patterns of the proposed MIMO antenna at (a) 2.4 GHz, (b) 5.2 GHz, and (c) 5.8 GHz.

in the notched bands is relatively large. That is because in the case of mismatch, the isolation plays a major role and has a large proportion in ECC.

In Fig. 14, the measured normalized radiation patterns of the three frequencies 2.4, 5.2, and 5.8 GHz for the proposed (MIMO) antenna are plotted. For measurements, during the excitation of portl, port 2 is cut off with a matched load of 50 Ω. The radiation patterns were measured for the total electric-field. It can be seen that, in *y-z* plane the antenna is omnidirectional and near-omnidirectional in *x-z* plane, which can receive the signals from all the directions. In addition, at the frequencies of 2.4, 5.2, and 5.8 GHz, the peak gains reach 1.56, 4.43, and 4.7 dBi, respectively, with the corresponding efficiencies of 70.1 %, 82.3 %, and 81.2 %.

5. Conclusions

The basic idea of this design is the development of a decoupling method for dual-band with a simple and compact geometry. In this article, a dual-band (MIMO) antenna for WLAN applications is exposed. A novel monopole element is used to provide high isolation for higher band, where radiating element and the isolation structure share the same structure. Moreover, the metal strip, which is the only additional decoupling structure, has occupied a small space and produces a significant enhancement of isolation at lower band.

Acknowledgments

This work was supported by the National Natural Science Foundation of China (No. 61172115 and No. 60872029), the High-Tech Research and Development Program of China (No. 2008AA01Z206), the Aeronautics Foundation of China (No. 20100180003), the Fundamental Research Funds for the Central Universities (No. ZYGX2009J037), Project 9140A07030513DZ02098 and the Fundamental Research Funds for the Central Universities (Grant No. ZYGX2012YB019).

References

[1] FOSCHINI, G. J., GANS, M. J. On limits of wireless communications in a fading environment when using multiple antennas. *Wireless Personal Communications*, 1988, vol. 6, no. 3, p. 311–335. DOI: 10.1023/A:1008889222784

[2] ZHENG, L., TSE, D.N.C. Diversity and multiplexing: A fundamental tradeoff in multiple-antenna channels. *IEEE Transactions on Information Theory*, 2003, vol. 49, no. 5, p. 1073 to 1096. DOI: 10.1109/TIT.2003.810646

[3] ZHANG, S., YING, Z., XIONG, J., et al. Ultrawideband MIMO/diversity antennas with a tree-like structure to enhance wideband isolation. *IEEE Antennas Wireless and Propagation Letters*, 2009, vol. 8, p. 1279–1282. DOI: 10.1109/LAWP.2009.2037027

[4] COULOMBE, M., KOODIANI, S. F., CALOZ, C. Compact elongated mushroom (EM)-EBG structure for enhancement of patch antenna array performances. *IEEE Transactions on Antennas and Propagation*, 2010, vol. 58, no. 4, p. 1076–1086. DOI: 10.1109/TAP.2010.2041152

[5] YANG, F., RAHMAT-SAMII, Y. Microstrip antennas integrated with electromagnetic band-gap EBG structures: A low mutual coupling design for array applications. *IEEE Transactions on Antennas and Propagation*, 2003, vol. 51, no. 10, p. 2936–2946. DOI: 10.1109/TAP.2003.817983

[6] GHOSH, S., TRAN, T., LE-NGOC, T. Dual-layer EBG-based miniaturized multi-element antenna for MIMO systems. *IEEE Transactions on Antennas and Propagation*, 2014, vol. 62, no. 8, p. 3985–3997. DOI: 10.1109/TAP.2014.2323410

[7] SHARAWI, M. S., NUMAN, A. B., KHAN, M. U., et al. A dual-element dual-band MIMO antenna system with enhanced isolation for mobile terminals. *IEEE Antennas and Wireless Propagation Letters*, 2012, vol. 11, p. 1006–1009. DOI: 10.1109/LAWP.2012.2214433

[8] LI, H., XIONG, J., HE, S. A compact planar MIMO antenna system of four elements with similar radiation characteristics and isolation structure. *IEEE Antennas and Wireless Propagation*

Letters, 2009, vol. 8, p. 1107–1110. DOI: 10.1109/LAWP.2009.2034110

[9] PARK, J., CHOI, J., PARK, J. Y., et al. Study of a T-shaped slot with a capacitor for high isolation between MIMO antennas. *IEEE Antennas and Wireless Propagation Letters,* 2012, vol. 11, p. 1541–1544. DOI: 10.1109/LAWP.2012.2226695

[10] ANDUJAR, A., AUGUERA, J. MIMO multiband antenna system combining resonant and nonresonant elements. *Microwave and Optical Technology Letters*, 2014, vol. 56, no. 5, p. 1076–1084. DOI: 10.1002/mop.28282

[11] ANDUJAR, A., AUGUERA, J. MIMO multiband antenna system with nonresonant elements. *Microwave and Optical Technology Letters*, 2015. vol. 57, no. 1, p. 183–190. DOI: 10.1002/mop.28810

[12] HSU, C., LIN, K., SU, H. Implementation of broadband isolator using metamaterial-inspired resonators and a T-shaped branch for MIMO antennas. *IEEE Transactions on Antennas and Propagation*, 2011, vol. 59, no. 10, p. 3936–3939. DOI: 10.1109/TAP.2011.2163741

[13] SU, S., LEE, C., CHANG, F. Printed MIMO-antenna system using neutralization-line technique for wireless USB-dongle applications. *IEEE Transactions on Antennas and Propagation*, 2012, vol. 60, no. 2, p. 456–463. DOI: 10.1109/TAP.2011.2173450

[14] SU, H., HSU, C., LIN, K. Dual-band insulator design by stacking capacitively loaded loops for MIMO antennas. *Electronics Letters*, 2010, vol. 46, no. 20, p. 1364–1365. DOI: 10.1049/el.2010.1756

[15] LI, G., ZHAI, H., MA, Z., et al. Isolation-improved dual-band MIMO antenna array for LTE/WiMAX mobile terminals. *IEEE Antennas and Wireless Propagation Letters*, 2014, vol. 13, p. 1128–1131. DOI: 10.1109/LAWP.2014.2330065

[16] LI, L., HUO, F., JIA, Z., et al. Dual zeroth-order resonance antennas with low mutual coupling for MIMO communications. *IEEE Antennas and Wireless Propagation Letters*, 2013, vol. 12, p. 1692–1695. DOI: 10.1109/LAWP.2013.2294686

[17] ZHAO, W., LIU, L., CHEUNG, S. W., et al. Dual-band MIMO antenna using double-T structure for WLAN applications. In *2014 International Workshop on Antenna Technology: Small Antennas, Novel EM Structures and Materials, and Applications (iWAT).* Sydney (Australia), 2014, p. 232–235. DOI: 10.1109/IWAT.2014.6958646

[18] RISCO, S., ANGUERA, J., ANDUJAR, A., et al. Coupled monopole antenna design for multiband handset devices. *Microwave and Optical Technology Letters*, 2010, vol. 52, no. 2, p. 359–364. DOI: 10.1002/mop.24893

[19] YAN, S., SOH, P. J., VANDENBOSCH, G. A. E. Dual-band textile MIMO antenna based on substrate-integrated waveguide (SIW) technology. *IEEE Transactions on Antennas and Propagation*, 2015, vol. 63, no. 11, p. 4640–4647. DOI: 10.1109/TAP.2015.2477094

About the Authors ...

Chao-ming LUO was born in Jingdezhen, China. He received his B.S. degree from ChengDu University of Information Technology in 2009. He is now working towards his Ph.D. degree in Radio Physics at the University of Electronic Science and Technology of China (UESTC). His interests include antenna technology and wireless communication technology.

Jing-song HONG received the B.S. degree in Electromagnetic from LanZhou University, China, in 1991, and the M.S. and Ph.D. degrees in Electrical Engineering from the University of Electronic Science and Technology of China (UESTC), in 2000 and 2005, respectively. He is now a professor with UESTC. His research interests include the use of numerical techniques in electromagnetic and the use of microwave methods for materials characterization and processing.

Muhammad AMIN was born in D.I.Khan, Pakistan. He received his Master degree from Gomal University D.I.Khan Pakistan in 2003.After that he worked at weather surveillance Radar D.I.Khan till 4th March 2011. Then he taught Physics in the Education Department KPK Pakistan till August 2014. Since September 2014 he is doing his Ph.D degree in Radio Physics at the University of Electronic Science and Technology of China (UESTC). His interests include antenna technology and wireless communication technology.

Directly Printable Organic ASK Based Chipless RFID Tag for IoT Applications

Ayesha HABIB [1], *Sohaira ANSAR* [1], *Adeel AKRAM* [1], *Muhammad Awais AZAM* [1], *Yasar AMIN* [1,2], *Hannu TENHUNEN* [2,3]

[1] ACTSENA Research Group, University of Engineering and Technology, 47050- Taxila, Pakistan
[2] iPack VINN Excellence Center, Royal Institute of Technology, SE-16440, Stockholm, Sweden
[3]TUCS, University of Turku, Turku-20520, Finland

ayesha.habib@uettaxila.edu.pk

Abstract. *A chipless RFID tag with unique ASK encoding technique is presented in this paper. The coding efficiency is enhanced regarding tag capacity. The amplitude variations of the backscattered RFID signal is used for encoding data instead of OOK. Strips of different widths are used to have amplitude variations. The ASK technique is applied using three different substrates of Kapton®HN, PET, and paper. To incorporate ASK technique, dual polarized rhombic shaped resonators are designed. These tags operate in the frequency range of 3.1–10.6 GHz with size of 70 × 42 mm^2. The presented tags are flexible and offer easy printability. The paper-based decomposable organic tag appears as an ultra low-cost solution for wide scale tracking. This feature enables them to secure a prominent position in the emerging fields of IoT and green electronics.*

Keywords

Internet-of-Things (IoT), Radio Frequency Identification (RFID), chipless, Amplitude Shift Keying (ASK), resistive strip, coupling, damping

1. Introduction

A fully connected network in which every entity, whether physical or virtual, can be accessed at any time, any place using any route defines IoT, the advancement in embedded systems and nanotechnology has provided automated decisions and all in one platform capability to the IoT [1]. IoT consists of smart devices of time like smartphones, tablets, surveillance systems, industrial equipment, field operation devices, automobiles, etc. [2], [3]. Realization of such smaller, powerful processing devices with onboard sensing capabilities makes IoT an important beacon of this tech-era. These smart devices are integrating the concept of data encoding and sensing resulting in an intelligent interaction. IoT allows automated systems to immediately response according to information provided by these smart devices [4], [6]. Now this information can be easily accessed through internet and decisions can be made remotely [7]. IoT has found its applications in different areas including health, education, industry, traveling, surveillance systems, home, business, geographical survey, mining, etc.

To implement IoT, the ubiquitous network is required to manage the whole network remotely. Wireless Sensor Network (WSN) provides the required medium for management and communication of smart nodes. Hence, on primitive level IoT works by implementing WSN techniques. Furthermore, the task of identification and autonomous tracking of smart nodes is carried out by RFID.

The RFID system works using radio waves as their communication medium. A general RFID comprises of RFID interrogators, RFID transponders/tags, and middleware system. The transponders are mounted on the objects being tracked or monitored. These tags in the vicinity of the reader are activated by the stimulus initiated by reader side. This stimulus is an electromagnetic (EM) signal from reader towards transponder for its activation. The required data is encoded by modulating this incident EM signal [8]. The encoded signal is backscattered towards interrogator. This received signal is forwarded to middleware software for further processing.

RFID tags can be majorly classified on the basis of the power facility and modulating entity. The availability of power supply classifies these tags into active and passive ones. Active tags offer long read range, possess on board battery but they are expensive. The cost of a tag is a matter of concern while choosing RFID as an identification technique. To replace the low-cost line-of-sight barcode technology, RFID should appear as cost effective solution. In this regard, the cost of a passive tag is quite low as compared to active ones due to the absence of power supply. So, while choosing a RFID tag for our system, a tradeoff is done between their prices and read range [9]. This feature of *passive tags* allows a cost effective solution for wide scale deployment of tags in the environment [10]. The classification of RFID tags can also be done on the basis of the modulator. The modulation of the incident signal is responsible for generating EM response of the tag. The properties of reflected signal are different from incident signal due to modulation. On the basis of this difference, reader interprets encoded message. In this perspective, the mounted tag over the object may be chip based or chipless.

In case of chip-based tags, the transponder device consists of two parts 1) scatterer, 2) encoder. Such tags use Application Specific Integrated Circuit or silicon chip as an encoder. While the transmission and reception of EM signals is done through scatterer or the antenna part. Silicon chip is responsible for encoding the information through modulation. Through backscattering phenomena, the reader receives its encoded message. On the other hand, chipless tags appear as a cost-effective solution that eliminates the need for costly silicon chip [11]. The antenna part performs the modulation of the signal as well as the transmission and reception of signals. Moreover, these energy efficient tags harvest RF energy from the transmitted signal and act like rectenna [12]. Hence, the chipless tag dually acts as scatterer as well as an encoder.

Some common examples of cost effective tags include embedding them in bank notes, passports, toll cards, clothes, shoe soles, on locomotives, helmets of miners, etc. where they are replacing the barcode technology efficiently. Chipless RFID tags are low cost, easily printable, lightweight, energy efficient and can bear high temperature and pressures. They can be easily embedded in different spying applications. They can be equipped with sensing materials and can act as very simple sensor nodes. These fascinating attributes prove chipless RFID tags to be an excellent sensor node for wireless WSN [13].

Printable chipless RFID tags may be retransmission based and backscattering based tags. Retransmission based tags are larger in size and require monopole antennas with resonators [14]. The backscattering based tags consist of only multi-resonators. Data is usually encoded in frequency, amplitude and phase, time of backscattered signal. The EM response of the backscattered signal greatly depends on the geometry and size of resonating structures [15]. On the basis of presence and absence of resonators, on-off keying (OOK) is observed as a spectral response. In this scenario, x resonators can produce 2^x combinations for encoding of data. To further enhance this coding efficiency, amplitude variation technique of the backscattered signal can be used. This will provide more combinations i.e. more unique IDs for identification. In Amplitude Shift Keying (ASK), the data is encoded as variations in the amplitude of backscattered encoded signal while OOK is a simpler form of ASK in which absence or presence of signal depicts the '0' and '1', respectively. Hence, by opting ASK technique coding capacity of tag can be increased.

In this research, a technique is explored which can increase the coding efficiency of broadband Chipless RFID tag. More precisely, Amplitude Shift Keying (ASK) has been studied and verified through simulations. Using ASK collective data rate has been enhanced up to 24-bits that is the highest among the previously reported work up to date [16]. The tag is easily printable on the object, flexible and supports green electronics. ASK technique has been applied on tag design and analysis is performed using different substrates.

2. Theory and Working Principle

Broadband antennas cannot be used for signature generation. The coding efficiency of such antennas can be increased by using ASK. To excite the tag, instead of using conventional methods, the incident plane wave (circularly polarized) is applied. The circular plane wave equation for E-field,

$$\mathbf{E}(x,y,z,t) = E_o e^{j(\omega t - kz)}\hat{\mathbf{x}} + E_o e^{j(\omega t - kz + \pi/2)}\hat{\mathbf{y}}. \quad (1)$$

For Radar Cross Section (RCS) measurement of chipless tag, the far-field distance is calculated by using (2), where D is maximum dimension of tag and λ is wavelength

$$d = \frac{2D^2}{\lambda}. \quad (2)$$

The probes are set at far-field distance to measure the RCS response. The far-field distance is calculated as 114 mm for the proposed design. The backscattered signal is observed by using RCS. RCS has the information of reflected wave, which tells us about the selectivity, phase, amplitude, velocity, surface area and resonance of tag. Detection range and a cross-section of the target can also be estimated with information of RCS. Frequency domain footprint based tags are frequently designed previously [17]. Mostly, OOK is used as the spectral modulation technique. Presence or absence of resonating RCS peaks determines binary '1' and '0' respectively. The capacity of encoded bits is dependent on the number of resonators. To increase data bits, the number of resonators needs to increase while maintaining a compact size. Concentric structures are also used by previous researchers to use small space [18] efficiently.

Some hybrid techniques like pulse position modulation and phase modulation are also used to increase data capacity. These methods have enhanced the capacity and greater number of objects can be tagged [19].

2.1 Amplitude Shift Keying (ASK)

In this research work amplitude of received signal is being observed. ASK is explored for six resonators. The overall capacity of the tag has been increased. Previously,

ASK has been applied for three resonators in credit card size [17].

The amplitude of receiving signal is varied by using resistive strips [20]. They are acting as bridging resistance [21] and the absence of resistive strip shows infinite resistance. These strips intervene in current paths and reduce pulse height. Amplitude variations are being observed for all six scatterers using these resistive strips. Mathematically, coding capacity is calculated by using formula

$$C = \log_2\left(\left(\frac{\Delta M_1}{dM_1}+1\right)\left(\frac{\Delta M_2}{dM_2}+1\right)\left(\frac{\Delta M_3}{dM_3}+1\right)\left(\frac{\Delta M_4}{dM_4}+1\right)\left(\frac{\Delta M_5}{dM_5}+1\right)\left(\frac{\Delta M_6}{dM_6}+1\right)\right) \quad (3)$$

Magnitude resolution $dM_{i=(1-6)}$ is a change in neighboring peaks due to the addition of resistive strip [18]. In this technique $dM_{i=(1-6)}$ is calculated by taking the difference of varying amplitudes values of neighboring peaks when flat strip width is applied and when no strip is present. Magnitude resolution for each resonator has been set after optimization and simulation observations. $\Delta M_{i=(1-6)}$ is the difference between the maximum and minimum amplitudes of the resonator on which resistive strip has been applied i.e. $\Delta M_i = M_{bi} - M_{ai}$. The layout of the tag design is shown in Fig. 1.

3. Tag Design and Optimization

The proposed tag is composed of six resonating structures. Dual rhombic loops are designed with silver conductive tracks having conductivity 9×10^6 S/m. The thickness of silver metal is kept 15 μm, for easy printing of the tag. The tag design having the same dimensions is analyzed for three substrates. In spite of achieving flexibility, the main purpose is to make cost effective, energy efficient tags also having admirable data encoding capacity. Resistive strips are kept 1 μm thick to attain resistance of 1000 Ω/sq. All tags are designed and simulated in CST STUDIO SUITE®. The overall dimension of tag design is 70×42 mm^2 shown in Fig. 1. The inkjet printing of tag is done using a table top printer based on Fujifilm Diamatix DMP2800 printer. The silver nano particle based conductive tracks are printed using Cabot conductive Ink CCI-300. We took nearly eighteen multiple passes and curing has been carried out during the printing process with the help of heat gun to avoid the misalignment between the passes. In the last two passes the resistive strips has been printed. Finally, the curing was carried out for 2 hr, at 150°C.

3.1 Realization on Kapton®HN Substrate

Using Kapton®HN substrate tag is designed as shown in Fig. 2. The thickness of the material is 175 μm with permittivity of 3.5 and loss tangent 0.004. The larger scatterer S1 has resonating frequency 3.5 GHz, the second scatterer

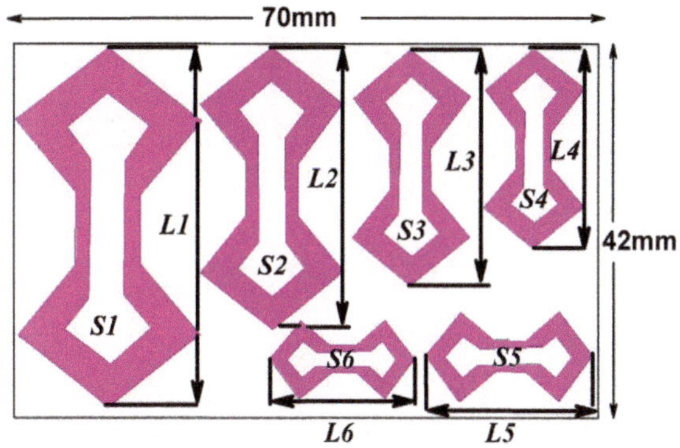

Fig. 1. Tag design dimensions. $L1$ = 41.21 mm, $L2$ = 32.48 mm, $L3$ = 27 mm, $L4$ = 22.94 mm, $L5$ = 19.97 mm, $L6$ = 17.33 mm.

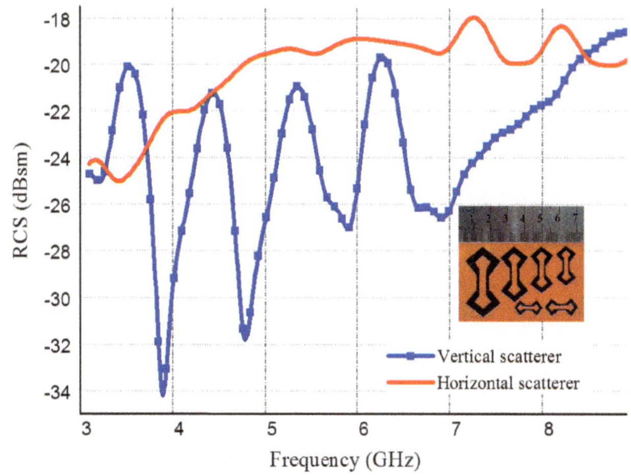

Fig. 2. Resonating peaks with no resistive strip (Kapton®HN).

Substrates	R1 (GHz)	R2 (GHz)	R3 (GHz)	R4 (GHz)	R5 (GHz)	R6 (GHz)
Kapton® HN	3.5	4.4	5.3	6.27	7.2	8.2
PET	3.7	4.72	5.7	6.8	7.9	9.1
Paper	3.42	4.26	5.23	6.21	7.08	8.13

Tab. 1. Resonating frequencies of 1-6 scatterers for three substrates.

S2 resonates at 4.4 GHz, the third scatterer S3 resonates at 5.3 GHz, and the fourth scatterer S4 resonates at 6.27 GHz, the fifth scatterer S5 resonates at 7.2 GHz and the sixth scatterer S6 resonates at 8.2 GHz as demonstrated in Tab. 1.

3.2 Realization on PET Substrate

The electrical properties of PET substrate comprise of permittivity and loss tangent having values of 2.9 and 0.0025, respectively. The thickness of material is kept to be 100 μm. All scatterers are designed to resonate in the band between 3.1 GHz and 10.6 GHz. S1 resonates at 3.7 GHz, S2 with resonating frequency of 4.72 GHz, S3 resonates at 5.7 GHz, S4 has the resonating frequency of 6.8 GHz, S5 resonates at 7.9 GHz and S6 resonates at 9.1 GHz as shown in Tab. 1.

Fig. 3. Realized printed tag using paper substrate.

3.3 Realization on Paper Substrate

The paper substrate used in the design has thickness of 250 μm, permittivity of 3.3 and loss tangent of 0.777 has been selected as an electrical variant. The printed tag structure using paper substrate is shown in Fig. 3. Six scatterers S1, S2, S3, S4, S5, and S6, resonate at frequencies of {3.42, 4.26, 5.23, 6.21, 7.08, and 8.13} GHz respectively as shown in Tab. 1.

4. Results and Discussions

In this section, the effect of adding a resistive strip on all resonators has been discussed. Coupling between resonators is avoided using position strategy [22]. All vertical resonators are not equally spaced from each other. A minimum distance of 0.8 mm between the 1st and the 2nd resonator is sufficient enough to decouple EM waves. The gap between the 2nd and the 3rd one is 1.9 mm while the 3rd and the 4th resonators are 2.2 mm. Also, the space between the 5th and 6th resonators is kept 1.3 mm. Resistive strips of different widths for horizontal and vertical resonators are applied to implement ASK.

Using Kapton®HN substrate, the width of the resistive strip has been varied from 0.1 mm to 7.5 mm (using 4-intervals) for S1, from 0.1 mm to 5 mm for S2, from 0.5 mm to 4 mm for S3 and from 0.5 mm to 3.5 mm for S4. For horizontal scatterers S5 and S6, a resistive strip of meager width is producing amplitude variations. So widths of 0.1 mm to 0.5 mm have been used for maximum amplitude variations. Amplitude variation from minimum to the maximum level for S1 is given in Tab. 2. The amplitude of resonating peak with resistive strip varies accordingly. Applying a resistive strip on the specific resonator also affects the amplitudes of neighboring peaks. In this way, amplitude variation analysis has been done for all resonators one by one. Damping of amplitude from maximum amplitude to minimum for the 1st scatterer can be easily observed in Fig. 4(a). The RCS vs. frequency plots for horizontal scatterers are also shown in Fig. 4(b). Also, at the largest strip width when the amplitude of the corresponding peak is almost deteriorated, a shift in neighboring peaks occurs which is shown in Fig. 4(a). It is deduced from observation that the change in thickness of the substrate is responsible for frequency shifting behavior.

Tags	Substrate	Max. Amplitude. of S1 M_a (dB)	Min. Amplitude. of S1 M_b (dB)	Data capacity of tag (bits)
Tag 1	Kapton®HN	–20.09	–24.6	16
Tag 2	PET	–19.89	–24.14	21
Tag 3	Paper	–21.22	–25.30	24

Tab. 2. Maximum and minimum amplitude variations and corresponding data capacity.

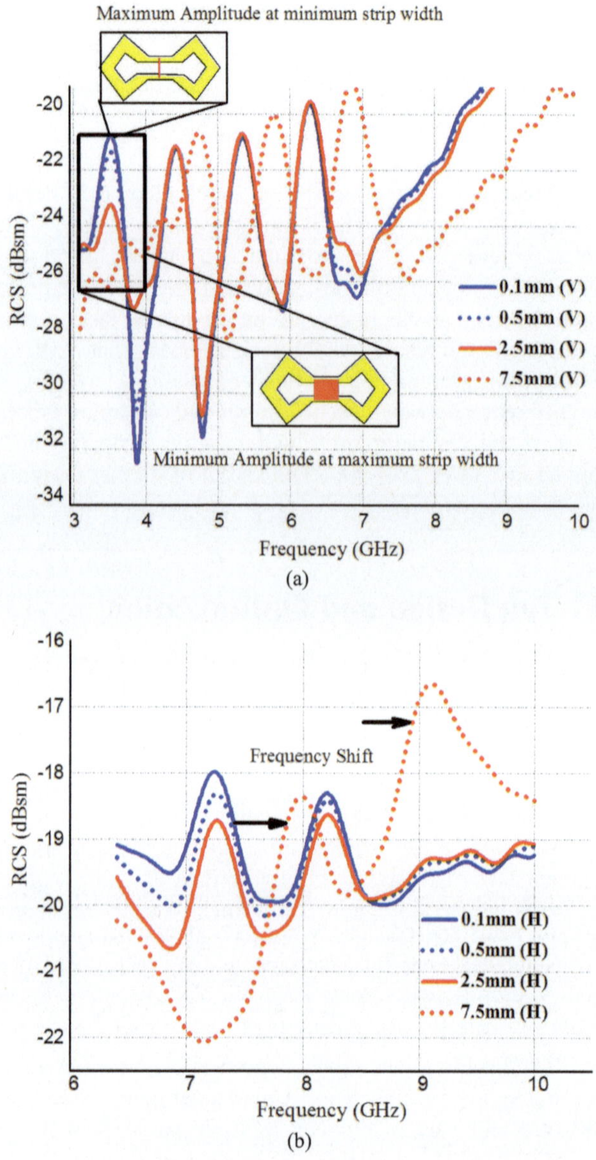

Fig. 4. Simulation results for ASK in S1 for Kapton®HN substrate. (a) Vertical scatterers showing ASK in S1. (b) Horizontal scatterers.

The surface current distribution for two different frequencies is being observed. Figure 5(a) shows the induced surface current corresponding to scatterer having the largest dimension and Figure 5(b) shows the mapping of EM waves on the scatterer having the smallest dimension for the lowest and highest frequencies, respectively.

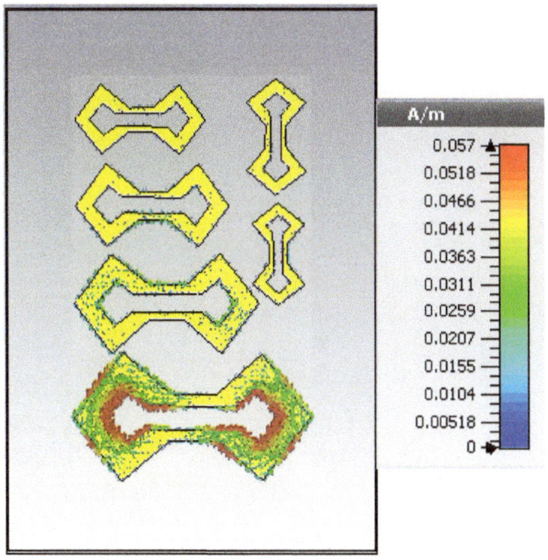

(a)

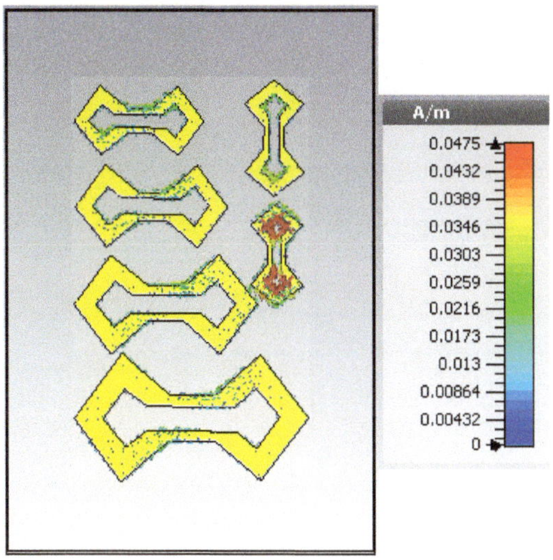

(b)

Fig. 5. Current distribution. (a) For the lowest frequency. (b) For the highest frequency.

For measuring RCS of a tag, the controlled testing environment is used. The whole setup comprises of two horn antennas, one for reception and other for transmission, Vector Network Analyzer R&S®ZVL13 and Chipless RFID tag. The measurement setup is in accordance with previously used standard measurement procedure as mentioned in detail elsewhere [23]. The tag design is manufactured and experimentally tested for five prototypes using each substrate for analysis of reliability parameters. In RCS values tag has tolerance of ±0.01 % and ±0.25 % in the desired frequency band as shown in Fig. 6.

Fully damped RCS response flattens the peak, and the width of the resistive strip at this point is known as flat peak strip width as shown in Tab. 3. For the maximum width of the resistive strip, the resistance would be low. Similarly for the minimum width of the resistive strip, the resistance becomes high and this can also be understood from (4)

$$R = \frac{\rho L}{A} \tag{4}$$

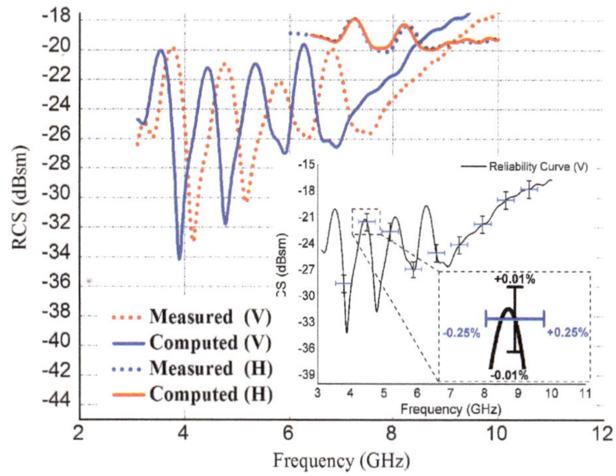

Fig. 6. Measured and RCS reliability graph of the tag.

Substrate	S1 (mm)	S2 (mm)	S3 (mm)	S4 (mm)	S5 (mm)	S6 (mm)
Kapton®HN	7.5	5	4	3.5	0.5	0.5
PET	7	3	3	2.5	0.5	0.4
Paper	6	2	2.5	2.5	0.4	0.4

Tab. 3. Maximum widths of the resistive strip for all scatterers giving flat peak response.

where L is the length of the resistive strip, ρ is resistivity of material and A is the cross-sectional area of the strip. It is experimentally found that the capacity of the tag increases by having more amplitude levels through resistive strip width variations. The amount of amplitude variations of neighboring peaks, i.e. magnitude resolution (dM_i) is dependent on the substrate and ultimately effects data coding capacity. All data values are collectively shown in Tab. 4 for mathematical calculations. By putting values in (3), capacity of Kapton®HN is calculated to be 16-bits.

$$C = \log_2[((\frac{(24.6-20.09)}{0.58}) + 1) \cdot ((\frac{(25.01-21.8)}{0.362}) + 1) \cdot ((\frac{(24.06-22.25)}{0.339}) + 1) \cdot ((\frac{(24.3-19.6)}{0.296} + 1) \cdot ((\frac{(19.08-17.9)}{0.45}) + 1) \cdot ((\frac{(19.4-18.3)}{1.244} + 1))] = 16\,\text{bits}.$$

In PET substrate-based tag, resistive strip widths are varied from 0.1 mm to 7 mm for S1, from 0.1 mm to 3 mm for S2, from 0.1 mm to 3 mm for S3, from 0.1 mm to 2.5 mm for S4, from 0.1 mm to 0.5 mm for S5 and from 0.1 mm to 0.4 mm for S6. From Fig. 7(a), ASK of the PET-based tag can be easily understood for S1.

Figure 7(b) shows the RCS response of horizontal scatterers using PET substrate. The resistive strip width and the size of the resonator are interlinked with each other. Strip widths are decreasing as the sizes of resonators decrease. For example the 1st scatterer (S1) at the width of 7 mm gives a flat peak response and the 6th scatterer (S6) gives a flat peak at 0.4 mm for PET. Table 3 shows a comparison of these values. It is being analyzed that using PET substrate there is no shift in neighboring peaks which re-

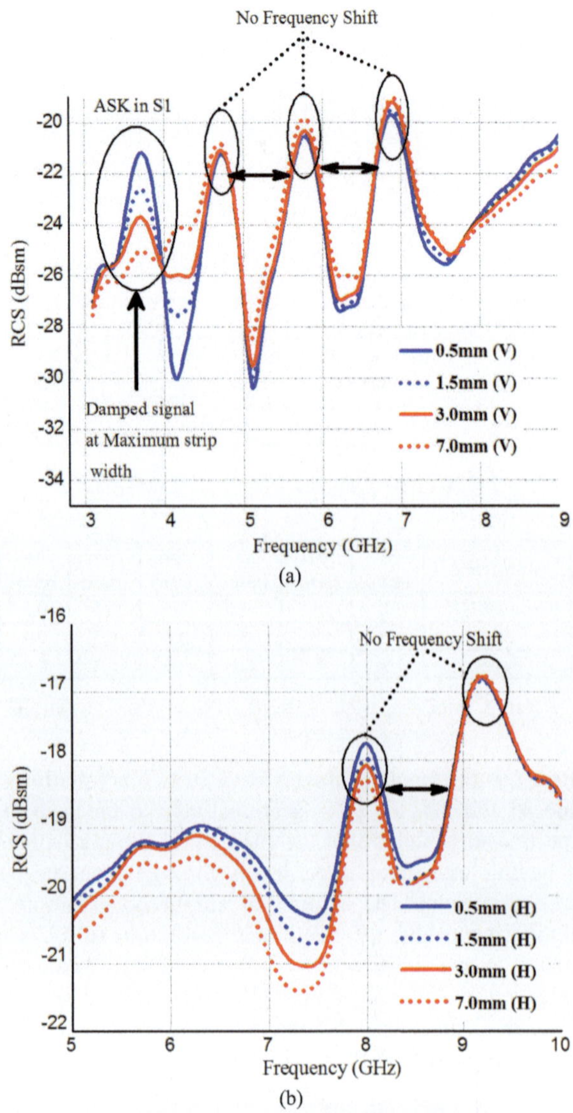

Fig. 7. Simulation results ASK implemented in S1 for PET substrate. (a) Vertical scatterers showing ASK in S1. (b) Horizontal scatterers.

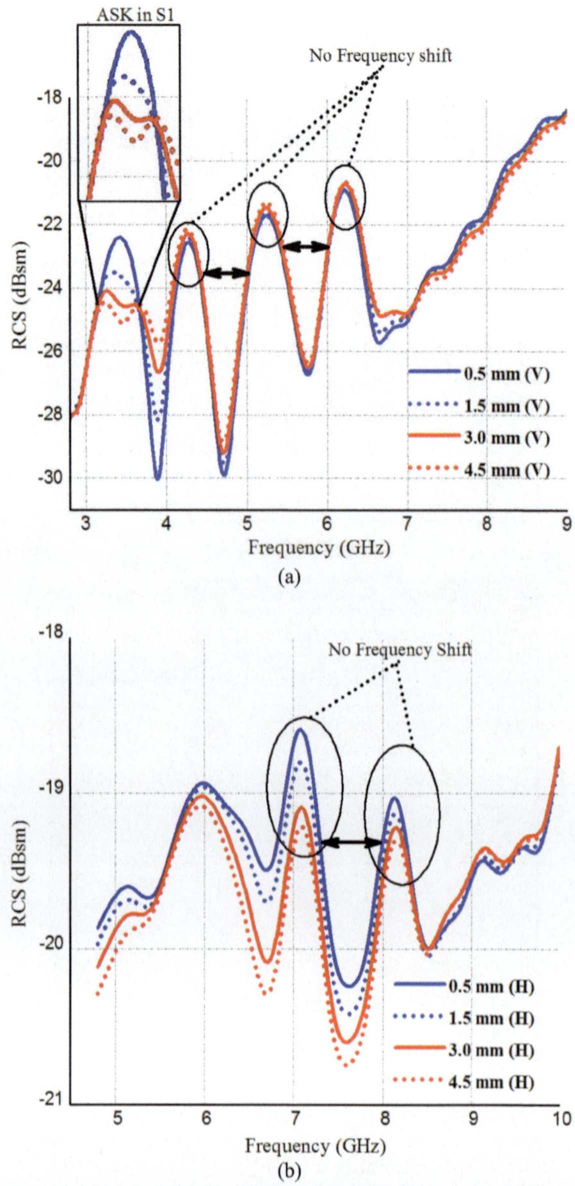

Fig. 8. Simulations results ASK implemented for scatterer 1 for the paper substrate. (a) Vertical resonators showing ASK in S1. (b) Horizontal scatterers.

duces the noise effects. Figures 7(a) and 7(b) show that using PET-based tag, neighboring peaks are less affected by noise; that's why over all tag capacity has been improved. Using values provided in Tab. 4 the calculated capacity of PET is 21 bits.

In the paper substrate, damping of resonating peaks is faster as compared to Kapton®HN and PET as shown in Fig. 8(a). It is observed that quality factor is affected, but still the tag is easily detectable [24], [25]. For the paper substrate the width of resistive strips are used 0.1 mm to 6 mm for S1, from 0.1 mm to 2 mm for S2, from 0.1 mm to 2.5 mm for S3, from 0.1 mm to 2.5 mm for S4, from 0.1 mm to 0.4 mm for both S5 and S6. Also, Figures 8(a) and 8(b) show that there is no frequency shift on neighboring peaks because of amplitude variation in the first scatterer. This behavior is responsible for overall tag capacity enhancement. So using paper substrate, the tag capacity is calculated to be 24 bits.

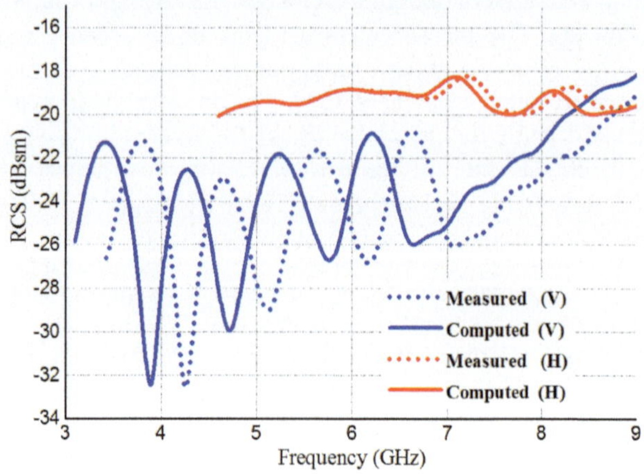

Fig. 9. Measured and computed results for the paper substrate.

Substrate	dM_1 (dB)	dM_2 (dB)	dM_3 (dB)	dM_4 (dB)	dM_5 (dB)	dM_6 (dB)
Kapton®HN	0.58	0.362	0.339	0.296	0.45	1.244
PET	0.56	0.17	0.15	0.37	0.53	0.1
Paper	0.51	0.82	0.089	0.09	0.11	0.08
Substrate	Ma_1 (dB)	M_{a2} (dB)	M_{a3} (dB)	M_{a4} (dB)	M_{a5} (dB)	M_{a6} (dB)
Kapton®HN	–20.09	–21.8	–22.25	–19.6	–17.9	–18.33
PET	–19.89	–21.18	–20.69	–19.89	–17.57	–16.96
Paper	–21.22	–22.5	–21.77	–20.8	–18.25	–18.92
Substrate	Mb_1 (dB)	M_{b2} (dB)	M_{b3} (dB)	M_{b4} (dB)	M_{b5} (dB)	M_{b6} (dB)
Kapton®HN	–24.6	–25.01	–24.06	–24.3	–19.08	–19.4
PET	–25.14	–25.43	–24.26	–24.22	–18.93	–17.8
Paper	–25.08	–26.08	–24.26	–24.09	–19.03	–19.59

Tab. 4. Magnitude resolution and M_a, M_b values for all substrates.

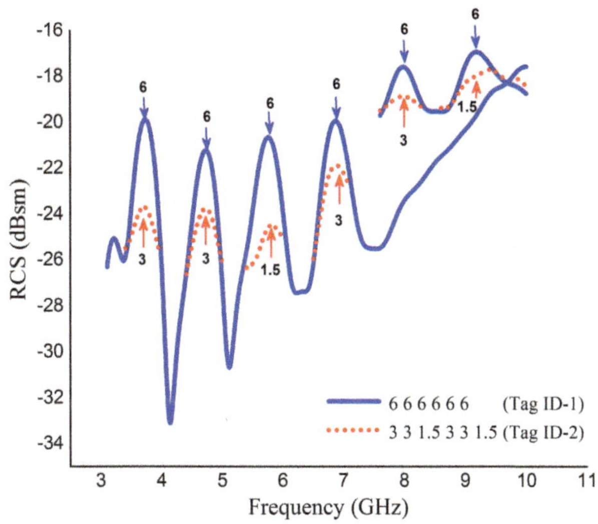

Fig. 10. ASK based coding principle.

The measured and computed results for a paper substrate with no resistive strip are shown in Fig. 9, which are in well coherence with computed values. The coding principle of the proposed tag configuration is based on ASK technique. Tag ID's for two different combinations are shown in Fig. 10. Likewise, multiple tag ID's can be generated to tag multiple objects.

5. Conclusions

Rhombic shaped broadband radiator's selectivity problem has been analyzed. For this purpose instead of conventional OOK, a unique Amplitude Shift Keying (ASK) has been applied. Keeping the size of chipless RFID within credit card format data encoding capacity has been increased. Environmental friendly 24-bit tag is realized. Scatterers are positioned in such a way that coupling effect is reduced to its minimum value. The tag design with three different substrates Kapton®HN, PET and paper with same encoding techniques have been verified. These tags provide not only flexibility, meager cost expenditure but also easy to manufacture by printing techniques. Especially paper substrate based tag is highly suitable for green electronics and IoT applications. Future work emphasizes on the equipping tag with sensing capabilities using ASK.

Acknowledgments

This work was financially supported by Vinnova (The Swedish Governmental Agency for Innovation Systems) and University of Engineering and Technology, Taxila, Pakistan through the Vinn Excellence Centers program and ACTSENA research group funding, respectively.

References

[1] WANT, R., SCHILIT, B. N., JENSON, S. Enabling the internet of things. *IEEE Journals and Magazines*, 2015, vol. 48, no. 1, p. 28 to 35. DOI: 10.1109/MC.2015.12

[2] GUBBI, J., BUYYA, R., MARUSIC, S., et al. Internet of Things (IoT): A vision, architectural elements, and future directions. *Future Generation Computer Systems*, 2013, vol. 29, no. 7, p. 1645–1660. DOI: 10.1016/j.future.2013.01.010

[3] TRUONG, H. L., DUSTDAR, S. Principles for engineering IoT cloud systems. *IEEE Cloud Computing*, 2015, vol. 2, no. 2, p. 68 to 76. DOI: 10.1109/MCC.2015.23

[4] ISLAM, M., KARMARKAR, N. A novel compact printable dual-polarized chipless RFID System. *IEEE Transactions on Microwave Theory and Techniques*, July 2012, vol. 60, no. 7, p. 2142–2151. DOI: 10.1109/TMTT.2012.2195021

[5] OPREA, A., COURBAT, J., BARSAN, N., et al. Temperature, humidity and gas sensors integrated on plastic foil for low power applications. *Sensors and Actuators B: Chemical*, 2009, vol. 140, no. 1, p. 227–232. DOI: 10.1016/j.snb.2009.04.019

[6] NOOR, T., HABIB, A., AMIN, Y., et al. High-density chipless RFID tag for temperature sensing. *Electronics Letters*, 2016, vol. 52, no. 8, p. 620–622.DOI: 10.1049/el.2015.4488

[7] ROSELLI, L., MARIOTTI, C., MEZZANOTTE, P., et al. Review of the present technologies concurrently contributing to the implementation of the Internet of Things (IoT) paradigm: RFID, green electronics, WPT and energy harvesting. In *2015 IEEE Topical Conference on Wireless Sensors and Sensor Networks (WiSNet)*. San Diego (CA, USA), January 2015, p. 1-3. DOI: 10.1109/WISNET.2015.7127402

[8] DOBKIN, D., WANDINGER, T. *The RF in RFID, UHF RFID in Practice.* 2nd edition. Waltham (USA): Newnes, 2013 (Chapter 3.) ISBN 9780123948304

[9] PRERADOVIC, S., KARMAKAR, N. Chipless RFID: Bar code of the future. *IEEE Microwave Magazine*, December 2010, vol. 11, no. 7, p. 87–97. DOI: 10.1109/MMM.2010.938571

[10] MCINTYRE, L., MICHAEL, K., ALBRECHT, K. RFID: Helpful new technology or threat to privacy and civil liberties. *IEEE Potentials*, September/October 2015, vol. 34, no. 5, p. 13–18. DOI: 10.1109/MPOT.2015.2410392

[11] PRERADOVIC, S., KARMAKAR, N .Design of fully printable planar chipless RFID transponder with 35-bit data capacity. In *Proceedings of the 39th European Microwave Conference*. Rome (Italy), 2009, p. 13–16. ISBN: 978-1-4244-4748-0.

[12] TEDJINI, S., DUROC, Y. From radiator to signal processing antenna. In *Asia-Pacific Microwave Conference 2011*. Melbourne (Australia), 2011, p.175–178. ISBN: 978-1-4577-2034-5

[13] FAN.P. F., ZHOU, G. Z. Analysis of the business model innovation of the technology of Internet of Things in postal logistics. In *2011 IEEE 18Th International Conference on Industrial Engineering and Engineering Management (IE&EM)*. Changchun (China), 2011, p. 532–536. ISBN: 978-1-61284-446-6. DOI: 10.1109/ICIEEM.2011.6035215

[14] BHUIYAN, S., KARMAKAR, N. Chipless RFID tag based on split-wheel resonators. In *2013 7th European Conference on Antennas and Propagation (EuCAP)*. Gothenburg (Sweden), 2013, p. 3054–3057. ISBN: 978-1-4673-2187-7

[15] REZAIESARLAK, R., MANTEGHI, M. Complex-natural-resonance-based design of chipless RFID tag for high-density data. *IEEE Transactions on Antennas and Propagation*. February 2014, vol. 62, no. 2, p. 898–904. DOI: 10.1109/TAP.2013.2290998

[16] VENA, A., BABAR, A. A., SYDÄNHEIMO, L., et al. A novel near-transparent ASK-reconfigurable inkjet-printed chipless RFID tag. *IEEE Antennas and Wireless Propagation Letters*, 2013, vol. 12, p. 753–756. DOI: 10.1109/LAWP.2013.2270932

[17] COSTA, F., GENOVESI, S., MONORCHIO, A. A chipless RFID based on multi resonant high-impedance surfaces. *IEEE Transactions on Microwave Theory and Techniques,* 2013, vol. 61, no. 1, p. 146–153. DOI: 10.1109/TMTT.2012.2227777

[18] VENA, A., PERRET, E., TEDJINI, S. Chipless RFID tag using hybrid coding. *IEEE Transactions on Microwave Theory and Techniques,* December 2011, vol. 59, no. 12, p. 3356–3364. DOI: 10.1109/TMTT.2011.2171001

[19] BORTFELDT, J. *The Floating Strip Micromegas Detector: Versatile Particle Detector for High Rate Applications.* 1st ed. Switzerland: Springer, 2015. ISBN 978-3-319-18892-8

[20] RANCE, O., SIRAGUSA, R., LEMAÎTRE-AUGER, P., et al. RCS magnitude coding for chipless RFID based on depolarizing tag. In *2015 IEEE MTT-S International Microwave Symposium.* Phoenix (AZ, USA), 2015, p. 1-4. DOI: 10.1109/MWSYM.2015.7166929

[21] IONESCU, C., SVASTA, P., VASILE, A., et al. Investigations on organic printed resistors based on PEDOT: PSS. In *2012 IEEE 18th International Symposium on Design and Technology in Electronic Packaging (SIITME).* Alba Iulia (Italy), 2012, p. 85–89. DOI: 10.1109/SIITME.2012.6384352

[22] HAVLICEK, J., SVANDA, M., MACHAC, J., et al. Improvement of reading performance of frequency-domain chipless RFID transponders. *Radioengineering,* June 2016, vol. 25, no. 2, p. 219 to 229. DOI: 10.13164/re.2016.0219

[23] LEI XU, KAMA HUANG. Design of compact trapezoidal bow-tie chipless RFID tag. *International Journal of Antennas and Propagation,* October 2014, vol.2015, Article ID 502938, 7 p. DOI: 10.1155/2015/502938

[24] FENG, Y., XIE, L., CHEN, Q., et al. Low-cost printed chipless RFID humidity sensor tag for intelligent packaging. *IEEE Sensors Journal,* 2015, vol. 15, no. 6, p. 3201–3208. DOI: 10.1109/JSEN.2014.2385154

[25] JAVED, N., HABIB, A., AMIN, Y., et al. Directly printable moisture sensor tag for intelligent packaging. *IEEE Sensors Journal,* June 2016, vol. 16, no. 16, p. 6147–6148. DOI: 10.1109/JSEN.2016.2582847

About the Authors...

Ayesha HABIB is working as a research scholar at Telecommunication Engineering Dept., University of Engineering and Technology Taxila, Pakistan. Currently, she is pursuing her Ph.D. in chipless RFID domain. She is a member of IEEE and IEICE.

Sohaira ANSAR received her B.Sc. degree in Telecom Engineering from the University of Engineering and Technology (UET), Taxila in 2014. In the same year, she joined UET, Taxila as Ms research scholar.

Adeel AKRAM is a Dean and Professor of Telecommunication & Information Engineering Dept., University of Engineering and Technology Taxila, Pakistan. He is leading a 5G Wireless Communication research group at UET Taxila, Pakistan.

Muhammad Awais AZAM received his Ph.D. degree in Pervasive and Ubiquitous Computing from London, UK in 2012. He is working as an Assistant Professor at the Dept. of Computer Engineering, UET Taxila, Pakistan.

Yasar AMIN is a Chairman and Associate Professor of Telecommunication Engineering Dept., University of Engineering and Technology Taxila, Pakistan. He is founder of ACTSENA Research Group at UET Taxila, Pakistan. Dr. Yasar is a member of IEEE, IET, ACM and ACES.

Hannu TENHUNEN is a Chair Professor of Electronic Systems at the Royal Institute of Technology (KTH), Stockholm, Sweden. Prof. Tenhunen has held professor position as full professor, invited professor or visiting honorary professor in Finland (TUT, UTU), Sweden (KTH), USA (Cornel U), France (INPG), China (Fudan and Beijing Jiatong Universities), and Hong Kong (Chinese University of Hong Kong), and has an honorary doctorate from Tallinn Technical University. Prof. Tenhunen is a member of the Academy of Engineering Science of Finland.

Compact Size UWB Monopole Antenna with Triple Band-Notches

Wael ALI [1], Ahmed A. IBRAHIM [2], Jan MACHAC [3]

[1] Dept. of Electronics & Comm. Engineering, College of Engineering, Arab Academy for Science, Technology and Maritime Transport (AASTMT), Alexandria, Egypt
[2] Dept. of Electrical Engineering, Faculty of Engineering, Minia University, Minia, Egypt
[3] Dept. of Electromagnetic Field, Faculty of Electrical Engineering, Czech Technical University in Prague, Czech Republic

wael_abd_ellatif@yahoo.com, ahmedabdel_monem@mu.edu.eg, machac@fel.cvut.cz

Abstract. *This paper presents triple band notched ultra wide band (UWB) monopole antenna with overall size of 36 × 32 mm² fed by microstrip transmission line. In order to achieve a good impedance matching from 2.7 GHz to 13.4 GHz, a tapered transition between the rectangular patch and the feeding line is utilized. The three notched frequency bands are accomplished by a defected microstrip structure (DMS) which is inserted in the microstrip feeding line and by an open loop slot etched in the radiating patch. The three band notches are 3.15–4 GHz, 5.7–6.3 GHz and 7.9–8.6 GHz. They prevent the receiving of the signals of IEEE 802.16 WiMAX band, WLAN band, and ITU applications respectively. The UWB antenna was designed and simulated, then fabricated and tested in order to investigate its impedance and radiation characteristics. Good agreement between the simulated and measured data is achieved. The obtained results show that the proposed antenna is convenient for UWB applications.*

Keywords

UWB monopole antenna, open loop resonant slot, defected microstrip structure, notched characteristic

1. Introduction

Recently, the rapid growth in the field of wireless communications and electronics has been extended. Ultra wide band (UWB) antennas play critical role in the UWB communication systems. The UWB antennas are designed to cover range 3.1–10.6 GHz assigned by FCC (Federal Communication Commission) [1]. The UWB technology has several advantages such as high data rate, low power consumption, simple hardware configuration in practical applications and good resistance for multipath [2], [3]. The UWB antenna is considered as one of the important components to realize the UWB communication systems. Therefore, UWB antenna has to achieve several requirements such as high and stable gain, omnidirectional radiation pattern, wide impedance matching through all frequency bands, low cost and compact size [4], [5].

Presently, printed UWB monopole antennas are used due to their merits such as small size, easy of fabrication, low cost and their compatibility with RF components [6–10]. However, there are many narrowband communication systems that interfere with the UWB systems such as WiMAX (3.3–3.7 GHz), IEEE 802.11/b/g (5.15 GHz to 5.825 GHz) WLAN band and X-band operating at 7.725 GHz to 8.275 GHz for ITU applications. In order to reduce the potential interferences from these undesired narrowband signals, UWB antennas with filtering behavior (notched characteristics) are designed. There is a number of antennas of this type presented in literature working with different elements assuring the notched characteristic – stop-bands. Some of them are: Paper [11] presents the antenna using the stepped impedance resonator and fork-shaped stubs. The antenna presented in [12] applies an annular strip etched above the radiating element and two slots, one C-shaped and one arc-shaped embedded in the radiating patch. Rather complex resonator enlarging the area taken by the antenna is used in [13]. Microstrip two-port filtering structure composed of a couple of modified capacitance loaded loop resonators is used in [14]. Each antenna designed in [15] comprises a half-circle shaped patch with an open rectangular slot and a half-circle shaped ground plane.

In this paper, a compact microstrip fed monopole UWB antenna with three stop bands and overall size of 36 × 32 mm² is presented. The notched frequency bands are achieved by employing open loop resonator etched in the patch and defected microstrip structure (DMS) stop-band filter inserted in the feeding line. First of all, a reference antenna is designed, which exhibits radiating characteristics in the frequency band 2.7–13.4 GHz. The first and the third notches are achieved by using DMS at frequency bands (3.15–4 GHz and 7.9–8.6 GHz). The second notch is achieved by using open loop slot at frequency band (5.7–6.3 GHz). The proposed antenna has been designed, simulated, fabricated and measured. The designed antenna performance has been validated using the electromagnetic full wave simulations. The commercial software HFSS was employed in the full wave simulations.

2. Design Procedures

2.1 UWB Antenna Configuration

The UWB antenna design is comprised of the steps: First of all, conventional patch antenna with partial ground plane is designed (case I) to achieve UWB behavior from 3.1 GHz to 10.6 GHz as shown in Fig. 1(a). The VSWR of this antenna plotted in Fig. 2 shows that the antenna operates from 3.5 GHz to 6.5 GHz with VSWR lower than 2. Matching of the UWB antenna can be improved by using a tapered transition in the radiator (case II) as illustrated in Fig. 1(b). The VSWR of the designed antenna illustrates that the antenna works from 3.1 GHz to 12 GHz (UWB applications). Finally to achieve good matching properties, simple rectangular slot is added in the ground plane (case III) as shown in Fig. 1(c). The case III is therefore used as our reference UWB antenna. The wide band operation of the UWB antenna has been achieved by optimizing the dimensions of the radiating patch and the partial ground plane. The UWB antenna is designed on RO4003 substrate with relative permittivity $\varepsilon_r = 3.38$, loss tangent equal to 0.0027 and thickness $h = 0.813$ mm. The monopole antenna is fed by 50 Ω microstrip line.

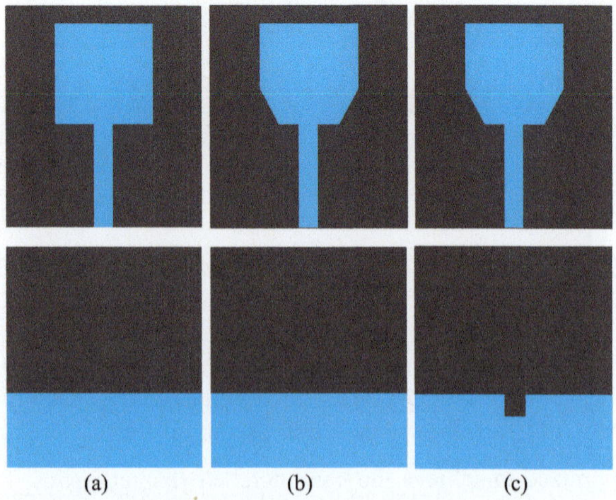

Fig. 1. UWB antenna layouts – top and bottom : (a) Case I, (b) Case II, (c) Case III.

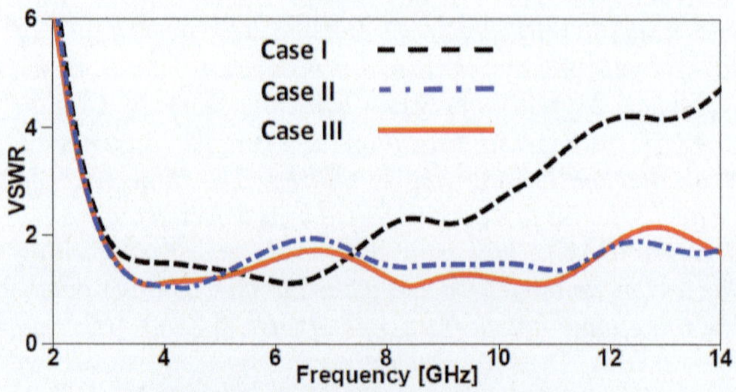

Fig. 2. VSWR of the UWB antenna.

2.2 DMS Band-Stop Filter Design

DMS plays an important role in designing filters due to several advantages such as size and complexity reduction. It also reduces the radiation loss which enhances the stop band attenuation. The 2-D layout of the DMS filter is shown in Fig. 3. It is considered as a simple slot inserted in the microstrip line. This slot changes the effective capacitance and inductance of the microstrip line which leads to band stop behavior. The dimensions of the DMS filter were optimized to achieve the desired resonant frequency. The simulated S-parameters (return loss and insertion loss) of the DMS filter are plotted in Fig. 4. It is clear that the filter has two rejected frequency bands. The first band is 3.15 to 4 GHz and the second band is 7.9–8.6 GHz.

Fig. 3. The layout of DMS band stop filter.

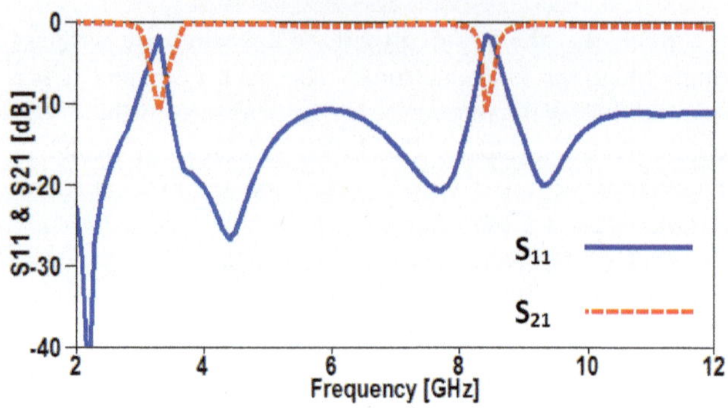

Fig. 4. S-parameters magnitude of the DMS filter.

2.3 Open Loop Resonant Slot

The simulated current distribution of the reference antenna at 6 GHz is plotted in Fig. 5 (a). It can be noticed that the surface current is concentrated near the input of the patch and decreases away from the center of the patch. Therefore, by inserting a slot half of wavelength in length at the place where there is the maximum of surface current, band notch features can be accomplished. The open loop resonant slot position and dimensions were selected and

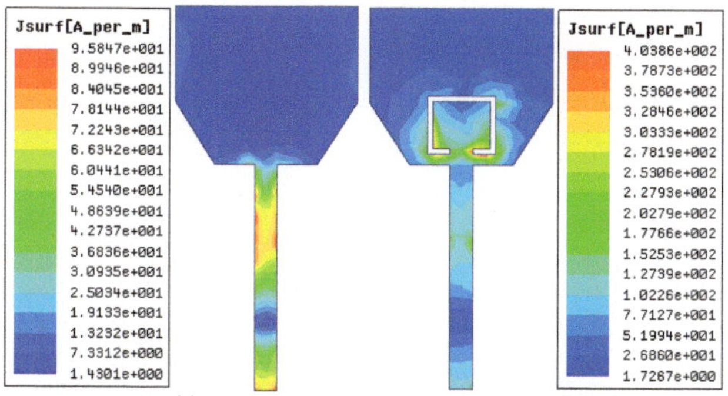

(a) (b)

Fig. 5. Simulated surface current distributions of antenna at 6 GHz: (a) without a slot, (b) with the slot.

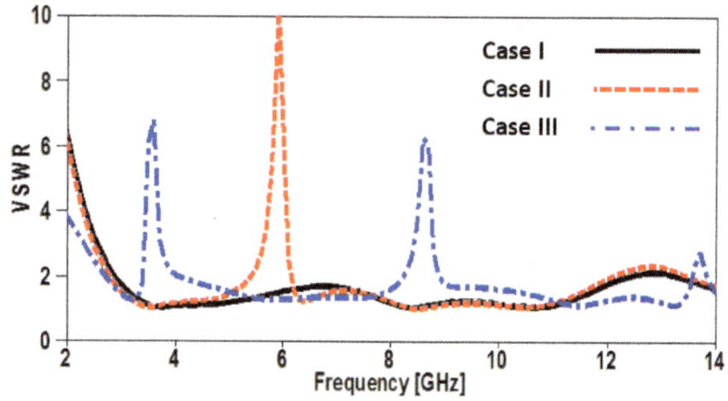

Fig. 7. The VSWR of the proposed antenna for three different cases, see the text.

optimized to achieve the band notched characteristics. The simulated current distribution at 6 GHz for monopole antenna with the loop resonator is shown in Fig. 5(b) and it is obvious that the surface current is concentrated around the slot, which means that the antenna does not radiate at this frequency band.

3. Proposed UWB Antenna Design

The 2-D layout of the proposed UWB monopole antenna is illustrated in Fig. 6. The detailed dimensions are stated in Tab. 1. The antenna consists of the previously investigated UWB monopole antenna with the open loop resonant slot etched in the patch and the DMS filter inserted in the microstrip line. The simulated VSWR of the

Parameter	L_1	L_2	L_3	L_4	L_5	L_6	L_7
Value (mm)	5.6	5.2	14	20	4.8	5.6	6
parameter	L_8	L_9	W_1	W_2	W_3	W_4	W_5
Value (mm)	2	19	16	6	5	3.5	1.6
parameter	W_6	W_7	W_8	W	L	d	
Value (mm)	1.2	2	2	32	36	0.2	

Tab. 1. Dimensions of the proposed antenna structure.

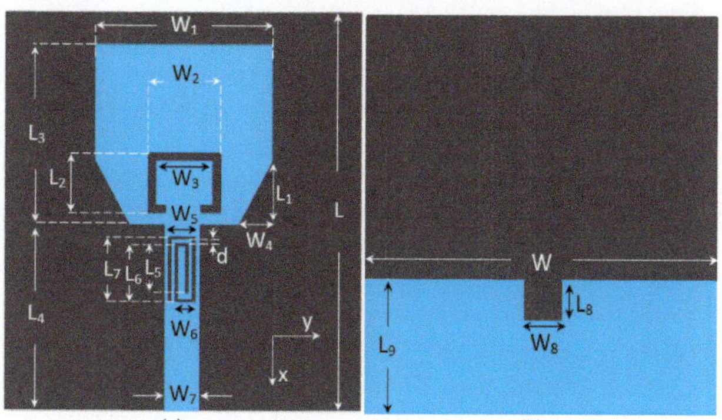

(a) (b)

Fig. 6. The geometry of the proposed UWB antenna: (a) Front view, (b) back view.

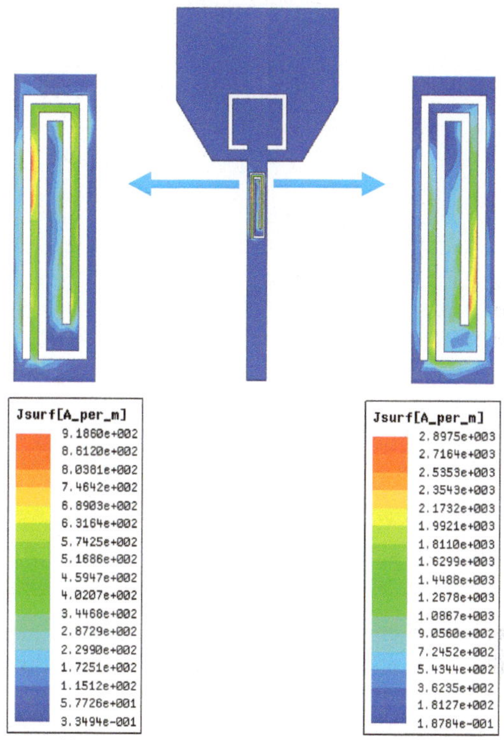

(a) (b)

Fig. 8. Simulated surface current distributions of the antenna with the DMS filter: (a) at 3.4 GHz, (b) at 8.4 GHz.

particular UWB antenna versions is shown in Fig. 7. Case I corresponds to the response of the reference antenna. Case II represents the antenna with the open loop resonator only. It is clear that the VSWR is lower than 2 except the notched frequency band 5.7–6.3 GHz. Case III represents the antenna with the DMS filter which is embedded in the feeding line only. It can be noticed from the plot that the VSWR is lower than 2 except the two notched frequency bands 3.15–4 GHz and 7.9–8.6 GHz.

The simulated current distributions of the UWB monopole antenna with DMS filter at 3.4 GHz and 8.4 GHz are demonstrated in Fig. 8. It is obvious from Fig. 8 that the current is concentrated around the DMS filter and practically no current passes through the patch. This confirms the fact that the DMS acts as dual band-stop filter and the antenna does not radiate at these two frequency bands.

4. Experimental Results

The proposed UWB antenna has been simulated, fabricated and then tested using Agilent N9918A vector network analyzer (VNA). The antenna was fabricated on RO4003 substrate with relative dielectric constant of 3.38, thickness of 0.813 mm and dielectric loss tangent equal to 0.0027. The photographs of the fabricated antenna are shown in Fig. 9. The VSWR of the antenna was measured and compared with the simulated one as demonstrated in Fig. 10. As seen from the plot, there is a good agreement between basic shapes of simulated and measured VSWR. The measured rejection level in the three notched band is not as good as predicted by simulations. This is with the highest probability caused by fabrication tolerances and by the used SMA connector. The results show that the proposed antenna is working as UWB antenna from 2.7 GHz to 13.45 GHz. However, there are three notches at 3.15–4 GHz, 5.7–6.3 GHz and 7.9–8.6 GHz with VSWR which is higher than 2.

A comparison between simulated and measured realized gain of the proposed triple band notched UWB monopole antenna is illustrated in Fig. 11(a). The average measured realized gain of the UWB antenna is nearly constant around 3 dBi, except the three notched frequency bands with values equal –2 dBi, –1.5 dBi and –1.1 dBi respectively. The radiation efficiency of UWB antenna is illustrated in Fig. 11(b). It is seen from Fig. 11(b) that the average efficiency of the UWB antenna equals around 85 %. However the efficiency of UWB antenna is reduced to 30 %, 20 % and 30 % at the three notched frequency bands.

The radiation patterns of the proposed antenna in the *x-z* plane, *y-z* plane and *x-y* plane (see the coordinate system in Fig. 6(a)) were measured inside anechoic chamber at 3.1 GHz, 5.3 GHz, and 9.1 GHz. The normalized simulated and measured radiation patterns in *x-z*, *y-z* and *x-y* planes of the proposed antenna are shown in Fig. 12 and Fig. 13, respectively. Position angle $\theta = 0$ in Figs. 12 and 13 is directed in the positive direction of *x*-axis in (a), *z*-axis in (b), and *x*-axis in (c). It follows from Fig. 12 that the UWB monopole antenna has nearly omnidirectional patterns in *y-z* plane and almost bi-directional patterns in *x-z* and *x-y* planes at the three different frequencies, which implies that the proposed antenna has a large and uniform coverage confirming the fact that the proposed antenna is convenient for UWB application.

Comparison between our proposed antenna and other in literature presented designs in terms of measured values is illustrated in Tab. 2. Table 2 presents notch frequencies, realized gain at these notches. It can be concluded from this comparison that the proposed antenna has compact size and higher bandwidth. The gain at centers of notched band is comparable, or somewhat higher.

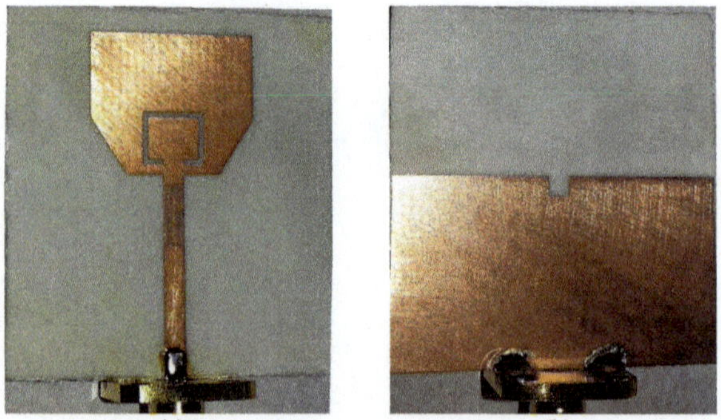

(a) (b)

Fig. 9. Photographs of the fabricated UWB antenna: (a) Front view, (b) back view.

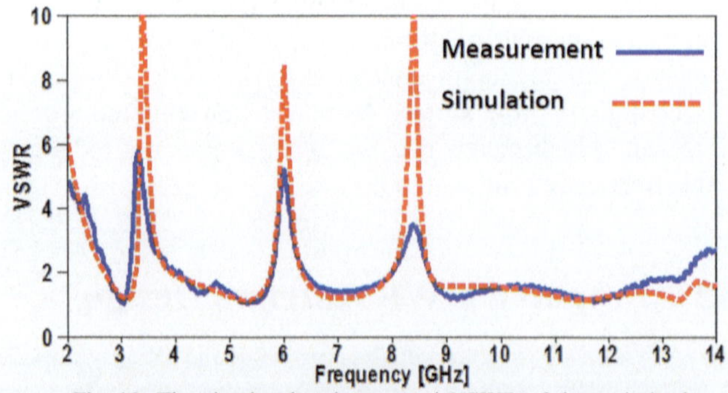

Fig. 10. The simulated and measured VSWR of the optimized structure against frequency.

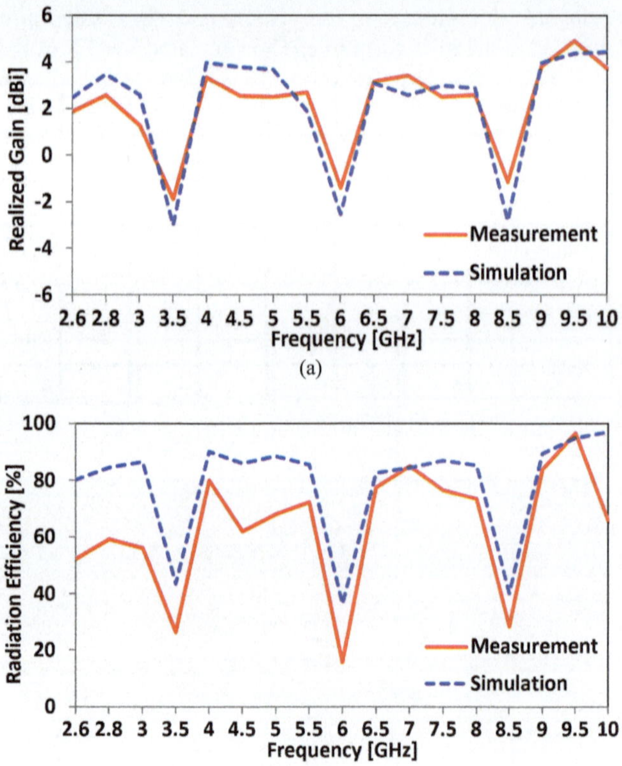

(b)

Fig. 11. The simulated and measured results of the optimized structure against frequency: (a) gain, (b) efficiency.

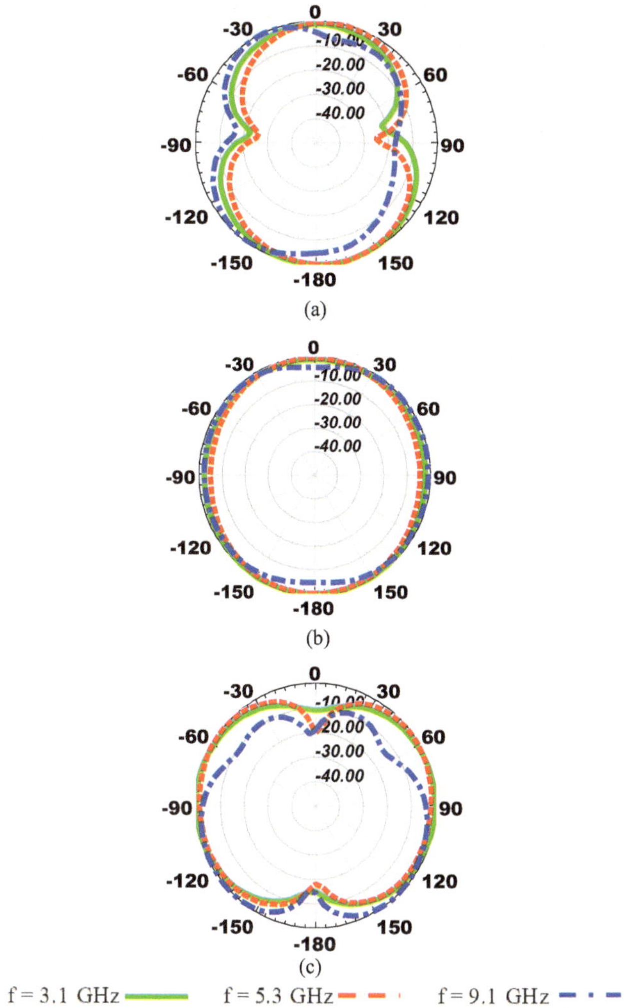

Fig. 12. The simulated normalized radiation patterns at f=3.1 GHz, 5.3 GHz and 9.1 GHz: (a) x-z plane, (b) y-z plane, (c) x-y plane.

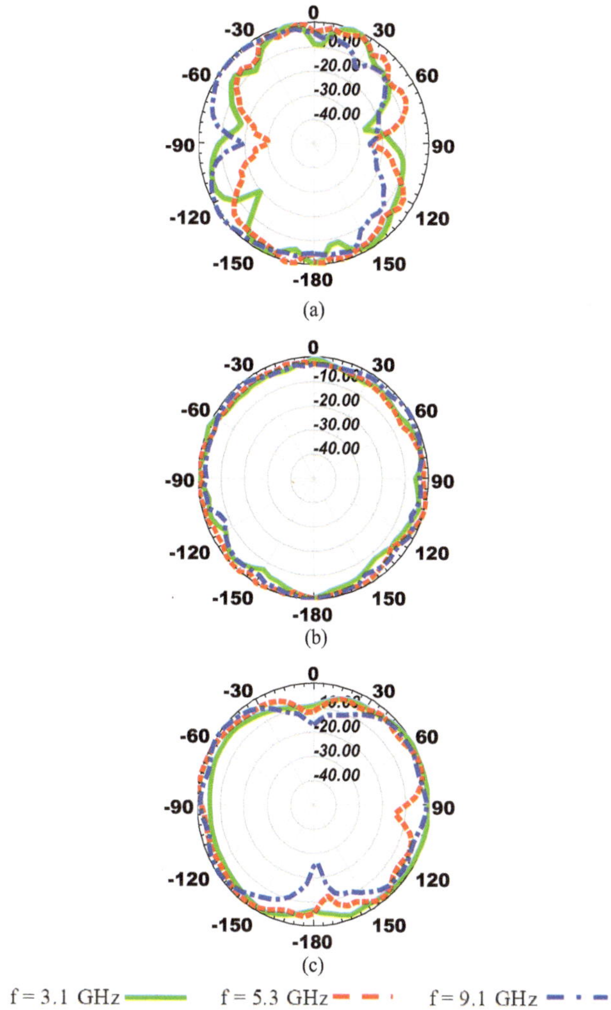

Fig. 13. The measured normalized radiation patterns at f = 3.1 GHz, 5.3 GHz and 9.1 GHz: (a) x-z plane, (b) y-z plane, (c) x-y plane.

	Size [mm^2]	Frequency range [GHz]	Notch frequencies [GHz]	Realized gain [dBi]
[16]	35 × 35	2.2-12.8	3.54	–5.6
			5.73	–0.9
			8.56	–3
[17]	44 × 12	2.2-11.3	2.4	–0.5
			3.8	–1.6
			5.5	–0.3
[18]	50 × 42	2-11	3.45	–5.5
			5.55	–5
			7.4	–4.3
[19]	45 × 40	3-11	4.1	–3.8
			5.3	–3.4
This work	36 × 32	2.7-13.4	3.3	–2
			6	–1.5
			8.4	–1.1

Tab. 2. Comparison between different UWB triple-band notched antenna designs.

5. Conclusion

The design of the compact size monopole UWB antenna with triple band notches has been presented. A tapered transition between the rectangular patch and the feeding line has been used to achieve good impedance matching from 2.7 GHz to 13.4 GHz. The defected microstrip structure and the open loop slot etched in the radiation patch have been used to produce triple band notches. The three band notches are 3.15–4 GHz, 5.7–6.3 GHz and 7.9–8.6 GHz for IEEE 502.16 WiMAX band, WLAN band, and ITU respectively. Moreover, good agreement has been achieved between the simulated and measured results. This confirms that the proposed antenna is suitable for UWB applications.

Acknowledgments

The authors would like to thank the National Telecommunication Institute (NTI) in Egypt for fabricating the proposed antenna and to thank Dr. Hadia Elhennawy and Eng. Abdel Hamid Hatem for their support in measuring the radiation characteristics of the fabricated antenna in the anechoic chamber of Ain Shams University's antenna laboratory in Egypt.

The finalizing of the work has been supported by the Grant Agency of the Czech Republic under project No. 17-00607S.

References

[1] FEDERAL COMMUNICATIONS COMMISSION. *First Report and order, Revision of part 15 of the commission's Rule Regarding Ultra-Wideband Transmission System FCC 02-48*. 2002.

[2] ALLEN, B., DOHLER, M., OKON, E., et al. (eds.). *Ultra Wideband Antennas and Propagation for Communications, Radar and Imaging*. Wiley, 2006. ISBN: 978-0-470-03255-8

[3] FONTANA, R. J. Recent system applications of short-pulse ultra-wideband (UWB) technology. *IEEE Transactions on Microwave Theory and Techniques*, 2004, vol. 52, no. 9, p. 2087–2104. DOI: 10.1109/TMTT.2004.834186

[4] WILLIAM, J., NAKKEERAN, R. A compact CPW-fed UWB slot antenna with cross tuning stub. *Progress In Electromagnetics Research C*, 2010, vol. 13, p. 159–170. DOI: 10.2528/PIERC10022505

[5] ABDELRAHEEM, A. M., ABDALLA, M. A., ELREGILY, H. A., et al. Coplanar UWB antenna for high speed communication systems. In *2012 International Conference on Engineering and Technology (ICET)*. Cairo (Egypt), 10-11 Oct. 2012, 5 p. DOI: 10.1109/ICEngTechnol.2012.6396155

[6] ALIPOUR, A., HASSANI, H. R. A novel omni-directional UWB monopole antenna. *IEEE Transactions on Antennas and Propagation*, 2008, vol. 56, no. 12, p. 3854–3857. DOI: 10.1109/TAP.2008.2007398

[7] ABDELREHEEM, A., ABDALLA, M. Compact curved half circular disc-monopole UWB antenna. *International Journal of Microwave and Wireless Technologies*, 2016, vol. 8, no. 2, p. 283 to 290. DOI: 10.1017/S1759078714001524

[8] BOUTEJDAR, A., ABD ELLATIF, W. A novel compact UWB monopole antenna with enhanced bandwidth using triangular defected microstrip structure and stepped cut technique. *Microwave and Optical Technology Letters*, 2016, vol. 58, no. 6, p. 1514–1519. DOI: 10.1002/mop.29820

[9] ZHAN, K., GUO, Q., HUANG, K. A miniature planar antenna for Bluetooth and UWB applications. *Journal of Electromagnetic Waves and Applications*, 2010, vol. 24, no. 16, p. 2299–2308. DOI: 10.1163/156939310793699109

[10] BOUTEJDAR, A., IBRAHIM, A. A., BURTE, E. P. *Novel Microstrip Antenna Aims at UWB Applications*. [Online] Cited 2016-08-31. Available at: http://mwrf.com/passive-components/novel-microstrip-antenna-aims-uwb-applications

[11] ZHANG CHAOZHU, JING ZHANG, LIN LI. Triple band-notched UWB antenna based on SIR-DGS and fork-shaped stubs. *Electronics Letters*, 2014, vol. 50, no. 2, p. 67–69. DOI: 10.1049/el.2013.2513

[12] MOHAMMED, H. J., ABDULLAH, A. S., ALI, R. S., et al. Design of a uniplanar printed triple band-rejected ultra-wideband antenna using particle swarm optimisation and the firefly algorithm. *IET Microwaves, Antennas & Propagation*, 2016, vol. 10, no. 1, p. 31–37. DOI: 10.1049/iet-map.2014.0736

[13] SUNG, Y. Triple band-notched UWB planar monopole antenna using a modified H-shaped resonator. *IEEE Transactions on Antennas and Propagation*, 2013, vol. 61, no. 2, p. 953–957. DOI: 10.1109/TAP.2012.2223434

[14] WANG, J., YIN, Y., LIU, X. Triple band-notched ultra-wideband antenna using a pair of novel symmetrical resonators. *IET Microwaves, Antennas & Propagation*, 2014, vol. 8, no. 14, p. 1154–1160. DOI: 10.1049/iet-map.2014.0239

[15] ZHU, F., GAO, S., HO, A. T.S., et al. Multiple band-notched UWB antenna with band-rejected elements integrated in the feed line. *IEEE Transactions on Antennas and Propagation*, 2013, vol. 61, no. 8, p. 3952–3960. DOI: 10.1109/TAP.2013.2260119

[16] SARKAR, D., SRIVASTAVA K. V., K. SAURAV, K. A compact microstrip-fed triple band-notched UWB monopole antenna. *IEEE Antennas and Wireless Propagation Letters*, 2014, vol. 13, p. 396 to 399. DOI: 10.1109/LAWP.2014.2306812

[17] LOTFI, P., SOLTANI, S., AZARMANESH, M. Triple band notched UWB CPW and microstrip line fed monopole antenna using broken ∩-shaped slot. *International Journal of Electronics and Communications*. 2011, vol. 65, no. 9, p. 734–741. DOI: 10.1016/j.aeue.2010.11.001

[18] JAGLAN, N., KANAUJIA, B. K., GUPTA, S. D., et al. Triple band notched UWB antenna design using electromagnetic band gap structures. *Progress In Electromagnetics Research C*, 2016, vol. 66, p. 139–147. DOI: 10.2528/PIERC16052304

[19] KALTEH, A. A., DADASHZADEH, G. R., NASER-MOGHADASI, M., et al. Ultra-wideband circular slot antenna with reconfigurable notch band function. *IET Microwaves, Antennas & Propagation*, 2012, vol. 6, no. 1, p. 108–112. DOI: 10.1049/iet-map.2011.0125

About the Authors ...

Wael ALI received his B.Sc. and M.Sc. in Electronics and Communications Engineering from Arab Academy for Science, Technology and Maritime Transport (AASTMT), Alexandria, Egypt in 2004 and 2007, respectively. He obtained his Ph.D. in Electronics and Communications Engineering from Alexandria University, Alexandria, Egypt in 2012. He is a lecturer in Arab Academy for Science, Technology and Maritime Transport (AASTMT), Alexandria, Egypt. His research interests include smart antennas, microstrip antennas, microwave filters, and metamaterials.

Ahmed A. IBRAHIM was born in 1986. He received the B.Sc. degree, with grade of very good, in Electrical Engineering from the Electronic and Communication Engineering Department, Elminia University, Elminia, Egypt in 2007. He was awarded the M.Sc. degree in Electronic and Communication Engineering from Elminia University in 2011 and the Ph.D. degree from Electronic and Communication Engineering from Minia University in 2014. He is now post-doctoral fellow in the Otto-von-Guericke-Universität Magdeburg, Germany. He is also a lecturer in Electronic and Communication Engineering Department. His research focused on the design and analysis of microstrip antennas, microstrip filters, and its application in wireless communications, metamaterial MIMO antennas and different metamaterial applications in microwave bands.

Jan MACHAC is a professor at the Department of Electromagnetic Field, Faculty of Electrical Engineering, Czech Technical University in Prague, Czech Republic. His main scientific interests are: investigation of planar passive elements and subsystems of microwave technology, planar antennas, planar microwave filters, propagation of electromagnetic waves in periodic structures, and metamaterials. Prof. Machac is an author or co-author of more than 250 publications in scientific journals and scientific international and national conferences. He is a Senior Member of IEEE Microwave Theory and Technique Society, member of the IEEE MTT-S Int. Microwave Symposium TPRC, since 2008, member of the TPC of the European Microwave Conference during 1995-1997, and since 2013, reviewer of: IEEE Transactions on Microwave Theory and Techniques, IEEE Microwave and Wireless Components Letters, Electronics Letters, and IET Microwave Antennas and Propagation.

11

Multifunctional Antenna with Reconfigurable Ultra-Wide Band Characteristics

Asha VERMA, Manoj SINGH PARIHAR

Dept. of Electronics & Communication Engineering, PDPM-Indian Inst. of Information Technology, Design and Manufacturing Jabalpur (PDPM-IIITDMJ), Jabalpur, India

asha.v04@gmail.com, mscareiit@gmail.com

Abstract. *In this paper a multifunctional antenna is presented which offers an ultra-wideband (UWB) operation, an UWB operation with two switchable notches and reconfigurable dual-band operation for WiMAX and WLAN applications, respectively. Total seven functions/states could be achieved from a single antenna using an electronic switching. The antenna uses dual slots on the ground plane to provide a wide bandwidth, ranging from 3.1 GHz to 10.6 GHz. U-shaped slot and C-shaped printed strip in the ground are used to generate two notches at 3.6 GHz (WiMAX) and 5.2 GHz (WLAN/ WiFi) bands, respectively. Moreover, four parasitic strips are added in the feed side to make antenna functional at either 3.6 GHz or 5.2 GHz or both. Total five PIN diodes are required to obtain seven operations from the proposed antenna. Seven structures are fabricated and measured to verify the seven states and results are found in good agreement with estimated results obtained from the simulation.*

Keywords

Reconfigurable antenna, UWB antenna, C-shaped strip, PIN diode, parasitic strip

1. Introduction

Reconfigurable antennas have received significant attention in recent year for modern wireless communication system due to their multi-functional properties. Multiband/wideband antennas are very much required in present communication systems for covering multiple applications such as WiMAX, WLAN, GPS, GSM, UMTS, DCS, PCS, and LTE within a single system. Conventionally, multiband applications require separate receiving/transmitting antennas at different operating frequencies with different radiation characteristics. For multi-band operation different techniques have been adopted such as adding parasitic strips to create additional resonances [1–3]. The use of multiple antennas to cover multiple bands would require a large physical area and subsequently system complexity and costwill increase manifold. A single wide band antenna could also serve the same purpose, but it may not provide a better selectivity between the adjacent bands which results in poor signal to noise ratio (SNR). Moreover, all applications might not be activated all the time, hence the wide band/multi-band antennas would create an electromagnetic interference (EMI) to other systems from those applications. Though, using a reconfigurability concept in an antenna, all applications could be accommodated into a single circuit capable of switching between different applications whenever they required/activated. Thus the concept of reconfigurability would make the system more compact, less complex and cost effective. To achieve reconfigurability, various switching techniques have been adopted such as PIN diodes, RF-MEMS switches, and photoconductive switches.

These techniques modify the physical shape of the antenna electronically; consequently the surface current distribution gets altered, which in turn modifies the operating characteristics of an antenna such as operating frequency and radiation pattern [4–6]. UWB antennas should have the properties of wide impedance bandwidth (3.1 GHz–10.6 GHz), constant gain, linear phase, stable radiation pattern, low spectral power density, which makes UWB system suitable for short range high speed communication [7], [8]. Microstrip fed wide slot antennas might be a good selection to obtain an ultra-wide impedance bandwidth due to less interaction with surface waves and could be easily integrated with active devices [9].

Due to wide bandwidth of UWB system, it is interfered by nearby existing communication systems such as WiMAX and WLAN. However, this interference could be minimized by creating notched response at these frequency bands. Notch structure could be realized by inserting U-shaped, C-shaped, π-shaped slot, stepped impedance stub (SIS), stepped impedance resonators (SIR), SIS loaded hexagon stepped impedance resonators (HSIR), rectangular split ring, defected microstrip structure band stop filter (DMS-BSF), open ended T-shaped stub, etching spur slot, and half and quarter-wavelength strip in the form of resonator, either in the ground plane or in radiating plane itself [10–13], [15–20]. In most of these works, authors have reported reconfigurable UWB operation with single notch or multiple notches. In [21], reconfigurability is obtained through physical rotation of an antenna to achieve reconfigurable dual frequency operation along with UWB operation without notches.

In this paper, an UWB reconfigurable antenna with three slots in the ground planes and parasitic strips on the feed side is proposed to provide seven operations/states in different bias conditions. The proposed antenna is excited by a 50 Ω microstrip feed line. To the best of our belief, seven modes of operation using a single reconfigurable antenna which could achieve UWB response with notches and dual resonating modes have not been reported so far.

The basic design taken from [14], consists of dual slots (wide and small rectangular) etched on the ground plane to provide a fixed lower ultra-wide band response (2.17–6.25 GHz) with dual notches in the WiMAX band. In this work the design has been modified substantially to cover the complete UWB response with reconfigurable characteristics.

The work is organized as follows: In Sec. 2, the design of the proposed reconfigurable antenna and its parametric analysis is presented. Section 3 shows simulated and measured reflection coefficients, radiation characteristics and gain of the proposed antenna. Finally the work is concluded in Sec. 4.

2. Antenna Design

The proposed layout of UWB antenna for reconfigurable dual notched and dual band operation is illustrated in Fig. 1, including DC biasing circuitry. It consists of rectangular dual slots (wide and smaller rectangular slots), U-shaped slot and C-shaped segmented metal strip in the ground plane. The wide rectangular slot resonates at a lower cutoff frequency of UWB due to its large size, which elongates current path. Similarly the small rectangular slot generates minimum current path thus operates at higher cutoff frequency. The areas of the wide and small rectangular slots control the lower and upper cutoff frequencies of the UWB response, respectively [14].

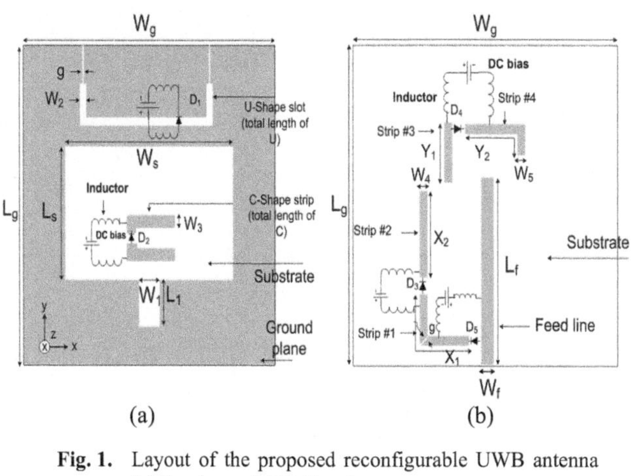

Fig. 1. Layout of the proposed reconfigurable UWB antenna (a) Top layer. (b) Bottom layer.

Ground plane (Top layer)	L_g	W_g	L_s	W_s	L_1	W_1	W_2	W_3	U_1	U_2, U_3
	40	40	14	27	6.4	4	1	2	24	6
Feed plane (Bottom)	L_f	W_f	X_1	X_2	Y_1	Y_2	W_4	W_5	g	C_1, C_2
	23.2	2.2	10	16	7	13	2	2	0.1	11

Tab. 1. Optimized design parameters.

As shown in Fig. 1(a), U-shaped slot provides rejection at 3.6 GHz frequency, whereas C-shaped segmented strip is used to reject WLAN (5.2 GHz) band. It is noted that the lengths of U-shaped slot and C-shaped strip are approximately $0.5\lambda_g$ and act as resonating circuit at respective notch frequencies (3.6 and 5.2 GHz). As shown in Fig. 1(b), four parasitic copper strips are used on the feed side to allow the same antenna to radiate in either 3.6 GHz (WiMAX) or 5.2 GHz (WLAN) band or in both bands. These strips generate additional current paths and act as radiating resonators at 3.6 GHz and 5.2 GHz. The overall size of the antenna (including substrate) is (40×40 mm^2). It is fabricated on an RT/Duroid5870 Rogers substrate with dielectric constant $\varepsilon_r = 2.55$ and thickness $h_S = 0.77$ mm.

The proposed structure is simulated in Finite Integral Technique (FIT) based CST Microwave Studio EM simulator. The optimized dimensions (mm) of the proposed antenna are revealed in Tab. 1. The length of slot 'U' could be varied to control the notch frequency of WiMAX band, while WLAN notch can be controlled by the length of segmented C-shaped metal strip. Frequency 3.6 GHz of single band antenna could be obtained by connecting metal strips $\#X_1$ and $\#X_2$ to the feed line. The WLAN band is achieved when only strip $\#X_1$ is connected (strip $\#X_2$ is disconnected) to the feed line. The role of strip Y_1 and Y_2 is to provide WiFi band in dual-band mode (strips $\#X_1$ and $\#X_2$ are connected to feed line for WiMAX band in this case). To validate the above statements parametric study is carried out using CST microwave studio. The deviation of lengths of U-shaped slot and C-shaped strip and their effect on frequency response is shown in Fig. 2. It is established that changing the lengths of U-shaped slot OR C-shaped strip, notch frequency could be varied as desired. Theoretically, the lengths U and C of slot and strip can be calculated as

$$U, C = \frac{c}{2 \cdot f_{3.6}, f_{5.2} \cdot \sqrt{\varepsilon_{\mathrm{eff}}}} \tag{1}$$

where f is the respective notch frequency, c is the velocity of light in free space, and $\varepsilon_{\mathrm{eff}}$ is the effective dielectric constant.

The effect of variation of length X on frequency response is shown in Fig. 3. Here the length X is considered as an addition of X_1 and X_2. It is seen that length X could be varied to shift the lower frequency (3.6 GHz) of the dual band antenna while to attain upper frequency band 5.2 GHz

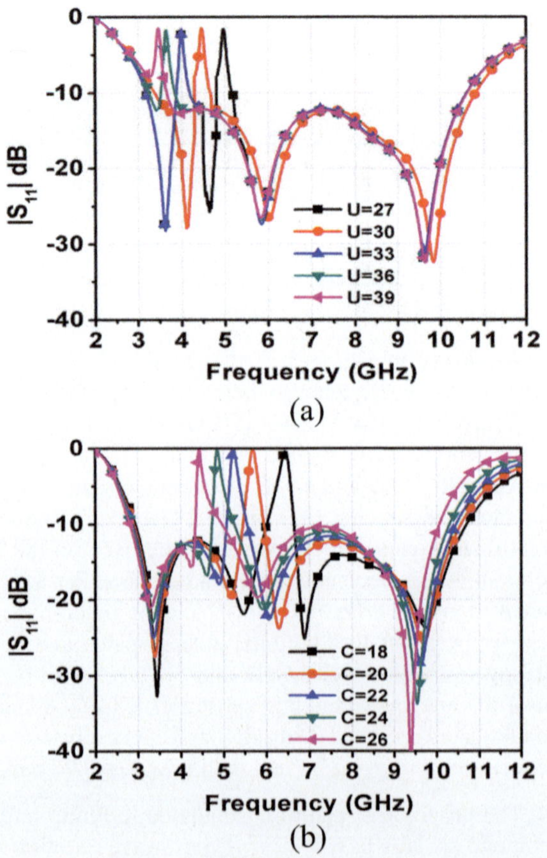

Fig. 2. Reflection coefficient versus frequency plot for variation in: (a) the lengths of U-shaped slot, (b) the length of C-shaped segmented metal strip.

the length X is altered as X_2. The second resonance which is not so strong, is observed at 7.0 GHz; it is because of loading strips Y_1 and Y_2, used to achieve 5.2 GHz operating band (WLAN).

To realize reconfigurability in the proposed antenna, five HPND-4005 PIN diodes (D_1-D_5), with appropriate biasing are embedded at different locations. All diodes are biased independently and RF to DC isolation is provided by creating a small gap g of 0.1 mm. In the EM simulation switches/diodes are realized using the metal strips. The ON state (ideal ON) of the diode is realized by inserting a metal strip of (0.4×0.2 mm^2) at the physical position of the diode and it is removed to realize diode's OFF state (ideal OFF). For full wave EM simulation, five discrete/lumped ports are placed in CST at the physical locations of diodes. After post processing, 6 × 6 S-matrix file generated and imported in Advanced Design System (ADS) for co-simulation because the CST simulation does not take into account the packaging effect of the diode or complex RLC model [4].

As shown in Fig. 4, spice model of actual HPND-4005 PIN diode will be connected to the five ports (2-6) of the model exported from an EM simulator. The first port will be used for RF input excitation. The spice parameters PIN diode model used in ADS are: V_{th}=1 V, R_S = 4.7 Ω, C_j = 0.017 pF. All diodes are biased through separate DC supplies.

The capacitors and inductors of 10 pF and 50 nH are used to provide DC and RF isolation, respectively [5], [6], [11–13]. To limit the DC current a series resistor of 100 Ω is used with all diodes. To activate each PIN diode, a bias

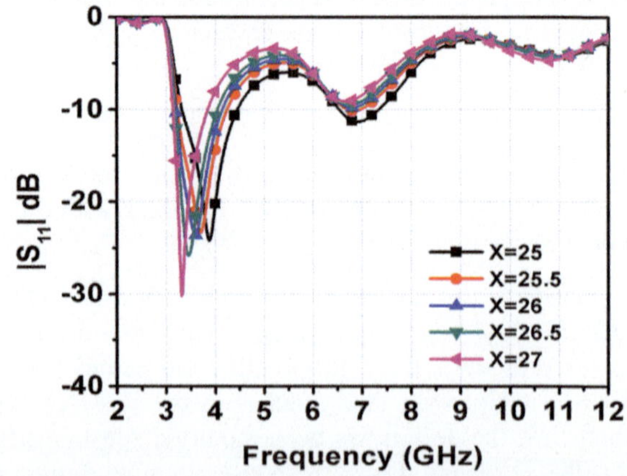

Fig. 3. Reflection coefficient versus frequency variation plot for different values of parasitic strip length X (X_1+X_2).

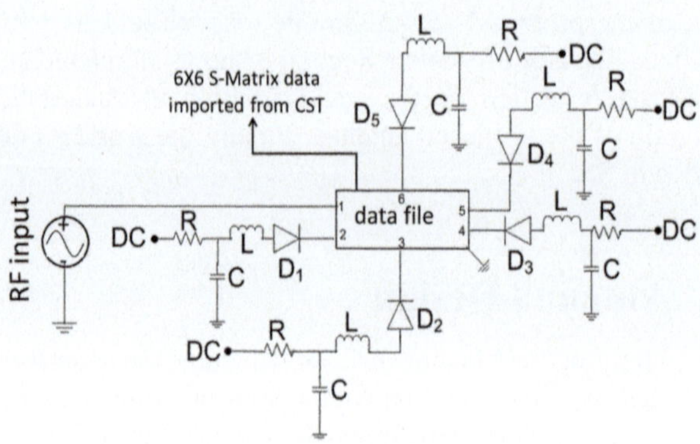

Fig. 4. Full wave circuit model of the proposed antenna.

State	D_1	D_2	D_3	D_4	D_5	Operation
State-1	ON	OFF	OFF	OFF	OFF	UWB (3.1-10.6 GHz)
State-2	OFF	OFF	OFF	OFF	OFF	UWB with single notch at 3.6 GHz (WiMAX)
State-3	ON	ON	OFF	OFF	OFF	UWB with single notch at 5.2 GHz (WLAN)
State-4	OFF	ON	OFF	OFF	OFF	UWB operation with dual notches at 3.6 GHz and at 5.2 GHz (WiMAX & WLAN)
State-5	ON	OFF	ON	OFF	ON	Resonance at 3.6 GHz (WiMAX)
State-6	ON	OFF	OFF	ON	ON	Resonance at 5.2 GHz (WLAN)
State-7	ON	OFF	ON	ON	ON	Dual-band response at 3.6 GHz and at 5.2 GHz (WiMAX & WLAN)

Tab. 2. Different combination of diode states.

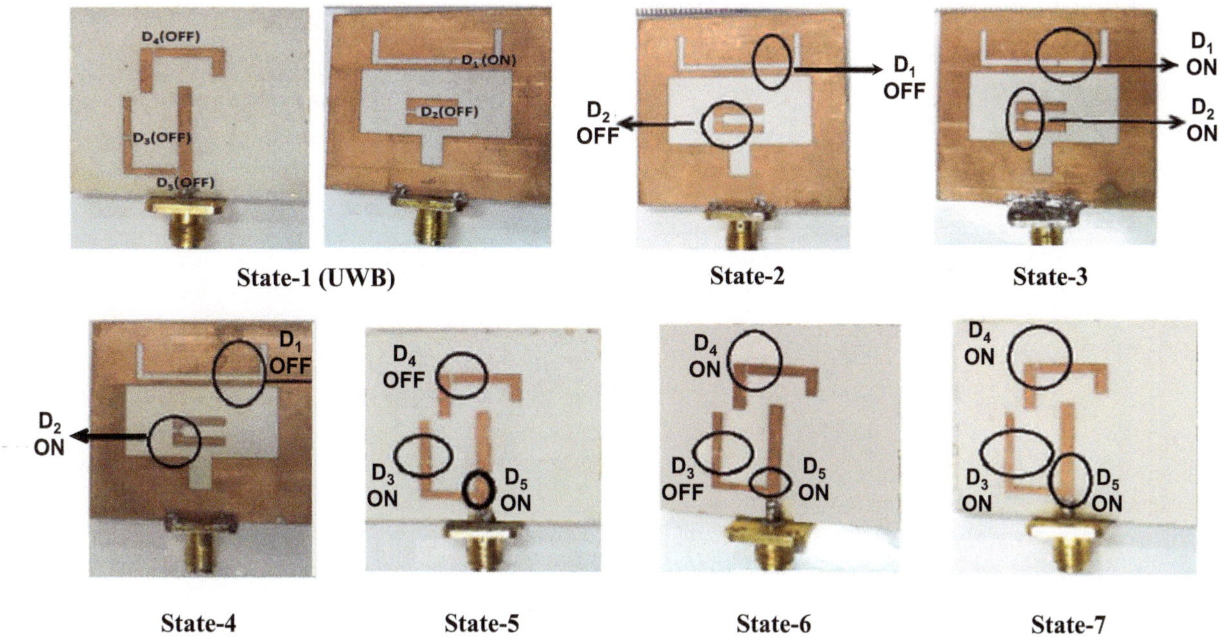

Fig. 5. Photographs of the reconfigurable antenna for seven different states.

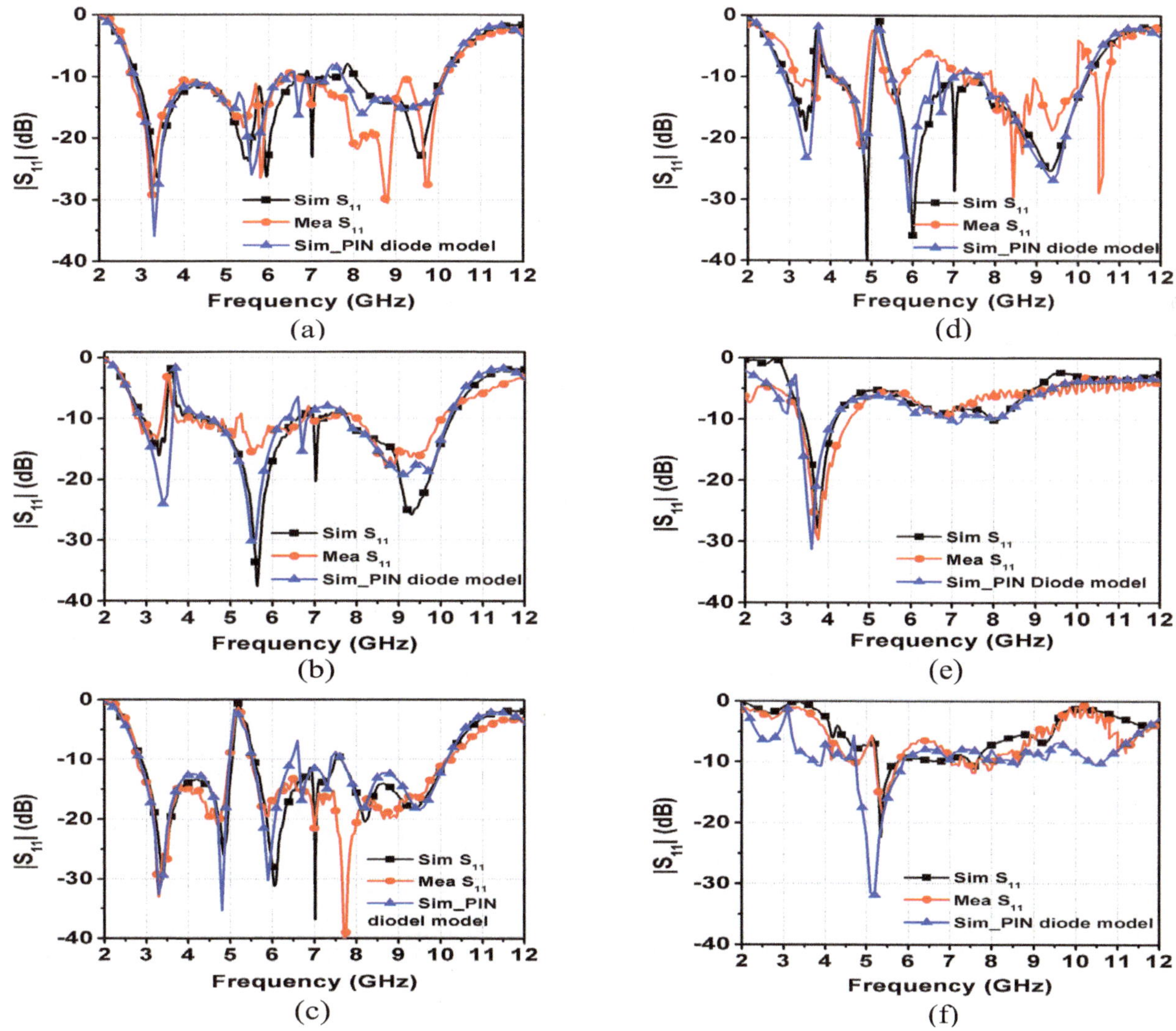

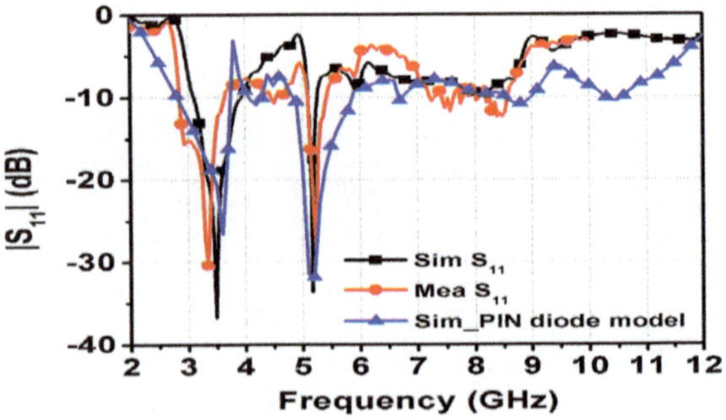

(g)

Fig. 6. Reflection coefficient of the proposed antenna: (a) State-1, (b) State-2, (c) State-3, (d) State-4, (e) State-5, (f) State-6, and (g) State-7.

current of 10 mA is applied. Seven states of the proposed antenna are summarized in Tab. 2, under different bias conditions. Due to unavailability of actual PIN diodes, seven distinct circuits are printed with ideal switches (metal strips) to verify the seven states of the proposed circuit and it is found that the operating principle is verified using ideal switches and the results obtained with the actual PIN diode model. Fabricated prototypes of seven states (state-1 to 7) are shown in Fig. 5. The circuits were fabricated using the LPKF PCB prototyping machine. Fabricated circuits were tested using Agilent's Vector Network Analyzer E5071C for its frequency response in seven states.

3. Simulation and Measured Results

Frequency responses of the simulated and measured reflection coefficient in seven states are shown in Figs. 6. Here measurement results are compared from two simulators: (a) from CST where metal strips are used to realize the ideal ON and they are removed in OFF states, (b) from ADS where spice model of HPND-4005 is used. It is depicted that measurement results are in good agreement with simulations in the lower UWB band. Though a small deviation is seen in the simulation results obtained from CST and ADS, which may be due to the difference in the switching methods being used. Additionally, the measured results in the Figs. 6 (a) and (d) are different with the simulated ones. The reason is that, as mentioned above, seven states of the proposed antenna are obtained with seven different printed antennas (Fig. 5).

One thing which is very common in the measured results shown in Figs. 6 (a) and (d) is that both states have used the same bottom layer and it was observed that the gaps to realize an OFF state of the switch/diode were not fabricated properly in this layer due to the conventional lithography process. It could also be observed that the difference is reasonably significant in the upper UWB band rather lower UWB region. This is quite obvious because at higher frequency any deviation in the physical dimensions becomes more critical. Furthermore, as shown in Fig. 6(a), the measured impedance bandwidth is 7.4 GHz (2.9 to 10.3 GHz) covering the UWB spectrum nearly. Figures 6(b-d) show the UWB response with single and dual notches. The first notch occurred at 3.5 GHz with rejection level of –1.9 dB. The second notch is obtained at 5.2 GHz with –1.1 dB rejection. Figures 6(e-g) show the responses of the proposed antenna in single mode and dual-band mode operations. The radiating single mode is achieved at 3.75 GHz with an impedance bandwidth (–10 dB) of 1 GHz which covers the entire WiMAX band. The second single radiating mode is obtained at 5.4 GHz with a bandwidth of 0.5 GHz, sufficient to cover WLAN band.

Figure 7 shows the surface currents in seven states. In state-1 shown in Fig. 7(a) at 3.1 GHz, large current is observed around the bigger slot that is quite understandable because it decides the lower cutoff for UWB response. In state-2 and 3, Figs. 7(b) and (c), current is mostly flowing in U-shaped slot and C-shaped metal strip which confirm their rejection behavior at respective frequencies. Figure 7(d) shows that in state-5 current is concentrated in the strip $X_1 + X_2$, responsible for single band operation at 3.6 GHz or 5.2 GHz, respectively. In Fig. 7(e) and (f) current is flowing largely in $X_1 + X_2$ and Y_1 and Y_2 which provides radiation at 3.65 GHz and 5.2 GHz respectively in dual-band mode (X_1 and X_2 are combined to radiate at 3.6 GHz in this mode).

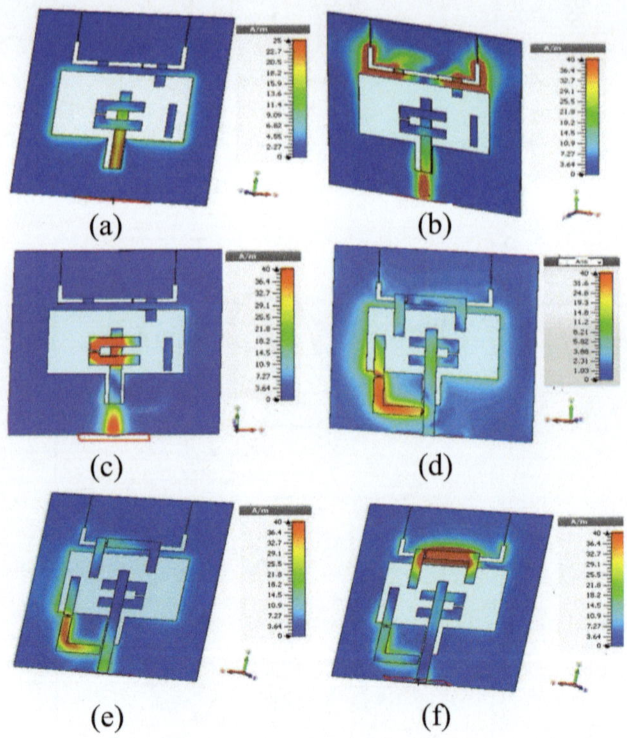

Fig. 7. Surface current distributions of the proposed antenna in different operating states: (a) State-1, at 3.1 GHz, (b) State-2, notch at 3.6 GHz, (c) State-3, notch at 5.2 GHz, (d) State-5, single band at 3.6 GHz, (e) State-7, single band at 3.56 GHz, (f) State-7, dual band at 5.2 GHz.

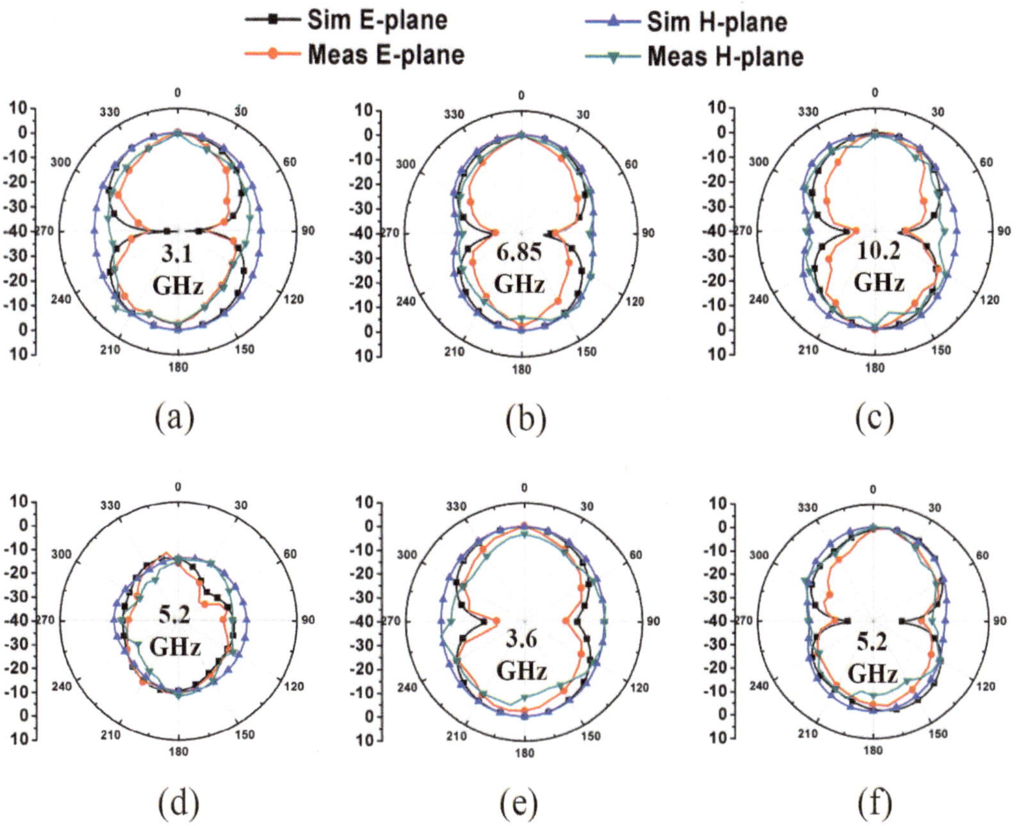

Fig. 8. Far-field radiation patterns in E-plane and H-plane: (a-c) UWB mode, (d) UWB with notch at 5.2 GHz, (e) single radiating mode at 3.6 GHz, (f) single radiating mode at 5.2 GHz.

Ref	Size (L×W) mm^2	ε_r/h	Feeding Technique	States/ switches	Results/Operations
[15]	32×24	2.65/1.6	CPW	NA	UWB band, UWB band with notch either at 5.5 or 8.5 GHz or both notches
[16]	32 × 24	2.65/1.6	CPW	4/2	UWB band, UWB band with notch either at 5.5 or 7 GHz or dual notches at both
[17]	55 × 45	2.2/1.6	Microstrip	8/3	UWB band , UWB band with notch either at 2.4 or 3.5 or 5.2 GHz or dual notches (2.4 & 3.5, or 2.4 & 5.2 or 3.5 & 5.2) or triple notches at 2.4, 3.5 and 5.2 GHz
[18]	28 × 12.3	2.56/1.6	ACS	2/1	UWB band, UWB band with notch either at 3.5 or 8.2 GHz or both
[19]	31.2 × 30	2.56/1.6	Microstrip	6/4	UWB band , UWB band with notch either at 6.8 or 8.7 GHz or dual notches (6 & 8.7, or 5.5 & 11.5) or triple notches at 5.5,6.8 and 11.5 GHz.
[20]	32 × 24	2.65/1.6	CPW	4/2	UWB band, UWB band with notch either at 3.5 or 8 GHz or both
[21]	58 × 65.5	2.2/1.6	Microstrip	3/ using physical rotation	UWB band, reconfigurable with reconfigurable dual band operation either at 4 or 6.65GHz
This work	40 × 40	2.55/0.77	Microstrip	7/5	UWB band, UWB band with reconfigurable notch either at 3.5 or 5.2 GHz or dual notches , resonating mode response either at 3.6 or 5.2 GHz or both modes

CPW: Coplanar Waveguide; ACS: Asymmetric Coplanar Strip-Fed; NA: Not Available

Tab. 3. Comparison with previous reported works.

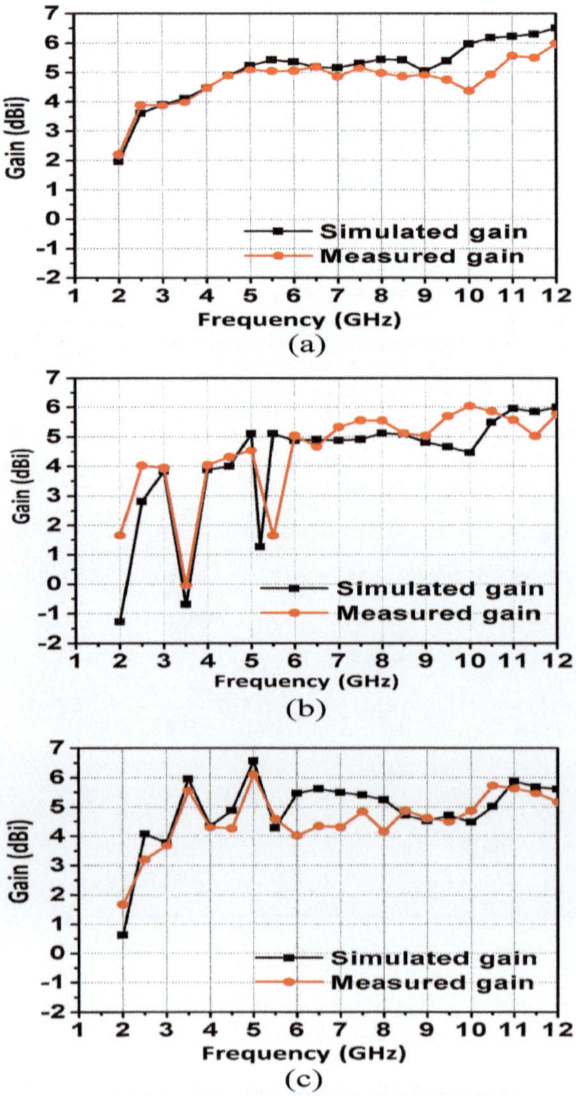

Fig. 9. Gain of the proposed antenna (a) UWB (State-1), (b) UWB with notches at 3.6 GHz & 5.2 GHz (State-5), (c) with parasitic strips for dual band at 3.6 GHz & 5.2 GHz (State-7).

Simulated and measured far-field radiation patterns in E-plane (yz- plane) and H-plane (zx-plane) at frequency 3.1 GHz, 5.85 GHz and 10.6 GHz obtained in state-1 (for UWB operation) are shown in Figs. 8(a-c). It is apparent that the antenna provides almost stable and omnidirectional patterns in the entire UWB range. In Fig. 8(d), E-plane and H-plane radiation patterns for UWB operation with notch at 5.2 GHz (state-3) are plotted. It is evident that at 5.2 GHz radiation is suppressed drastically (–18 dB) from all the direction as desired to avoid any interference in UWB band from WLAN band. Same type of behavior may be expected with notch at 3.6 GHz (state-2). Figures 8(e) and (f) show the radiation patterns for dual band reconfigurable antenna resonating at 3.6 GHz (state-5) and 5.2 GHz (state-6). It is clearly shown that parasitic strips contribute for radiation of desired pattern in 3.6 GHz and 5.2 GHz band.

Figure 9 shows simulated and measured gain of the proposed antenna in three states: UWB operation (state-1), UWB with a dual notched (state-4), and dual band operation (state-7).

As shown in Fig. 9(a), a maximum gain of 5 dBi is realized with a variation of < 2 dBi in the whole UWB frequency range. Figure 9(b) shows the simulated and measured gain of UWB antenna with dual notches, the gain is significantly reduced to 0 dBi and 1.5 dBi at notch frequency 3.6 GHz and 5.2 GHz, respectively.

The gain values are not much lower at the notch frequencies as expected. This might be due to parasitic coupling between different elements of the antenna. Figure 9(c) shows the gain of the antenna in the dual band mode. Maximum achieved gain is 5.5 dBi and 6 dBi at frequency of 3.65 GHz and 5.3 GHz, respectively. Although the out-of-band gain rejection (4-5 dBi) in state-7 is not as much as good and needs to be improved further.

Comparison of the proposed multifunctional reconfigurable UWB antenna with earlier reported works in similar domain is shown in Tab. 3. It is found that, in addition to UWB operation with multiple notches at 3.6 GHz and 5.2 GHz, the proposed antenna could also be used to provide the resonating modes at aforesaid frequencies.

4. Conclusion

A new compact size (40×40 mm^2) UWB antenna with reconfigurable dual-notched and dual-band characteristics is proposed. Switching mechanism is achieved by using a metal strip to realize ON and OFF states. The proposed antenna could offer seven modes of operation by activating/deactivating five PIN diodes. The proposed antenna is simulated, fabricated and tested for different modes of operation and shows good agreement between measured and simulated return loss, radiation and gain characteristics. The proposed antenna may be an appropriate candidate for UWB operation, UWB operation with WiMAX and WLAN bands rejection capability and for WiMAX/ WLAN applications.

References

[1] MAK, A. C. K., ROWELL, C. R., MURCH, R. D. et al. Reconfigurable multiband antenna designs for wireless communication devices. *IEEE Transactions on Antennas and Propagation*, July 2007, vol. 55, no. 7, p. 1919–1928. DOI: 10.1109/TAP.2007.895634

[2] PARK, Y. K., SUNG, Y. A reconfigurable antenna for quad-band mobile handset applications. *IEEE Transactions on Antennas and Propagation*, June 2012, vol. 60, no. 6, p. 3003–3006. DOI: 10.1109/TAP.2012.2194672

[3] ZHANG, T., LI, R., JIN, G. et al. A novel multiband planar antenna for GSM/UMTS/LTE/Zigbee/RFID mobile devices. *IEEE Transactions on Antennas and Propagation*, Nov. 2011, vol. 59, no. 11, p. 4209–4214. DOI: 10.1109/TAP.2011.2164201

[4] LIM, J. H., BACK, G. T. , KO, Y. I., et al. A reconfigurable PIFA using a switchable PIN-diode and a fine-tuning varactor for USPCS/WCDMA/m-WiMAX/WLAN. *IEEE Transactions on Antennas and Propagation*, July 2010, vol. 58, no. 7, p. 2404 to 2411. DOI: 10.1109/TAP.2010.2048849

[5] LI, Y., ZHANG, Z., ZHENG, J. A compact hepta-band loop-inverted F reconfigurable antenna for mobile phone. *IEEE Transactions on Antennas and Propagation*. Jan. 2012 vol. 60, no. 1, p. 389–392. DOI: 10.1109/TAP.2011.2167949

[6] LEE, J. H., SUNG, Y. A simple penta-band reconfigurable antenna for mobile device applications. *Microwave and Optical Technology Letters,* 2014, vol. 56, no. 3, p. 769–774. DOI: 10.1002/ mop.28139

[7] Revision of Part 15 of the Commission's Rules Regarding Ultra-Wideband Transmission Systems. *FCC, Washington, DC, Tech. Rep. ET-Docket*,8, Apr. 2002, p. 1–118. DA/FCC: FCC-10-151

[8] AZIM, R., ISLAM, M. T., MISRAN, N. Compact tapered-shape slot antenna for UWB applications. *IEEE Antennas and Wireless Propagation Letters*, 2011, vol. 10, p. 1190–1193. DOI: 10.1109/LAWP.2011.2172181

[9] KAHRIZI, M., SARKAR, T. K., MARICEVIC, Z. A. Analysis of a wide radiating slot in the ground plane of a microstrip line. *IEEE Transactions on Microwave Theory and Techniques*, Jan. 1993, vol. 41, no. 1, p. 29–37. DOI: 10.1109/22.210226

[10] TASOUJI, N., NOURINIA, J., GHOBADI, C., et al. A novel printed UWB slot antenna with reconfigurable band-notch characteristics. *IEEE Antennas and Wireless Propagation Letters,* 2013, vol. 12, p. 922–925. DOI: 10.1109/LAWP.2013.2273452

[11] SRIVASTAVA, G., DWARI, S., KANAUJIA, B. K. A compact UWB antenna with reconfigurable dual notch bands. *Microwave and Optical Technology Letters,* 2015, vol. 57, no. 12, p. 2737 to 2742. DOI: 10.1002/mop.29424

[12] VALIZADE, A., GHOBADI, CH., NOURINIA, J., et al. A novel design of reconfigurable slot antenna with switchable band notch and multiresonance functions for UWB applications. *IEEE Antennas and Wireless Propagation Letters,*2012 vol. 11, p. 1166–1169. DOI: 10.1109/LAWP.2012.2218271

[13] BADAMCHI, B., NOURINIA, J., GHOBADI, C., et al. Design of compact reconfigurable ultra-wideband slotantenna with switchable single/dual band notchfunctions. *IET Microwave, Antennas and Propagation,* 2013, vol. 8, no. 8, p. 541–548. DOI: 10.1049/iet-map.2013.0311

[14] CHEN, W. S., LEE, B. Y., CHANG, P. Y. A compact microstrip-line-fed slot antenna with dual-band notched for WiMAX operation. *Progress In Electromagnetics Research C,* 2010, vol. 16, p. 13–23. DOI: 10.2528/PIERC10080803

[15] LI, Y., LI, W., YU, W. A switchable UWB slot antenna using SIS-HSIR and SIS-SIR for multi-mode wireless communications applications. *Applied Computational Electromagnetics Society Journal*, 2012, vol. 27, no. 4, p. 340–351.

[16] LI, Y., LI, W., YE, Q. A reconfigurable wide slot antenna integrated with sirs for UWB/multiband communication applications. *Microwave and Optical Technology Letters,* 2013, vol. 55, no. 1, p. 52–55. DOI: 10.1002/mop.27253

[17] AL-HUSSEINI, M., RAMADAN, A., EL-HAJJ, A. et al. Design based on complementary split-ring resonators of an antenna with controllable band notches for UWB cognitive radio applications. In *Proceedings of the IEEE International Symposium on Antennas and Propagation and USNC/URSI National Radio Science Meeting (APSURSI '11),* July 2011, p. 1120–1122. DOI: 10.1109/APS.2011.5996479

[18] LI, Y., LI, W., YE, Q. Miniaturization of asymmetric coplanar strip-fed staircase ultrawideband antenna with reconfigurable notch band. *Microwave and Optical Technology Letters,* vol. 55, p. 1467–1470. DOI: 10.1002/mop.27634

[19] LI, Y., LI, W., YE, Q. A reconfigurable triple-notch-band antenna integrated with defected microstrip structure band-stop filter for ultra-wideband cognitive radio applications. *International Journal of Antennas and Propagation*, 2013, 13 p. DOI: 10.1155/2013/472645

[20] LI, Y., LI, W., MITTRA, R. A compact CPW-fed circular slot antenna with reconfigurable dual band-notch characteristics for UWB communication applications. *Microwave and Optical Technology Letters*, 2014, vol. 56, no. 2, p. 465–468. DOI: 10.1002/mop.28087

[21] TAWK, Y., CHRISTODOULOU, C. G. A new reconfigurable antenna design for cognitive radio. *IEEE Antennas and Wireless Propagation Letters*, 2009, vol. 8, p. 1378–1381. DOI: 10.1109/LAWP.2009.2039461

About the Authors ...

Asha VERMA received her Master of Technology (M.Tech.) from PDPM-IIITDM Jabalpur, India, in 2015. Her research interests include UWB technologies and reconfigurable antennas.

Manoj S. PARIHAR received his Ph.D. from the Indian Institute of Technology Delhi, India, in 2012. Presently, he is working as an assistant professor in PDPM-IIITDM Jabalpur, India. His current research interests are reconfigurable circuits, microwave integrated circuits, EMC/EMI, RFID, RF MEMS, RF energy harvesting, and UWB technology.

Substrate Integrated Waveguide Cavity Filters: Miniaturization and New Materials for IoT Applications

Cristiano TOMASSONI[1], *Maurizio BOZZI*[2]

[1] Department of Engineering, University of Perugia, Perugia, Italy
[2] Department of Electrical, Computer and Biomedical Engineering, University of Pavia, Pavia, Italy

cristiano.tomassoni@unipg.it, maurizio.bozzi@unipv.it

Abstract. *This paper presents an overview of the current research trends in the field of substrate integrated waveguide (SIW) technology, with particular emphasis on the issues related to the emerging applications in the framework of the Internet of Things (IoT) and the fifth generation of mobile communication (5G). More specifically, different techniques adopted to miniaturize SIW cavities are described, with the aim of reducing the footprint of SIW components and filters. Moreover, the use of innovative materials, like paper, textile and 3D printed dielectric substrates, is presented and discussed, and the implementation of ecofriendly, wearable, and fully 3D structures is illustrated.*

Keywords

Substrate integrated waveguide (SIW), filters, paper-based electronics, wearable microwave components, additive manufacturing

1. Introduction

The rapid evolution of new applications in the framework of the Internet of Things (IoT) [1], [3] and the fifth generation of mobile communication (5G) [4], [5] poses a set of requirements for the developers of microwave wireless systems. Besides the low cost, the fast design process, and the simple fabrication technology, the future generation of microwave systems needs to exhibit easy integration, compact size, as well as the possibility to be implemented on unconventional materials (like paper, textile, and 3D printed materials).

The future generation of microwave systems needs to integrate in a single device the microwave circuitry, sensing capability, and wireless transmission of data and (possibly) power. In order to keep the design and fabrication costs as low as required by the market, the use of a single, suitable technology is mandatory.

Moreover, specific applications require the use of innovative materials that are not commonly used in the fabrication of microwave circuits: wearable components and antennas, for instance, can be conveniently integrated on garments, by adopting the textile as the substrate material. Ecofriendly applications, in the field of agriculture and environmental monitoring, can benefit from the use of paper (instead of plastic) for the dielectric substrate.

In this scenario, the substrate integrated waveguide (SIW) technology [6], [7] represents a particularly suitable candidate, due to its capability to implement passive components, active devices, and antennas by using a single technology and manufacturing process. High performance components in terms of losses are usually realized by using waveguide technology. A lot of effort has been devoted to mass and weight reduction for this technology [8]–[11], even in combination with high dielectric materials [12], [13]. The main problems remain their integrability. SIW is the integrated version of this technology and is very competitive from the point of view of losses. In fact, the SIW technology allows to implement interconnects and cavities by using a standard printed-circuit board (PCB) process, based on a laminated dielectric substrate and metal vias that provide the lateral confinement. SIW technology presents several advantages over other planar technologies (e.g., microstrip line and coplanar waveguide), such as complete shielding, and low loss, while preserving a low fabrication cost and a well-established manufacturing process.

One of the major limitations of SIW structures, however, is related to the dimensions, which depend on the operation frequency and the dielectric material. While using dielectric substrates with high permittivity allows reducing interconnect and cavity size, it has typically a detrimental effect on losses, and consequently it leads to poor quality factors. For this reason, several solutions have been proposed to reduce the size of SIW structures, including the substrate integrated folded waveguide [8] and the half-mode substrate integrated waveguide [15].

This paper presents an overview of recent achievements in the implementation of novel SIW cavity filters for IoT applications, with particular emphasis on the miniaturization and the use of new materials. The use of folded, half-mode and quarter-mode SIW cavities will be discussed

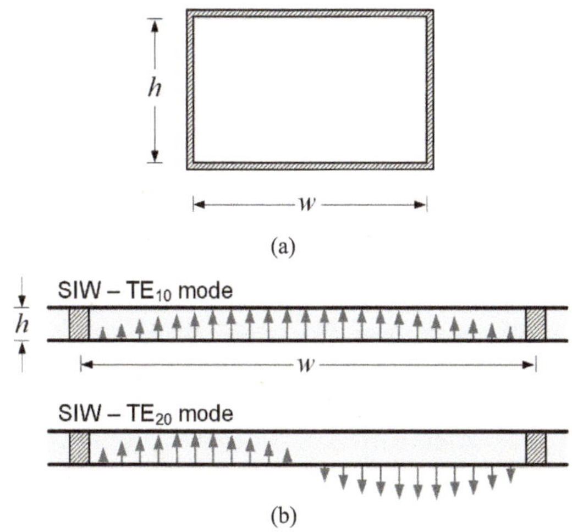

Fig. 1. Geometry comparison of a standard and an integrated waveguide: (a) Cross section of a standard rectangular waveguide; (b) Cross section of a substrate integrated waveguides, and electric field of the first two modes.

and several filters based on these structures will be presented. Moreover, the implementation of SIW filters on paper (for eco-friendly applications), on textile (for wearable systems), and by 3D printing (to exploit the maximum versatility and to tailor the material properties) will be presented and discussed through several examples.

2. Miniaturized SIW Cavities

The substrate integrated waveguide represents an integrated version of the classical rectangular waveguide [6], [7]. As shown in Fig. 1, the main difference in terms of dimensions resides in the height: in the standard rectangular waveguide the height h is usually about a half of the width w, while in SIW the dimension h is much smaller than w and it corresponds to the thickness of the dielectric substrate, used in the PBC fabrication.

2.1 Folded SIW Cavity

The basic idea of folded SIW is to reduce the footprint by folding the structure around a metal septum [8]. This solution allows reducing the footprint of the structure by a factor two.

A folded SIW can be obtained by using a two-layer topology (Fig. 2). The metal septum between the two layers is connected to the side posts on one side, whereas it leaves a gap on the opposite side. This leads to a U-shaped waveguide. The electric field of the fundamental mode of the folded SIW is depicted in Fig. 2 (above) and in practice corresponds to the fundamental TE_{10} mode of the standard rectangular waveguide, once the waveguide is folded around the metal septum. Conversely, the electric field of the second mode of the folded SIW is illustrated in Fig. 2 (below): this mode corresponds to the TE_{20} mode of the standard rectangular waveguide, after folding the structure around the metal septum.

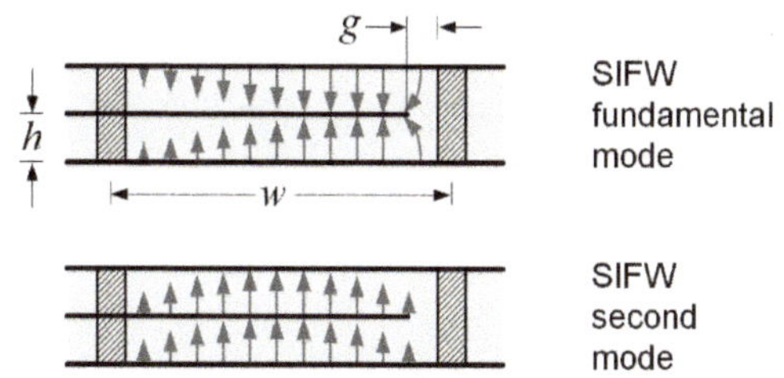

Fig. 2. Geometry of a folded SIW and representation of the electric field of the first two modes.

Note that the fundamental mode exhibits an odd symmetry with respect to the middle plane between the two dielectric layers, while the second mode shows an even symmetry.

Because of its topology, this structure can be easily fed by stripline and, thanks to the modal symmetry, the stripline excites the fundamental mode but not the second (TE_{20}) mode, thus resulting in a wider spurious free behavior.

The gap g between the metal septum and the side posts (Fig. 2) affects the resonance of the folded waveguide by changing the electrical length of the structure. In particular, the narrower the metal septum, the higher is the waveguide cutoff frequency. This effect can be exploited for obtaining filtering structures as that sketched in Fig. 3 [16]. In this case, the waveguide sections with a large septum alternate to waveguide sections with a narrow septum (which defines the insets). Large-septum waveguide sections behave as resonators, narrow-septum waveguide sections behave instead as impedance inverters. This leads to the classical equivalent circuit shown in Fig. 4, consisting in transmission lines and admittance inverters.

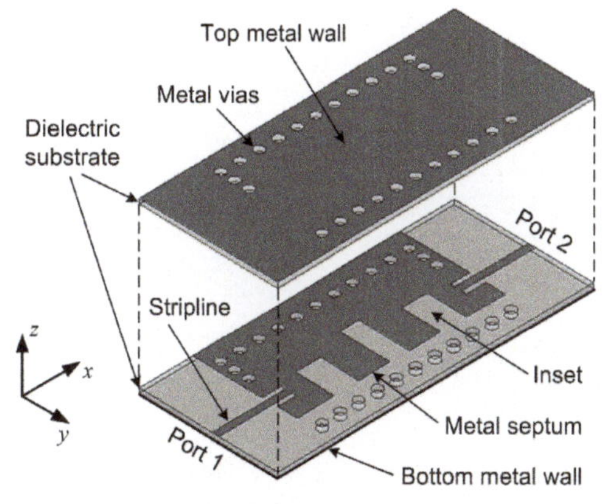

Fig. 3. Geometry of a filter based on the folded SIW cavity (from [16]).

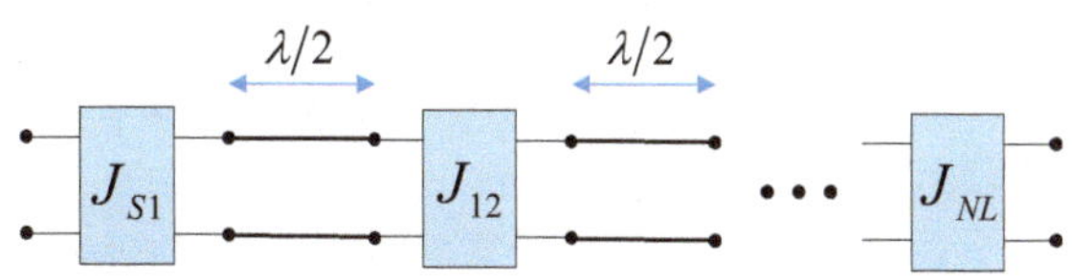

Fig. 4. Equivalent circuit of folded SIW filter.

By using the proposed topology, a two-pole filter centered at 4.5 GHz with 1.4 GHz bandwidth has been designed and manufactured [16]. The filter design can be performed in two different ways: the classical way is based on the properly dimensioning of resonator lengths and insets, to realize the filtering response of the equivalent circuit in Fig. 4. The second technique, described in [16], is based on the tuning of the resonant modes of the cavity, by modifying the size of the insets.

The filter has been manufactured by milling technique, by adopting two layers of Taconic TLX-9 laminate, with dielectric permittivity $\varepsilon_r = 2.5$, loss tangent $\tan\delta = 0.0019$, and thickness $h = 0.76$ mm. The photograph of the prototype is shown in Fig. 5a. The comparison between measured and simulated scattering parameters is reported in Fig. 5b.

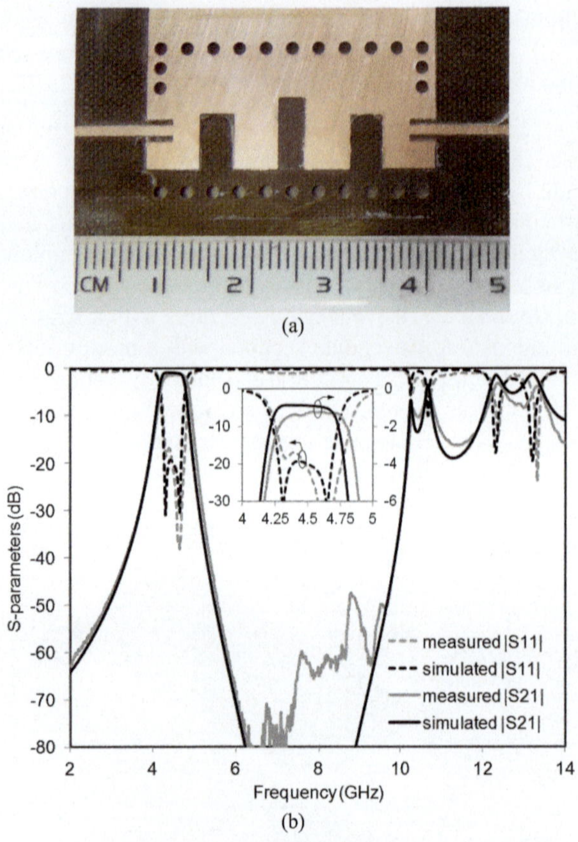

Fig. 5. Second order filter based on a dual-mode folded SIW cavity (from [16]): (a) Photograph of the bottom layer; (b) Comparison between simulated and measured scattering parameters of the filter.

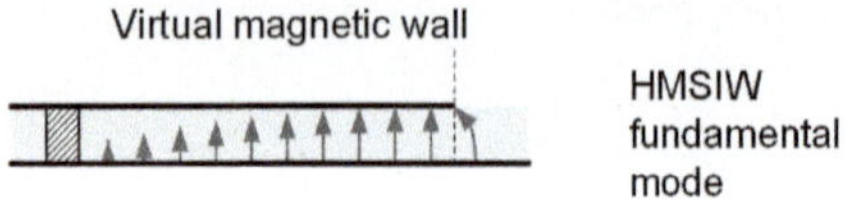

Fig. 6. Geometry of a half-mode SIW and representation of the electric field of the fundamental mode.

2.2 Half-Mode SIW Cavity

As shown in Fig. 1b, the fundamental mode of the SIW structures exhibits a magnetic wall in the vertical symmetry plane. This suggests that it is possible to use just a half of the SIW if the cut is closed with a magnetic wall.

Unfortunately, on the contrary of the electric wall that can be created simply by a metallic material, the magnetic wall is more difficult to realize. The behavior of the magnetic wall can be approximated by exploiting a high impedance load. According to Fig. 6, an abrupt interruption of the upper SIW wall results in a high impedance discontinuity. This effect can be considered a sort of virtual magnetic wall, and this allows to obtain the half-mode substrate integrated waveguide (HMSIW) [15].

The same concept can be exploited in the design of cavity resonators, to be adopted in the implementation of filters. Of course, the fact that the impedance is high but not infinite leads to an electromagnetic field that is not completely confined inside the SIW. Therefore, a small leakage due to radiation is present. This effect decreases the Q-factor. In any case, lower SIW height h or higher dielectric constant allow the implementation of a better virtual magnetic wall (with a higher Q-factor).

By using the HMSIW technology, it is possible to realize a filter with a geometry very similar to that used for folded waveguide, as firstly proposed in [17]. In fact, the fundamental mode of the HMSIW can be excited by a microstrip line, same as the fundamental mode of the folded SIW can be excited by a stripline. According to Fig. 7, wide HMSIW sections can be exploited as resonators, while narrower sections can be used to implement admittance inverters. This leads to the implementation of the circuit of Fig. 4, where narrower HMSIW sections correspond to lower values of impedance inverters.

By adopting the proposed geometry, a four-pole filter with central frequency at 4.5 GHZ and 1.4 GHz bandwidth has been designed and manufactured in [17]. In this case, the fabrication required the processing of one single layer of Taconic TLX-9 laminate, with a complexity reduction compared to the folded SIW filter proposed in the previous section. The photograph of the prototype is shown in Fig. 8a, and the comparison between measured and simulated scattering parameters is reported in Fig. 8b.

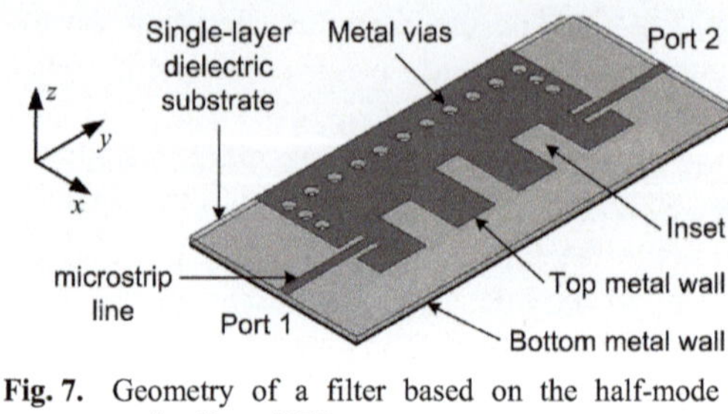

Fig. 7. Geometry of a filter based on the half-mode SIW cavity (from [17]).

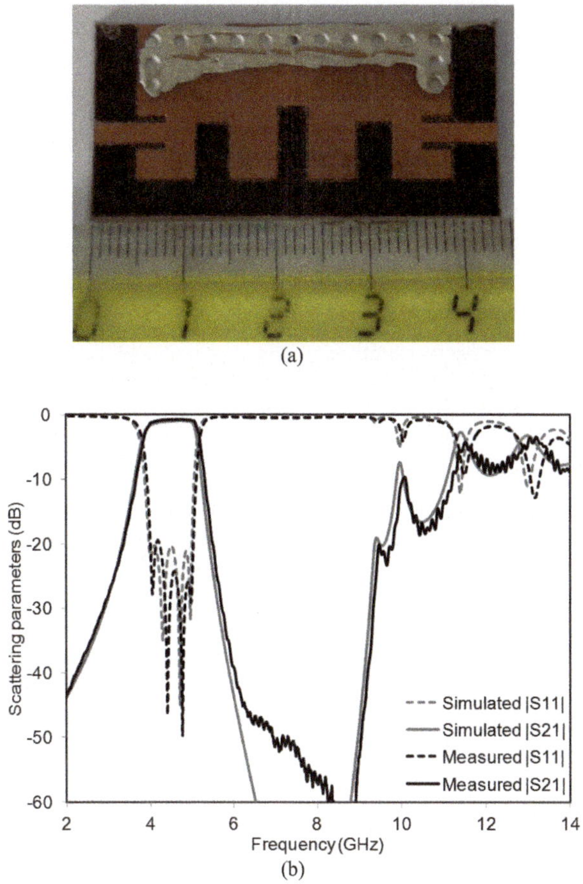

Fig. 8. Four-pole filter based on a half-mode SIW cavity (from [17]): (a) Photograph of the prototype; (b) Comparison between simulated and measured scattering parameters of the filter.

2.3 Quarter-Mode SIW Cavity

The concept of HMSIW cavity can be further extended to obtain a quarter mode SIW cavity (QMSIW). As shown in Fig. 9, a quarter mode-SIW cavity can be obtained by cutting a rectangular SIW cavity into four parts along the symmetry planes [18]. This is obtained by removing the top metal wall and metal vias of three quarters of the cavity, thus reducing the footprint of the cavity by 75%.

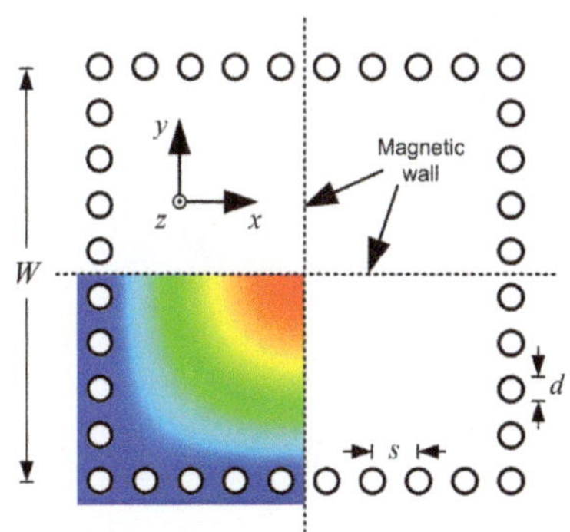

Fig. 9. Geometry of a quarter-mode SIW cavity and representation of the electric field of the fundamental mode (from [18]).

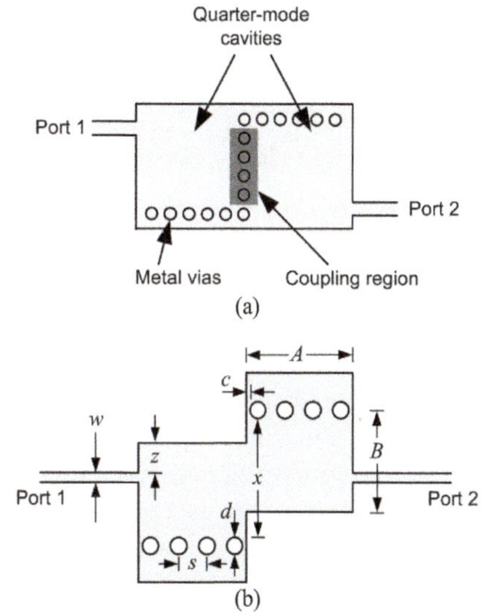

Fig. 10. Coupling mechanisms in a two-pole filter based on quarter-mode SIW cavity (from [18]): (a) Coupling is obtained by removing the common posts; (b) Coupling is controlled by shifting the two cavities.

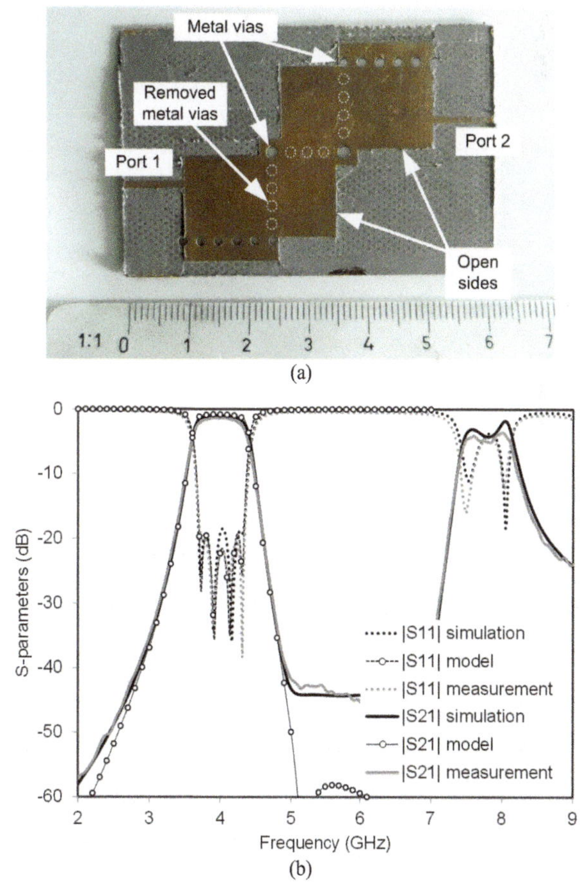

Fig. 11. Four-pole filter based on a quarter-mode SIW cavity (from [18]): (a) Photograph of the prototype; (b) Comparison between simulated and measured scattering parameters of the filter.

The equivalent circuit of Fig. 4 can be implemented also by using QMSIW cavities. According to Fig. 10a, the coupling between two adjacent QMSIW cavities is obtained by removing the row of common posts. The control of the coupling is obtained by side shifting the two cavities, as shown in Fig. 10b. The smaller the shift x, the lower is the admittance inverter value. The coupling to input and output transmission lines is instead controlled by the value of z: the larger the shift z, the lower is the admittance inverter value. In fact, small values of z mean that the microstrip line is connected to the point where the electric modal field is large (see Fig. 9), thus determining a large external coupling.

A four-pole filter centered at 4 GHz and with an FBW of 16% has been designed and manufactured in [18]. A dielectric substrate Taconic RF-35, with $\varepsilon_r = 3.5$, $\tan\delta = 0.0018$, and thickness $t = 0.508$ mm, was adopted for the fabrication. A photograph of the manufactured prototype is shown in Fig. 11a. Measured results are plotted in Fig. 11b, where also the simulated results and the response of the equivalent circuit are reported.

3. SIW Based on New Materials

The use of new materials and fabrication technologies is expected to be a key point of the next generation of wireless systems. The spread of wearable systems will lead to the use of fabric-based components, where the microwave structures are embedded in the garments. Examples of SIW components and antennas in textile have been reported in [19] and [20]. The need to implements eco-sustainable wireless systems is expected to be a strong requirement for some applications, for instance in the field of agriculture and environmental sensing. SIW circuits based on paper substrates and implemented by different techniques have been described in [21]–[23]. Finally, the additive manufacturing represents an emerging area of research, where the microwave components are fabricated by 3D printing, in a single-pass, low-cost manufacturing process. The fused-deposition modeling has been adopted for the implementation of SIW components and antennas [24], [25], by exploiting the possibility to synthetize the required dielectric material by properly adjusting the infill percentage during the printing.

3.1 Paper-based SIW Filters

The use of paper for the implementation of microwave components is particularly interesting, as paper is a cheap material, it is flexible and therefore potentially suitable to the implementation of conformal circuits and antennas, and it is eco-friendly, for those applications (agriculture, environmental monitoring, …) where reducing the ecological impact is critical.

SIW technology is particularly suitable to the implementation on paper, and two different techniques can be adopted. The use of inkjet printing was proposed in [22]: an inkjet printer, that uses a special ink with silver particles, realizes the top and bottom metal layers of the SIW, and the vias consist of brass rivets. Another approach was adopted in [23]: aluminum foils are pasted at the top and bottom sides of a thick paper sheet by using epoxy glue, to define the top and bottom metallization of the SIW structure; subsequently, a classical milling technique is adopted, and the metal vias are realized by conductive paste. The latter solution avoids the sintering process of the conductive ink and reduces the fabrication to a well establish PCB process.

The filters presented in the previous section and based on classical plastic substrates can be implemented on paper. The filters presented in this section have been fabricated by using the technology described in [23], based on the standard milling processing of a paper layer with aluminum foils. The paper substrate exhibits a thickness of 500 μm, relative dielectric permittivity $\varepsilon_r = 2.2$, and loss tangent $\tan\delta = 0.04$.

Two-pole filter on paper, based on a half-mode SIW cavity and operating at the central frequency of 4.5 GHz, was proposed in [20]. The photograph of the prototype is shown in Fig. 12a: the filter is quite compact, with a length less than 40 mm, including the input/output microstrip transitions. The simulated and measured frequency response is reported in Fig. 12b, showing a measured insertion loss is 5.0 dB at the frequency of 4.5 GHz.

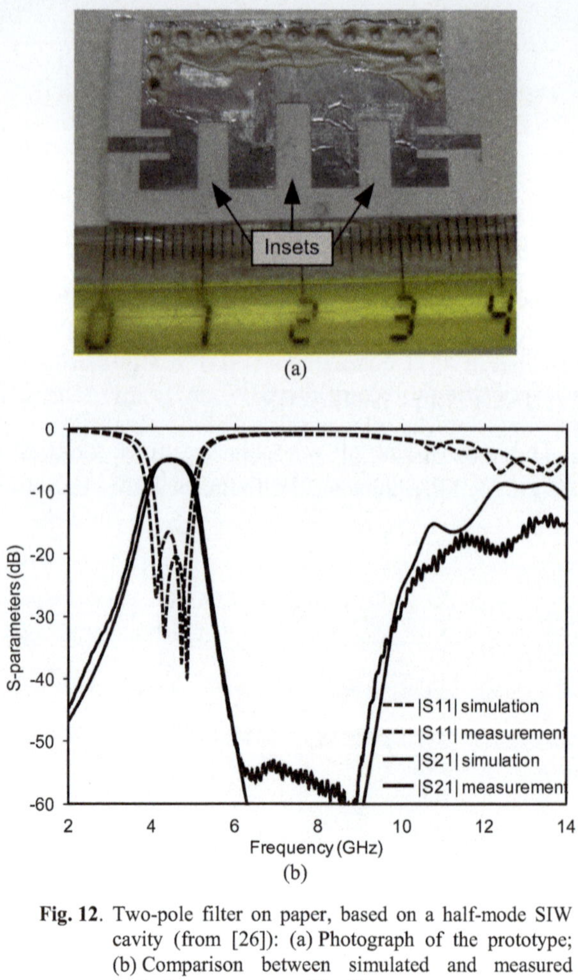

Fig. 12. Two-pole filter on paper, based on a half-mode SIW cavity (from [26]): (a) Photograph of the prototype; (b) Comparison between simulated and measured scattering parameters of the filter.

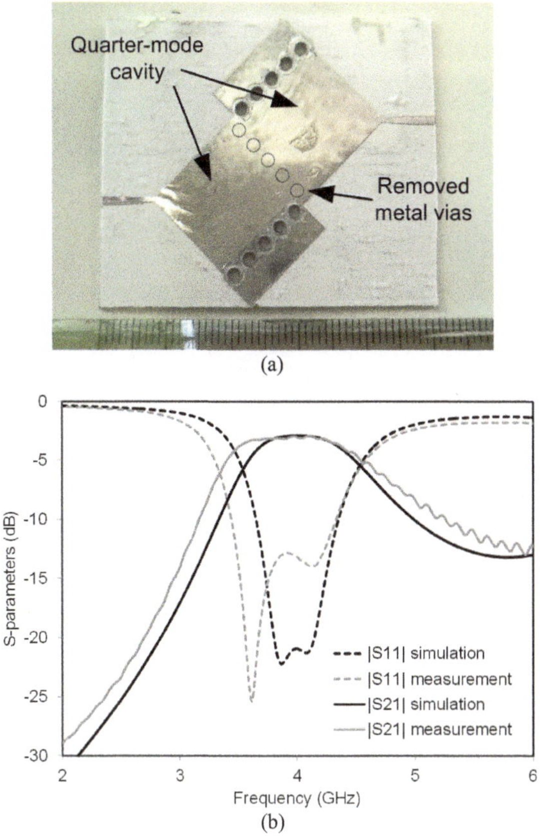

Fig. 13. Quarter-mode SIW filter on paper, fabricated by milling technique (from [23]): (a) Photograph of the prototype; (b) Comparison between simulated and measured scattering parameters of the filter.

A quarter-mode SIW filter on paper, operating at the frequency of 4 GHz, was firstly presented in [23]: it consists of two side-coupled quarter-mode SIW cavities, fed in the corners by microstrip lines with tapered transitions. Fig. 13 shows the photograph of the prototype and the simulated and measured frequency response of the filter.

3.2 SIW Filters on Textile

The deployment of wearable electronic devices is gaining increasing popularity, with applications in a variety of different fields, ranging from healthcare to rescue of workers in harsh environments [28]. Among the different solutions for the implementation of wearable devices, the direct use of textile as the substrate for the microwave components and antennas represents a particularly interesting solution, with the results of a perfect integration of the electronics in the garments.

SIW structures have been implemented on textile [19], [20], by using a layer of expanded cell rubber as the dielectric substrate, a conductive fabric (specifically, electro textile) for the metal layers, and brass rivets for the metal vias. Multilayer topologies are possible, by stacking different dielectric layers, with conductive fabric glued in between. The dielectric textile material adopted in this work exhibits a dielectric permittivity $\varepsilon_r = 1.575$ and a loss tangent $\tan\delta = 0.0238$, and the single-layer textile with a thickness of $h = 3.94$ mm. The photograph of a folded SIW structure is shown in Fig. 14a.

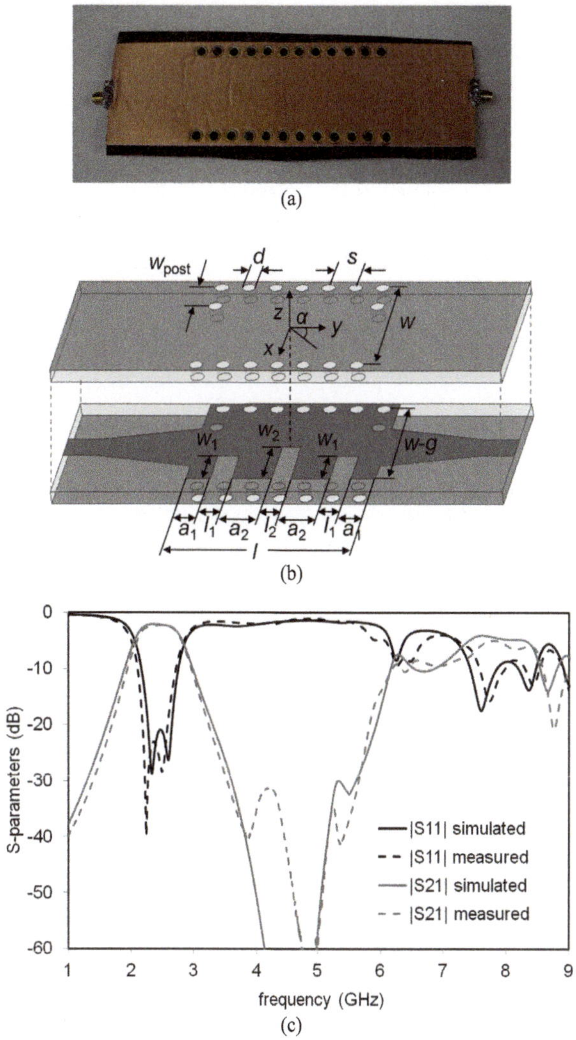

Fig. 14. Folded SIW filter on textile (from [20]): (a) Photograph of a folded SIW waveguide on textile; (b) Geometry of the filter; (c) Comparison between simulated and measured scattering parameters.

A two-pole SIW filter on textile, based on a folded SIW cavity with insets in the middle metal septum, was proposed in [20]. It was designed to operate at the frequency of 2.45 GHz, in the industrial, scientific and medical (ISM) band. The geometry of the filter is shown in Fig. 14b, and its simulated and measured frequency response is reported in Fig. 14c. The 3 dB bandwidth of the filter is 725 MHz and the insertion loss is 2.3 dB at 2.45 GHz.

3.3 SIW Filters by 3D Printing

Additive manufacturing techniques are becoming very popular in the microwave community, because they allow a great versatility in terms of shapes, materials, and fabrication process.

SIW structures based on 3D printing has been proposed in last few years, starting from the first 3D waveguide [24] to microfluidic sensors based on a 3D printed SIW cavity [29], and broadband slab SIW structures [25].

Among the variety of 3D printing techniques, the fused deposition modeling (FDM) is particularly interesting, as it allows to print structures with different material density (infill) from 100% to roughly 10% [30]. This feature permits to control the dielectric permittivity and loss tangent. In fact, in FDM process, the material is heated and extruded through a nozzle, and deposited to form a two-dimensional pattern. Stacked 2D layers results into the final 3D printed structure. One unique feature of the 3D printing is the capability of changing the density of the printed object, by varying the spacing during the filament deposition.

The flexibility of FDM technique can be conveniently exploited in the design of SIW cavity filters. In [27], ABS filament with dielectric permittivity $\varepsilon_r = 2.7$ and loss tangent $\tan\delta = 0.02$ was adopted. By using 40% infill factor, the dielectric permittivity is reduced to $\varepsilon_r = 1.6$ and the loss tangent is simultaneously reduced to $\tan\delta = 0.008$. The effects are a larger size of the circuit dimensions, as well as an increase in the quality factor of the resonant cavities. Based on this concept, two examples of three-pole filters were designed in [27]: the former with 100% infill (Fig. 15a) and the latter with 40% infill (Fig. 15b). While the smaller infill leads to increased component dimensions (Fig. 15c), it also guarantees an increased Q-factor and lower insertion loss, as demonstrated in [27].

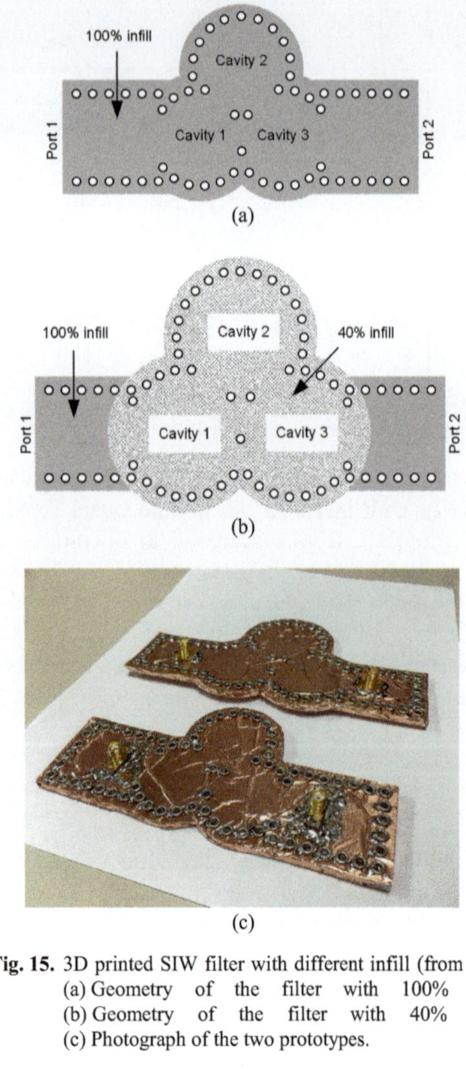

Fig. 15. 3D printed SIW filter with different infill (from [27]): (a) Geometry of the filter with 100% infill; (b) Geometry of the filter with 40% infill; (c) Photograph of the two prototypes.

Acknowledgments

The authors wish to acknowledge Prof. Luca Perregrini and several current and former students at the University of Pavia, Italy, Prof. Manos Tentzeris and his group at the Georgia Institute of Technology, Atlanta, USA, and Prof. Hendrik Rogier and his group at Ghent University, Belgium.

References

[1] GIUSTO, D., IERA, A., MORABITO, G., ATZORI, L. (Eds.) *The Internet of Things*. Springer, 2010. DOI: 10.1007/978-1-4419-1674-7

[2] Special issue: The internet of things. *IEEE Wireless Communications*, 2010, vol. 17, no. 6. ISSN: 1536-1284

[3] Special issue: The internet of things. *IEEE Communications Magazine*, 2011, vol. 49, no. 11. ISSN: 0163-6804

[4] RAPPAPORT, T. S., SUN, S., MAYZUS, R., et al. Millimeter wave mobile communications for 5G cellular: It will work! *IEEE Access*, 2013, vol. 1, no. 1, p. 335–349. DOI: 10.1109/ACCESS.2013.2260813

[5] BOCCARDI, F., HEATH, R. W., LOZANO, A., MARZETTA, T. L., POPOVSKI, P. Five disruptive technology directions for 5G. *IEEE Communications Magazine*, 2014, vol. 52, no. 2, p. 74–80. DOI: 10.1109/MCOM.2014.6736746

[6] BOZZI, M., GEORGIADIS, A., WU, K. Review of Substrate Integrated Waveguide (SIW) circuits and antennas. *IET Microwaves, Antennas and Propagation*, 2011, vol. 5, no. 8, p. 909–920. DOI: 10.1049/iet-map.2010.0463

[7] GARG, R., BAHL, I., BOZZI, M. *Microstrip Lines and Slotlines*. Artech House, 2013. ISBN: 9781608075355

[8] BASTIOLI, S., TOMASSONI, C., SORRENTINO, R. A new class of waveguide dual-mode filters using TM and nonresonating modes. *IEEE Transactions on Microwave Theory and Techniques*, 2010, vol. 58, no. 12, p. 3909–3917. DOI: 10.1109/TMTT.2010.2086068

[9] BASTIOLI, S., MARCACCIOLI, L., TOMASSONI, C., SORRENTINO, R. Ultracompact highly-selective dual-mode pseudoelliptic filters. *IET Electronics Letters*, 2010, vol. 46, no. 2, p. 147–149. DOI: 10.1049/el.2010.2971

[10] BASTIOLI, S., TOMASSONI, C., SORRENTINO, R. TM dual-mode pseudoelliptic filters using nonresonating modes. In *IEEE MTT-S International Microwave Symposium Digest*, p. 880–883. Anaheim (CA, USA), May 2010. DOI: 10.1109/MWSYM.2010.5518108

[11] TOMASSONI, C., BASTIOLI, S., SORRENTINO, R. Generalized TM dual-mode cavity filters. *IEEE Transactions on Microwave Theory and Techniques*, 2011, vol. 59, no. 12, p. 3338–3346. DOI: 10.1109/TMTT.2011.2172622

[12] TOMASSONI, C., BASTIOLI, S., SNYDER, R. V. Propagating waveguide filters using dielectric resonators. *IEEE Transactions*

on Microwave Theory and Techniques, 2015, vol. 63, no. 12, p. 4366–4375. DOI: 10.1109/TMTT.2015.2495284

[13] TOMASSONI, C., BASTIOLI, S., SNYDER, R. V. Compact mixed-mode filter based on TE_{101} cavity mode and $TE_{01\delta}$ dielectric mode. *IEEE Transactions on Microwave Theory and Techniques*, 2016, vol. 64, no. 12, p. 4434–4443. DOI: 10.1109/TMTT.2016.2623714

[14] GRIGOROPOULOS, N., IZQUIERDO, B. S., YOUNG, P. R. Substrate Integrated Folded Waveguides (SIFW) and filters. *IEEE Microwave and Wireless Components Letters*, 2005, vol. 15, no. 12, p. 829–831. DOI: 10.1109/LMWC.2005.860027

[15] HONG, W., LIU, B., WANG, Y. Q., et al. Half mode substrate integrated waveguide: A new guided wave structure for microwave and millimeter wave application. In *Proceedings of Joint 31st International Conference on Infrared Millimeter Waves and 14th International Conference on Terahertz Electronics*. Shanghai (China), Sept. 18-22, 2006, p. 219. DOI: 10.1109/ICIMW.2006.368427

[16] MORO, R., MOSCATO, S., BOZZI, M., PERREGRINI, L. Substrate integrated folded waveguide filter with out-of-band rejection controlled by resonant-mode suppression. *IEEE Microwave and Wireless Components Letters*, 2015, vol. 25, no. 4, p. 214–216. DOI: 10.1109/LMWC.2015.2400927

[17] DELMONTE, N., SILVESTRI, L., BOZZI, M., PERREGRINI, L. Compact half-mode SIW cavity filters designed by exploiting resonant mode control. *International Journal of RF and Microwave Computer-Aided Engineering,* 2016, vol. 26, no. 1, p. 72–79. DOI: 10.1002/mmce.20940

[18] MOSCATO, S., TOMASSONI, C., BOZZI, M., PERREGRINI, L Quarter-mode cavity filters in substrate integrated waveguide technology. *IEEE Transactions on Microwave Theory and Techniques*, 2016, vol. 64, no. 8, p. 2538–2547. DOI: 10.1109/TMTT.2016.2577690

[19] MORO, R., AGNEESSENS, S., ROGIER, H., BOZZI, M. Wearable textile antenna in substrate integrated waveguide technology. *IET Electronics Letters*, 2012, vol. 48, no. 16, p. 985–987. DOI: 10.1049/el.2012.2349

[20] MORO, R., AGNEESSENS, S., ROGIER, H., DIERCK, A., BOZZI, M. Textile microwave components in substrate integrated waveguide technology. *IEEE Transactions on Microwave Theory and Techniques*, 2015, vol. 63, no. 2, p. 422–432. DOI: 10.1109/TMTT.2014.2387272

[21] KIM, S., COOK, B., LE, T., et al. Inkjet-printed antennas, sensors and circuits on paper substrate. *IET Microwaves, Antennas and Propagation*, 2013, vol. 7, no. 10, p. 858–868. DOI: 10.1049/iet-map.2012.0685

[22] MORO, R., KIM, S., BOZZI, M., TENTZERIS, M. Inkjet-printed paper-based Substrate Integrated Waveguide (SIW) components and antennas. *International Journal of Microwave and Wireless Technologies*, 2013, vol. 5, no. 3, p. 197–204. DOI: 10.1017/S1759078713000494

[23] MOSCATO, S., MORO, R., PASIAN, M., BOZZI, M., PERREGRINI, L. An innovative manufacturing approach for paper-based substrate integrated waveguide components and antennas. *IET Microwaves, Antennas and Propagation*, 2016, vol. 10, no. 3, p. 256–263. DOI: 10.1049/iet-map.2015.0125

[24] MOSCATO, S., BAHR, R., LE, T., et al. Additive manufacturing of 3D substrate integrated waveguide components. *IET Electronics Letters*, 2015, vol. 51, no. 18, p. 1426–1428. DOI: 10.1049/el.2015.2298

[25] MASSONI, E., SILVESTRI, L., ALAIMO, G., et al. 3D-printed substrate integrated slab waveguide for single-mode bandwidth enhancement. *IEEE Microwave and Wireless Components Letters*, 2017, vol. 27, no. 6, p. 536–538. DOI: 10.1109/LMWC.2017.2701323

[26] MOSCATO, S., DELMONTE, N., SILVESTRI, L., et al. Compact Substrate Integrated Waveguide (SIW) components on paper substrate. In *Proceedings of the 45th European Microwave Conference (EuMC 2015),* Paris (France), 2015, p. 24–27. DOI: 10.1109/EuMC.2015.7345690

[27] TOMASSONI, C., BAHR, R., TENTZERIS, M., BOZZI, M., et al. 3D printed substrate integrated waveguide filters with locally controlled dielectric permittivity. In *Proceedings of the 46th European Microwave Conference* (*EuMC 2016*). London (UK), 2016, p. 253–256. DOI: 10.1109/EuMC.2016.7824326

[28] HERTLEER, C., ROGIER, H., VALLOZZI, L., et al. A textile antenna for off-body communication integrated into protective clothing for firefighters. *IEEE Transactions on Antennas and Propagation*, 2009, vol. 57, no. 4, p. 919–925. DOI: 10.1109/TAP.2009.2014574

[29] MOSCATO, S., PASIAN, M., BOZZI, M., et al. Exploiting 3D printed substrate for microfluidic SIW sensor. In *Proceedings of the 45th European Microwave Conference (EuMC 2015).* Paris (France), 2015, p. 28–31. DOI: 10.1109/EuMC.2015.7345691

[30] MOSCATO, S., BAHR, R., LE, T., PASIAN, M., et al. Infill dependent 3D-printed material based on ninjaflex filament for antenna applications. *IEEE Antennas and Wireless Propagation Letters*, 2016, vol. 15, p. 1506–1509. DOI: 10.1109/LAWP.2016.2516101

About the Authors ...

Cristiano TOMASSONI was born in Spoleto, Italy. He received the Ph.D. degree in Electronics Engineering from the University of Perugia, Perugia, Italy, 1999. In 1999, he was a Visiting Scientist with the Lehrstuhl für Hochfrequenztechnik, Technical University of Munich, Munich, Germany, where he was involved with the modeling of waveguide structures and devices by using the generalized scattering matrix (GSM) technique. In 2001, he was a Guest Professor with the Fakultät für Elektrotechnik und Informationstechnik, Otto-von-Guericke University, Magdeburg, Germany. During that time, he was involved with the modeling of horn antennas having nonseparable cross sections by using hybrid methods combining three different techniques: the finite-element method (FEM), mode-matching technique (MMT), and generalized multipole technique (GMT). During his career, he was also involved in the modeling of low-temperature co-fired ceramics (LTCC) by using the method of moments (MoM). Furthermore he studied new analytical methods to implement boundary conditions in the transmission-line matrix (TLM) method and he modeled aperture antennas covered by dielectric radome by using spherical waves. Since 2007, he has been an Assistant Professor with the University of Perugia, where he currently teaches the 'Aerospace Microwave and Radiofrequency Systems' course. His main area of research concerns the modeling and design of waveguide devices and antennas. His research interests also include the development of reduced-size cavity filters, reconfigurable filters, and printed reconfigurable antenna arrays. Cristiano Tomassoni is a member of the MTT-8 Filters and Passive Components Technical Committee. He was the recipient of the 2012 Microwave Prize presented by the IEEE Microwave Theory and Technique Society (IEEE MTT-S), recognized to the author of the paper, published in any official IEEE publication during the calendar year, that is judged to be the most significant contribution in the field of interest of the MTT Society.

Maurizio BOZZI was born in Voghera, Italy, in 1971. He received the Ph.D. degree in Electronics and Computer Science from the University of Pavia, Italy in 2000. He is currently an Associate Professor of the University of Pavia and a Guest Professor of the Tianjin University, China. He held research positions with various universities worldwide, including the Technische Universität Darmstadt, Germany, the Universitat de Valencia, Spain, and the École Polytechnique de Montréal, Canada. He has authored or co-authored more than 100 journal papers and 270 conference papers, and co-authored the book *Microstrip Lines and Slotlines* (Artech House, 2013). His main research interests concern the development of numerical methods for the electromagnetic modeling and design of microwave and millimeter-wave components with a particular focus on frequency-selective surfaces and substrate integrated waveguides. Prof. Bozzi was the General Chair of the IEEE MTT-S International Microwave Workshop Series-Advanced Materials and Processes (IMWS-AMP 2017), of the IEEE International Conference on Numerical Electromagnetic Modeling and Optimization (NEMO2014), and the General Chair of the IEEE MTT-S International Microwave Workshop Series on Millimeter Wave Integration Technologies (IMWS 2011). Prof. Bozzi is an Elected Member of the Administrative Committee of the IEEE Microwave Theory and Techniques Society (MTT-S) for term 2017–2019. He was the Secretary of IEEE MTT-S for year 2016 and a member of the General Assembly (GA) of the European Microwave Association (EuMA) for the term 2014–2016. Moreover, he is an Associate Editor of the IEEE Microwave and Wireless Components Letters, the IET Microwaves, Antennas and Propagation, and the IET Electronics Letters. He received several awards, including the 2015 Premium Award for the Best Paper in IET Microwaves, Antennas & Propagation, the 2014 Premium Award for the Best Paper in Electronics Letters, the Best Young Scientist Paper Award of the XXVII General Assembly of URSI in 2002, and the MECSA Prize at the Italian Conference on Electromagnetics (RiNEm) in 2000.

Compact Microstrip Antenna with Pattern-Reconfigurable Characteristic

Wei-Dong MA, Guang-Ming WANG, Ya-Wei WANG, Bin-Feng ZONG

Air and Missile Defense College, Air Force Engineering University, Shaanxi Province 710051, China

maweidong607@126.com, wgming01@sina.com

Abstract. *In this paper, a compact microstrip antenna with radiation pattern reconfigurable characteristic employing MEMS switches is proposed and investigated. This novel antenna mainly includes four parts: a circular-shaped patch, six MEMS switches, six fan-shaped coupling cells and six fan-shaped radiation cells. Through controlling the states of MEMS switches, radiation pattern selectivity of the proposed antenna can be achieved. When switching operation states of the antenna from state 1 to state 12 sequentially, the main radiation beam direction in azimuth planes could rotate 30° from 0° to 330° with each switch. Meanwhile, the radiation beam directions in elevation plane and the operation frequencies of the proposed antenna for different operation states are stable at about 35° and 4 GHz, respectively. With the consideration of MEMS switches actual size, the antenna structure and dimensions were optimized and adjusted. The simulated and measured results were analyzed and compared, indicating this compact antenna can be widely used for multifunctional applications and modern wireless communication systems.*

Keywords

Reconfigurable antenna, compact microstrip antenna, radiation pattern reconfigurability, MEMS switch

1. Introduction

Reconfigurable antenna (RA) for the function of changing its electrical parameters dynamically in respond to various environments has been studied extensively [1]. RA can satisfy the diverse requirements and strict standards of advanced wireless communication systems due to its numerous merits such as reducing size, expanding capacity, improving security and lessening interference [2], [3]. According to the alterations of electrical parameters, RAs can be mainly divided into four categories: frequency-reconfigurable antenna (FRA), polarization-reconfigurable antenna (PoRA), pattern-reconfigurable antenna (PaRA), and multiparameter-reconfigurable antenna (MRA) [4].

In recent years, the PaRA is attracting more and more attention as it is very useful for systems which suffer from multipath interference [5]. On account of its pattern selectivity, the PaRA is capable of enhancing system security, improving communication reliability and reducing energy loss to better the comprehensive performance of whole system. The textbook approach to antenna pattern reconfiguration is using phased-array antenna with fast response. However, the phased array antenna usually has a large volume and complex feed-network. Therefore, a lot of substantial research work has been carried out to PaRAs with electrical tuning. Electronic switch components, such as field effect transistors (FETs), micro-electromechanical systems (MEMSs), PIN diodes and varactors, are often used to reconfigure the radiation patterns of antennas [6–9].

Compared with traditional and classical switches, MEMS switch made on a common silicon substrate is tiny enough for integration. Due to its advantageous properties such as high isolation, good linearity, wide band and low loss, RF MEMS switches have been used in reconfigurable antennas for many years [10–12].

Designs resorted MEMS switch to achieve radiation pattern reconfigurability can be found in many literatures [13–15]. A reconfigurable microstrip antenna presented in [13] uses MEMS switches for pattern selectivity realization. The reconfigurable antenna provides three radiation-pattern modes by controlling the states of four MEMS switches. In [14], the reconfigurable rectangular spiral antenna could radiate at different directions because that the four MEMS switches can switch the connection or disconnect of five discrete arms. However, each resonant state can only cover a narrow range. The antenna shown in [15] can realize the operation frequency, radiation pattern and polarization mode reconfiguration simultaneously by switching MEMS switches. And the impacts of bias circuit network and feed-network on the efficiency of antenna are important to the design and need to be taken into account. In these designs, MEMS switches are selected to reconfigure the radiation patterns of the antennas mainly because that they can provide superior performance than active switches such as PIN diodes, and be integrated in these designs conveniently.

In this paper, a compact microstrip antenna with radiation pattern reconfigurability using MEMS switches is proposed and investigated. Six MEMS switches are used to respectively connect or disconnect the fan-shaped coupling cells and fan-shaped radiations cells of the proposed antenna, changing the current distribution and achieving radiation pattern diversity. Twelve basic operation states (state 1 to state 12) are obtained for the proposed antenna. Sequentially switching operation states from state 1 to state 12 can rotate the main radiation beam by 30° from 0° to 330° in azimuth plane, maintaining 35° in elevation plane. Moreover, this novel PaRA can operate around 4 GHz stably with all operation states, indicating that it is a good candidate for advanced communication systems of, for example, satellite C band (3.7 GHz ~ 4.2 GHz). Moreover, the proposed PaRA possesses of many merits such as low cost, compact structure and stable radiation performance.

2. Antenna Design and Analysis

2.1 Antenna Configuration

The top and side views of the proposed PaRA are illustrated in Fig. 1 detailedly. This novel PaRA designed on a FR4 substrate (dielectric constant $\varepsilon_r = 4.3$, loss tangent $\tan\delta = 0.02$ and thickness $H = 3$ mm), mainly consists of four parts. The first (red area) part is a circular-shaped patch located at the center of the proposed antenna for connecting a 50-Ω coaxial cable. The second (blue area) and third (yellow area) parts are six fan-shaped coupling cells and six fan-shaped radiation cells respectively, which distribute around the antenna center symmetrically. Six RF switches, the fourth part, are used to connect or disconnect the coupling cells and radiation cells, hence change the radiation structure. When the switch shifts to ON-state, the radiation cell connected to the corresponding coupling cell will be activated. Contrarily, the radiation cell will not work when the switch shifts to OFF-state. With controlling the states of switches, the proposed PaRA can operate at different states and achieve radiation pattern reconfigurable characteristic.

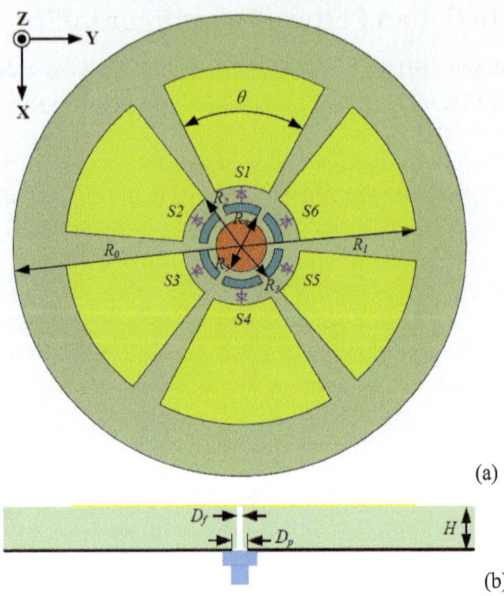

Fig. 1. Geometry of the proposed PaRA: (a) top, (b) side view.

2.2 MEMS Switch and Bias Circuit

Six Radant MEMS RMSW101 switches (S1 ~ S6) with advantageous properties of small size, high isolation, low loss and high linearity, are selected as electrical components to reconfigure the radiation pattern of the proposed PaRA. For MEMS switch actuation, 90 V DC voltage between the source and gate is required, which will pull the free end of the cantilever into contact with the drain. The data book of the used type of MEMS switch shows that at the operation frequency 4 GHz, the insertion loss is about 0.26 dB in the ON-state which brings in RF continuity. Also the isolation is 21 dB in the OFF-state which results in RF discontinuity.

The bias circuit network for MEMS switch actuation is designed according to the data book of MEMS switch. The source and gate of MEMS switch are respectively connected the pads in the bias line for DC actuation voltage loading. Moreover, the source and drain of the MEMS switch is required to connect to the DC ground plane. For circuit simplicity, the RF ground plane is used for the DC ground plane. And the DC continuity between antenna and ground can be achieved with stub line through via structure. Besides, DC biasing lines are properly designed to reduce the effect of bias current to radiation characteristic and matching characteristic of antenna. 100 kΩ SMD (Surface Mounted Devices) resistances and 270 nH SMD inductances are applied in biasing lines to reduce the loss as well as the influence on antenna performance from biasing lines. In order to avoid the radiators are blocked by biasing lines in measurement and performance degradation of the antenna, the biasing lines are brought from the ground by via structure to connect the DC source.

2.3 Parameters Analysis

The parameters of the antenna determine its performance. The exradius R_1 of fan-shaped radiation cell is a major influence upon the operation frequency of the proposed antenna. Figure 2 plots the simulated operation frequency versus R_1. It displays that the operation frequency gradually decreases from 4.27 GHz to 3.76 GHz as R_1 increases from 19 mm to 22 mm with step increment of 1 mm. $R_1 = 20.6$ mm closed to a half guided-wave wavelength is selected as the appropriate parameter for that the

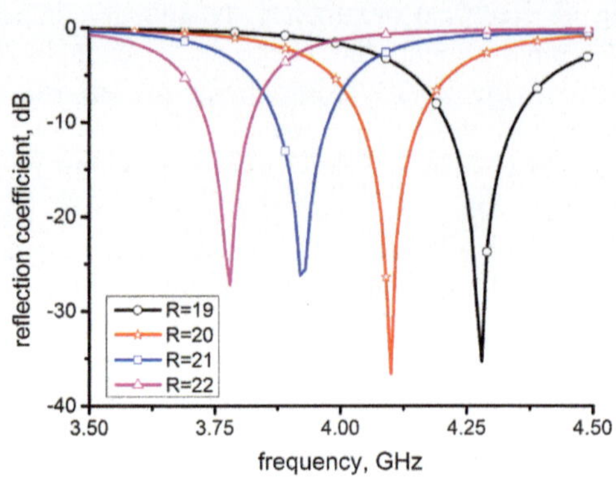

Fig. 2. Simulated operation frequency versus R_1.

Parameters	R_0	R_1	R_2	R_3	R_4
Dimensions	62 mm	20.5 mm	5 mm	4 mm	3.5 mm
Parameters	R_5	D_f	D_p	H	θ
Dimensions	3 mm	1.4 mm	4.1 mm	3 mm	50°

Tab. 1. Dimensions of the proposed PaRA.

novel PaRA could operate at 4 GHz exactly. Meanwhile, $\theta = 50°$ is obtained for good radiation and low-coupling. With consideration of the actual size of MEMS switch, the gap ($R_2 - R_3$) between fan-shaped coupling cell and fan-shaped radiation cell is chosen as 1 mm for easy-fabrication. Furthermore, the radius R_5 of circular-shaped patch mainly affects the antenna's efficiency. And simulations show the efficiency arrive at its maximum when $R_5 = 3$ mm. In order to get good impedance match and high gain, the exradius R_3 = 4 mm and inradius R_4 = 3.5 mm of fan-shaped coupling cell are acquired respectively. Lastly, the radius R_0 = 62 mm and height H = 3 mm of the entire antenna is the optimized results for compact structure and low-profile ultimately.

The proposed PaRA is designed and analyzed with a popular EM simulated tool HFSS.14. The detailed antenna dimensions after optimization are listed in Tab. 1.

2.4 Operation Principles

With controlling the states of switches, different fan-shaped radiation cell of the antenna can be fed selectively by the circular-patch through corresponding coupling cell. As a result, the surface current distribution of the antenna will be changed, realizing the deflection of the radiation direction of the antenna. Moreover, the diverse operation states of the proposed PaRA corresponding to different states of six MEMS switches are obtained. The amply simulated information of different operation states are summarized in Tab. 2. It shows that the proposed antenna has 12 basic operation states (one or two-adjacent ON-state MEMS switches), accordingly radiating at 12 different directions (from 0° to 330° with step increment of 30°) in azimuth planes and maintaining 35° in elevation plane. Meanwhile, the operation frequencies of the proposed

Operation State	MEMS Switches*						Resonance Frequency (GHz)	Beam Direction (°)	Gain (dBi)
	S1	*S2*	*S3*	*S4*	*S5*	*S6*			
state1	1	0	0	0	0	0	3.94 ~ 4.05	**0**	5.16
state2	1	1	0	0	0	0	3.91 ~ 4.03	**30**	5.75
state3	0	1	0	0	0	0	3.95 ~ 4.05	**60**	5.16
state4	0	1	1	0	0	0	3.91 ~ 4.03	**90**	5.75
state5	0	0	1	0	0	0	3.95 ~ 4.05	**120**	5.16
state6	0	0	1	1	0	0	3.91 ~ 4.03	**150**	5.75
state7	0	0	0	1	0	0	3.95 ~ 4.05	**180**	5.16
state8	0	0	0	1	1	0	3.91 ~ 4.03	**210**	5.75
state9	0	0	0	0	1	0	3.95 ~ 4.05	**240**	5.16
state10	0	0	0	0	1	1	3.91 ~ 4.03	**270**	5.75
state11	0	0	0	0	0	1	3.95 ~ 4.05	**300**	5.16
state12	1	0	0	0	0	1	3.91 ~ 4.03	**330**	5.75

* 0 means ON-state and 1 means OFF-state of MEMS switch

Tab. 2. The operation states of the antenna.

antenna with different operation states all come around 4 GHz and the 10-dB impedance bandwidths fluctuated at the range of 3.91 GHz to 4.05 GHz slightly. Similarly, the gains, as well as other performances of the proposed PaRA with different operation states are almost same with each other.

Note that other operation states (two-nonadjacent or more ON-state MEMS switches) of the proposed antenna almost have the similar radiation patterns with the 12 basic operation states (one or two-adjacent ON-state MEMS switches) in Tab. 2, respectively. However, the antenna performances of other operation states decrease observably though its operation frequencies change a little. After removing the redundant and invalid operation states with optimization and analysis, there are 12 basic operation states as shown in Tab. 2 are obtained. And in the following sections, these 12 basic operation states are mainly discussed, measured and summarized.

3. Experimental Results of the PaRA

A prototype shown in Fig. 3 was fabricated with commercial MEMS switches according to the optimized dimensions in Tab. 1. The experimental results of the proposed PaRA of 12 basic operation states are displayed as follows. Due to its symmetrical configuration, return losses of state 1, state 3, state 5, state 7, state 9 and state 11 are identical. Similarly, so do the return losses of state 2, state 4, state 6, state 8, state 10 and state 12.

For briefness, Figure 4 shows the simulated and measured $|S_{11}|$ in dB of state 1 and state 2, respectively. Others can be obtained in the light of symmetry. It can be found that the simulated 10-dB impedance bandwidth (3.95 GHz ~ 4.05 GHz) of state 1 is approximately the same with the bandwidth (3.91 GHz ~ 4.03 GHz) of state 2. Moreover, the measured and simulated results agree very well.

3.1 Radiation Pattern Reconfigurability

When different MEMS switch shifting to ON-state, different fan-shaped radiation cell will be activated, making antenna radiation pattern diversity into realization. The simulated and measured 2-D radiation patterns of the proposed PaRA in azimuth plane for 12 basic operation states

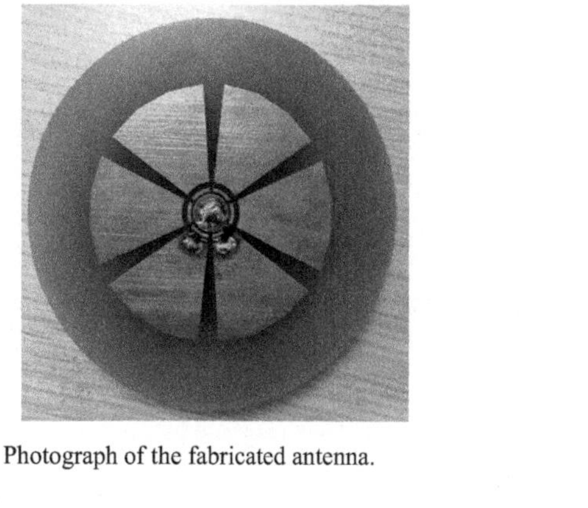

Fig. 3. Photograph of the fabricated antenna.

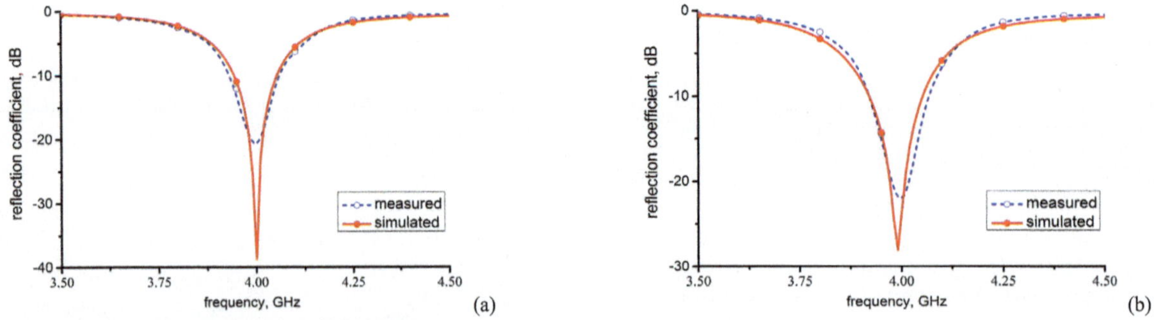

Fig. 4. Simulated and measured $|S_{11}|$ in dB of the proposed antenna (a) state 1, (b) state 2.

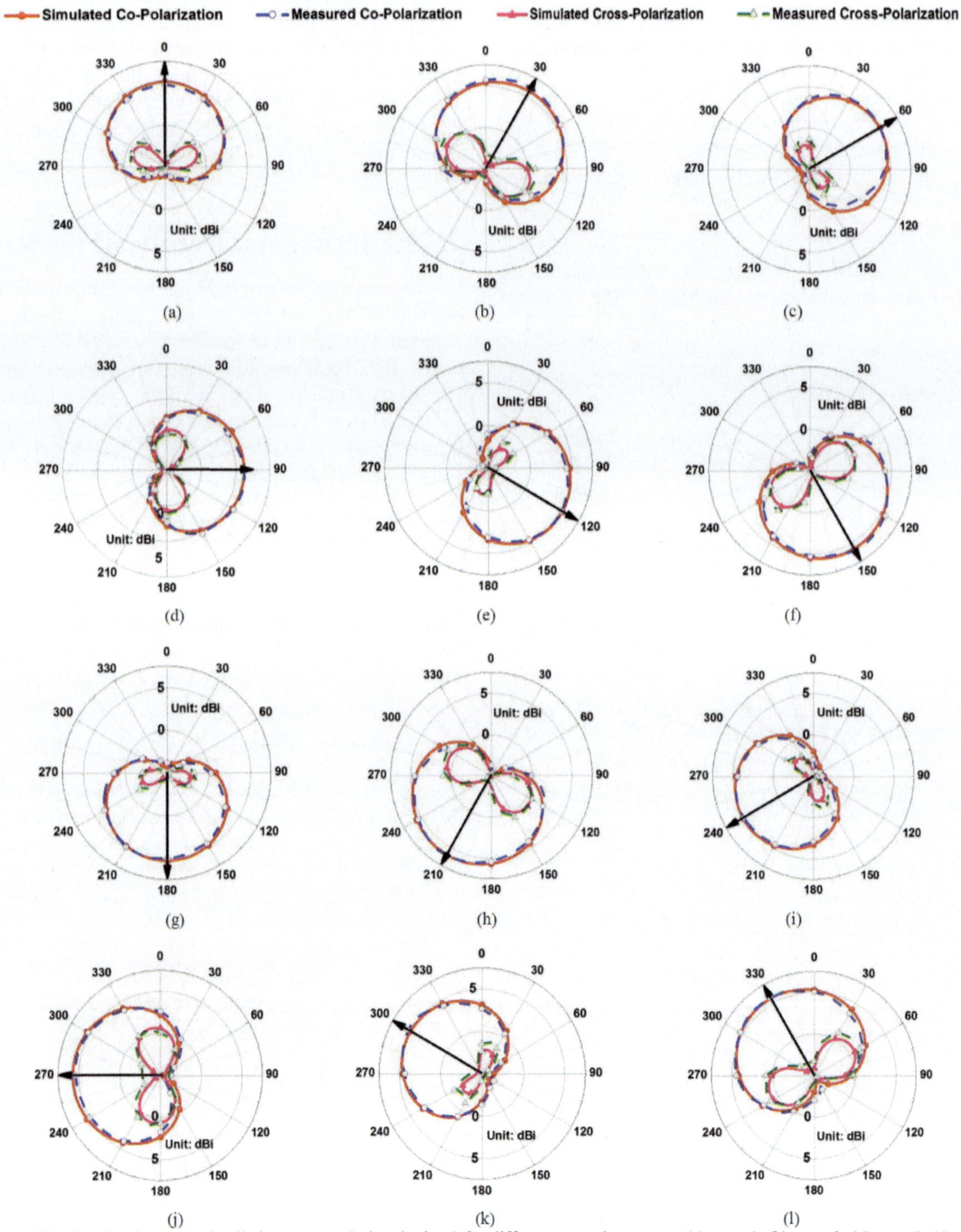

Fig. 5. Simulated and measured radiation patterns (azimuth plane) for different operation states : (a) state 1, (b) state 2, (c) state 3, (d) state 4, (e) state 5, (f) state 6, (g) state 7, (h) state 8, (i) state 9, (j) state 10, (k) state 11, (l) state 12.

are shown in Fig. 5 in detail. From state 1 to state 12, the main beam direction of the proposed PaRA varies from 0° to 330° with step increment of 30°. It indicates that this novel design can cover the whole azimuth plane. It also manifests that each operation state provides corresponding main-beam direction and demonstrates the radiation pattern reconfigurable characteristic of the proposed PaRA.

3.2 Antenna Performance

For the 12 basic operating states of the proposed PaRA, its performances keep stable except the main beam direction in azimuth plane varies. Experimental results from measurement and simulation as follow demonstrate the performances of the proposed PaRA.

Figure 6 displays the measured 2-D radiation patterns of the proposed PaRA from state 1 to state 4 for briefness in elevation plane. For other operation states, similar results had been obtained. It can be observed from Fig. 6 that the radiation patterns in elevation are quite similar and the main beam directions all point at about 35°. Slight shifts appeared as a result of the fabricated errors and influences from MEMS switches and their biasing circuit network. The cross-polarizations of the radiation patterns in elevation plane are too small to be shown and so omitted.

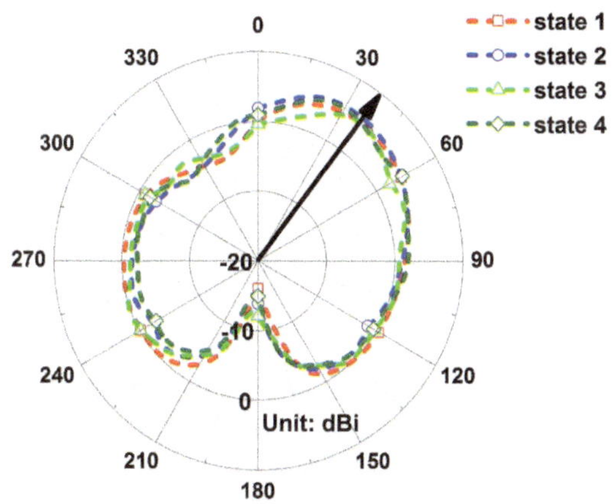

Fig. 6. Measured 2-D radiation patterns in elevation plane.

Operation State	state1	state2	state3	state4
Gain (dBi)	5.02	5.60	5.11	5.54
Efficiency	85.7%	84.7%	86.1%	85.3%
Operation State	state5	state6	state7	state8
Gain (dBi)	5.05	5.63	5.03	5.69
Efficiency	85.9%	86.6%	85.3%	87.1%
Operation State	state9	state10	state11	state12
Gain (dBi)	5.14	5.61	5.09	5.55
Efficiency	86.9%	85.4%	85.6%	84.9%

Tab. 3. Measured gains and efficiencies of the proposed PaRA.

The measured realized gains and efficiencies of the proposed PaRA with 12 different operation states are summarized in Tab. 3. It can be discovered that the realized gains for different operating states are all more than 5 dBi and efficiencies for different operating states are mostly up to 85%. Moreover, the experiment results of measurement and simulation are compared in detail. The results not only showed a good agreement but also demonstrated the stable and good performances of the proposed PaRA.

4. Conclusion

A reconfigurable microstrip antenna with radiation pattern selectivity using MEMS switches is designed and investigated. The novel PaRA has compact structure with circle patch, radiation cells, coupling cells and MEMS switches. With controlling the states of MEMS switches, the proposed PaRA can operate at 12 basic different states and achieve radiation pattern reconfigurable characteristic. When the operation states of the proposed PaRA steer sequentially, the main radiation beam direction can rotate by 30° from 0° to 330° in azimuth plane. Meanwhile, the main beam directions in elevation plane and the operation frequencies of the PaRA with different states undulate slightly at 35° and 4 GHz, respectively. Moreover, the realized gains, efficiencies as well as other performances of the PaRA are all very nice and stable. Besides, the novel PaRA possesses of many merits, such as low cost, compact structure and stable radiation performance, which indicate that it is well-suitable for multifunctional applications and modern communication systems.

Acknowledgements

This work is supported by the National Natural Science Foundation of China under Grant Nos. 61372034.

References

[1] HAUPT, R. L., LANAGAN, M. Reconfigurable antennas. *IEEE Antennas and Propagation Magazine,* 2013, vol. 55, no. 1, p. 49–61. DOI: 10.1109/MAP.2013.6474484

[2] COSTANTINE, J., TAWK, Y., BARBIN, S. E., CHRISTODOULOU, C. G. Reconfigurable antennas: Design and applications. *Proceedings of IEEE,* 2015, vol. 103, no. 9, p. 425 to 437. DOI: 10.1109/JPROC.2015.2396000

[3] CHENG, Y.-J. Substrate integrated waveguide frequency-agile slot antenna and its multibeam application. *Progress In Electromagnetics Research, PIER,* 2012, vol. 130, p. 153–168. DOI: 10.2528/PIER12061602

[4] RODRIGO, D., CETINER, B. A., JOFRE, L. Frequency, radiation pattern and polarization reconfigurable antenna using a parasitic pixel layer. *IEEE Transaction on Antennas and Propagation,* 2014, vol. 62, no. 1, p. 3422–3427. DOI: 10.1109/TAP.2014.2314464

[5] NGUYEN V.-A., JEONG. M.-H., DAO. M.-T, PARK. S.-O. Four-port beam reconfigurable antenna array for pattern diversity systems. *IET Microwaves, Antennas and Propagation*, 2015, vol. 6, no. 10, p. 1179–1186. DOI: 10.1049/iet-map.2011.0606

[6] XIAO-LIN YANG, JIAN-CHENG LIN, GANG CHEN, FANG-LING KONG. Frequency reconfigurable antenna for wireless communications using GaAs FET switches. *IEEE Antennas and Wireless propagation Letters,* 2015, vol. 14, p. 807–810. DOI: 10.1109/LAWP.2014.2380436

[7] KAGAN TOPALLI, CIVI, O. A., DEMIR, S., et al. Dual-frequency reconfigurable slot dipole array with a CPW based feed network using RF MEMS technology for X- and Ka- band applications. In *IEEE Antennas and Propagation Society International Symposium.* Honolulu (HI, USA), 2007, p. 825–828. DOI: 10.1109/APS.2007.4395621

[8] LIN, X., YANG, X., KONG, F. A frequency monopole antenna with switchable stubbed ground structure. *Radioengineering*, 2015, vol. 24, no. 2, p. 449–454. DOI: 10.13164/re.2015.0449

[9] BEHEDAD, N., SARABANDI, K. A varactor-tuned dual-band slot antenna. *IEEE Transaction on Antennas and Propagation,* 2006, vol. 54, no. 2, p. 401–408. DOI: 10.1109/TAP.2005.863373

[10] YASHCHYSHYN, Y. Reconfigurable antennas by RF switches technology. In *5th International Conference on Perspective Technology and Methods in MEMS Design (MEMSTECH 2009).* Polyana-Svalyava (Ukraine), 2009, p. 155–157.

[11] ZOHUR, A., MOPIDEVI, H., RODRIGO, D., et al. RF MEMS reconfigurable two-band antenna. *IEEE Antennas of Wireless Propagation Letters,* 2013, vol. 12, p. 73–75. DOI: 10.1109/LAWP.2013.2238882

[12] GRAU, A., ROMEU, J., LEE, M.-J., et al. A dual-linearly-polarized MEMS-reconfigurable antenna for narrowband MIMO communication systems. *IEEE Transaction on Antennas and Propagation,* 2010, vol. 58, no. 1, p. 4–17. DOI: 10.1109/TAP.2009.2036197

[13] ZHANG, S., HUFF, G. H., FENG, J., et al. A pattern reconfigurable micro-strip parasitic array. *IEEE Transaction on Antennas and Propagation,* 2004, vol. 52, no. 10, p. 2773–2776. DOI: 10.1109/TAP.2004.834372

[14] JUNG, C. W., LEE, M., LI, G. P., et al. Reconfigurable scan-beam single-arm spiral antenna integrated with RF-MEMS switches. *IEEE Transaction on Antennas and Propagation,* 2006, vol. 54, no. 2, p. 455–463. DOI: 10.1109/TAP.2005.863407

[15] GRAU BESOLI, A., DE FLAVIIS, F. A multifunctional reconfigurable pixeled antenna using MEMS technology on printed circuit board. *IEEE Transaction on Antennas and Propagation,* 2011, vol. 59, no. 12, p. 4413–4424. DOI: 10.1109/TAP.2011.2165470

About the Authors ...

Wei-Dong MA was born in Shaanxi province, People's Republic of China. He received his B.S. from the China University of Geoscience, Wuhan, China, in 2012, and M.S. degrees from the Air Force Engineering University, Xi'an, China, in 2014. His research interests include smart antenna, reconfigurable antenna, composite right/left-handed (CRLH) transmission lines, and metamaterial-based antennas.

Guang-Ming WANG was born in Anhui province, People's Republic of China. He received his B.S. and M.S. degrees from the Missile Institute, Air Force Engineering University, Xi'an, China, in 1982 and 1990, respectively, and his Ph.D. degree from the Electronic Science and Technology University, Chengdu, China, in 1994. Then, he joined the Air Force Engineering University as a professor in 2000 and is now the head of the Microwave Laboratory Center in it. His current interest includes microwave circuits, antenna and propagation, and also the new structures include EBG, PBG, smart antenna, metamaterials and fractals, etc.

Ya-Wei WANG was born in Anhui province, People's Republic of China. He received his B.S. and M.S. degrees from the Air Force Engineering University, Xi'an, China, in 2008 and 2011, respectively. His research interests are mainly spiral antennas, metamaterial-based antennas, etc.

Bin-Feng ZONG was born in Jiangsu province, People's Republic of China. He received his B.S. and M.S. degrees from the Air Force Engineering University, Xi'an, China, in 2010 and 2012, respectively. His research interests include microstrip resonators, composite right/left-handed (CRLH) transmission lines and metamaterial-based antennas, etc.

14

Low-Cost and High-Gain SIW Circularly Polarized Circular-Horn-Loaded Antenna for Broadband Millimeter-Wave Applications

Ming DU [1], Jun XU [1], Yuliang DONG [2], Xiao DING [1]

[1] School of Physical Electronics, University of Electronic Science and Technology of China, Chengdu, China
[2] National Institute of Measurement and Testing Technology of China, Chengdu, China

mingdu_uestc@163.com, xujun@uestc.edu.cn, dongyull@163.com, xding@uestc.edu.cn

Abstract. *A wideband, low-cost and high-gain circularly polarized (CP) circular-horn-loaded antenna based on substrate integrated waveguide (SIW) technology operating at Ka band is presented. The proposed antenna, which is built on a single-layer substrate, consists of five parts: a short-ended SIW, a center-symmetric wide slot, an L-shaped probe, a circular horn and a transition from SIW to air-filled rectangular waveguide for measurement. The slot is etched on the upper ground of the SIW, while the L-shaped probe for generating CP wave is printed inside the slot and connected to the SIW. A circular horn is also loaded on the surface of the SIW slot for high gain. Then, the proposed antenna with a dimension of $45 \times 45 \times 24.16$ mm^3 was fabricated and measured. The measured results show that the antenna has a wide impedance bandwidth of 28.6% from 30 to 40 GHz for $|S_{11}| \leq 10$ dB and a wide axial ratio (AR) bandwidth of 22.2% from 32 to 40 GHz for AR ≤ 3 dB. The measured maximum gain is 15.6 dBi at 36 GHz with slight fluctuations over the 30–40-GHz frequency range.*

Keywords

Substrate integrated waveguide (SIW), circularly polarized (CP), horn, low cost, high gain, broadband millimeter-wave (mmW) antenna

1. Introduction

For the merit of low loss [1], the substrate integrated waveguide (SIW) technology has received much attention in millimeter-wave (mmW) wireless systems [2], such as front-end systems [3], [4], filters [5], antennas [6], [7], etc. As the indispensable components of mmW wireless systems, antennas play an important role in information transmission. Meanwhile, there has been an increasing interest in the investigation of circularly polarized (CP) antennas because of their advantages of suppressing the multi-path interference and reducing the polarization mismatch [8]. Therefore, over past twenty years, many kinds of CP antennas based on SIW technology have been investigated [9–13].

The SIW slot covered by a rotated strip or a CP patch may be one common method for generating CP radiation [9], [10]. For example, a low-temperature co-fired ceramic integrated SIW slot antenna covered by a rotated strip was proposed in [9]. This antenna achieves a 3-dB axial ratio (AR) bandwidth of 11.6%. [10] presents a SIW slot array antenna covered by corner-truncated CP array patches. The measured –15-dB $|S_{11}|$ and 3-dB AR bandwidths of this antenna are about 6.4 % and 2.86 %, respectively. This method can achieve good CP performance. However, the multi-layer technology is always needed. The multi-layer technology merits high degree of freedom of designing and beam forming capability, but the cost is high.

The multiple feeding is also a good method for generating CP radiation, and this kind of antenna always shows wide bandwidth. In [11], a differential-fed CP planar aperture antenna based on SIW technology operating at 60 GHz was proposed. There are two feeding ports in one antenna element. The proposed four-element array could achieve wide 3-dB AR bandwidth of 16.7% (56 to 66.2 GHz) within its –10-dB $|S_{11}|$ bandwidth and stable gain throughout the operating bandwidth with a maximum of 14.6 dBi. The differential feeding is useful in differential mmW monolithic integrated circuits and can achieve stable and symmetric radiation patterns. Nevertheless, for other applications such as phased arrays, the feeding network is complex.

Single-fed CP SIW antennas are always research focus because of low cost, simple structure and easy fabrication. A CP circular ring-slot antenna embedded in single-layered SIW operating at X band was proposed in [12]. The measured results exhibit that the antenna has a broadband impedance bandwidth of 18.74% for VSWR ≤ 2 and a 3-dB AR of 2.3%. In [13], the SIW cavity-backed CP antennas using two different feeding transitions were presented. The measured results prove that broadband impedance bandwidth of 17.32% and 14.42% under the criteria of less than VSWR 2:1 and 1.5:1, respectively, were obtained and a 3-dB AR of 2.34% with a maximum gain of 7.79 dBic was accomplished by using the proposed antenna

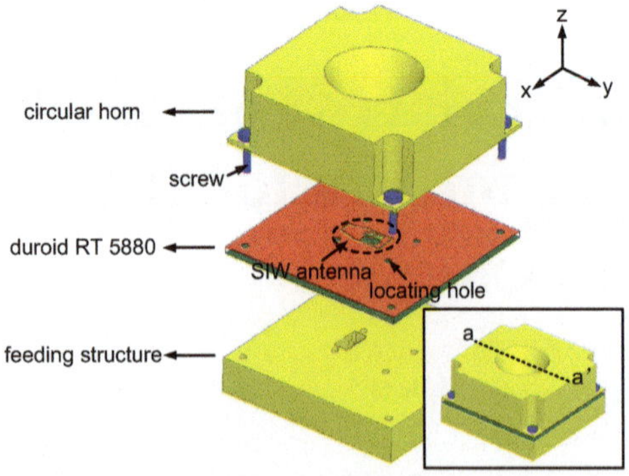

Fig. 1. Proposed antenna configuration.

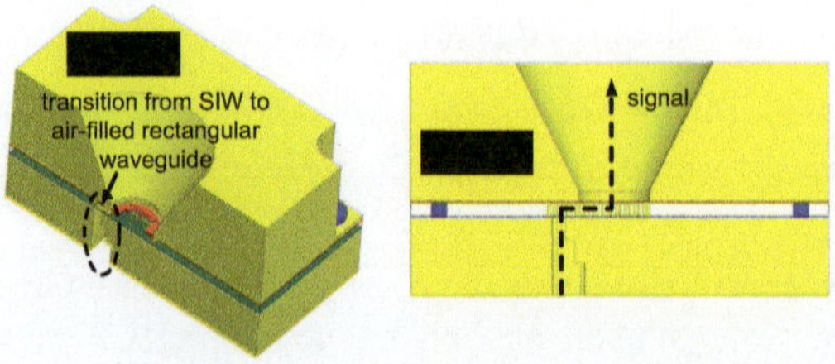

Fig. 2. 3D and side views of the a-a' profile of the proposed antenna.

with coax-to-SIW transition operating at the X band of 10 GHz center frequency. Both the two antennas are single-fed, low-cost and can achieve good CP radiation pattern, however the 3-dB AR bandwidth is relatively narrow.

Therefore, the design of a low-cost single-fed SIW CP antenna with both wide impedance and AR bandwidths is not an easy task.

In addition, high gain is also necessary for mmW applications because of the high transmission and propagation losses at mmW. Forming an array may be the most common method to achieve high gain. This method merits easy integration with other planar circuits. However, for parallel-fed arrays, because of the large width of the SIW, the radiating elements and feeding network of this kind of array are not easy to be built on a single substrate at the same time in a simple and tighten way, especially when the dielectric permittivity of the substrate is small, the radiating element needs multiple feeding and a square array with large number of elements is needed. As a result, the design of the feeding network is not so easy, the feeding network is complex which will cause extra loss and limitation of the number of elements of a square array. The series-fed array can solve this problem, but the bandwidth is usually narrow [10].

To address the problems above, this paper presents a low-cost single-fed SIW CP antenna built on single-layer substrate with both wide impedance and AR bandwidths. A circular horn is also loaded on the surface of the proposed SIW CP antenna to forming a CP horn antenna for high gain. This kind of antenna will decrease the chance of a dense integration but it needs no feeding network. Therefore, the structure is simple and the loss caused by the feeding network is low. And compared with the multi-layer technology, the cost of the extra metal horn is low.

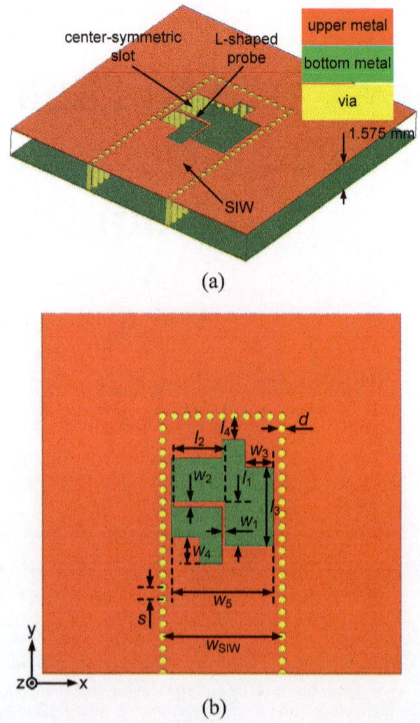

Fig. 3. (a) 3D and (b) top views of the proposed SIW antenna.

Owing to the above advantages, the proposed antenna is suitable for low-cost, high-gain and broadband mmW applications.

2. Antenna Design

2.1 Total Antenna Configuration

The proposed antenna configuration is shown in Fig. 1. It consists of a duroid RT 5880 substrate, a metal circular horn, and a metal feeding structure for measurement. The substrate used is duroid RT 5880 with dielectric permittivity $\varepsilon_r = 2.2$, loss tangent $\tan\delta = 0.0009$ and thickness of 1.575 mm. For seeing the proposed antenna more clearly, the 3D and side views of the a-a' profile of the proposed antenna are also plotted in Fig. 2.

2.2 SIW Antenna

As the SIW antenna is the excitation of the circular horn, we start our design from the SIW antenna. Figure 3 shows the 3D and top views of the SIW antenna. As shown, a center-symmetric wide slot is etched on the upper ground of a short-ended SIW. In order to generate CP wave, an L-shaped probe is printed inside the slot and connected to the SIW. The detailed dimensions are: $d = 0.3$ mm, $s = 0.5$ mm, $w_{SIW} = 5$ mm, $w_1 = 0.1$ mm, $w_2 = 0.2$ mm, $w_3 = 1.2$ mm, $w_4 = 1.1$ mm, $w_5 = 4.3$ mm, $l_1 = 1.81$ mm, $l_2 = 2.14$ mm, $l_3 = 3.25$ mm and $l_4 = 0.975$ mm. The evolution of the proposed antenna is depicted in Fig. 4 to explain the CP performance of the SIW antenna. All the three antennas have been discussed here, .i.e., antenna 1 (ant 1), antenna 2 (ant 2), and antenna 3 (ant 3). Ant 1 is the SIW

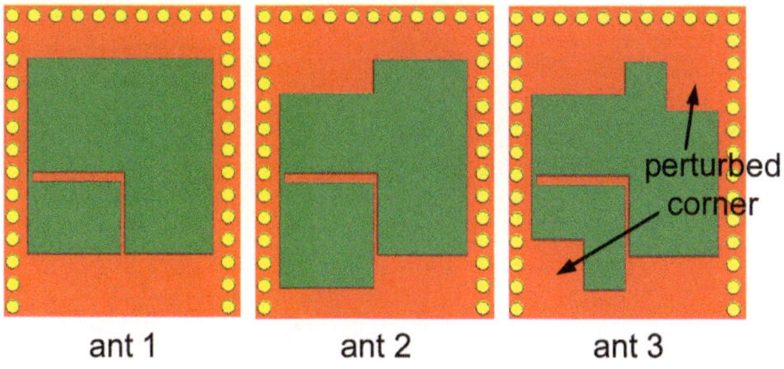

Fig. 4. Evolution of the SIW antenna.

antenna with a rectangular slot. Ant 3 is the proposed SIW antenna with a corner-perturbed center-symmetric slot. And ant 2 is the SIW antenna with a basic center-symmetric slot.

The effects of the each step of all three antennas are described in Fig. 5. The commercial electromagnetic software HFSS was used to obtain all the simulated results. It can be seen that ant 1 has the worse $|S_{11}|$. Ant 3 has a wide impedance bandwidth of 23.8% from 31.5 to 40 GHz for $|S_{11}| \leq -10$ dB. The impedance bandwidth of ant 2 is almost the same with that of ant 3 but shifts to lower frequencies. The AR is improved gradually from ants 1 to 3. The AR bandwidth of ant 1 with the highest center frequency is the narrowest. The 3-dB AR bandwidth of ant 3 is 21.3% from 31.5 to 39 GHz.

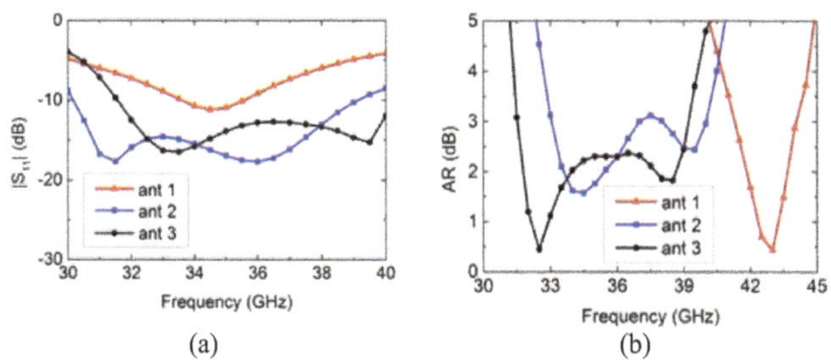

Fig. 5. Simulated (a) $|S_{11}|$s and (b) ARs for antennas 1-3.

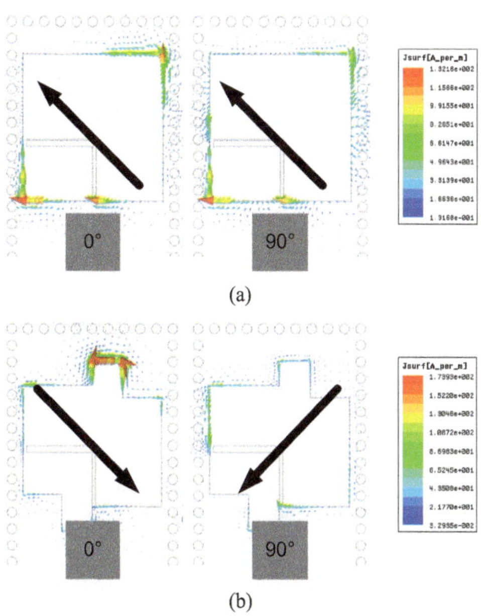

Fig. 6. Simulated currents on the surface of the SIW slot of (a) ant 1 and (b) ant 3 at 35 GHz and different phases.

The shape of the slot is the key for achieving good CP performance. By shaping the slot into central symmetry with perturbed corner, the currents on the SIW slot will be improved because of the rotating edges. For illustrating this point clearly, the simulated currents on the surface of the SIW slot of ants 1 and 3 at 35 GHz and different phases are plotted in Fig. 6. It can be seen that the currents of ant 1 are in the same direction at 0° and 90°, therefore the SIW slot of ant 1 generates the linearly polarized wave. On the contrary, the currents of ant 3 rotate clockwise from 0° to 90° which results in left-handed CP (LHCP) radiation. Right-handed CP (RHCP) radiation can be obtained by adopting a mirror image of the presented configuration. Therefore, the LHCP will be the co-polarization, while the RHCP will be the cross-polarization.

CP is generated by two orthogonal E vectors with equal amplitude and 90° phase difference. The two orthogonal E vectors are generated by the horizontal and vertical parts of the L-shaped probe in the proposed SIW antenna. The rectangular SIW slot is influenced by the probe differently, so the SIW slot cannot generate CP wave. By shaping the slot into central symmetry with perturbed corner, the SIW slot will be influenced almost equally. Therefore, the SIW slot can generate CP wave.

The simulated E-field distributions on the surface of the SIW slot of ant 3 at 35 GHz and different phases are also plotted in Fig. 7. The directions of the E-field rotate clockwise from 0°to 270° which also demonstrates that the LHCP will be the co-polarization, while the RHCP will be the cross-polarization.

The simulated radiation patterns of ant 3 at 35 GHz are shown in Fig. 8. It can be seen that good broadside radiation is achieved. The LHCP components in both planes are the co-polarization while the RHCP components

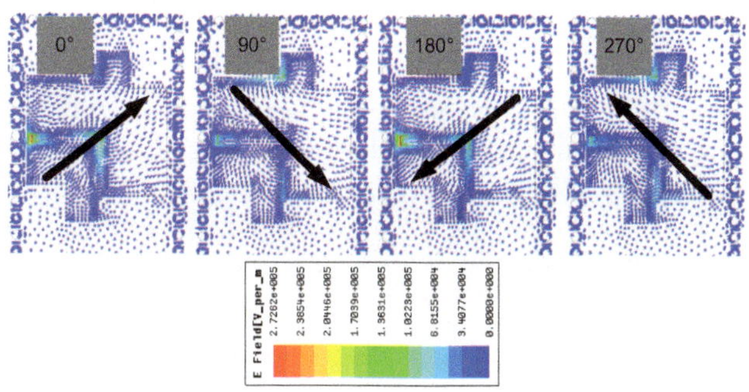

Fig. 7. Simulated E-field distributions on the surface of the SIW slot of ant 3 at 35 GHz and different phases.

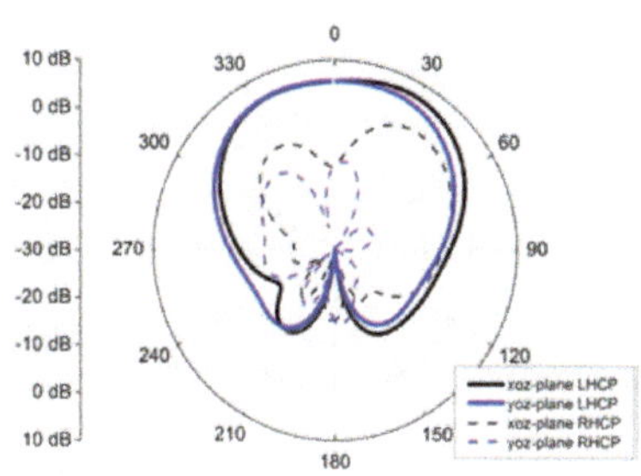

Fig. 8. Simulated radiation patterns of ant 3 at 35 GHz.

in both planes are the cross-polarization. The gain of ant 3 is 5.55 dBi. Meanwhile, the proposed SIW antenna merits good front-to-back ratio (FBR) 35.55 dB.

2.3 Feeding Structure

The transition from SIW to air-filled rectangular waveguide proposed in [14] is used to feed the SIW antenna. The configuration of the transition with detail dimensions is shown in Fig. 9. The simulated $|S_{11}|$ and $|S_{12}|$ of the transition is shown in Fig .10. It can be seen that the simulated return loss and insertion loss of the back-to-back transition are better than 10 and 0.5 dB from 30 to 40 GHz.

2.4 Circular-Horn-Loaded SIW Antenna

In order to obtain high gain, a circular horn is loaded on the surface of the SIW antenna to forming a CP horn antenna.

Traditionally, the horn antenna is LP and it can be turned into a CP antenna by using a circular-wave polarizer [15], [16]. However, this type of CP horn antenna has large size. A dual CP rectangular horn antenna with a chiral metamaterial was presented in [17]. The antenna achieves LHCP performance in the frequency range of 12.4 GHz to 12.5 GHz, and RHCP performance in the range of 14.2 GHz to 14.4 GHz. The circular polarization also can

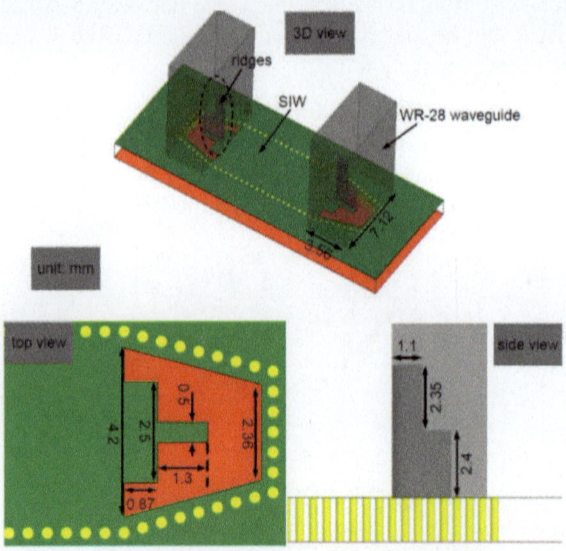

Fig. 9. Configuration of the back-to-back transition from SIW to air-filled rectangular waveguide with detail dimensions.

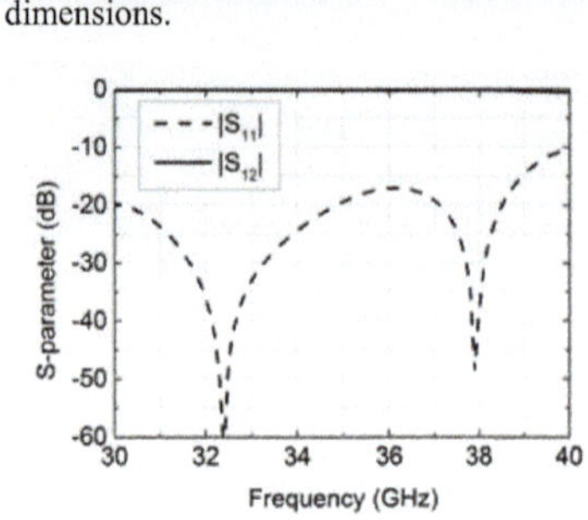

Fig. 10. Simulated $|S_{11}|$ and $|S_{12}|$ of the back-to-back transition.

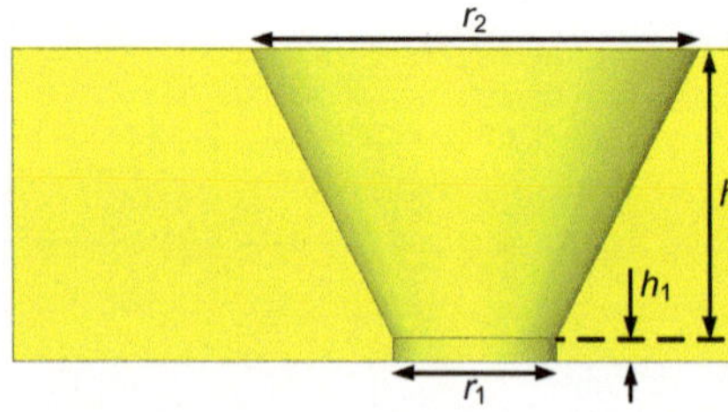

Fig. 11. Configuration of the circular horn.

be obtained by truncating the horn properly [18]. A wideband CP circular horn antenna fed by an L-probe was proposed in [19]. It is a good method to achieve circular polarization owing to the simple structure and low cost.

The CP horn antenna can be obtained using all the above methods, but all of them are fed by a rectangular waveguide or a coaxial cable. Therefore, extra transition is needed to connect the antenna to the planar circuit which will introduce extra loss. The CP horn antenna proposed in this paper is fed by the SIW antenna directly which can solve this problem.

Figure 11 shows the configuration of the circular horn. The detail dimensions are: r_1 = 0.78 mm, r_2 = 21.42 mm, h_1 = 1.12 mm and h_2 = 13.4 mm.

In order to achieve the best performance, some values of the parameters of the SIW antenna are tuned slightly. The tuned parameters are (original value to present value): w_2 = 0.2 to 0.1 mm, w_3 = 1.2 to 0.82 mm, w_4 = 1.1 to 0.75 mm, l_1 = 1.81 to 1.75 mm, l_2 = 2.14 to 2.1 mm, l_3 = 3.25 to 3.69 mm and l_4 = 0.975 to 1 mm. The values of the other parameters are not changed.

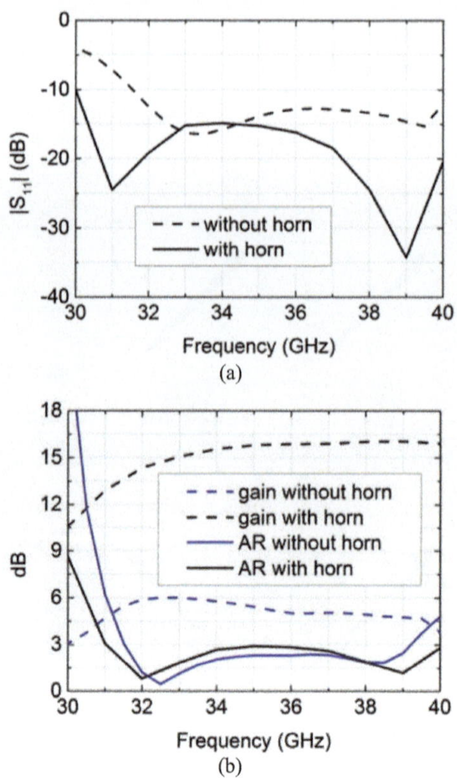

Fig. 12. Simulated (a) $|S_{11}|$s, (b) ARs and gains of the SIW antenna with and without the circular horn.

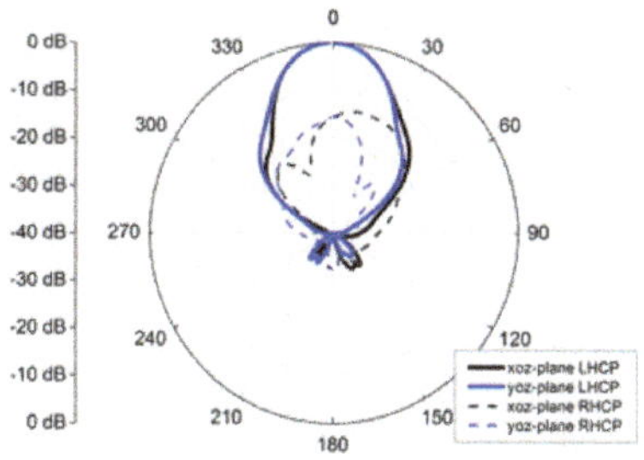

Fig. 13. Simulated radiation patterns of the SIW antenna with the circular horn at 35 GHz.

For illustrating the effect of the circular horn on the SIW antenna, the simulated $|S_{11}|$s, ARs and gains of the SIW antennas with and without the circular horn are shown in Fig. 12. As observed, the $|S_{11}|$ is improved by the loaded circular horn. The impedance bandwidth becomes 28.6% for $|S_{11}| \leq -10$ dB from 30 to 40 GHz. The maximum gain is increased greatly from 6 to 16 dBi. And the AR bandwidths of the SIW antennas with and without the circular horn are almost the same.

The simulated radiation patterns of the SIW antenna with the circular horn at 35 GHz are shown in Fig. 13. It can be seen that good broadside radiation is also achieved. The FBR is increased to 47.8 dBi.

3. Measured Results and Discussions

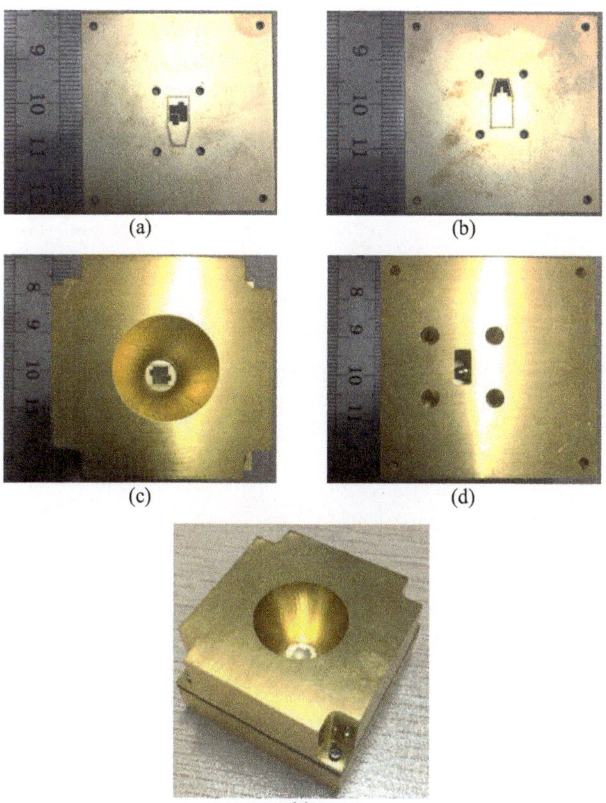

Fig. 14. Photographs of the antenna prototype: (a) Top and (b) back views of the SIW antenna, and (c) top, (b) back and (e) 3D views of the total antenna.

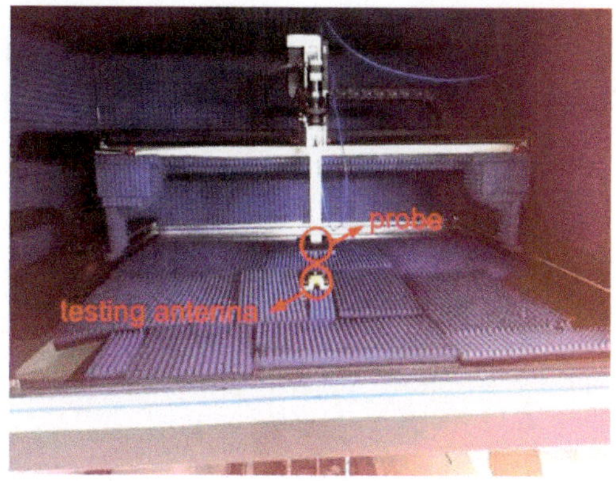

Fig. 15. Photo of the measurement setup.

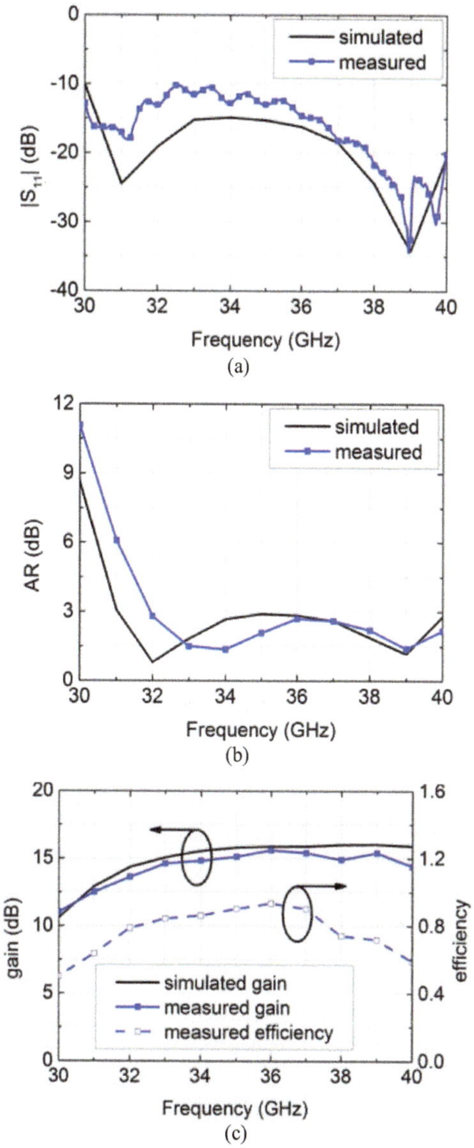

Fig. 16. Simulated and measured $|S_{11}|$s (a), ARs (b) and gains (c).

To verify the simulations above, an antenna prototype was built and measured. Photographs of the antenna prototype are given in Fig. 14. The measurement setup is shown in Fig. 15. The measurement system is a near-field one. The distance between the antenna prototype and the probe is 35 mm and the detection range is 120 mm × 120 mm with the detection step of 2 mm. It should be noted that only the $|S_{11}|$, AR, gain below 40 GHz and the radiation patterns in the upper-half plane are measured due to the limitation of the experimental equipment.

The simulated and measured $|S_{11}|$s ARs and gains are given in Fig. 16. It can be seen that the measured results agree well with the simulated ones. The antenna prototype has a wide impedance bandwidth of 28.6% from 30 to 40 GHz for $|S_{11}| \leq -10$ dB and a wide axial 3-dB AR bandwidth of 22.2% from 32 to 40 GHz. The measured maximum gain is 15.6 dBi at 36 GHz with slight fluctuations over the 30–40-GHz frequency range. The efficiency is about 80% within the working band and the maximum efficiency is 93% at 36 GHz.

Simulated and measured radiation patterns of the proposed array at 32, 35, and 39 GHz are indicated in Fig. 17. It can be seen that good agreements are achieved between the simulated and measured radiation patterns. The LHCP components in both planes are generated as the co-polarization while the RHCP components in both planes are generated as the cross-polarization.

The slight discrepancy between measurements and simulations may be due to fabrication and assembling errors, and uncertainties, such as changes in the dielectric constant and thickness of the used substrate, metallization losses, and surface roughness.

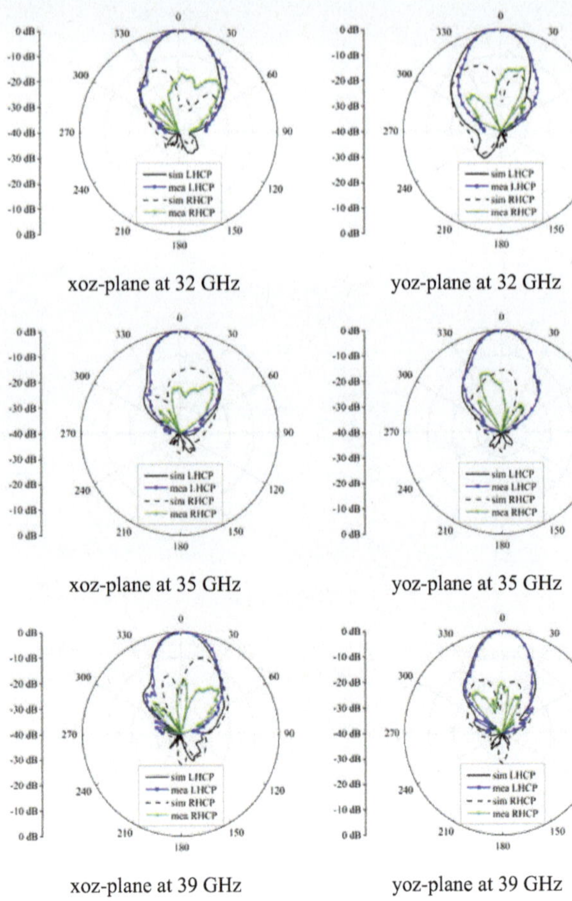

Fig. 17. Simulated and measured radiation patterns.

4. Conclusion

A low-cost circular-horn-loaded CP antenna based on SIW technology operating at Ka band is presented and demonstrated. The measured results show reasonable agreement with the simulated ones. This kind of CP antenna, which is single-fed, low-cost and high gain, has both wide impedance and 3-dB AR bandwidth. Owing to its advantages, this CP antenna can be a good candidate for the mmW applications.

Acknowledgments

This work was supported by the National Natural Science Foundation of China (No. 61401065), and China Scholarship Council (CSC).

References

[1] GUPTA, K. C., GARG, R., BAHL, I., BHARTIA, P. *Microstrip Lines and Slotlines*. 3rd ed., rev. London (UK): Artech House, 1996. ISBN: 9780890067666

[2] BOZZI, M, GEORGIADIS, A, WU, K. Review of substrate-integrated waveguide circuits and antennas. *IET Microwaves, Antennas and Propagation*, 2011, vol. 5, no. 8, p. 909–920. DOI: 10.1049/IET-MAP.2010.0463

[3] LI, Z., WU, K. 24GHz FMCW radar front-end system on substrate. In *2007 IEEE Radio and Wireless Symposium*. Long Beach (CA, USA), 2007, p. 233–236. DOI: 10.1109/RWS.2007.351810

[4] LI, Z, WU, K. 24-GHz frequency-modulation continuous-wave radar front-end system-on-substrate. *IEEE Transactions on Microwave Theory and Techniques*, 2008, vol. 56, no. 2, p. 278–285. DOI: 10.1109/TMTT.2007.914363

[5] WONG, S. W., WANG, K., CHEN, Z. N., CHU, Q. X. Design of millimeter-wave bandpass filter using electric coupling of substrate integrated waveguide (SIW). *IEEE Microwave and Wireless Components Letters*, 2014, vol. 24, no. 1, p. 26–28. DOI: 10.1109/LMWC.2013.2288177

[6] DING, Y., WU, K. A 4×4 ridge substrate integrated waveguide (RSIW) slot array antenna. *IEEE Antennas and Wireless Propagation Letters*, 2009, vol. 8, p. 561–564. DOI: 10.1109/LAWP.2009.2021006

[7] MALLAHZADEH, A. R., ESFANDIARPOUR, S. Wideband H-plane horn antenna based on ridge substrate integrated waveguide (RSIW). *IEEE Antennas and Wireless Propagation Letters*, 2012, vol. 11, p. 85–88. DOI: 10.1109/LAWP.2012.2183110

[8] JOHNSON, D.R., VOLAKIS. L. *Antenna Engineering Handbook*. 4nd ed., rev. New York (USA): McGraw-Hill, 2007. ISBN-13: 978-0071475747

[9] LI, Y., CHEN, Z.N., QING, X., et al. Axial ratio bandwidth enhancement of 60-GHz substrate integrated waveguide-fed circularly polarized LTCC antenna array. *IEEE Transactions on Antennas and Propagation*, 2012, vol. 60, no. 10, p. 4619–4626. DOI: 10.1109/TAP.2012.2207343

[10] LI, T., WANG, B., DOU, W. Substrate integrated waveguide slot array antenna covered by circularly polarised array patches. *Electronics Letters*, 2015, vol. 51, no. 21, p. 1634–1635. DOI: 10.1049/EL.2015.2206

[11] BISHARAT, D. J., LIAO, S., XUE, Q. High gain and low cost differentially fed circularly polarized planar aperture antenna for

broadband millimeter-wave applications. *IEEE Transactions on Antennas and Propagation*, 2016, vol. 64, no. 1, p. 33–42. DOI: 10.1109/TAP.2015.2499750

[12] KIM, D., LEE, J. W., CHO, C. S., LEE, T. K. X-band circular ring-slot antenna embedded in single-layered SIW for circular polarisation. *Electronics Letters*, 2009, vol. 45, no. 13, p. 668–669. DOI: 10.1049/EL.2009.0901

[13] KIM, D. Y., LEE, J. W., LEE, T. K., CHO, C. S. Design of SIW cavity-backed circular-polarized antennas using two different feeding transitions. *IEEE Transactions on Antennas and Propagation*, 2011, vol. 59, no. 4, p. 1398–1403. DOI: 10.1109/TAP.2011.2109675

[14] DONG, J., YANG, Z., PENG, H., YANG, T. Full Ka-band right-angle transition from substrate integrated waveguide to air-filled rectangular waveguide. *Electronics Letters*, 2015 vol. 51, no. 22, p. 1796–1798. DOI: 10.1049/EL.2015.2462

[15] CHEN M, TSANDOULAS G. A wide-band square-waveguide array polarizer. *IEEE Transactions on Antennas and Propagation*, 1973, vol. 21, no. 3, p. 389–391. DOI: 10.1109/TAP.1973.1140486

[16] HUANG, Y., GENG, J., JIN, R., et al. A novel compact circularly polarized horn antenna. In *2014 IEEE Antennas and Propagation Society International Symposium (APSURSI)*. Memphis (TN, USA), 2014, p. 43–44. DOI: 10.1109/APS.2014.6904353

[17] MA, X., HUANG, C., PAN, W., et al. A dual circularly polarized horn antenna in Ku-band based on chiral metamaterial. *IEEE Transactions on Antennas and Propagation*, 2014, vol. 62, no. 4, p. 2307–2311. DOI: 10.1109/TAP.2014.2301841

[18] GAN, T. H., TAN, E. L. Design of waveguide fed broadband circular polarization truncated horn antenna for high power applications. In *2011 Asia-Pacific. IEEE Microwave Conference Proceedings (APMC)*. Melbourne (Australia), 2011, p. 1194–1197.

[19] FUKUSAKO, T, SHAFAI, L. Design of broadband circularly polarized horn antenna using an L-shaped probe. In *IEEE Antennas and Propagation Society International Symposium*. Albuquerque (USA), 2006, p. 3161–3164. DOI: 10.1109/APS.2006.1711281

About the Authors ...

Ming DU was born in China in 1989. He received the B.S. degree in Electrical Engineering from the University of Electronic Science and Technology of China (UESTC) Chengdu, China, in 2011, and is currently pursuing the Ph.D. degree in UESTC. His research interests include microwave wave circuits and antennas.

Jun XU was born in China in 1963. He received the B.S. degree in 1984, and M.S. degree in 1990 from the University of Electronic Science and Technology of China (UESTC) Chengdu, China. He engaged in basic research in the field of radio physics subject and teaching work. He was hired as an associate professor in 1997, then as a professor in 2000 in UESTC. His main research interests include microwave theory and technology, millimeter-wave hybrid integrated technology, millimeter-wave communication and radar radio frequency technology. He has published over 80 papers in refereed journals.

Yuliang DONG was born in Sichuan, China, in 1972. He received the B.S. degree in Electronic Engineering from the Northwestern Polytechnical University, Xi'an, China, in 1993, and Ph.D. degree from Beihang University, Beijing, in 2005. He is now working in the Institute of Applied Physics, University of Electronic Science and Technology of China, No.4, Section 2, North Jianshe Road, Chengdu, Sichuan, 610054, China. His research interests include microwave wave circuits, passive components, antennas and microwave CAD technology.

Xiao DING was born in Sichuan, China, in 1982. He received the Ph.D. degree in Radio Physics from the University of Electronic Science and Technology of China (UESTC), Chengdu, China, in 2013. He joined UESTC in 2014, where he is currently an associate professor. He has been a Visiting Scholar at the South Dakota School of Mines and Technology, SD, USA, a Visiting Fellow with the University of Houston, TX, USA. His research interests include wide-angle scanning phased array, LTCC millimeter wave array and reconfigurable antenna and its applications.

15

Passive UHF RFID Tags with Specific Printed Antennas for Dielectric and Metallic Objects Applications

Katherine SIAKAVARA, Sotirios GOUDOS, Argiris THEOPOULOS, John SAHALOS

Dept. of Physics, Aristotle University of Thessaloniki, 54124 Thessaloniki, Greece

skv@auth.gr, ssoti@physics.auth.gr, atheo@physics.auth.gr, sahalos@auth.gr

Abstract. *Design process and respective results for the synthesis of specific Radiofrequency Identification (RFID) tag antennas, suitable for dielectric and metallic objects, are presented. The antennas were designed for the UHF (865 MHz-869 MHz) band and their basic configuration is that of the printed spiral type. Six modification steps to the classical spiral layout are proposed and it was proved that they can lead to tags with high readability and reading distances up to 10 m when designed for dielectric object and up to 7 m in the case of metallic objects. The results of the measurements of the fabricated tags are explained via theoretical evaluations which take into account reflection phenomena, which are present in a real environment at which the tags are used.*

Keywords

Printed antennas, RFID tags, spiral antennas

1. Introduction

The concept of communicating by means of reflected electromagnetic power is not new as can be traced back to more than six decades of years. Over this time period, the idea has evolved and been realized as modern, efficient and thus worldwide applied technology which exploits the capability of receiving and storing data by RF modulated and backscattered waves and in this way providing automatic wireless identification and tracking capability of things and personnel. Nowadays, applications of RFID systems span from inventory tracking, security and ticketing to electronic payment. Frequencies inside various bands of electromagnetic spectrum as the Low Frequency (LF) band, the High Frequency (HF), the Ultra High Frequency (UHF) and Microwave Frequency (MW) (2.4 GHz to 2.4835 GHz and 5.8 GHz) bands are dedicated to RFID applications. Among these, the UHF band is preferred for most of applications since UHF passive tags are capable of offering relatively fast reading speeds and can be read over long distances [1], [2].

The main grades of an RFID system are: a) the reader (interrogator) supplied with an antenna, b) the tags (transponders) which are microchips combined with an antenna in a compact package, c) a host computer, and d) a middleware including software and data base. Concisely, a passive RFID system operates in the following way: RFID reader transmits a modulated RF signal towards the RFID tag. The chip receives power from the tag's antenna and responds by varying its input impedance modulating, in this way, the backscattered signal. Modulation type, often used in RFIDs, is Amplitude Shift Keying (ASK). At ASK the chip impedance switches between two states: at the one of them it is matched to the antenna (chip collects power in that state) and at the other one it is strongly mismatched, backscattering the incident power [1], [3].

Basic requirement of the RFID forward-backward communication, is the ensuring of sufficient power at the input of the tag's IC in order that the IC to work. Two are the key factors for this to be obtained. The first one is the use of a suitable antenna capable of gathering satisfactory amount of power from the environment, as at a passive tag its antenna is the sole source of energy for the chip. The second key is the transfer of a high, as possible, fraction of this power to the IC [4]. The power selected via the tag's antenna depends inversely from the distance between reader and tag, the antenna's directive gain and these parameters along with the IC's sensitivity determine the maximum permitted distance between reader and tag at which the tag can be read. So, the designer of a tag has to solve two main problems: To design an antenna small in size but with high gain and to match the antenna to the IC. The majority of the commercial chips have complex input impedance with small real part and imaginary part of high capacitive value. Consequently in order the conjugate matching between antenna and IC to be obtained, the antenna has to operate out of resonance having an inductive input reactance. These specific requirements make the antenna's design and generally of the whole tag not an easy task.

In accordance to the literature, various configurations have been employed for the design of passive tag antennas, which aim at both satisfactory gain and matching to the IC be achieved: antennas, of simple dipole or monopole type [5], [6], of meandered line shape [7–10], of fractal shape [11–14], of metamaterial-inspired type [15–17] or based on

Split Ring Resonator (SRR) concept [18]. Recently, published works for RFID tag antenna design propose coupled shorted patches, configured as step-impedance sections and being suitable for dielectric and metallic hosts [19]. Coupled shorted patches textured by comb-notches for efficient matching to the IC and utilization as an on-body sensors have been also proposed in [20]. Articles [21–25] suggest antennas of spiral shape which are properly modified via optimization techniques, for the required specifications of matching and gain levels to be obtained.

The target of the present work was the synthesis of passive tag antennas operating in the European RFID UHF band of 865 MHz–869 MHz, with maximum, as possible, reading distance being suitable for application at dielectric and metallic objects. For the design, the spiral shape was adopted and ICs with power threshold of –18 dBm were considered. The design focuses not only to the matching condition between the antenna and the chip but mainly to the maximization of the gain and the distribution of the maximum gain values inside a wide space range around the tag, thus ensuring high readability, independently of the orientation of the tag with respect to the reader's antenna. These being the targets, the novel contribution of the work was the way by which the classical spiral shape of the printed antenna would be modified and the additional elements that would potentially be incorporated, to the desired performance be obtained.

The rest of the article's content is divided in three sections. In Sec. 2 the process of the tag antenna's design and the power budget of the RFID link are analytically presented. Section 3 includes the presentation of the designed tags for dielectric and metallic objects. Analytical information about their structure is given, and their records and effectiveness are shown via results received by simulation and measurements. Comparison of the designed and fabricated tags' performance with commercial tags or other ones found in the literature is also made. Section 4 includes the general conclusions.

2. Proposed Design and Link Budget

2.1 The Spiral's Shape Modification Process

In the present work, as explained in the previous section, the spiral shaped antenna printed on a dielectric slab was selected as initial geometry for the design of the RFID tags. The first reason for this selection was the basic attribute of the spiral layout to fill the area it occupies. In this way, a strip antenna with long length would be created and due to this long length, the antenna would be capable of operating at UHF band, without having large size. The second reason comes from the spiral geometry which has several geometrical parameters, as the number and the length of the spiral branches, the distance between them, the width of the printed lines, etc. This fact would potentially help in finding a proper set of parameter values which would lead to an antenna with enhanced performance [21], [22]. Furthermore, the spiral geometry offers two additional advantages. One of them is the easy theoretical calculation which can be preceded for the calculation of the input impedance, the radiated field and the gain, applying a hybrid combination of the Transmission Line Model and the low value of electromotive force induced to the spiral along with the Green function method [21]. The other advantage is that the branches of the spiral can be modified as meandered strip lines [23], or meandered and simultaneously loaded with small patches at the end or along branches [24], [25]. Although these additional modifications increase the number of geometrical parameters making, at a first glance, the antenna design more complicated, actually, in the published works it was proved that it turns to an advantage because offers more degrees of freedom which can be handled via optimization algorithms. In this way, effective tags can be designed [21–25].

As mentioned above and due to the aforementioned reasons, in the present work the spiral shape was selected for the design of the tag antennas. The IC SL3S1203_1213/UCODE G2iL, of sensitivity –17.6 dBm was selected for the design and fabrication of the tags. As it is well known, there are not mathematical formulas in closed form, which would make easy the design and the modification of the layout of the majority of printed antennas. So, the enhancement of the antenna's performance was approached by a new reconfiguration process of its shape and without any mathematical formulas or any standard optimization technique to be applied. At first, it was supposed that the spiral comes from an open terminated transmission line of two conductors, bent around the input terminals of the line, at which the chip (IC) element is connected conductively (Fig. 1a). Then six modifications of its layout were made. They are divided in two types, in accordance to the target at which they aimed. The modifications of the first type aimed mainly at the matching of the antenna to the IC, namely at the reduction of the signal reflection coefficient value, Γ, between the antenna and the IC, as it is defined in Sec. 2.2. The modifications of the second type aimed mainly at the configuration of the radiation pattern. However when modifications of one type were being made, e.g. a modification for a desired radiation pattern to be achieved, a check about a low reflection coefficient value was also made, via the simulation, and vice versa. So, the final proposed tag (Fig. 1b) is a result of many repeated reconfigurations of its layout, which substantially were applied in combination, in order to lead to the final tag with the satisfactory performance.

The parameters undergone to modifications of the first type are as follow.

a) The overall length and the number and distances among the branches were changing step by step.

b) The branch line 1 was cut at the position A. The remaining branch lines were bent at the corners.

c) The initial input terminals of the spiral at which the chip

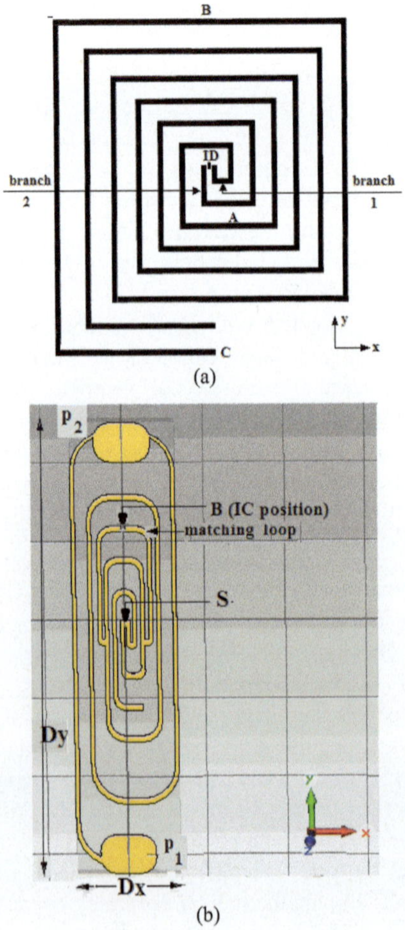

Fig. 1. The dual-branch spiral shape: a) the initial scheme, b) the final prototype after the modifications (Size: 115 mm × 26 mm × 3.38 mm).

would be connected were short-circuited (point S, Fig. 1b). This modification offered the potential to change the position of the chip connection as explained in the next step.

d) A planar printed matching loop was connected to the spiral and the IC was adapted to the loop (point B, Fig. 1b). This loop, in accordance to the theory, increases the inductive imaginary part of the antenna's input impedance and obtains to counterbalance the strongly capacitive input impedance of the IC, which comes from its energy storage stage and is about two to three hundred Ohms. As explained, the position and the size of the loop were found during the process of the repeated layout's modifications and the respective simulation results.

The modifications of the second type were as follows:

a) Scaling of the modified spiral along the x and y directions with different scale factors, at each one, was made. In accordance to the antenna theory, the elongation of an antenna along a specific direction, would make its radiation pattern more narrow on the planes including this direction and simultaneously would increase the gain value. This relation between the width of the main lobe and the gain values of the radiation pattern is an issue that has to be taken into account when an antenna for a specific application is designed. At the proposed antenna the elongation was made along the y direction and the final ratio of the antenna length at the directions y and x was made step by step, the target being the value of the gain to be increased but simultaneously the width of the main lobe of the radiation pattern around the tag to be not very narrow. A very narrow lobe would lead to very high gain but this fact, in the case of the RFID application, would be a disadvantage, because it would increase very much the space area from which and toward which the tag could not effectively receive or transmit signals. The synthesized prototype exhibits a wide space zone, around it, inside which the gain values remain almost constant and high. This fact would enhance the readability of the tag, giving to it the potential to be tracked when having random orientations with respect to the reader's antenna.

b) Small metallic patches with bent corners at the positions B and C of branch line 2 were placed. The patches work as two 'islands' of current constituting a two element array along the y-direction with distance a little less than D_y. Their radiation is added to those of the spiral's currents and due to the arrangement of the two element array along the y axis it enhances the gain of the antenna on the xz plane. On the other hand, the incorporation of the patches in the spiral branches affects the reflection coefficient between the chip and the spiral. So, the total size and the bending of them were found step by step checking simultaneously the radiation pattern, the gain and the reflection coefficient.

2.2 The Link Budget

During the process of tag design and in all cases, the criterion of its effective performance was the power that can be ensured at the IC input, which has to be greater than its specified power threshold. The forward link budget can be obtained by using free space Friis equation as [1]

$$P_{\text{tag}} = P_{\text{EIRPrd}} \left(\frac{\lambda}{4\pi R_{\text{rd-tag}}} \right)^2 D_{\text{tag}} e_{\text{tag}} \tag{1}$$

where P_{tag} is the available power at the input of the tag antenna, coming from the incident signal, $P_{\text{EIRPrd}} = P_{\text{Trd}} G_{\text{rd}}$ is the effective isotropically radiated power from the reader, P_{Trd} is the power radiated from the reader, G_{rd} is the gain of the reader's antenna, D_{tag}, e_{tag} are the directivity and the efficiency of the tag antenna, respectively, and $R_{\text{rd-tag}}$ is the distance between the tag and the reader. In the ideal backscattering ASK modulation, the response of the tag switches between the states '0' and '1'. At the state of '1' the impedance, seen by the antenna towards the tag, is zeroed via a shorting conductor and the gathered power is totally backscattered. The state of '0' is the absorbing state at which the backscattered power is minimized, as the IC absorbs part of the gathered power, P_{tag} [1]. This absorbed power is

$$P_{\text{in,chip}} = (1-|\Gamma|^2)P_{\text{tag}}np \qquad (2)$$

where factors n and p describe the package losses of the IC and the potential polarization mismatching between the reader's and the tag's antennas respectively, $\Gamma = (Z_{\text{in,chip}} - Z_a^*) / (Z_{\text{in,chip}} + Z_a)$ is the reflection coefficient due to the mismatching, and $Z_{\text{in,chip}}$, Z_a are, correspondingly, the input impedances of the chip and the antenna. In accordance to the theoretical analysis, the design of the antenna focuses to the achievement of sufficiently high gain level and simultaneously low value of Γ. In this way, high, as possible, reading distance would be obtained.

In the case of an RFID link, not occurring in absolutely free space, the impact of various scattering mechanisms of the signal in the environment has to be taken into account when P_{tag} is evaluated. In this work and in order to compare the theoretically expected $P_{\text{in,chip}}$ with the corresponding values received via measurements in Open Area Test Site (OATS), equation (1) was modified as

$$P_{\text{tag}}^{R_{\text{V,H}}} = P_{\text{EIRPrd}}\left(\frac{\lambda}{4\pi R_{\text{rd-tag}}}\right)^2 D_{\text{tag}}e_{\text{tag}}R_{\text{V,H}} \qquad (3)$$

where by the coefficient $R_{\text{V,H}}$ (4), the reflection on the ground of OATS for vertically or horizontally polarized propagating signals is taken into account, considering the contribution of only one reflection and not surface waves or secondary effects of the ground [26], [27]

$$R_{\text{V,H}} = 1 + C_{\text{V,H}}\exp(j\Delta\varphi). \qquad (4)$$

In (4), in accordance to the literature, $C_{\text{V,H}}$ is coefficient depending on the polarization of the wave and the angle of incidence to the ground, as well as the ground's characteristics. Term $\Delta\varphi$, stands for the phase difference between the reflected and the direct path from the transmitting to the receiving antenna, and depends on the height of antennas over the ground, the distance between them and the wavelength. In accordance to the above, equation (2) is generalized to (5)

$$P_{\text{in,chip}}^{R_{\text{V,H}}} = (1-|\Gamma|^2)P_{\text{EIRPrd}}\left(\frac{\lambda}{4\pi R_{\text{rd-tag}}}\right)^2 D_{\text{tag}}e_{\text{tag}}R_{\text{V,H}}np. \qquad (5)$$

3. Designed Tags and Results

3.1 Tag_DO :Tag for Dielectric Objects

Starting from spirals of original shape and applying the proposed process, the repeated modifications to their layout provided the tag illustrated in Fig. 1b and the respective photo of the fabricated and measured prototype is presented in Fig. 2a. It is termed as Tag_DO and is the best of the obtained tags.

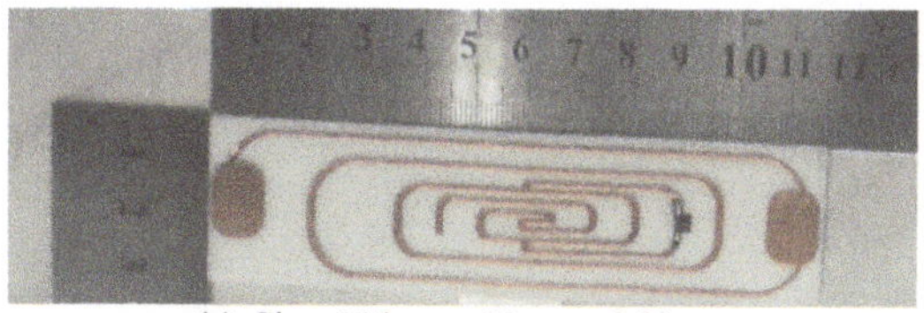

(a) Size: 115mm x 26mm x 3.38mm

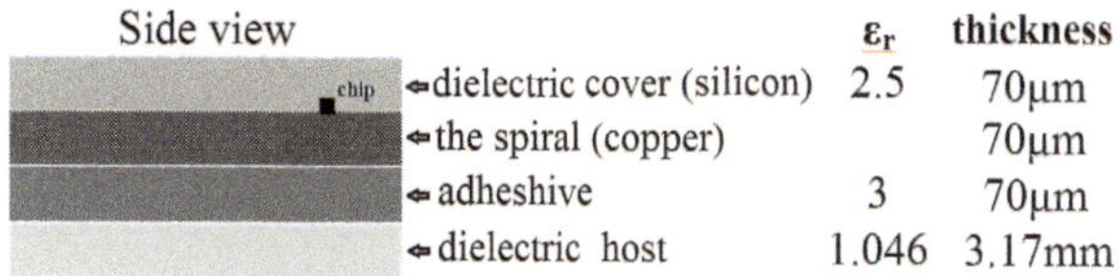

(b)

Fig. 2. Tag_DO : a) The fabricated prototype of the scheme of Fig. 1.b, b) the side view.

In detail, the changes made to the ordinary spiral shape were : a) the layout was changed and the ratio of the lengths of its two sides was made greater than 1, namely $D_y/D_x = 4.42$; b) the initial input terminals of the spiral were short-circuited(point S); c)planar printed matching loop was connected to the spiral and the IC was adapted to this loop (point B); d) two printed metallic patches, p_1 and p_2 (12.5 mm × 0.94 mm) were incorporated into the spiral at proper positions.

Firstly, the performance of the tag was judged via simulation of its operation and the results are presented in Fig. 3 and 4. In Fig. 3, the real and imaginary parts of the antenna input impedance are illustrated. It is shown that at the frequency of 867 MHz the value of the impedance is $Z_a = (36 + j220)$ Ohms. This value is close enough to the IC' s input impedance $Z_{\text{in,chip}} = (25 - j237)$ Ohms, and yields a satisfactory low reflection coefficient $|\Gamma| = -12.5$ dB.

Figure 4 depicts, indicatively, the 3D directivity pattern of the antenna at 867 MHz. A wide range with high directive gain around the tag is ascertained and it is in agreement with the respective target of the design. Detailed results for the directivity and the efficiency at the central frequency and at the limits of the frequency band of operation are presented in Tab. 1 and 2

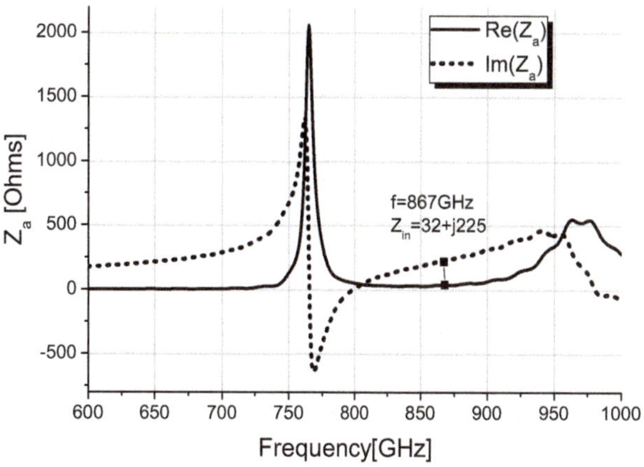

Fig. 3. Tag_DO: Real and imaginary part of the input impedance.

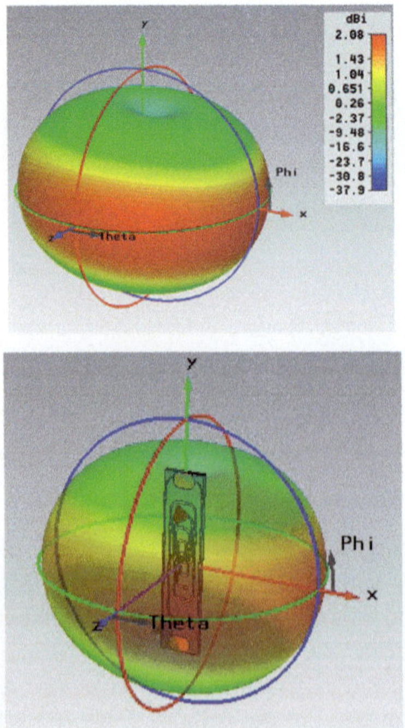

Fig. 4. Tag_DO: 3D directivity pattern of the tag at 867 MHz.

Plane	D_{max}[dB]	D_{min}[dB]
-XY-	1.94	–2.26
-XZ-	2.075	–0.938
-YZ-	2.02	–12.21

Tab. 1. Tag_DO: Directivity at 867 MHz.

865 MHz	867 MHz	869 MHz
–0.4 dB	–0.33 dB	–0.25 dB

Tab. 2. Tag_DO : Efficiency.

To validate the tag's performance, and so the robustness of the design procedure, the tag was fabricated (Fig. 2a) and measured. The fabricated tag was tested using the reader MTI Wireless Edge/MT–242032/NRH. The reader antenna characteristics were: emitted power 1 W, gain 7 dBi, and circular polarization. The measurements were received in OATS (Fig. 5a) at the roof of the building which hosts the laboratory. The floor of the area was covered by aluminum sheet. The tag was positioned in parallel and perpendicularly to the horizontal floor plane(*-xz-* plane), was rotated around the *-x-*, *-y-* and *-z-* axes and 12 measurements were received at *-yz-* , *-xz-*, *-xy-* planes, respectively. What was checked, was the readability or not of the tag versus its distance from the reader. Figure 6 shows the statistical results by plane and over the total of 36 measurements. The presented results regard to the reading efficiency, namely the number of the times the tag was read over the total times its readability was checked. The rotation of the tag was judged necessary in order a general conclusion to be formed for its efficiency, because in practice, the tag, when attached to the host object, would potentially see the reader, or be shown by it, via a direction different from that of its maximum directivity.

In Fig. 6a it is observed that the reading efficiency varies with the orientation of the tag, reaching to 100 % and holding this portion even at 10 m away from the reader

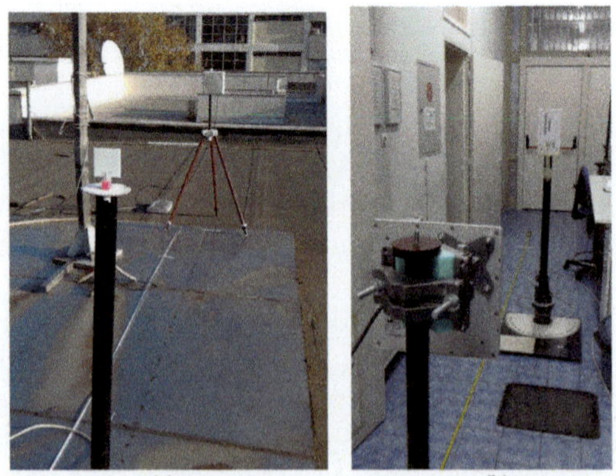

(a) (b)

Fig. 5. a) The OATS at which the measurements of Fig. 6 were received. b) The interior of the laboratory and the experimental arrangement by which the measurements of Fig. 8 were received.

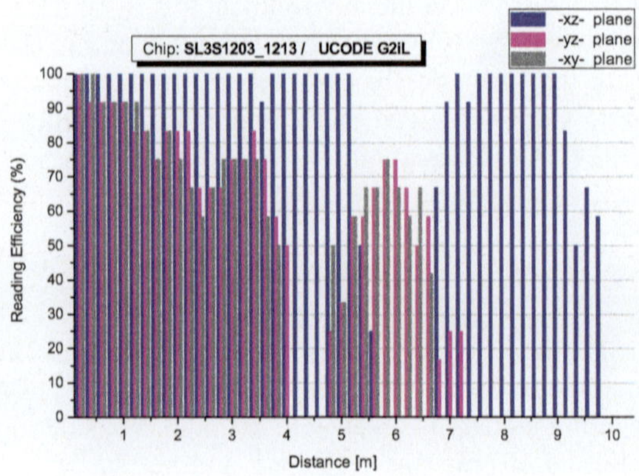

(a)

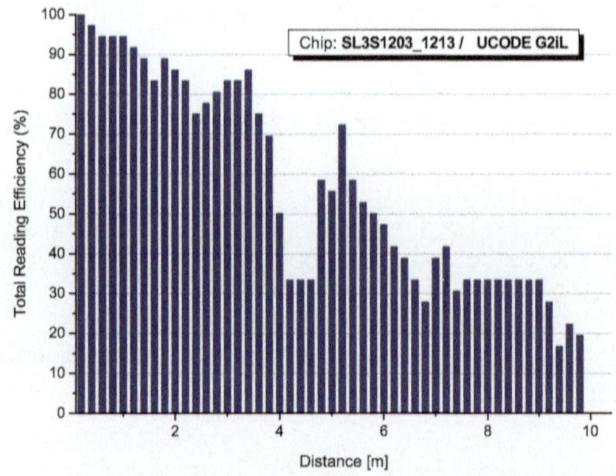

(b)

Fig. 6. Tag_DO: Measured statistical results for the reading efficiency versus the distance from the reader: a) by plane, b) over all three planes.

when the tag is oriented with its large dimension parallel to *-y-* axis (*xz* plane measurements). This axis was considered perpendicular to ground plane of the OATS. Satisfactory results are also obtained when the tag was positioned with its small side perpendicular to the ground (*yz* plane measurements). The reading efficiency gets much smaller when the tag is positioned in parallel to the ground. The results give, substantially, information about how the tag would be used. The results of Fig. 6b present the reading efficiency over all three orientations of the tag. A noticeable observation is that the reading efficiency is minimized at specific distances from the reader, while increases again with further increment of this distance, appearing a local maximum and then tending gradually to zero. It was also observed that the distances of minimization differ from plane to plane. It was estimated that this performance is due to the reflection of the reader's signal at the floor of the OATS. In detail, it is due to the component of the electric intensity of the incident field from which the tag draws energy and also the reflection coefficient for each component, the point of reflection and consequently the distance at which the reflected ray field component cancels the directly propagating one. To validate this estimation, theoretical calculation of the input power of the IC was made via (5). The tag draws power from the vertical or the horizontal component of the reader's incident wave, depending on its orientation. Moreover, its radiation pattern exhibits maximum and minimum directivity per main plain as shown in Tab. 1. The evaluations were made taking into account these issues. Indicative results are depicted in Fig. 7.

Concisely, Figure 7 depicts the input power to the IC, evaluated by four ways:

a) Graph P1: pure theoretically via 5, taking into account the reflection

b) Graph P2: theoretically but via utilization of measured field values (in presence of reflection). Namely, the electric field intensity was measured via a prototype antenna, at the positions in space, away from the reader, at which the tag was tested. Then, the power resulted from these experimental field values, replaced the terms $P_{EIRPrd} R_{V,H} / (4\pi(R_{rd\text{-}tag})^2)$ of (5).

c) Graph P3: reduction to free space, of the results of graph P2 was made, dividing the results presented in this graph by the theoretical reflection coefficients.

d) Graph P4: pure theoretically in free space (2). All the calculations were made using polarization loss factor $p = 0.5$, as the fabricated antenna has linear and that of the reader, circular polarization. The IC package loss factor was $n = 0.5$.

It was validated that the reading efficiency is in agreement with the theoretical evaluations as at the distances with maximum readability, up to 100 %, the input IC power exceeds the IC's sensitivity threshold even if the tag is oriented towards the reader with the direction of its minimum directivity. It happens for distances up to ~5.2 m. Correspondingly, the readability is minimized or zeroed when the tag is positioned at distances at which the reader's signal is minimized, due to the reflection mechanism, and simultaneously the IC's input power is less than its threshold level, even if the tag is oriented to the reader with the direction of its maximum directivity. It is valid between ~5.2 m to ~6.2 m. At distances from 6.2 m to 9.5 m the readability reaches again 100 % as the power input to the chip becomes again greater than its threshold. At these distances perhaps the reflection enhances the reader's signal instead of decreasing it.

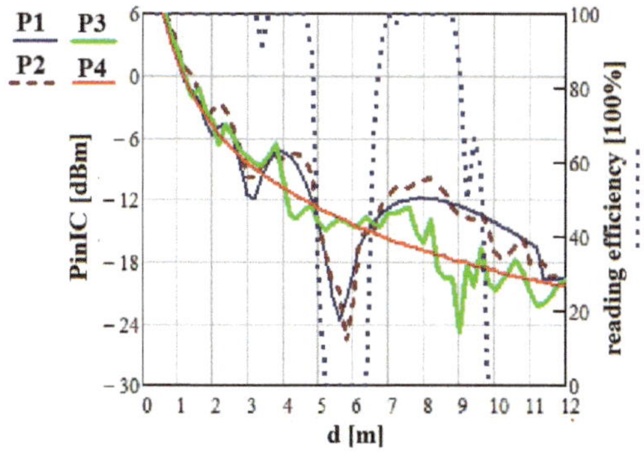

(a)

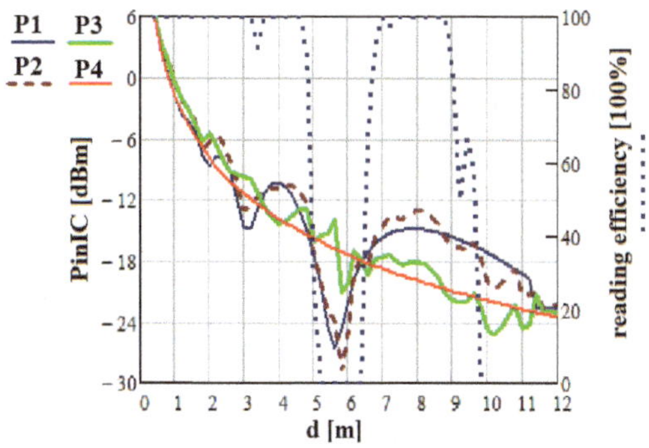

(b)

Fig. 7. Chip input power [dBm] versus the distance from the reader. Estimation via: (a) the maximum and (b) the minimum gain of the tag antenna. Plotted results: pure theoretical (5) taking into account the reflection (P1), theoretical but via the measured field values (P2), theoretical via the measured field values and by reduction to free space (P3), pure theoretical (2) in free space (P4).

The detecting range of the tag is judged very sufficient as it exhibits maximum reading distances comparable or superior to those of several ones, commercial or presented in the literature, with IC of equal sensitivity. Indicatively: a) the tag NXP GP-33[28] of size: 33 mm × 33 mm × 3.38 mm, and the tag DogBone (Monza 4) of the Smartrac Com [29], of size: 86 mm × 24 mm, were measured in OATS (Fig. 5a) and exhibited reading distances 4.7 m and 9.5 m, respectively. Tags, with size 51 mm × 43 mm, proposed in [8] obtain detection range about 8.6 m. In [10] tags of size 48 mm × 48 mm, exhibit detection range about 6 m. In [18] tags of size 30 mm × 30 mm have detection range 9.3 m and in [20] tags of size 10.4 mm × 5.9 mm × 0.76 mm have the range of 7.3 m.

Beyond the experimental tests in OATS indoor tests, with the tag attached on paper box considered as a dielectric host, were made. The applied procedure was similar to that in OATS and the respective measurements were received in the interior of the laboratory (Fig. 5b). The statistical results of total reading efficiency are presented in Fig. 8. In this case, due to the presence of the paper box, as well as the various mechanisms of signal scattering at the indoor environment, decrease of readability with respect to those in OATS (Fig. 6b) is ascertained.

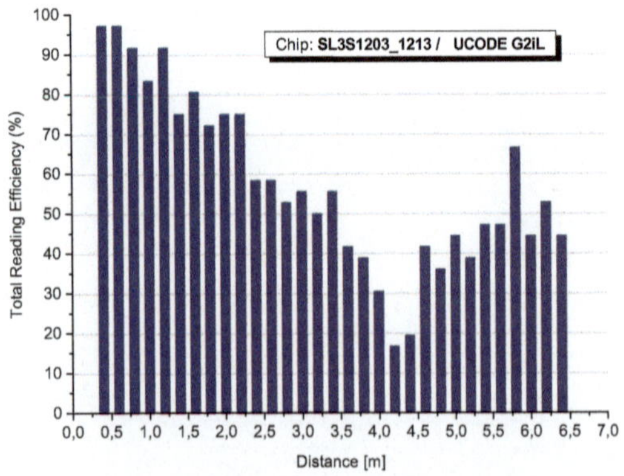

Fig. 8. Tag_DO attached to a paper box including printer toner: Measured statistical results, over all three planes, for the reading efficiency versus the distance from the reader.

3.2 Tag_MO: Tag for Metallic Objects

The design of a tag for metallic objects is of increased difficulty as the tag antenna has to meet the requirements of the satisfactory gain level, the proper distribution of the radiated power in the space, the matching to IC and all of them have to be ensured in the presence of a metallic object being in very close proximity to the tag antenna.

The metallic surface of the object due to its electromagnetic behavior, affects strongly the performance of the antenna and this influence increases by the proximity of the antenna to the surface. In the case of the RFID tag, this proximity is the maximum possible as the tag is attached to the object. In the literature, various configurations of antennas destined for metallic hosts have been proposed as, inverted-F type [30], antennas modeled by loop and open and shorted printed strip lines [31], bow-tie patches connected through metallic pins to the back plane [32] or bow-tie slot type combined with Artificial Magnetic Conductor (AMC) [33], meander-type [34] or metamaterial structured antennas [35], patches, inset fed by open stub printed line at which the IC is mounted [36], patch antenna with proximity-coupled feed with line shorted at its end [37] and coupled shorted patches fed by printed loop for the IC hosting [38]. The microstrip type is superior to the printed type due to the fact the effectiveness of the printed models are strongly reduced when they are positioned very close to metallic surfaces and moreover large shifting of their resonance frequency occurs. To these disadvantages, the large length of the lines in order to resonate at the UHF RFID band has to be added. A disadvantage of the microstrip type proposed for RFID tags is that satisfactory performance is obtained by incorporating short walls or metallic pins between the radiating element and the ground plane, thus making the fabrication more difficult.

In the present work, for the design of the tag antennas destined for metallic objects, the spiral shape was also adopted along with the modifications made for the design of Tag_DO, except one of them. The spiral shape, as already has been mentioned, has the advantage of large line length but small overall size and moreover shorting pins

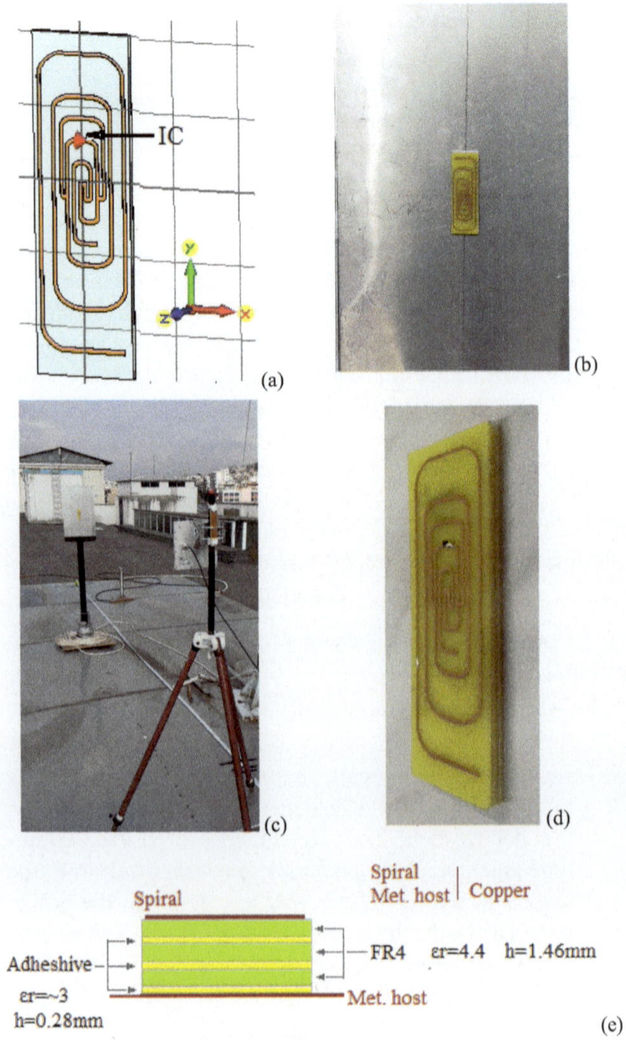

Fig. 9. Tag_MO: a) The layout of the tag; b) the tag on a metallic sheet; c) the fabricated prototype on the metallic sheet at the OATS where the measurements of Fig. 11 were received; d) tag's profile and details of its structural parameters.
Size of the tag: 29.32 mm × 94.07 mm × 5.29 mm.
Size of the metallic host: 50 cm × 30 cm × 0.1 cm.

are not necessary. The IC SL3S1203_1213 /UCODE G2iL and, as a substrate, the dielectric FR4 with $\varepsilon_r = 4.4$ was also used. The design of the antenna, and also its fabrication, were made using 3 dielectric layers, each one of the thickness of 1.46 mm, and adhesive between them. The three layer substrate keeps the metallic spiral at a distance from the metallic surface greater than that of a single layer and thus its effectiveness is enhanced, however at the cost of a high side profile.

Moreover, at this tag, termed as Tag_MO, printed patches were not incorporated to the spiral because, in contrast to Tag_DO, their presence was degrading the performance of the antenna instead to enhance it, for all the use of three layer dielectric. Perhaps it was due to the trapping of power at the cavities they created with the metallic host. The synthesis and simulation of the antenna was made considering the tag attached to planar thin metallic layer of size 50 cm × 30 cm × 0.1 cm standing for

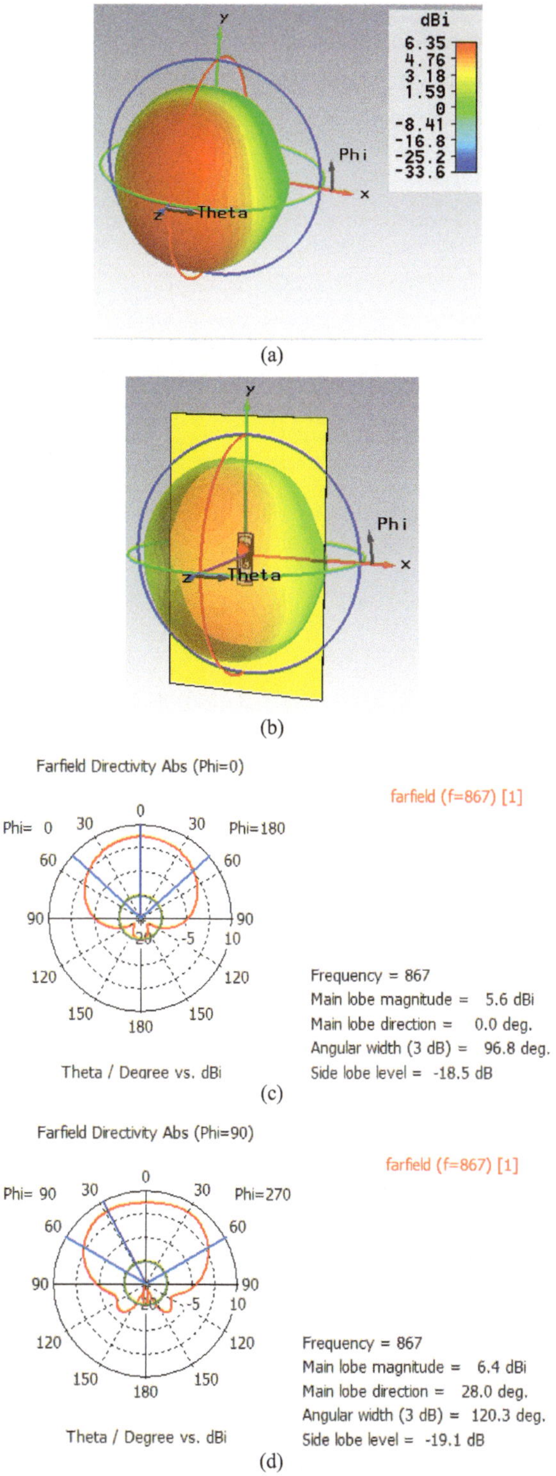

Fig. 10. Tag_MO: (a) and (b) 3D directivity patterns at 867 MHz received by simulation in the presence of the metallic sheet; (c) and (d) polar directivity patterns on *-xz-* and *-yz-* planes respectively.

Plane	D_{max} [dB]	D_{min} [dB]
-XZ-	5.59	–2.82
-YZ-	6.36	–0.9

Tab. 3. Tag_MO: Directivity at 867 MHz.

865 MHz	867 MHz	869 MHz
–0.612 dB	–0.159 dB	–0.4 dB

Tab. 4. Tag_MO: Efficiency.

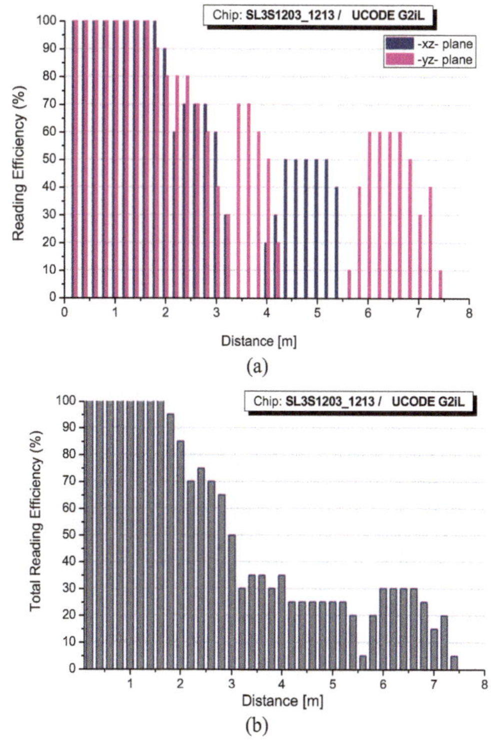

Fig. 11. Tag_MO: Measured statistical results for the reading efficiency versus the distance from the reader: a) by plane, b) over all three planes.

the metallic host. Details about the geometry and the materials of the tag Tag_MO, as well as the fabricated prototype and the area used for measurements (OATS) are presented in Fig. 9. The OATS was that, used for the measurements of Tag_DO.

From the experimental results, statistically processed and shown Fig. 11a, a satisfactory readability of the Tag_MO is ascertained which reaches the values of ~5.5 m or ~7.5 m depending on the orientation of the tag. The distances of minimizations and local maximizations that appear, are different at the two planes, as in the case of Tag_DO. The overall performance (Fig. 11b) is judged as sufficient. The proposed antenna has advantages as a) simple structure compared with other ones with metamaterial [33] or AMC lattices [31], shorting pins [29], [30], [35], [36] etc.; and b) a performance as good as other tags presented in the literature. Indicatively in [19] tags of size 105 mm × 60 mm × 0.76 mm exhibit reading distance of 7.3 m, in [36] tags of size 111.13 mm × 23.56 mm have the reading distance of 5 m, in [37] reading distance of 8.9 m is obtained with tags of size 72.1 mm × 25.5 mm × 3.2 mm and in [38] tags with size 88 mm × 25 mm × 3.5 mm have reading distance of 5 m.

4. Conclusion

The target of the work was the synthesis of antennas suitable for UHF RFID passive tags with high readability and large reading distances. The spiral shape was selected as the initial antenna geometry and a new approach, consisting in modification of the spiral layout, was adopted in order that the target to be obtained. The proposed process focused, besides the ordinary matching between the antenna and the IC, to the proper configuration of the antenna radiation patterns. The aim was, the patterns to demonstrate wide space areas, inside which the gain values are high. In this way the readability of the tag would increase as, in practice, the plane of the tag would have various orientations with respect to the reader's antenna. The tags were fabricated and measured and also theoretical results for the link budget in a real environment, taking into account phenomena of reflection, were received. The agreement between theoretical and experimental results is satisfactory. The tag for dielectric objects appears readability even 100 % at distances up to 10 m, when it is properly oriented. The readability reduces for random orientation, however remaining greater than 50 % up to distance of 6 m. The tag designed for metallic hosts has also satisfactory performance demonstrating readability greater than 50 % up to 7 m, when properly oriented. For random orientation, the above value reduces being greater than 30 % up to 7 m.

Acknowledgments

This research has been co-financed by the European Union (European Social Fund – ESF) and Greek national funds through the Operational Program "Education and Lifelong Learning" of the National Strategic Reference Framework (NSRF) - Research Funding Program: THALES: Reinforcement of the interdisciplinary and/or inter-institutional research and innovation (MIS 377271).

References

[1] CURTY, J. P., DECLERCQ, M., DEHOLLAIN, C., JOEHL, N. *Design and Optimization of Passive UHF RFID Systems*. Springer, 2007. ISBN: 978-1841270715

[2] BROWN, D. E. *RFID Implementation.* Mc Graw Hill, 2007. ISBN: 978-0-07-226324-4

[3] RAO, K. V. S., NIKITIN, P. V., LAM, S. F. Antenna design for UHF RFID tags: A review and a practical application. *IEEE Transactions on Antennas and Propagation*, 2005, vol. 53, no. 12, p. 3870–3876. DOI: 10.1109/TAP.2005.859919

[4] MARROCCO, G. The art of UHF RFID antenna design: Impedance-matching and size-reduction techniques. *IEEE Antennas and Propagation Magazine*, 2008, vol. 50, no. 1, p. 66 – 79. DOI: 10.1109/MAP.2008.4494504

[5] PANDA, J. R, SALADI, A. S. R, KSHETRIMAYUM, S. R. A compact printed monopole antenna for dual-band RFID and WLAN applications. *Radioengineering*. 2011, vol. 20, no. 2, p. 464–467.

[6] JANKOWSKI-MIHUŁOWICZ, P., KAWALEC, D., WĘGLARSKI, M. Antenna design for semi-passive UHF RFID transponder with energy harvester. *Radioengineering*. 2015, vol. 24, no. 3, p. 722–728. DOI: 10.13164/re.2015.0722

[7] MARROCCO, G. Gain-optimized self-resonant meander line antennas for RFID applications. *IEEE Antennas and Wireless Propagation Letters*, 2003, vol. 2, p. 302–305. DOI: 10.1109/LAWP.2003.822198

[8] ABDULHADI, A. E., ABHARI, R. Dual printed meander monopole antennas for passive UHF RFID tags. In *Proceedings of the IEEE International Symposium on Antennas and Propagation.* Washington (USA), 2011. p. 988–991. DOI: 10.1109/APS.2011.5996444

[9] RIDA, R. A., YANG, L., BASAT, S. S., FERRER-VIDAL, A., NIKOLAOU, S., TENTZERIS, M. M. Design, development and integration of novel antennas for miniaturized UHF RFID tags. *IEEE Transactions on Antennas and Propagation*, 2009, vol. 57, no. 11, p. 3450–3457. DOI: 10.1109/TAP.2009.2027347

[10] PAREDES, F., ZAMORA, G., HERRAIZ-MARTÍNEZ, F. J., MARTÍN, F., BONACHE, J. Dual-band UHF-RFID tags based on meander-line antennas loaded with spiral resonators. *IEEE Antennas and Wireless Propagation Letters*, 2011, vol. 10, p. 768–771. DOI: 10.1109/LAWP.2011.2162716

[11] AIVAZIS, E., SIAKAVARA, K., SAHALOS, J. N. Design and analysis of efficient fractal antennas for UHF RFID passive tags. In *Proceedings of the 16th European Conference on Antennas and Propagation (EuCAP)*. Prague (Czech Republic), 2012, p. 3021 to 3025. DOI: 1109/EuCAP.2012.6206102

[12] GHALIBAFAN, J., KASHANI, F. H. A circularly polarized fractal microstrip antenna for RFID applications. In *Proceedings of IEEE International Symposium on Radio-Frequency Integration Technology*. Singapore, 2009. p. 319–322. DOI: 10.1109/RFIT.2009.5383721

[13] RUSU, M., HIRVONEN, M., RAHIMI, H., ENOKSSON, P., RUSU, C., PESONEN, N., VERMESAN, O., RUSTAD, H. Minkowski fractal microstrip antenna for RFID tags. In *Proceedings of the 38th European Microwave Conference (EuMC)*. Amsterdam (The Netherlands), 2008. p. 666–669. DOI: 10.1109/EUMC.2008.4751540

[14] SULTAN, Q. H., SABAAWI, A. M. A. Design a new fractal loop antenna for UHF RFID tags based on a proposed fractal curve. In *Proceedings of the 2nd International Conference on Computer Technology and Development (ICCTD)*. Cairo (Egypt), 2010, p. 6 to 9. DOI : 10.1109/ICCTD.2010.5646055

[15] MITTRA, R., HOENSCHEL, A. Some novel designs for RFID antennas and their performance enhancement with metamaterials. In *IEEE International Workshop on Anti-counterfeiting, Security, Identification.* Xiamen (China), 2007. DOI: 10.1109/IWASID.2007.373681

[16] STUPF, M., MITTRA, R., YEO, J., MOSIG, J. R. Some novel design for RFID antennas and their performance enhancement with metamaterials. In *Proceedings of the IEEE Antennas and Propagation Society International Symposium*. Albuquerque (NM, USA), 2006, p. 1023–1026. DOI: 1109/APS.2006.1710707

[17] KIM, S., KAWAHARA, Y., GEORGIADIS, A., COLADO, A., TENTZERIS, M. M. Low-cost inkjet-printed fully passive RFID tag using metamaterial-inspired antennas for capacitive sensing applications. In *Proceedings of the IEEE MTT-S International Microwave Symposium Digest (IMS)*. 2013. DOI: 10.1109/MWSYM.2013.6697644

[18] ZUFFANELLI, S., ZAMORA, G., AGUILA, P., PAREDES, F., MARTIN, F., BONACHE, J. Analysis of the split ring resonator (SRR) antenna applied to passive UHF-RFID design. *IEEE Transactions on Antennas and Propagation*, 2016, vol. 64, no.3, p. 856–864. DOI : 10.1109/TAP.2015.2513084

[19] SVANDA, M., POLIVKA, M. Stepped impedance coupled-patches tag antennas for platform-tolerant UHF RFID applications. *IEEE Transactions on Antennas and Propagation,* 2015, vol. 63, no. 9, p. 3791–3797. DOI: 10.1109/TAP.2015.2447034

[20] SVANDA, M., POLIVKA, M. Matching technique for an on-body low-profile coupled-patches UHF RFID tag and for sensor antennas. *IEEE Transactions on Antennas and Propagation,* 2015, vol. 63, no. 5, p. 2295–2301. DOI : 10.1109/TAP.2015.2403399

[21] GOUDOS, S. K., SIAKAVARA, K., SAHALOS, J. N. Novel spiral antennas design using swarm intelligence for passive UHF RFID tags. In *Proceedings of the 7th European Conference on Antennas and Propagation (EuCAP).* Gothenburg (Sweden), 2013. p. 3223–3227.

[22] GOUDOS, S. K., SIAKAVARA, K., SAHALOS, J. N. Novel spiral antenna design using artificial bee colony optimization for UHF RFID applications. *IEEE Antennas Wireless Propagation Letters,* 2014, vol. 13, p. 528–531. DOI: 10.1109/LAWP.2014.2311653

[23] GOUDOS, S. K., SIAKAVARA, K., SAHALOS, J. N. Modified spiral RFID tag antenna optimal design using artificial bee colony optimization. In *Proceedings of the 43rd European Microwave Conference (EuMC).* Nuremberg (Germany), 2013, p. 1255–1258. DOI: 10.23919/EuMC.2013.6686892

[24] GOUDOS, S. K., SIAKAVARA, K., SAHALOS, J. N. Design of load-ended spiral antennas for RFID UHF passive tags using improved artificial bee colony algorithm. *International Journal of Electronics and Communications (AEU),* 2015, vol. 69, no. 1, p. 206–214. DOI: 10.1016/j.aeue.2014.09.008

[25] GOUDOS, S. K., SIAKAVARA, K., THEOPOULOS, A., VAFIADIS, E. E., SAHALOS, J. N. Application of Gbest-guided artificial bee colony algorithm to passive UHF RFID tag design. *International Journal of Microwave and Wireless Technologies,* 2016, vol. 8, no. 3, p. 537–545. DOI: 10.1017/S1759078715000902

[26] JAKES, W. C. *Microwave Mobile Communications.* John Wiley & Sons, New York, 1974. ISBN: 0780310691

[27] JANASWAMY, R. *Radiowave Propagation and Smart Antennas for Wireless Communications.* Kluwer Academic Publishers, 2001. ISBN: 978-1475775105

[28] www.rfidtagworld.com/products/RFID-Tag-Antenna-NXP-UHF-Manufacturer1241.html

[29] www.smartrac - group.com/files/content/Products_Services/PDF/DogBone_M4.pdf

[30] HIRVONEN, M., PURSULA, P., JAAKKOLA, K., LAUKKANEN, K. Planar inverted-F antenna for radio frequency identification. *Electronic Letters,* 2004, vol. 40, p. 848–850. DOI: 10.1049/el:20045156

[31] LU, J. H., ZHENG, G. T. Planar broadband tag antenna mounted on the metallic material for UHF RFID system. *IEEE Antennas and Wireless Propagation Letters,* 2011, vol. 10, p. 1405–1408. DOI: 10.1109/LAWP.2011.2178997

[32] LIN, K. H., CHEN, S. L., MITTRA, R. A looped-bowtie RFID tag antenna design for metallic objects. *IEEE Transactions on Antennas and Propagation,* 2013, vol. 61, no. 2, p. 499–505. DOI: 10.1109/TAP.2012.2224082

[33] HADARIG, R. C., DE COS GOMEZ, M. E., ALVAREZ, Y., LAS-HERAS, F. Novel bow-tie-AMC combination for 5.8 GHz RFID tags usable with metallic objects. *IEEE Antennas and Wireless Propagation Letters,* 2010, vol. 9, p. 1217–1220. DOI: 10.1109/LAWP.2010.2100358

[34] VISHNUPRIVA, V., ANJU, M., AJU, J, THOMASKUTTY, M. Meander dipole RFID tag for metallic objects. In Proceedings of the 2nd *International Conference on Electronics and Communication Systems (ICECS).* Coimbatore (India), 2015, p. 1027–1029. DOI: 10.1109/ECS.2015.7124734

[35] UGARTE-MUÑOZ, E., HERRAIZ-MARTÍNEZ, F. J., GONZÁLEZ-POSADAS, V., SEGOVIA-VARGAS, D. Patch antenna based on metamaterials for a RFID transponder, *Radioengineering,* 2008, vol. 17, no. 2, p. 66–71.

[36] MO, L., QIN, CH. Planar UHF RFID tag antenna with open stub feed for metallic objects. *IEEE Transactions on Antennas and Propagation,* 2010, vol. 58, no. 9, p. 3037–3043. DOI: 10.1109/TAP.2010.2052570

[37] SON, H.-W., JEONG, S.-H. Wideband RFID tag antenna for metallic surfaces using proximity-coupled feed. *IEEE Antennas and Wireless Propagation Letters,* 2011, vol. 10, p. 377–380. DOI: 10.1109/LAWP.2011.2148151

[38] PENG H. YANG, YAN LI, I, LIJUN JIANG, ET AL. Compact metallic RFID tag antennas with a loop-fed method. *IEEE Transactions on Antennas and Propagation,* 2011, vol. 59, no. 12, p. 4454–4462. DOI: 10.1109/TAP.2011.2165484

About the Authors ...

Katherine SIAKAVARA was born in 1954. She received the B.Sc. degree in Physics in 1977 and the M.Sc. degree in Electronics, in 1979, from the Aristotle University of Thessaloniki, Greece. She was awarded the Ph.D. degree in Electrical Engineering from the same university in 1982. She is currently a Professor in the Department of Physics of the University of Thessaloniki, Greece. Dr. Siakavara is author or co-author of 49 international journal papers, 45 papers in proceedings of international conferences, two book chapters in English, and two books in Greek. Her research interests are in the areas of applied electromagnetism, analysis and design of antenna systems, microwaves, and radio-wave propagation and radio-communications.

Sotirios K. GOUDOS was born in 1968. He received the B.Sc. degree in Physics in 1991 and the M.Sc. of Postgraduate Studies in Electronics in 1994, both from the Aristotle University of Thessaloniki. In 2001, he received the Ph.D. degree in Physics from the Aristotle University of Thessaloniki and in 2005 the Master in Information Systems from the University of Macedonia, Greece. In 2011, he obtained the Diploma degree in Electrical and Computer Engineering from the Aristotle University of Thessaloniki. He joined the Department of Physics, Aristotle University of Thessaloniki, in 2013 and is currently an Assistant Professor of the Department. He is author or co-author of more than 90 papers of international scientific journals and conferences and his research interests include antenna and microwave structures design, electromagnetic compatibility of communication systems, evolutionary computation algorithms, and mobile communications.

Argiris THEOPOULOS was born in 1978. He received the B.Sc. degree in Physics in 2005 and the M.Sc. degree in Electronic Physics in 2008 from the Aristotle University of Thessaloniki. He has been working as an Electronics Engineer at Radiocommunications Laboratory of Aristotle University of Thessaloniki since 2013. His main fields of interest are design and development of embedded systems.

John N. SAHALOS was born in 1943. He received his B.Sc. and Ph.D. degrees in Physics. He also received the Diploma (BCE + MCE) in Civil Engineering and the professional Diploma of postgraduate studies in Radio-Electrology. He worked at the ElectroScience Laboratory, the Ohio State University, Columbus, as a Postdoctoral University Fellow. He was a Professor at the University of Thrace and a Professor at the AUTH. Dr. Sahalos is now Professor at the ECE Department of the University of Nicosia, Cyprus and director of the Radio & Telecommunications Laboratory (RTeLab). He was a visiting Professor at several universities in USA and Europe. He is the author of four books (three of them in Greek), seven book chapters, and more than 400 articles published in the scientific literature.

An Integrated Frequency Reconfigurable Antenna for Cognitive Radio Application

Sonia SHARMA, Chandra Charu TRIPATHI

Dept. of Electronics and Communication Engineering, UIET, Kurukshetra University, Kurukshetra, India

sonia990@gmail.com, tripathiuiet@gmail.com

Abstract. *A novel integrated antenna that consists of spectrum sensing and frequency reconfigurable antenna, printed on a single substrate is presented here. The integration of both antennas is based on the fact that reasonably large metallization area of UWB antenna can be used as a space for printing another narrowband antenna. Prominently, the second antenna should be placed on a region of large UWB antenna where there is less distribution of E field. The proposed antenna senses the spectrum over a wide frequency range from 2 GHz to 12 GHz via a U shape monopole having partial ground structure. After sensing the spectrum condition, the slot antenna with reconfigurable feeding structure can switch its operating bands among six narrowbands: 5.65 GHz, 3.6 GHz, 5 GHz, 2.94 GHz, 4.5 GHz, 2.15 GHz and two dual bands at 2 GHz, 5.48 GHz and 1.7 GHz, 5 GHz. In addition to this, the proposed antenna can also reconfigure its frequency in wide-bandwidth using two ground stubs. Since wide band response for spectrum sensing and each of eight frequency state works independently, therefore, there is no interference among various signals.*

Keywords

Cognitive radio, frequency reconfiguration, slot antenna, spectrum sensing, wideband antenna

1. Introduction

The next generation smart systems should have capability to automatically detect the spectrum condition and accordingly change the radio operating parameters to modify system performance such as maximize throughput, mitigate interference, facilitate interoperability etc. In addition, the physical size of these devices must follow the downsizing trend of the system unit. A smart Cognitive Radio (CR) compatible antenna with smaller physical size meets the essential requirement for these systems. Such antenna has a spectrum-sensing antenna for detecting the spectrum condition and reconfigurable antenna for communicating purpose [1], [2]. Many antenna designs have been already reported in the literatures, which are compatible with CR [3–6]. Further to decrease the system size, these two antennas must be optimally design on a single substrate. Only a handful of techniques based on integration concept of these antennas, on a single substrate are reported in literature [7–15]. In these techniques, small narrowband (NB) antenna are integrated in UWB antenna by using various topology viz. using large metallization area of UWB antenna as ground plane for a smaller antenna [7], [8], by utilizing the empty space in between the radiating element, and in between two tapered slots [9], by adding reconfigurable part in the feeding area of Vivaldi antenna [10], by adding reconfigurable band pass filter in feed line [11], by adding JK inverter in the switchable feed line structure [12], by placing the NB antenna at a point where electric field does not disturb the UWB antenna [13], by adding UWB antenna above the reconfigurable antenna [14], by reusing the printed area in different operating modes [14], etc. In all the above designs the numbers of frequency reconfiguration states are limited because of size constraint so there is underutilization of available area. Further up gradation of these antennas to increase the reconfiguration states is quite challenging because of limited space constraints on board. This may compromise the goal of combining the wideband and reconfigurable antenna together for CR systems.

Therefore, in the present work, UWB antenna and reconfigurable antenna are integrated in such a way that they do not disturb the characteristics of each other. The challenging goal in this work is to use the available space optimally for multiple purposes. For space management, the

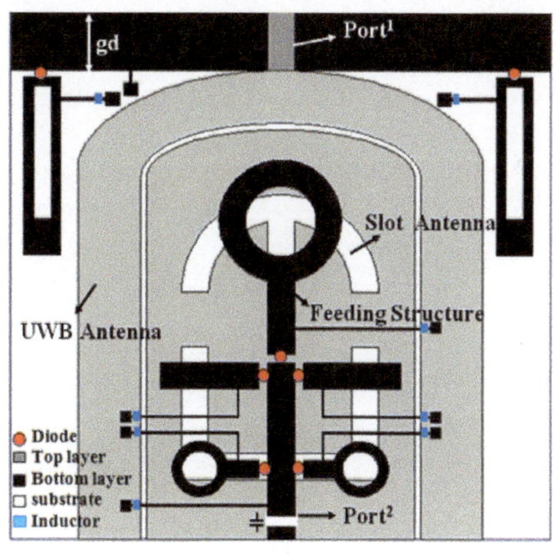

Fig. 1. Layout of the proposed integrated antenna.

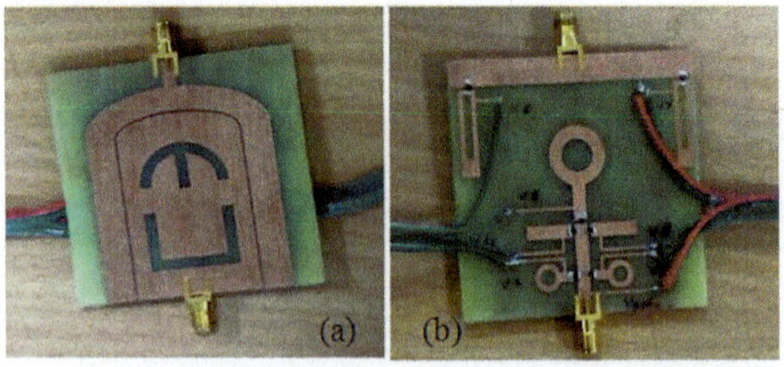

Fig. 2. Fabricated prototype: (a) Top view. (b) Bottom view.

large metal area of UWB antenna can be used for printing slot antenna and the partially etched ground area can be used for printing reconfigurable feeding structure for the slot antenna. Proper isolation of high field and low field area ensure the least interaction between two antenna and efficient utilization of space for multipurpose.

2. Design Strategy/Design Evolution

For CR systems, an antenna consists of two parts: a UWB antenna as a spectrum sensor and a reconfigurable NB antenna for communicating purpose. So, a U shape monopole, which is excited by port[1], is chosen as a spectrum sensor, over a wide spectrum range from 2 to 12 GHz. A reconfigurable slot antenna, which is excited by port[2], is used as communicating antenna in an operating range between 1 GHz to 6 GHz. The target is to integrate these two individual antennas on a single substrate to use the available space optimally such that space of one antenna can be used for printing two antennas. Importantly, the NB antenna should be positioned in a region of the UWB antenna where there is less concentration of electric field intensity. The high and low field areas on the UWB antenna are isolated by etching a thin slit in order to have space for the NB antenna. This ensures less interaction between the two antennas. The corresponding antenna topology is shown in Fig. 1 and the fabricated prototype is shown in Fig. 2. It consists of two layers. The top layer is the large U shape antenna and the bottom layer consists of partial ground plane and reconfigurable feeding structure. The design steps adopted for complete integration topology are explained below.

2.1 Integration Methodology

Here, the methodology adopted to integrate a NB antenna into the large UWB antenna is explained thoroughly. Firstly a U shape monopole antenna namely configuration I is designed on $6 \times 6\ cm^2$ substrate as shown in Fig. 3(a). The proper selection of monopole shape and partial ground plane size optimizes the impedance bandwidth of UWB antenna. Figure 5 shows the effect of partial ground size on the return loss characteristics in UWB range. The result clearly show that the UWB behavior of antenna improved as we increased the ground plane size from 5 mm to 7 mm. Further increasing the ground size, the UWB response starts degrading, so the optimum size of partial ground plane is chosen as 7 mm for UWB response from 2 to 12 GHz. Figure 4(a) shows the E field distribution in UWB antenna for configuration I. It is observed that the electric field intensity is very high on the outer periphery and very low intensity is found on the central area of the antenna. If one can separate the high E field area from low E field area

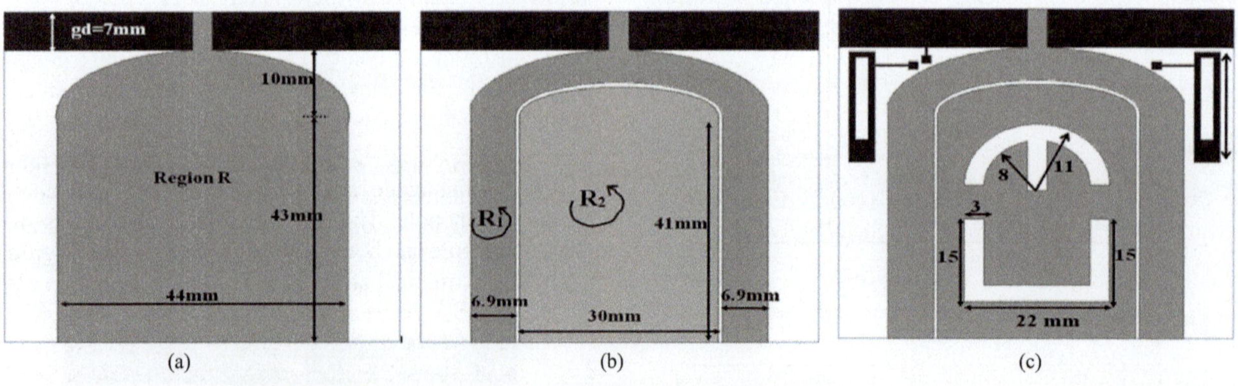

Fig. 3. Layout of UWB antenna for (a) Configuration I, (b) Configuration II, (c) Configuration III.

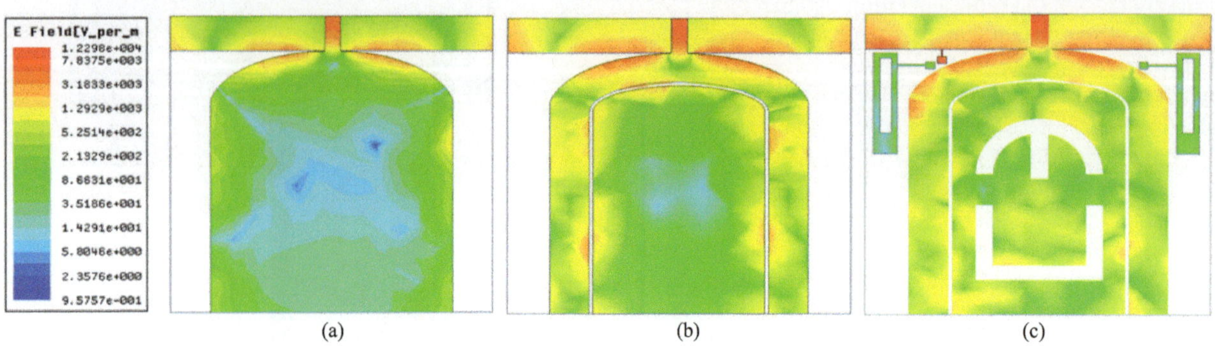

Fig. 4. E field distribution in (a) Configuration I (b) Configuration II, (c) Configuration III.

in UWB monopole without much affecting the S_{11} response, than this space can be reused to integrate another NB antenna. Therefore, in configuration-II a very thin slit is etched in the UWB antenna, which separates the region R into two parts R_1 as high E field area, R_2 very low E field area as shown in Fig. 3(b). The E field distribution for configuration-II is supporting the statement that region R_2 can act as nearly passive area for integrating another antenna. Figure 6 shows that S_{11} parameter from configuration I to II does not disturb so much and is acceptable of UWB operation from 2 GHz to 12 GHz.

Now, in configuration III, two slot antennas (U shape, inverted E shape) are successfully integrated in region R_2 without disturbing the UWB behavior. This configuration is used for spectrum sensing mode and its functioning is explained in detail later. All simulations are done using HFSS ver. 16. Figure 4 shows the E field distribution for all the three configurations and Figure 6 shows the comparisons of S_{11} parameter.

2.2 Mutual Coupling between Ports

In the proposed design, the UWB and the reconfigurable antenna are integrated together but work independently because both antennas are excited by different ports, one at a time. Both operating modes work separately without interfering each other's characteristics. Further, design optimization of integrated antenna is done so that passive antenna does not interrupt the operating state of active antenna. To prove that both antennas do not interfere with each other's characteristics, coupling coefficient for both the ports are studied.

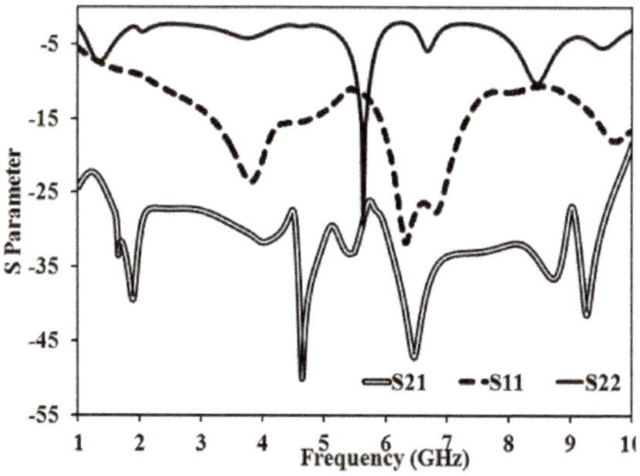

Fig. 7. S parameter (dB) vs. frequency for port isolation.

Figure 7 shows the simulated S parameter. The response clearly shows that S_{21} is better than 25 dB from 2 to 12 GHz, which proves that there is no transmission of RF power from port[1] to port[2]. So there is no need to design extra port isolation circuit because there is good isolation in both operating modes and this ensures no interaction of both the antennas.

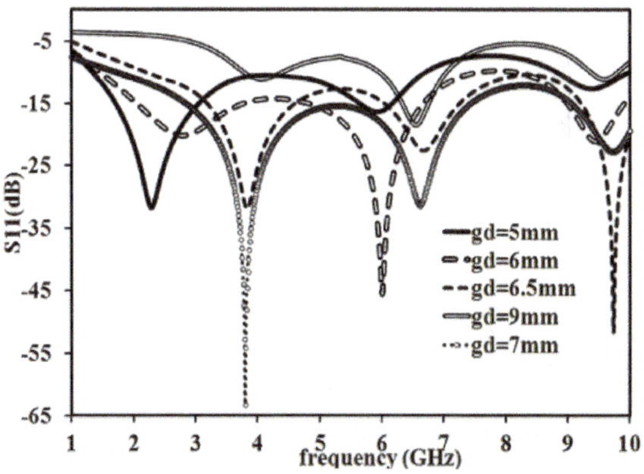

Fig. 5. Effect of ground plane size on UWB characteristics.

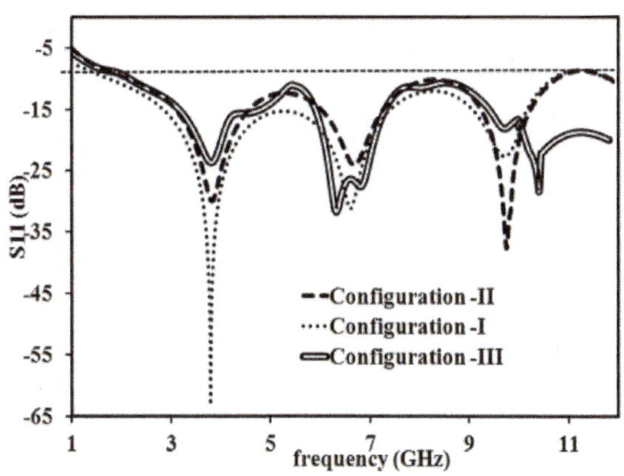

Fig. 6. Comparison of S_{11} vs. freq. for Configuration I, II, III.

3. Antenna Architecture and Working Operation

Now antenna architecture, geometry description and working for each mode is explained in detail below.

3.1 UWB Antenna Design

Figure 8 shows the active portion of integrated antenna when operated in spectrum sensing mode (configuration III). In this mode, antenna is fed by port[1] to excite the U shape monopole. The top layer structure consists of U shape monopole and bottom layer consists of partial ground and two symmetrical stubs (S_6, S_7). The two ground stubs namely S_6, S_7 are used to reconfigure the frequency in wideband (WB) mode. Diode D_6 and D_7 are used to control the state of stubs S_6 and S_7 independently. Three modes

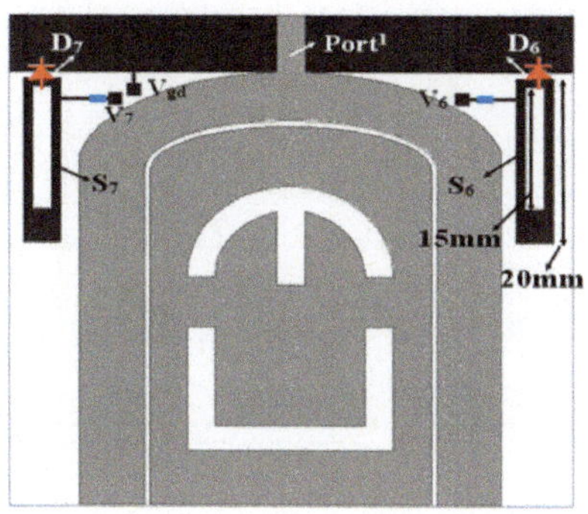

Fig. 8. Active area of integrated antenna in UWB mode.

have been studied using different combination of D_6 and D_7. When both diodes are in OFF state then UWB antenna operates in Spectrum Sensing mode from 2 to 12 GHz. When either one of the diode D_6 or D_7 is ON by applying +5 V on V_6 or V_7 pad then UWB antenna operates in WB mode-I. If both diodes D_6 and D_7 are forward biased by applying +5 V onV_6 and V_7 DC pad then UWB antenna operates in WB mode-II.

3.2 Frequency Reconfigurable Antenna Design

Figure 9 shows the active portion of the integrated antenna when it operates in frequency reconfigurable mode. The two slot antennas are imprinted such that they use switchable feed line structure for their active operation. Figure 9(b) shows feeding structure which is composed of two feed lines (F_1, F_2), four stubs (S_1, S_2, S_3, S_4), 5 PIN Diodes (D_1, D_2, D_3, D_4, D_5) and 6 DC pads (V_1, V_2, V_3, V_4, V_5, V_{gd}). The heart of the proposed design is electronically switchable feeding structure, which is specially designed for frequency reconfiguration using micro strip open stubs. These stubs are placed at proper position, such that these stubs match the antenna's input impedance at different desired frequencies. Five PIN diodes are used to control the activation of different stubs independently. The diode positioning along with DC biasing network are given in Fig. 9(b). The orientations of all PIN diodes are such that they share single ground pad. Applying 5 voltages on their respective DC pad and ground signal on V_{gd} pad activate the respective PIN diodes.

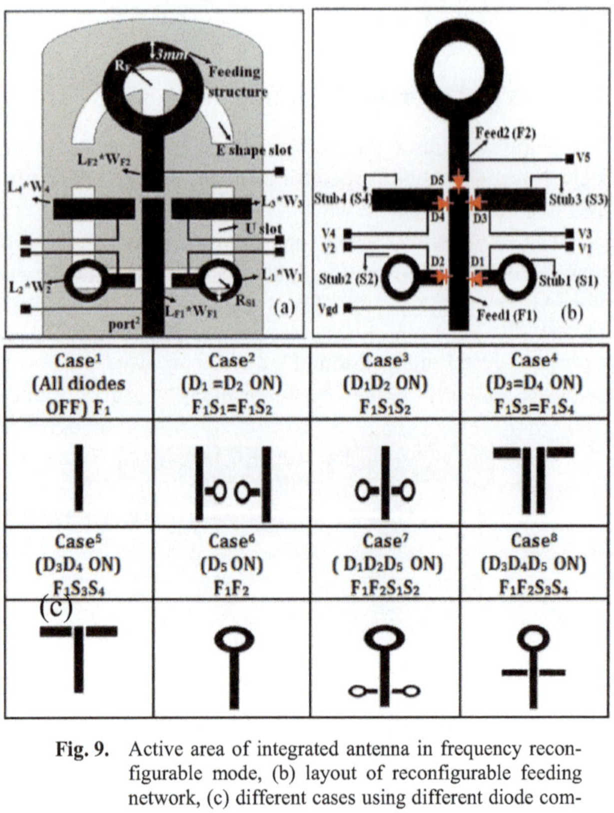

Fig. 9. Active area of integrated antenna in frequency reconfigurable mode, (b) layout of reconfigurable feeding network, (c) different cases using different diode combinations.

F_1	$Stub^1$=$Stub^2$	$Stub^3$=$Stub^4$	F_2	R_{S1}, R_F
$L_{F1} \times W_{F1}$ 20 × 3	$L_1 \times W_1$ = $L_2 \times W_2$ = 4.56 × 3	$L_3 \times W_3$ = $L_4 \times W_4$ = 12 × 3	$L_{F2} \times W_{F2}$ 9.23 × 3	2, 3

Tab. 1. Geometrical parameter of the feeding structure (mm).

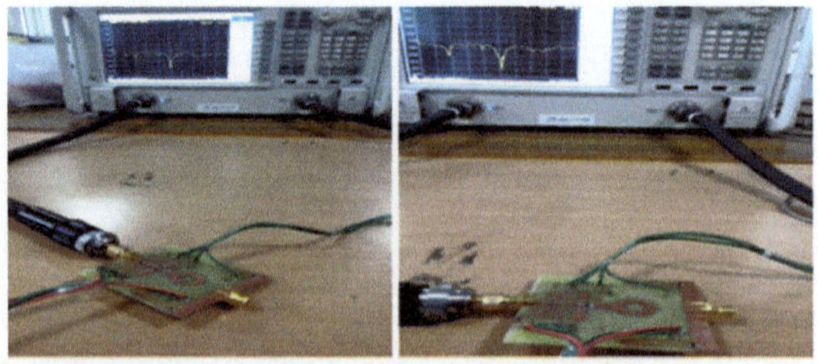

Fig. 10. Experimental setup for the integrated antenna.

A number of cases for different combination of these diodes can be envisaged. Figure 9(c) shows the diode state and pictorial view of active portion of feeding structure under different cases. When no diode is forward biased than feed line portion F_1 excites the lower U slot antenna (case[1]). Now by activating different stubs combination with feed line F_1, lower U slot antenna can reconfigure its resonant frequency over 4 different bands (case[2] - case[5]). When diode D_5 is forward biased than feed line portion F_1F_2 also excites the upper E shaped slot antenna (case[6]). For this combination two cases are also discussed, i.e. when stub S_1 and S_2 are attached with feed line F_1F_2 (case[7]) and stub S_3 and S_4 are attached with the feed line portion F_1F_2 (case[8]). The geometrical parameters for the reconfigurable feeding structure are given in Tab. 1.

4. Results and Discussion

To validate the proposed design, the prototype has been fabricated and results are verified experimentally here. Figure 10 shows the experimental setup for integrated antenna. In the prototype, Infineon BAR-6402 diodes are used here. In Ansoft HFSS 16, PIN diodes are model using lumped RLC boundary to add the effect of actual diodes.

4.1 Spectrum Sensing Mode/ WB Frequency Reconfigurable Mode

Antenna works in normal mode or spectrum sensing mode, when both stubs (S_6, S_7) are not attached with the partial ground structure (D_6 D_7 -00). Figure 11 shows the simulated and measured S_{11} vs. frequency for normal mode. It is clear that simulated and measured S_{11} is better than −10 dB in UWB bandwidth from 2 GHz to 10 GHz. Figure 11 also shows the very low electric field distribution is observed in the central part of the antenna and maximum on the outer edges of the U shape monopole.

The gain of the antenna is 2.6 dB at 5 GHz as shown in Fig. 12(a). The E and H plane pattern is nearly omnidirectional except a discontinuity at $\theta = 180°$. Figure 12(b) shows the simulated and measured E plane pattern at 5 GHz.

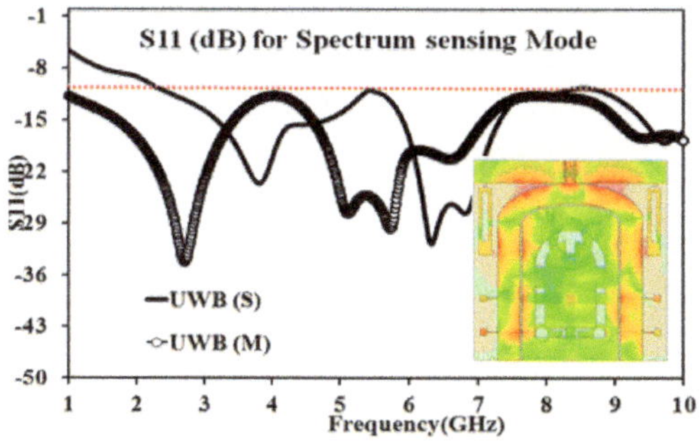

Fig. 11. S_{11} vs. frequency in normal mode.

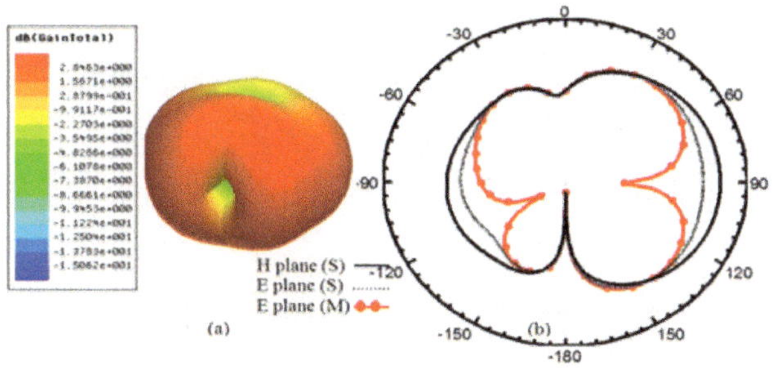

Fig. 12. (a) Gain polar plot. (b) Observed radiation pattern at 5 GHz.

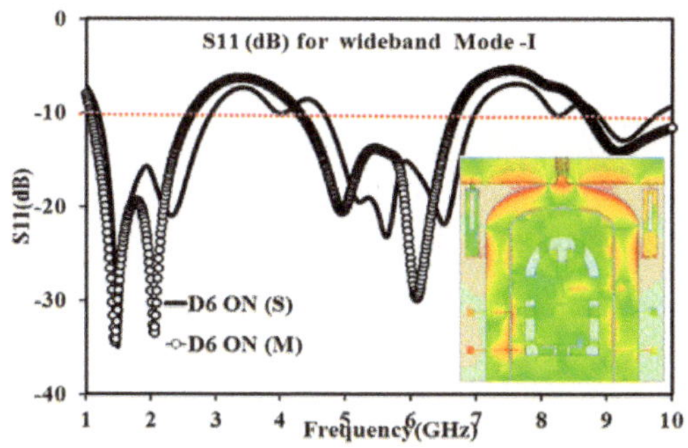

Fig. 13. S_{11} vs. frequency in wideband mode- I.

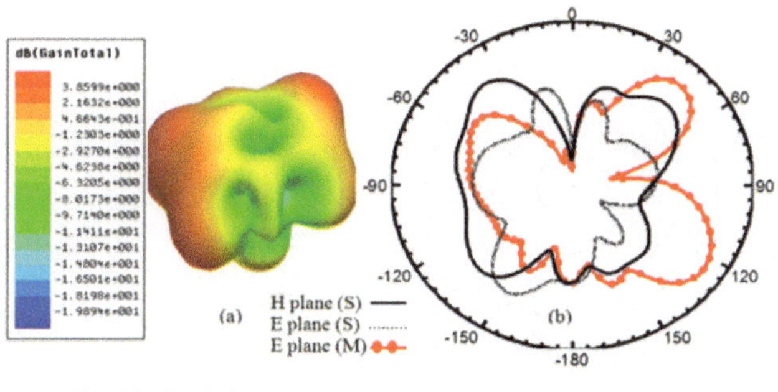

Fig. 14. (a) Gain polar plot, (b) Observed radiation pattern at 5 GHz.

When diode D_6 is forward biased, then stub S_6, reconfigures the operating bandwidth and the antenna works for WB-I. Now the antenna has two WBs from 1 to 3 GHz and from 5 to 7 GHz as shown in Fig. 13. Simulated and measured S_{11} is better than –10 dB for 1–3 GHz and 5–7 GHz. Figure 13 shows that E field redistributes under UWB antenna and stub S_6. The observed value of gain is 2.4 dB at 5 GHz as shown in Fig. 14(a). Simulated and measured radiation pattern at 5 GHz has distorted shape as shown in Fig. 14(b). If both diodes D_6 and D_7 are forward

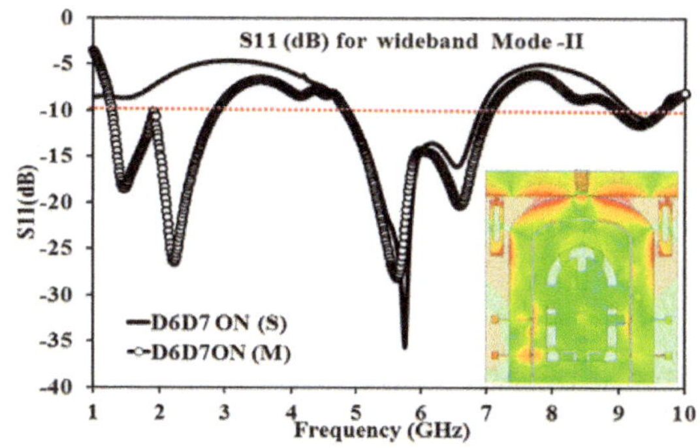

Fig. 15. S_{11} vs. frequency in wideband mode-II.

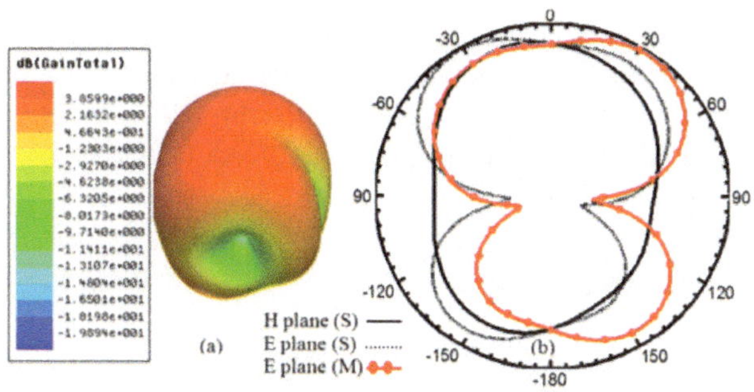

Fig. 16. (a) Gain polar plot, (b) Observed radiation pattern at 5 GHz.

biased then stub S_6 and S_7 are attached with the partial ground to tune the operating bandwidth.

In this case, the antenna operates in WB-II. Simulated and measured S_{11} is better than –10 dB from 5 to 7 GHz. Figure 15 also shows that E field redistributes and the maximum field is found under UWB antenna and stubs S_6 and S_7. In this case, the maximum value of gain is 3.6 dB as shown in 3D polar plot of gain in Fig. 16(a). The shape of H plane is nearly omni-directional and E plane is nearly dumbbell-shaped as shown in Fig. 16(b).

4.2 NB Frequency Reconfigurable Mode

In this mode, different combinations of stubs are activated at different times to achieve frequency reconfiguration. All the cases are discussed by comparing simulated S_{22} with the measured value and analyzing electric field distribution. The 3D polar plot for gain and simulated and measured radiation pattern for each mode is also given for each frequency state. In case[1], when all the diodes are in OFF condition then feedline section F_1 excites the U slot structure. Figure 17(a) shows the simulated and measured S_{22} vs. frequency for this case. The simulated value of return loss (RL) is 30 dB at resonating frequency 5.65 GHz whereas the measured value of RL = 28 dB at 5.68 GHz. Electric field distribution for case[1] is also given in Fig. 17(a) which clearly shows that maximum E field is distributed under F_1 portion. The observed value of gain is 3.43 dB and comparison of simulated and measured radiation pattern is shown in Fig. 17(b-c) respectively. In case[2], either of diode D_1 or D_2 is forward biased by applying +5 V on V_1 or V_2 pad then antenna resonates at 3.6 GHz as shown in Fig. 18. In this case, the U slot antenna is fed by feedline section F_1 in conjunction with stub[1] or stub[2]. It should be noted here that resonating frequency for feedline structure F_1S_1 or F_1S_2 is found the same because geometry of stub[1] and stub[2] are identical so they are symmetrically placed across feedline section F_1. In this case, the simulated frequency is 3.6 GHz with $RL = 28.3$ dB whereas the measured frequency is 3.69 GHz with $RL = 40$ dB as shown in Fig. 18. The maximum E field intensity is found under F_1S_1 structure. The observed value of gain is 2.8 dB and simulated and measured radiation patterns are shown in Fig. 18(c) respectively.

In case[3], when diode D_1 and D_2 is forward biased by applying +5 V on DC pad V_1 and V_2 then active feed portion is F_1 along with stub[1] and stub[2]. In this case, the simulated frequency is 5 GHz with $RL = 30$ dB whereas the measured frequency is 5.15 GHz with $RL = 30$ dB as shown in Fig. 19. In this case, electric field is maximum under the $F_1S_1S_2$ structure and at the corner of the U slot antenna. The observed value of gain is 2.7 dB. The simulated and measured radiation patterns are shown in Fig. 19(c).

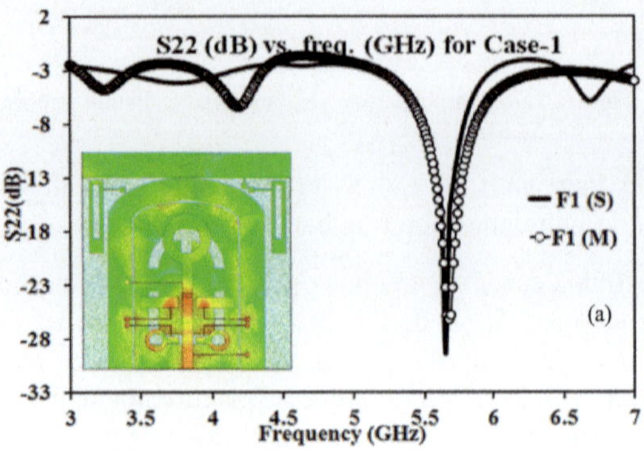

Fig. 17(a) S_{22} vs. frequency for case[1].

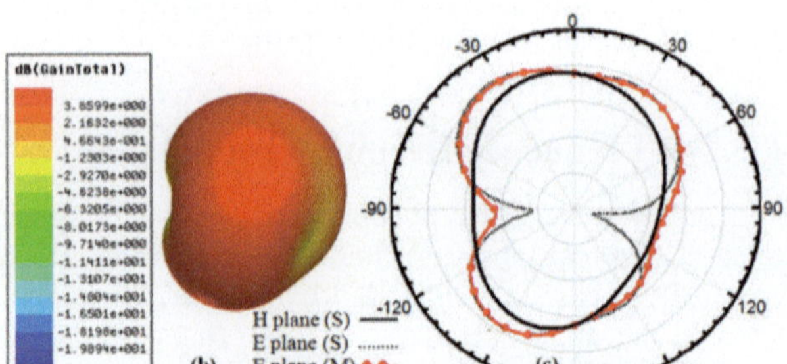

Fig. 17(b) Gain polar plot. **(c)** Observed radiation pattern for case[1].

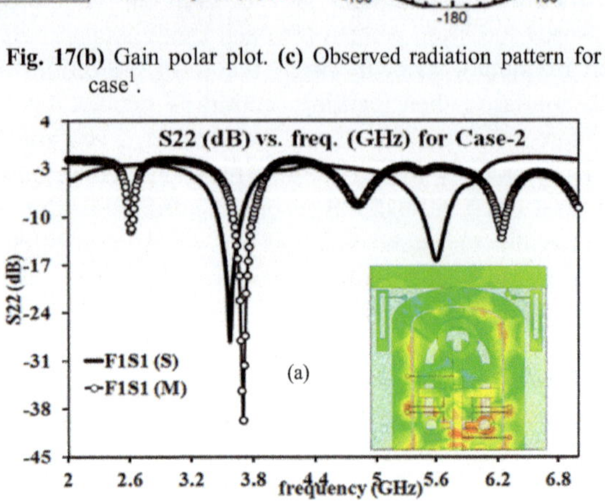

Fig.18(a) S_{22} vs. frequency for case[2].

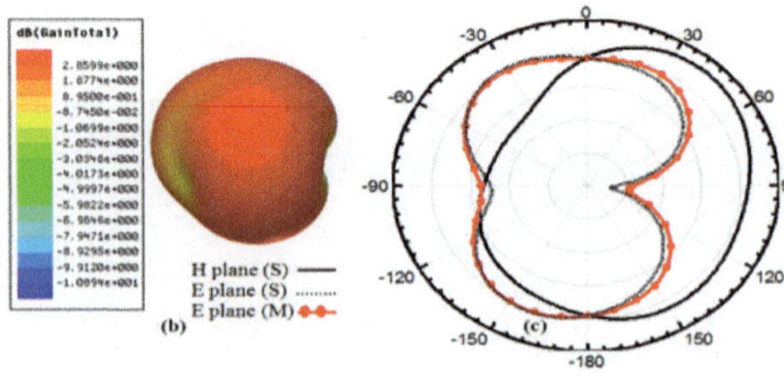

Fig. 18(b) Gain polar plot. **(c)** Observed radiation pattern for case[2].

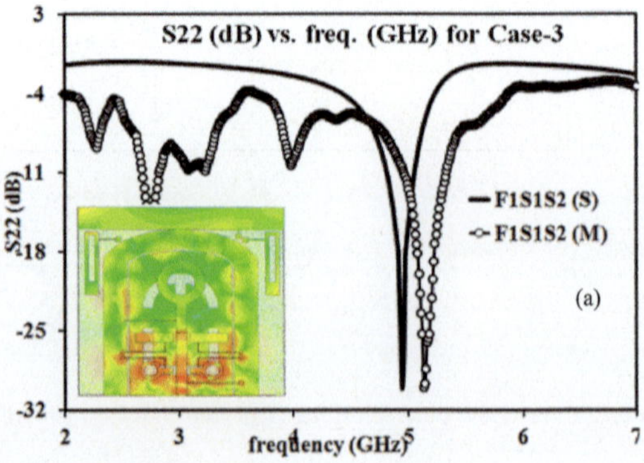

Fig. 19(a) S_{22} vs. frequency for case[3].

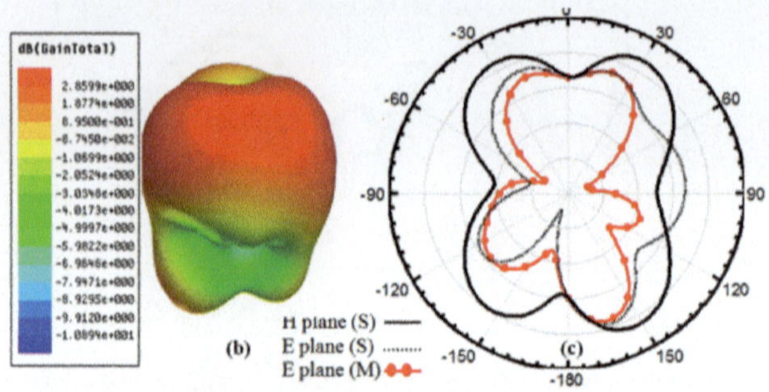

Fig. 19(b) Gain polar plot. **(c)** Observed radiation pattern for case[3].

In case[4], when either of diode D_3 or D_4 is forward biased then either stub[3] or stub[4] is selected to tune the resonating frequency for dual band. Again in this case, due to the geometrical symmetry of F_1S_3 or F_1S_4 structure because of the same geometrical parameters of stub[3] and stub[4], their resonating frequency is found the same for both cases. So only the results for F_1S_3 are given here. In this case, simulated dual band is found at 2 GHz and 5.48 GHz with $RL = 36$ dB, and 35 dB respectively and the measured dual band is obtained at 1.89 and 5.38 GHz with $RL = 27$ and 38 dB, respectively, as shown in Fig. 20(a). The observed value of gain is 2.2 dB and the simulated and measured radiation patterns are shown in Fig. 20(c), respectively. The discrepancies between the simulation and measurement radiation patterns can be attributed to the presence of the diodes, in close proximity to the antenna during the pattern measurements.

In case[5], when diode D_3 and D_4 are forward biased by applying +5 V on DC pad V_3 and V_4, then the U slot is excited by T shape feedline structure ($F_1S_3S_4$). In this case, the simulated resonating frequency is 2.94 GHz with $RL =$

34 dB whereas the measured frequency is 3 GHz with *RL* = 35 dB as shown in Fig. 21(a). The electric field is found high under $F_1S_3S_4$ structure as shown in Fig. 21(a). The observed value of gain is 3.7 dB and radiation patterns are shown in Fig. 21(c).

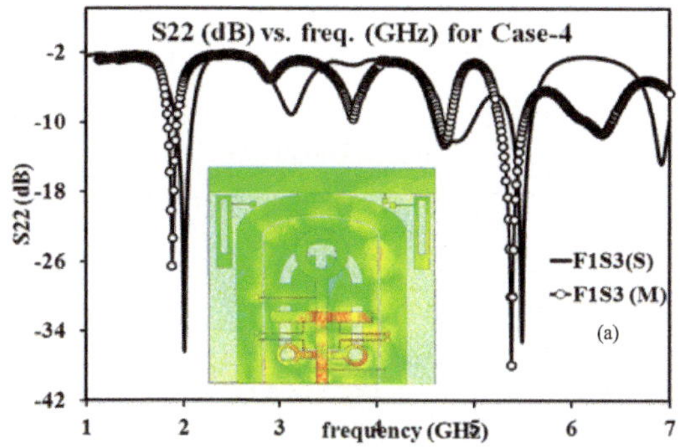

Fig. 20(a) S_{22} vs. frequency for case[4].

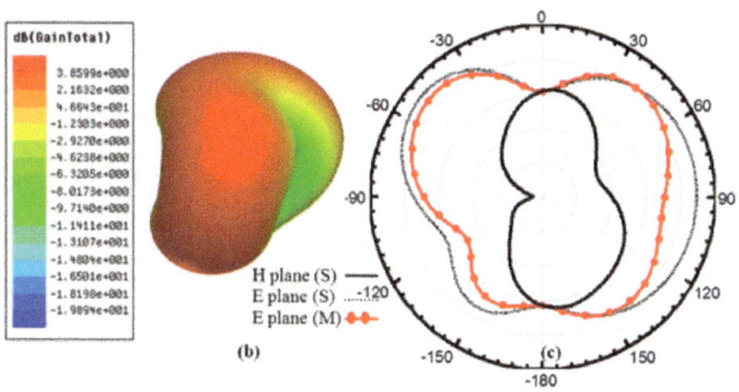

Fig. 20(b) Gain polar plot. **(c)** Observed radiation patterns for case[4].

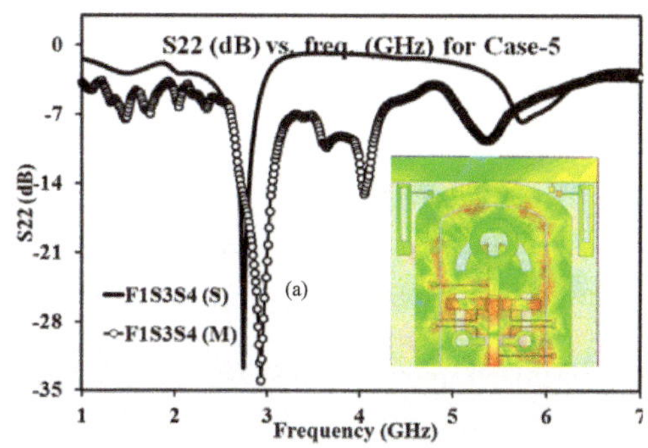

Fig. 21(a) S_{22} vs. frequency for case[5].

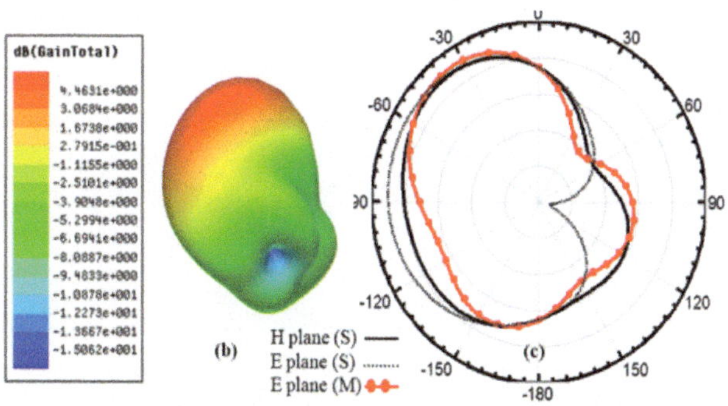

Fig. 21(b) Gain polar plot. **(c)** Observed radiation pattern for case[5].

In case[6], upper inverted E shaped slot is excited by feed line F_1F_2 by forward biasing the diode D_5. The antenna this time resonates at 4.48 GHz with *RL* = 36 dB whereas the measured resonating frequency is 4.5 GHz with *RL* = 40 dB as shown in Fig. 22(a). The observed value of gain is 1.8 dB and radiation patterns are shown in Fig. 22(b-c).

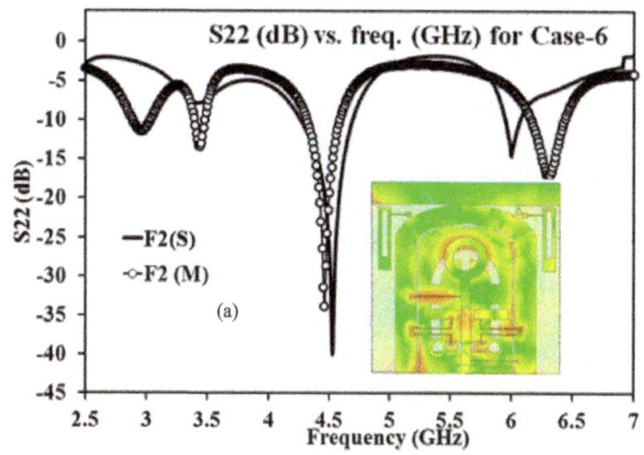

Fig. 22(a) S_{22} vs. frequency for case[6].

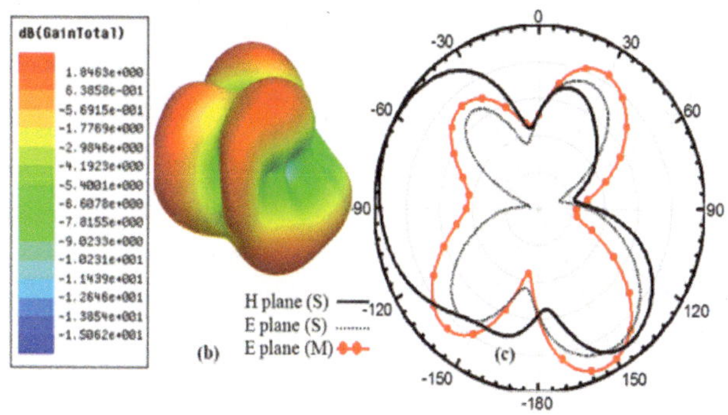

Fig. 22(b) Gain polar plot. **(c)** Observed radiation patterns for case[6].

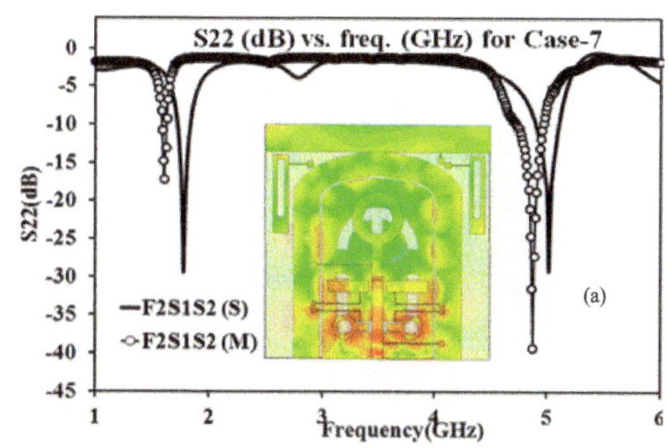

Fig. 23(a) S_{22} vs. frequency for case[7].

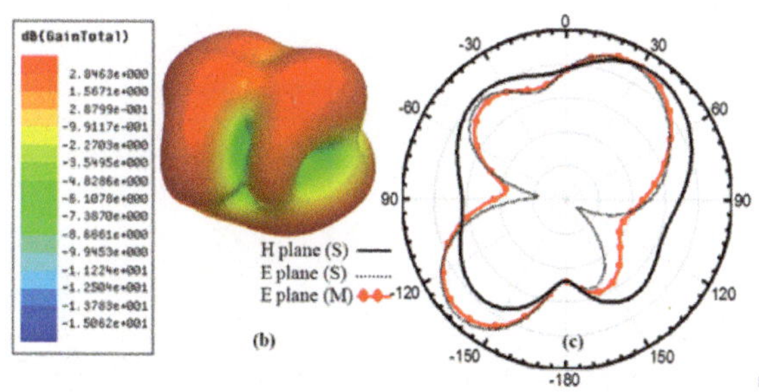

Fig. 23(b) Gain polar plot. (c) Observed radiation patterns for case[7].

In case[7], when diode D_1, D_2, D_5 is forward biased then the slot antenna is excited by feed line section F_1F_2 in conjunction with stub[1] and stub[2]. The electric field distribution for this case is given in Fig. 23(a), which clearly indicates that maximum electric intensity is found under $F_1F_2S_1S_2$ structure. The simulated dual band is found at 1.7 GHz and 5 GHz with the measured RL = 30 dB and 31 dB respectively whereas the measured dual band is at 1.66 GHz and 4.88 GHz with RL = 18 dB and 31 dB respectively as shown in Fig. 23(a). The observed value of gain is 2.7 dB and the simulated and measured radiation patterns are shown in Fig. 23(c).

In case[8], feed line section $F_1F_2S_3S_4$ excites the slot antenna by forward biasing the diodes D_3, D_4, D_5. In this case, the antenna resonates at 2.15 GHz with RL = 38 dB whereas the measured RL = 27 dB is at 2 GHz as shown in Fig. 24(a). E field distribution also shows that maximum field is distributing under $F_1F_2S_3S_4$ structure. The observed value of gain is 1.7 dB and the simulated and measured radiation patterns are shown in Fig. 24(c). Simulated and measured parameters are also tabulated in Tab. 2.

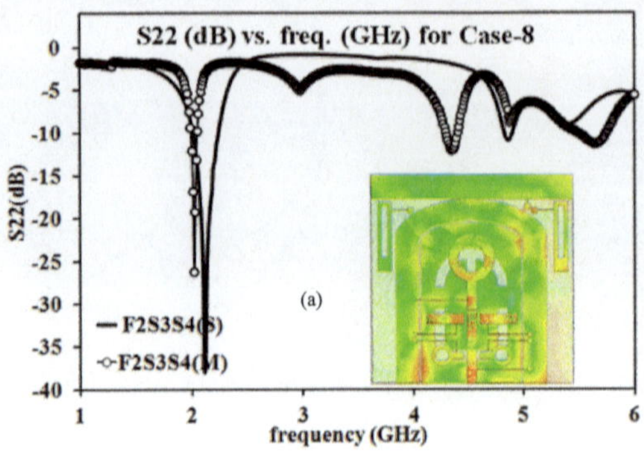

Fig. 24(a) S_{22} vs. frequency for case[8].

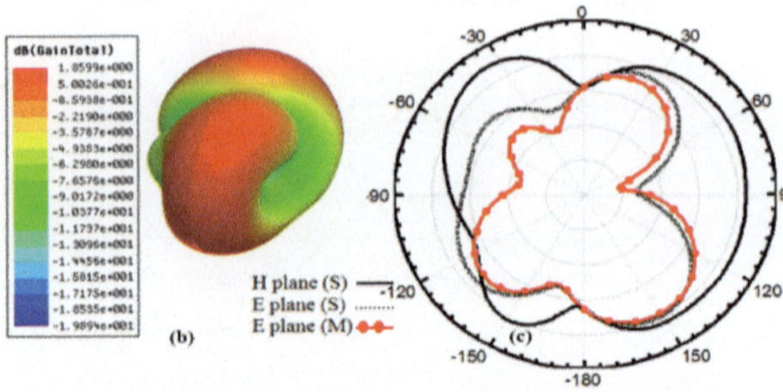

Fig. 24(b) Gain polar plot. **(c)** Observed radiation patterns for case[8].

	Freq. (GHz)	*BW* (MHz)	*RL* (dB)	Eff. η%
Case [1]	5.65	340	30	81
Case [2]	3.6	220	28.3	72
Case [3]	5	260	30	74
Case [4]	2 5.48	160 210	36 35	69 69.8
Case [5]	2.94	190	34	79
Case [6]	4.5	390	40	70
Case [7]	1.7 5	110 210	30 31	71 69
Case [8]	2.15	220	38	67

Tab. 2. Simulated parameters.

5. Conclusion

In this work, we integrate the NB antenna into a large UWB antenna without affecting each performance. The UWB antenna, slot antenna was printed onto the top side of the substrate and the reconfigurable feeding structure for the U slot antenna was printed onto the bottom side of the substrate. This has the advantage of reducing space requirements and making the two antennas lying in the same plane required for cognitive radio communication. The arrangement has potential for various CR applications.

References

[1] HUSSAIN, R., SHARAWI, M. S. A cognitive radio reconfigurable MIMO and sensing antenna system. *IEEE Antennas and Wireless Propagation Letters,* 2015, vol. 14, p. 257–260. DOI: 10.1109/LAWP.2014.2361450

[2] TAWK, Y, CONSTANTINE, J., CHRISTODOULOU, C. G. Cognitive-radio and antenna functionalities: A tutorial. *IEEE Antennas and Propagation Magazine*, 2014, vol. 56, no. 1, p. 231 to 243. DOI: 10.1109/MAP.2014.6821791

[3] SHARMA, S., TRIPATHI, C. C. Frequency reconfigurable U-slot antenna for SDR application. *Progress in Electromagnetics Research Letters,* 2015, vol. 55, p. 129–136. DOI: 10.2528/PIERL15071304

[4] ABOUFOUL, T., ALOMAINY, A., PARINI, C. Reconfiguring UWB monopole antenna for cognitive radio applications using GaAs FET switches. *IEEE Antennas and Wireless Propagation Letters,* 2012, vol. 11, p. 392–395. DOI: 10.1109/LAWP.2012.2193551

[5] TAWK, Y., BKASSINY, M., EL HOWAYEK, G., et al. Reconfigurable front-end antennas for cognitive radio applications. *IET Microwaves, Antennas and Propagation,* 2011, vol. 5, no. 8, p. 985–992. DOI: 10.1049/iet-map.2010.0358

[6] WU, G. T., LI, R. L., EOM, S. Y., et al. Switchable quad-band antennas for cognitive radio base station applications. *IEEE Transactions on Antennas and Propagation*, 2010, vol. 58, no. 5, p. 1468–1476. DOI: 10.1109/TAP.2010.2044472

[7] EBRAHIMI, E., KELLY, J. R., HALL, P. S. Integrated wide-narrowband antenna for multi-standard radio. *IEEE Transactions on Antennas and Propagation*, 2011, vol. 59, no. 7, p. 2628–2635. DOI: 10.1109/TAP.2011.2152353

[8] MANTEGHI, M. A switch-band antenna for SDR applications. *IEEE Antennas and Wireless Propagation Letters*, 2009, vol. 8, p. 3–5. DOI: 10.1109/LAWP.2008.2005256

[9] AUGUSTIN, G., DENIDNI, T. A. An integrated ultra wideband /narrow band antenna in uniplanar configuration for cognitive radio systems. *IEEE Transactions on Antennas and Propagation,* 2012, vol. 60, no. 11, p. 5479–5484. DOI: 10.1109/TAP.2012.2207688

[10] HAMID, M. R., GARDNER, P., HALL, P. S., et al. Vivaldi antenna with integrated switchable band pass resonator. *IEEE Transactions on Antennas and Propagation,* 2011, vol. 59, no. 11, p. 4008–4015. DOI: 10.1109/TAP.2011.2164197

[11] QIN, P. F., WEI, F., JAY GUO, Y. A wideband-to-narrowband tunable antenna using a reconfigurable filter. *IEEE Transactions on Antennas and Propagation,* 2015, vol. 63, no. 5, p. 2282–2285. DOI: 10.1109/TAP.2015.2402295

[12] SHARMA, S., TRIPATHI, C. C. Wideband to concurrent tri-band frequency reconfigurable microstrip patch antenna for wireless communication. *International Journal of Microwave and Wireless Technologies*, 2016, vol. 9, no. 4, p. 915–922. DOI: 10.1017/S1759078716000763

[13] SHARMA, S., TRIPATHI, C. C. A wide spectrum sensing and frequency reconfigurable antenna for cognitive radio. *Progress in Electromagnetics Research C*, 2016, vol. 67, p. 11–20. DOI: 10.2528/PIERC16070803

[14] SHARMA, S., TRIPATHI, C. C. A versatile reconfigurable antenna for cognitive radio. In *Proceedings of the Asia Pacific Microwave Conference (APMC 2016)*. New Delhi (India), 2016, 4 p. DOI: 10.1109/APMC.2016.7931320

[15] SHARMA, S., TRIPATHI, C. C., A novel reconfigurable antenna with spectrum sensing mechanism for CR system. *Progress in Electromagnetics Research C*, 2017, vol. 72, p. 187–196. DOI: 10.2528/PIERC17010901

About the Authors ...

Sonia SHARMA received her M.Sc. degree in Electronics Science from Maharshi Dayanand University, Rohtak and M.Tech. in Electronics and Communication Engineering from UIET, Kurukshetra University, Kurukshetra. She is pursuing her Ph.D. in the area of reconfigurable antenna from UIET now.

Chandra Charu TRIPATHI received the B.Sc.(Honors-Physics) and M.Sc. (Solid State Electronics Devices) degrees from Banaras Hindu University, Varanasi, followed by Master in Engineering (Microelectronics) from BITS, Pilani, India. Presently, he is a Professor in the Department of Electronics and Communication Engineering at UIET, Kurukshetra University, Kurukshetra. He has more than 50 papers in refereed journals and 40 in national/ international conferences to his credit.

FPGA-based Low Latency Inverse QRD Architecture for Adaptive Beamforming in Phased Array Radars

Raafia IRFAN[1], *Haroon ur RASHEED*[1], *Waqas Ahmed TOOR*[2]

[1] Dept. of Electrical Engineering, Pakistan Inst. of Eng. and Applied Sciences (PIEAS), Nilore, Islamabad-45650, Pakistan
[2] Dept. of Electrical Engineering, Capital University of Science and Technology (CUST), Express Way, Zone 5, Islamabad-45650, Pakistan

fac2013.pieas@gmail.com

Abstract. *The main objective of this paper is to facilitate the adaptive beamforming which is one of the most challenging tasks in phased array radars receivers. Recursive least square (RLS) is considered as the most well suited adaptive algorithm for the applications where beamforming is mandatory, because of its good numerical properties, high convergence rate and low misadjustment. In this paper, some RLS variants are discussed and the most numerically suitable algorithm Inverse QR decomposition (IQRD) is selected for efficient adaptive beamforming. A novel architecture for IQRD RLS is also presented, which offers low latency and low area occupation for Field Programmable Gate Array (FPGA) implementation. This approach reduces the computations by utilizing the standard pipelining methodology. Hence, efficient adder and multipliers and Look Up Table (LUT) based solution for square root and division, has highly enhanced the performance of the algorithm. The proposed IQRD RLS architecture has coded in Verilog and analyzed its performance in terms of throughput, hardware resources and efficiency.*

Keywords

Adaptive filters, beamforming, phased-array radars, Recursive Least Square (RLS), Inverse QR Decomposition (IQRD), Field Programmable Gate Array (FPGA)

1. Introduction

Radar signal processing is one of the active area of research due to its vast applications in radar systems like target tracking, navigation. Most of the radar systems are composed of multiple transmitting and receiving phased array antennas, known as phased array radars. These phased array antennas are designed to transmit and receive the spatially propagating signals and to focus on the arriving direction of desired signals as well as suppressed the interfering signals [1], [2]. Beamforming is an important signal processing technique in phased array radars for directional signal transmission or reception [2]. Figure 1 shows a generalized block diagram of four-channel adaptive beamformer receiver having a uniform linear array structure (ULA) and an adaptive weight calculation (AWC) unit. In ULA, phased array radars, all the antenna elements are arranged along a line with uniform spacing *s*, to attenuate interferences coming from different directions along with desires signal [3]. AWC unit is based on an over determined system of equations in many cases and various algorithms has been developed to solve these equations. Such a system is also known as adaptive antenna system. AWC usually based on gradient based adaptive algorithms like Least Mean Square (LMS), Normalized Least Mean Square (NLMS), Recursive Least Square (RLS), Kalman filter, assuming all the input signals to ULA system are uncorrelated with each other [3], [4]. But, RLS is considered as the most commonly used algorithm as it has the fast convergence rate to find an appropriate solution and good numerical properties, but it has complex mathematical operations like matrix inversion. Hence, it limits its application in real time systems [5]. In numerical methods, a well-known method known as QR decomposition (QRD) is used to decompose any matrix $\mathbf{X}$ into orthogonal matrix $\mathbf{Q}$ ($\mathbf{QQ}^T = \mathbf{I}$), and upper triangular matrix $\mathbf{R}$ ($\mathbf{X} = \mathbf{QR}$). So, to overcome the computational complexity, various QR decomposition based RLS variants (QRD, Inverse QRD, Fast QRD) were developed, which are widely used in variety of applications, which leads to more efficient architectures, stability and accuracy [5], [6]. In the paper, the comparisons of few RLS variants are performed and select IQRD RLS as the least computationally expensive algorithm for hardware application.

The primary goal of this paper is to propose and develop a high-throughput architecture of the selected RLS variant, by minimizing its latency and device area occupation for adaptive beamforming application. The speed of the beamformer is limited due to recursive operation based AWC unit, therefore, minimizing latency is a critical issue [4], [7]. In this paper, a novel architecture is proposed for IQRD RLS to reduce the computational complexity by using the standard pipelining methodology and mapped into Virtex5 FPGA because it provides the powerful com-

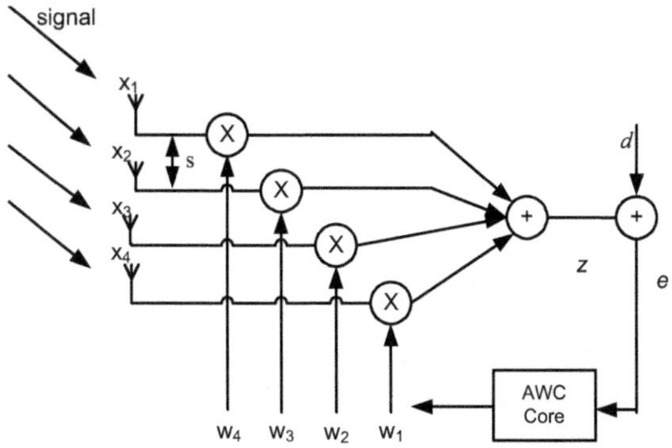

Fig. 1. Adaptive beamformer receiver with ULA structure.

putational architecture features for floating point arithmetic.

The paper is organized as follows: first of all, a brief introduction of adaptive beamforming is presented in Sec. 2. Further, in Sec. 3, a thorough study is performed to select IQRD as the most suitable RLS variant. Section 4 describes the conventional systolic array architecture and the proposed new architecture based on low latency modules. Finally, in Sec. 5, AWC unit implementation in Virtex 5 FPGA is discussed to confirm the advantages of the new structure.

2. Adaptive Beamforming

The core of adaptive beamformer receiver is to detect and estimate the target signal at the input of adaptive antenna array. Such systems are usually consisted of an antenna array along with adaptive processor (AWC) to adjust its weights in real time, such that the main lobe of the radar is focused on the arriving direction of target signal and suppresses undesired signals [3], [6].

The output of beamformer in Fig. 1 is given as:

$$\boldsymbol{e}(k)=\boldsymbol{d}(k)-\boldsymbol{z}(k) \tag{1}$$

where z be the vector array of estimated signal given as

$$\boldsymbol{X}\boldsymbol{w}=\boldsymbol{z} \tag{2}$$

where $\boldsymbol{w}$ ($m \times 1$) be weights vector, $\boldsymbol{X}$ ($m \times N$, $m > N$) is the input data matrix, $\boldsymbol{d}$ ($N \times 1$) is the desired training signal and $\boldsymbol{e}$ ($N \times 1$) is the additive noise error vector. The primary goal of the adaptive beamformer is to minimize the mean square error, such that $z(k) \approx d(k)$ and find out the optimized values of unknown weight vector $\boldsymbol{w}$.

$$\xi(k)=\sum_{i=0}^{k}\lambda^{k-i}\left[\boldsymbol{d}(i)-\boldsymbol{w}^{\mathrm{T}}(i)\boldsymbol{x}(k)\right]^2=\left\|\boldsymbol{e}(k)^2\right\| \tag{3}$$

where λ is a forgetting factor in conventional RLS [4].

3. RLS Variants for Adaptive Beamforming

In this section, a detail study of RLS variants is performed to select the most efficient one for the AWC unit in adaptive beamformer. But, first of all a brief introduction of QR decomposition based on Givens Rotation (GR) method is given. A GR rotation matrix $\boldsymbol{G}=\begin{bmatrix} c & s \\ -s & c \end{bmatrix}$ is used to premultiply a two-row matrix and transformed it into upper triangular matrix. Let two-row matrix be given as:

$$\boldsymbol{A}=\begin{bmatrix} \sqrt{\lambda}r_1 & \sqrt{\lambda}r_2 & \dots & \sqrt{\lambda}r_m \\ x_1 & x_2 & \dots & x_m \end{bmatrix}. \tag{4}$$

Premultiplying matrix $\boldsymbol{A}$ with $\boldsymbol{G}$ we get

$$\begin{bmatrix} c & s \\ -s & c \end{bmatrix}\begin{bmatrix} \sqrt{\lambda}r_1 & \sqrt{\lambda}r_2 & \dots & \sqrt{\lambda}r_m \\ x_1 & x_2 & \dots & x_m \end{bmatrix}=\begin{bmatrix} r_1' & r_2' & \dots & r_m' \\ 0 & x_2' & \dots & x_m' \end{bmatrix} \tag{5}$$

where

$$r_1'=\sqrt{\lambda r_1^2+x_1^2}\ , \tag{6}$$

$$c=\frac{\sqrt{\lambda}r_1}{\sqrt{\lambda r_1^2+x_1^2}}, \tag{7}$$

$$s=\frac{x_1}{\sqrt{\lambda r_1^2+x_1^2}}. \tag{8}$$

In [7], a numerically stable RLS variant QRD RLS, is described which avoids the complex matrix inversion operations. Moreover, it has the possibility of implementation in systolic array architecture (pipelined architecture). Hence, many researchers considered QRD RLS as the most suitable adaptive algorithm for hardware implementation [8–10]. The QRD decomposes input data matrix $\boldsymbol{X}$ through an orthogonal triangularization approach like Givens rotation (GR), Householder transformation or Gram Schmidt orthogonalization [11]

$$\boldsymbol{X}=\boldsymbol{Q}\boldsymbol{R}. \tag{9}$$

In most of the cases GR is considered as the most numerically well-conditioned method. $\boldsymbol{Q}$ be the unitary matrix ($\boldsymbol{Q}\boldsymbol{Q}^{\mathrm{T}}=\boldsymbol{I}$) generated as the sequence of Givens rotation matrices in [11], and $\boldsymbol{R}$ is the upper triangle matrix which is also the Cholesky factor of autocorrelation matrix $\boldsymbol{R}=\boldsymbol{X}^{\mathrm{T}}\boldsymbol{X}$ [11]. So, equation (9) can be written as:

$$\boldsymbol{Q}\boldsymbol{X}=\boldsymbol{R} \tag{10}$$

Premultiplying (1) with $\boldsymbol{Q}$ we get

$$\boldsymbol{Q}\boldsymbol{e}=\boldsymbol{Q}\boldsymbol{d}-\boldsymbol{Q}\boldsymbol{X}\boldsymbol{w}. \tag{11}$$

Let $\boldsymbol{e}'=\boldsymbol{Q}\boldsymbol{e}$ and $\boldsymbol{d}'=\boldsymbol{Q}\boldsymbol{d}$, so equation (11) can be written as

$$\boldsymbol{e}'=\boldsymbol{d}'-\boldsymbol{R}\boldsymbol{w}. \tag{12}$$

From (12) the weight update equation can be written as

$$\boldsymbol{w}=\boldsymbol{R}^{-1}\boldsymbol{d}',\quad \boldsymbol{e}'\approx 0. \tag{13}$$

The above equation is the solution of least square problem which is obtained by back substitution process. Hence, the QRD RLS algorithm is composed of two steps: first the QR

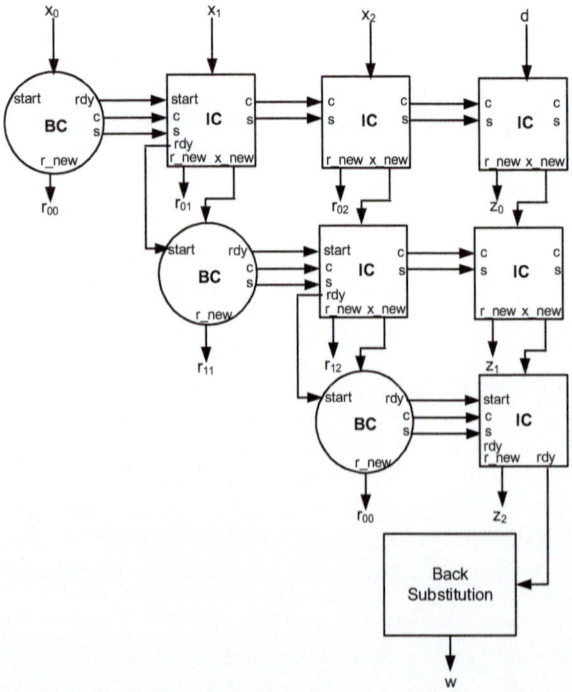

Fig. 2. QRD RLS systolic array.

decomposition of data matrix $\boldsymbol{X}$ (9) is performed and has systolic realization. In step 2, back substitution process begins for weight calculation (13), as shown in Fig. 2.

IQRD RLS is another important variant of RLS proposed by Alexander and Ghirnikar [12], to avoid the numerically tedious back substitution process for weight calculation. The core idea is to update the inverse matrix $\boldsymbol{R}$ (13) in each iteration of systolic array. So, the weights of filter are computed within systolic array. In [11], a new weight update equation is derived which is given as:

$$\boldsymbol{w}(k) = \boldsymbol{w}(k-1) + \boldsymbol{e}(k)\boldsymbol{R}^{-1}(k)\boldsymbol{R}^{-\mathrm{T}}(k)\boldsymbol{x}(k) \quad (14)$$

where the term $\boldsymbol{R}^{-1}(k)\boldsymbol{R}^{-T}(k)\boldsymbol{x}(k)$ is known as the Kalman gain. The systolic realization of IQRD RLS will be discussed in Sec. 3. Hence, IQRD RLS is less computationally expensive as compared to QRD RLS.

In Fast QRD (FQRD) RLS, matrix update equation is replaced by vector update equations. In [11] detail derivations of these equations are given, which further reduces the numerical complexity as compared to QRD RLS and IQRD RLS algorithms. Although, FQRD RLS is numerically less complex, but it has no weight update procedure, so it is limited to applications which try to estimate the output error vector [13].

In this section, three important RLS variants and their properties are discussed and summarized in Tab. I. It is worth mentioning that QRD and IQRD are computationally more complex than FQRD, but FQRD has no weight update procedure. So, we cannot consider it for adaptive beamforming application. On the other hand, QRD involves more computations (two step procedure) so it is also not preferable. Due to above reasons, IQRD RLS is the well-suited algorithm for hardware implementation.

Algorithm	Computational Complexity	Weight Update Procedure
QRD RLS	$O(N^2)$	Back Substitution
IQRD RLS	$O(N^2)$	Within Systolic array
FQRD RLS	$O(N)$	No Procedure

Tab. 1. Properties of RLS variants.

3.1 Performance Analysis of Adaptive Beamforming in Terms of Radiation Pattern

The performance analysis of adaptive beamforming, using RLS adaptive algorithm is done by varying different beamformer parameters. Researchers have performed the analysis of adaptive beamformers for smart antennas based on adaptive algorithms Least Mean Squares (LMS), Sample Matrix Inversion (SMI), Recursive Least Squares (RLS) and Conjugate Gradient Method (CGM). The analysis is done by varying the inter element spacing and the number of antenna elements for each algorithm and studying the radiation patterns, amplitude response, mean square error and absolute weights for adaptive beamforming algorithms [14], [15]. We assume a ULA beamformer with an operating frequency of 1 GHz. It receives an echo signal from the single target at azimuth of 30° and an interference signal at –45°. Simulations show that the RLS adaptive algorithm rejects the interference at –45° and enhanced the target signal at 30°. The following procedures are adopted for further analysis of RLS based adaptive beamformer:

- Change the number of antenna elements.
- Change the antenna elements inter space.

Figure 3 depicts the performance of beamformer with different number of antenna elements. The selected numbers of antennas are 10, 15 and 20. It is clear from the radiation pattern that when the number of antenna increases the main lobe of the beamformer is shrinked and pointed accurately towards the target angle and the side lobes are much suppressed. On the other hand, the perfor-

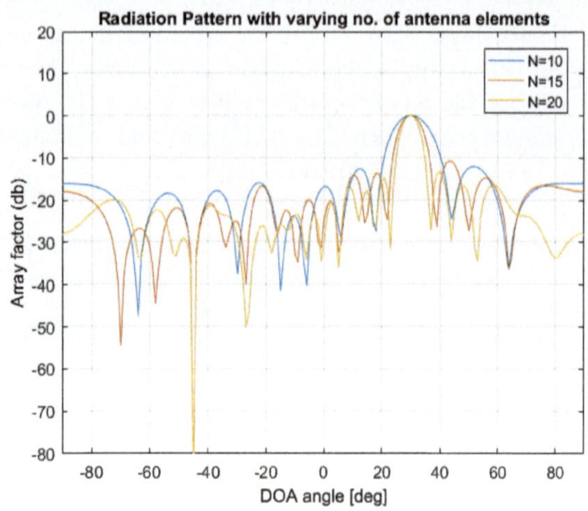

Fig. 3. Radiation pattern with different number of antennas N.

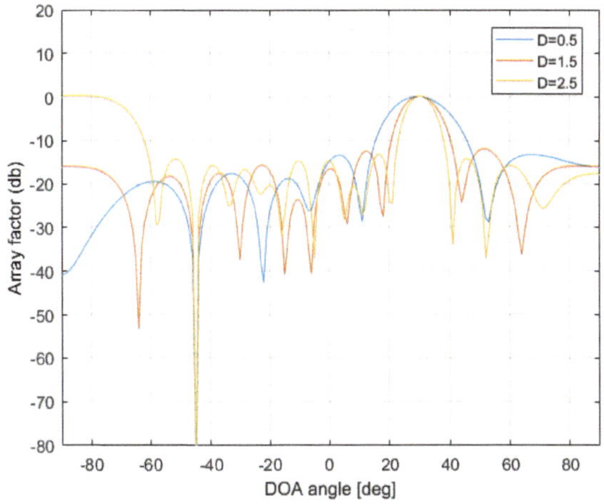

Fig. 4. Radiation pattern with different inter antenna element space *D*.

mance is also judged in terms of changing antenna elements inter space. The selected inter spaces are 0.5, 1.5 and 2.5 mm. Figure 4 is the resulting radiation pattern in which the performance of the adaptive beamformer is improved as the distances between the antenna elements are increased. The side lobes are also suppressed in a much better way.

Hence, the RLS based adaptive beamformer has a good performance and accuracy as the numbers of antenna elements are increased as well as the distance between each antenna elements is also increased.

4. IQRD Systolic Array Architecture

Systolic arrays are the well-known concept for matrix triangularization and inversion [16], [17] and the same idea is adopted for IQRD-RLS algorithm. Figure 5 shows the conventional systolic realization of the IQRD-RLS algorithm. The IQRD systolic triangular array consists of four types of processing cells; Boundary Cell (BC), Internal Cell (IC), Inverse Internal Cell (IIC) and Final Cell (FC).

A Boundary Cell determines the rotation parameters cos α and sin α to perform the rotations on input complex data rows $\boldsymbol{x}$, from data matrix $\boldsymbol{X}$ and update the corresponding row values of $\boldsymbol{R}$ matrix which is an upper triangular matrix. It also sends the rotation parameters to the right neighboring Internal Cells [16], [18]. The corresponding equations (6), (7) and (8) are modified as:

$$\cos\alpha = c = \frac{\sqrt{\lambda}\ \ r}{\sqrt{\lambda r^2 + |\boldsymbol{x}|^2}}, \tag{15}$$

$$\sin\alpha = s = \frac{x}{\sqrt{\lambda r^2 + |\boldsymbol{x}|^2}}\cdot \tag{16}$$

BC is more complex than IC, IIC and FC because it performs square-root and inverse operations. All the other cells perform only multiplication and inverse operations. Researchers have performed a lot of work to find a square-root and division free BC unit for systolic arrays. In [19], the following data transformations are applied to seek the square-root and division free expressions for transformed data.

$$r_j = \frac{1}{\sqrt{\alpha_1}} a_j,\quad x_j = \frac{1}{\sqrt{\beta_1}} b_j,\quad r'_j = \frac{1}{\sqrt{\alpha_2}} a'_j,\quad j = 1,2,...,m, \tag{17}$$

$$x'_j = \frac{1}{\sqrt{\beta_2}} b'_j,\qquad j = 2,3,...,m\cdot \tag{18}$$

However, the number of multiplications is increased due to these transformations, but it is a good approach to improve the system latency and reduce number of computations. In [8], another single cycle look-up table (LUT) based approach is used to replace the square root and inverse operations and to optimize the QRD RLS latency. This approach gives much improvement in overall performance of QRD RLS algorithm.

An Internal Cell (IC) received the rotation parameters cos α and sin α from the Boundary Cell (BC) in each row, and performed the rotations on the input values of data matrix $\boldsymbol{X}$ and updated the x and r values. The corresponding update equations are given as follows [16], [17]:

$$r_n = \sqrt{\lambda} c\cdot r + s\cdot x, \tag{19}$$

$$x_n = c\cdot x + \sqrt{\lambda}\overline{s}\cdot r. \tag{20}$$

Inverse Internal Cells receive the sine and cosine value of the rotation angle from the internal cell in the same row. Some part of the inverse internal cell operation is similar to the internal cell; i.e. it rotates the input data by multiplying it with input rotation angles. Along with this operation, Inverse internal nodes of IQRD-RLS use $\lambda^{(-½)}$ in (12) and (13). In this paper, we consider the case when $\lambda = 1$, so that IIC is equivalent to IC units.

Final Cell (FC) performs two functions: it received the rotation parameters cos α and sin α to rotate the input data and generate the updated filter weights using the following relation [10]

$$w = (-1)\cdot\gamma\cdot\sqrt{\lambda r^2 + |\boldsymbol{x}|^2} \tag{21}$$

where γ is computed using the following relation

$$\gamma(k) = \prod_{i=0}^{N}\cos\alpha_i(k)\cdot \tag{22}$$

A design methodology has been proposed to optimize the throughput and latency of the IQRD-RLS architecture. In Fig. 6, a block diagram of the proposed architecture is presented, which utilizes the standard pipelining methodologies and has 50% reduction in number of BC, IC and IIC units (Tab. 2). We have further optimized the IQRS RLS architecture, by omitting the square-root and division operations in BC, in terms of single cycle single cycle look up table (LUT) as shown in Fig. 7. Generally, a look up table

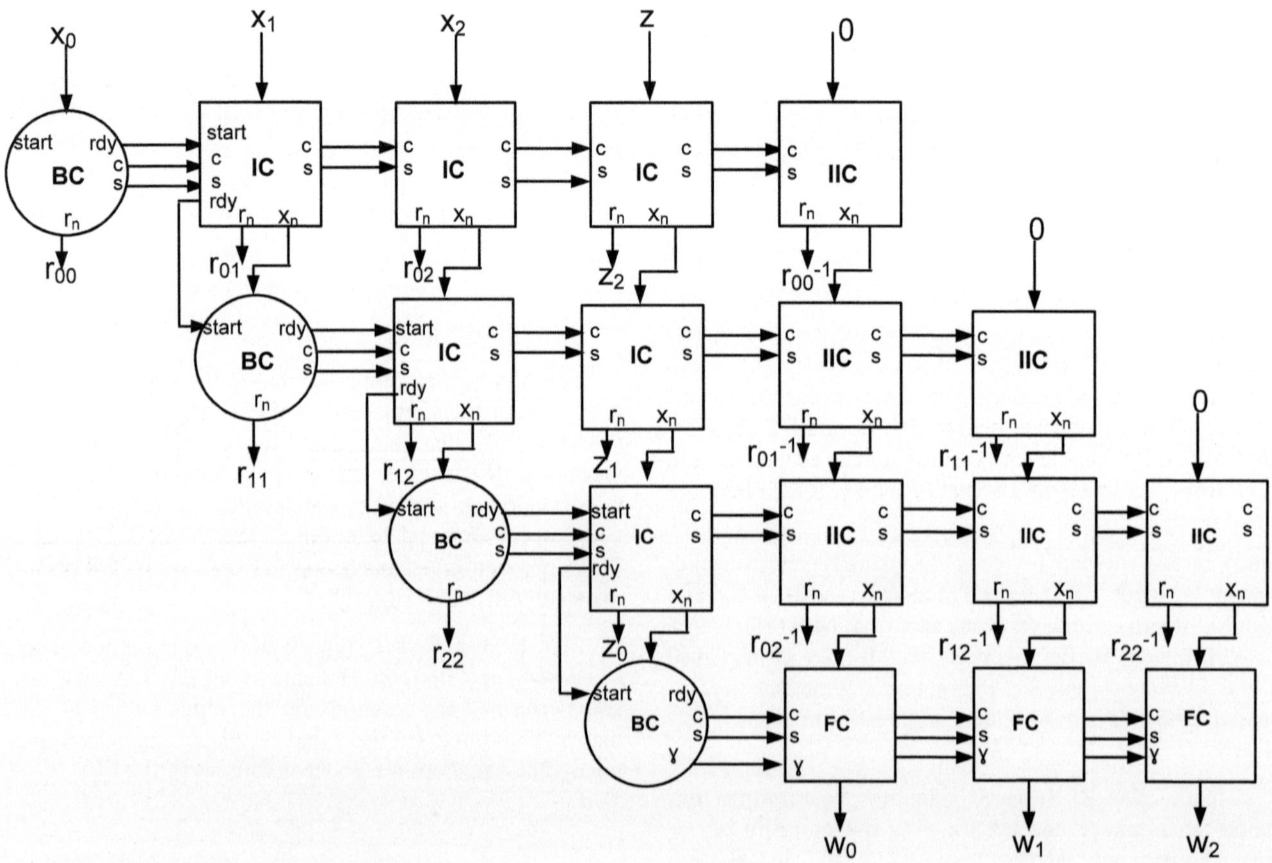

Fig. 5. Conventional IQRD systolic array.

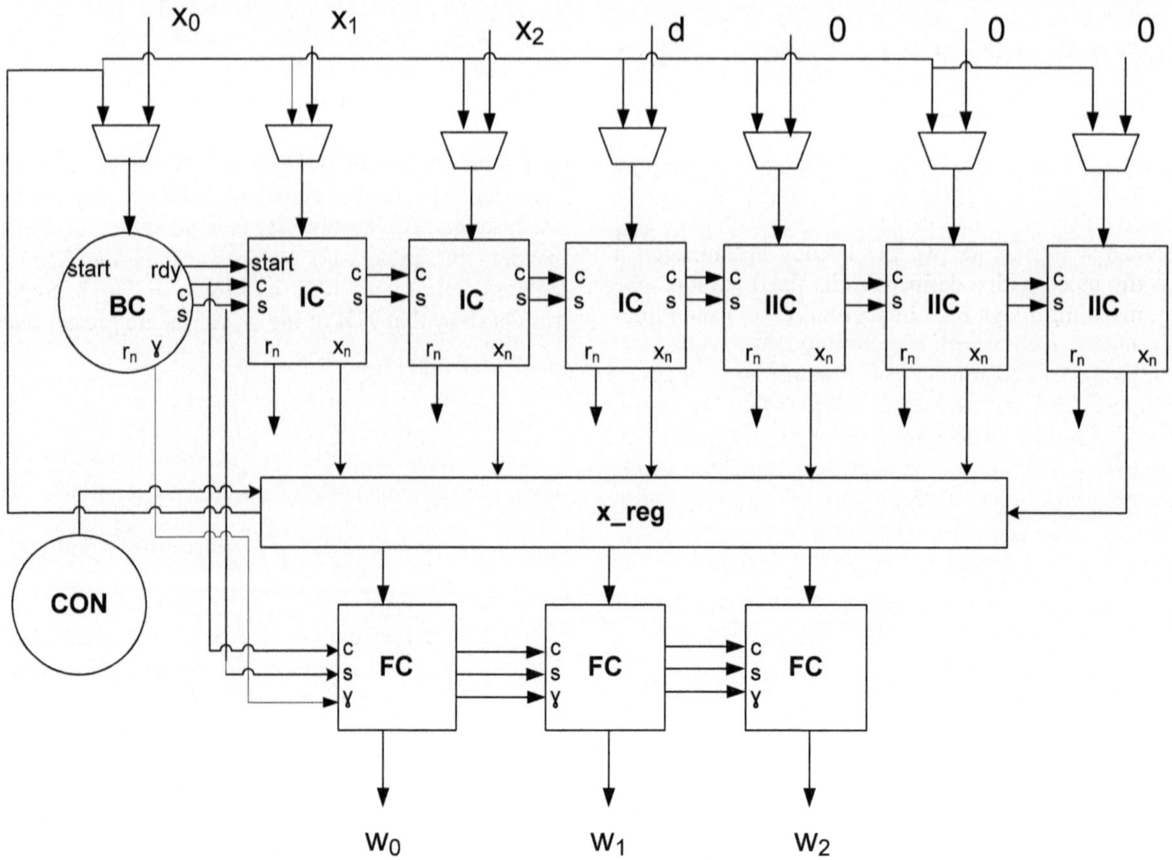

Fig. 6. The proposed IQRD architecture.

(LUT) is a table that determines what the output is for any given input. In FPGA truth tables are used to implement any digital logic. In [8], the same methodology is adopted for QRD RLS algorithm. This scheme avoids the complex hardware implementation with considerable decrease in latency and improves the speed of overall system. In some previous work the conventional implementation of BC is based on CORDIC algorithm, for non-recursive applications [20], [21]. But, digital beamforming is a recursive procedure, so, the conventional approach would not pay off because it leads to considerable latency. In the proposed architecture, the pipelined registers (R_reg, x_reg, Rinv_reg) are introduced which not only holds the updated values, but also feedback them to BC, IC and IIC cells. The controller CON handles the control signals as well as feedback timings. Hence equations (15) and (16) can be modified as:

$$A_s = \lambda r^2 + |\mathbf{x}|^2 = \lambda r^2 + x_i^2 + x_r^2, \tag{23}$$

$$A_{ss} = \sqrt{\lambda r^2 + |\mathbf{x}|^2}, \qquad A_i = \frac{1}{A_{ss}}, \tag{24}$$

$$c = \sqrt{\lambda} r \times A_i, \tag{25}$$

$$s_i = x_i \times A_i, \qquad s_r = x_r \times A_r. \tag{26}$$

The IC unit received the c, s_i and s_r values from neighboring BC to compute (19) and (20) and update the values of x_{ni}, x_{nr} and r_n. These computations required only adder and multiplier units as shown in Fig. 8. The updated values x_{ni}, x_{nr} are stored in pipelined register x_reg and also feedback to IC module, and updated r_n is stored in internal memory of IC. The c, s_i and s_r are also sent to the neighboring IC (Fig. 8).

One of the major differences between the LUT implemented in [8] and in this paper, is the floating-point representation format. In [8], the latency optimization of LUT is discussed with a defined N bits fixed point number format, including 1 sign bit, i integer bits ($i > 1$) and f fraction bits ($f > 1$). But, the floating-point multiplier and adder are not discussed with the new defined number format.

In this paper, we have utilized the same LUT approach for IQRD RLS architecture with single precision IEEE 754 format due to its wider range over fixed point. The single-precision numbers are stored in 32 bits: 1 for the sign, 8 for the exponent, and 23 for the fraction [22]. In the proposed architecture, the computation of A_i and A_{ss} are handled in single LUT, whereas A_s is the 32-bit input address field to the LUT as shown in Fig. 7. Simulation results show that 97% of the A_s values are greater than 1 as

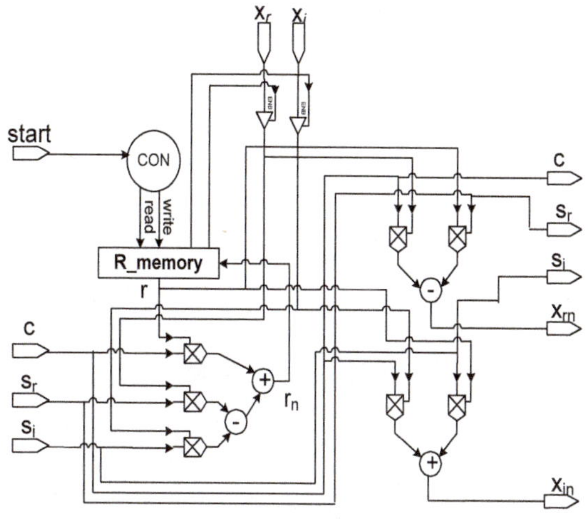

Fig. 8. Internal cell (IC) architecture.

Data matrix X [size]	No. of BC		No of IC		No of IIC	
	Conventional Architecture	Proposed Architecture	Conventional Architecture	Proposed Architecture	Conventional Architecture	Proposed Architecture
3×3	4	1	6	3	6	3
4×4	5	1	10	4	10	4
5×5	6	1	15	5	15	5

Tab. 2. Comparison between the proposed and conventional IQRD architecture.

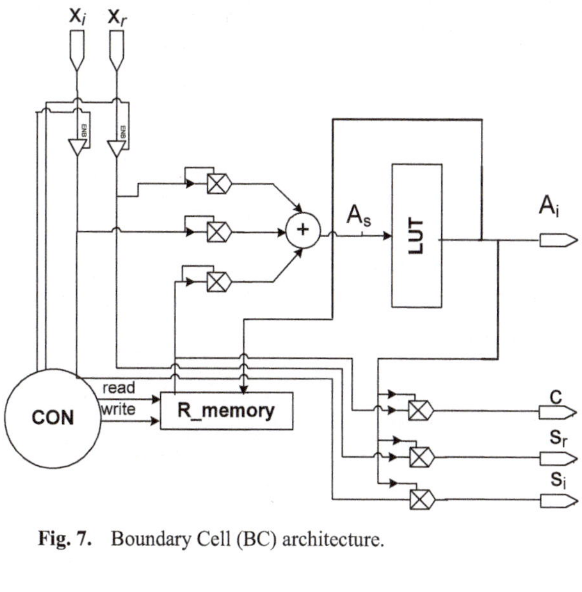

Fig. 7. Boundary Cell (BC) architecture.

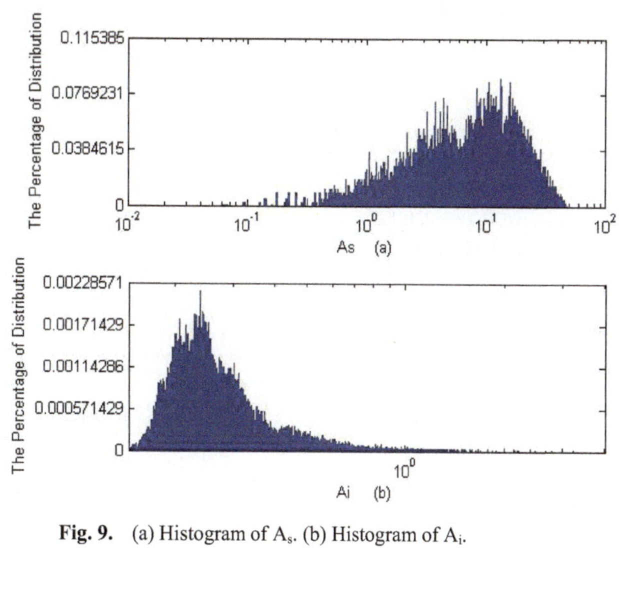

Fig. 9. (a) Histogram of A_s. (b) Histogram of A_i.

shown in Fig. 9a, when the all input variables in (23) have a typical uniform distribution within the range of [–4, 4], and their corresponding output values (A_i) are very small (Fig. 9b). Hence, probability of occurring a value less than 1 is very low thus reduces the range of input values A_s for LUT. The output of the LUT A_i has 32-bit bandwidth, which is used to compute the c, s_i and s_r values, for the IC units.

5. Implementation and Results

In this paper, we have optimized the BC module of systolic array in terms of complex mathematical operations. Many researchers also performed a lot of work in single precision floating point multiplier which is also a complex mathematical operation in hardware [23–25], but we keep our focus in divider and square-root operations and use the multiplier core of the FPGA. Xilinx Virtex 5 FPGA is the selected platform for hardware implementation.

As we have already discussed, adaptive algorithm is a basic part of the beamforming system and we select the IQRD-RLS algorithm as the most suitable algorithm. Three types of architectural modules are designed to judge the performance of algorithm in terms of latency, throughput (1sample/clock Time×clock cycles), error and efficiency (throughput/number of slices). The details are:

1. MD_0 (module 0), a *conventional* systolic array architecture with single precision floating point multipliers, dividers and a square root module (generated with Coregen software).
2. MD_1 (module 1), a *proposed* systolic array architecture with single precision floating point multipliers, dividers and a square root module (generated with Coregen software).
3. MD_2 (module 2), a *proposed* systolic array architecture with LUT based BC for divider and square root operation.

where MD stands for module. A radar target detection scenario is simulated in MATLAB to generate the data points for the testing of MD_0, MD_1 and MD_2. First, a radar received signal is modelled having one target at azimuth angle 30°, and two interferences at azimuth 15° and 45°, respectively. Assume that both targets and radar lie at the same; i.e. elevation angle is equal to zero. Figure 10 shows the radiation pattern. The received signal is impinged on 3 element ULA beamformer based phased array radar to generate almost 10^4 data points.

In adaptive beamforming application, latency is an important performance metric which affects the overall system throughput. To judge the LUT effects on algorithm performance we compare the latencies and throughput of the architectural modules. Table 3 details the latency of each unit in system individually, and throughput of adaptive filter (3-tap) for IQRD-RLS algorithm with three different architectural modules; i.e. MD_0, MD_1, MD_2. These modules are modelled using Verilog and their behavior description is placed and routed on Virtex5 device by ISE14.1i tool. It is clear from Tab. 3, the proposed architecture based module MD_2 module has offered 9 times reduction in latency as compared to MD_0 and MD_1 modules because of its optimized single cycle LUT based architecture. On the other hand, MD_2 module of IQRD-RLS has a higher throughput as compared to QRD-RLS M2 module. It is again important to notify that in [8], QRD-RLS is implemented using a new floating point format, whereas we implement the IQRD-RLS with standard IEEE-754 single precision floating point representation. Hence, it is again proved that IQRD-RLS is the most suitable algorithm for adaptive beamforming application.

Algorithm accuracy of the proposed architecture and in presence of LUT is found out by measuring the difference between the weight coefficients of the ideal method and the proposed method in MD_2 with 10^4 simulated samples. This is also known as weight error. Results show that the weight error is less than 0.00216 for 96% samples. Figure 11 illustrates the filter weights of the ideal and proposed methods for 3 tap beamformer (For clarity of the image we plot only 200 iterations).

The device utilization details or hardware summary are given in Tab. 4. It is noteworthy that MD_2 outperforms in terms of area efficiency and throughput. The conventional architecture MD_0 has the highest area efficiency, while MD_1 occupies much lesser area as compared to MD_0. The hardware efficiency and throughput are found out to draw a reasonable comparison between the architectures. To the best of our knowledge, this is the first implementation of IQRD RLS algorithm in FPGA, so, the performance of MD_2 is also measured, after it is synthesized for different target FPGAs as shown in Tab. 5. The throughput of the IQRD proposed architecture in all FPGAs is better than the QRD [8]. Hence, simulation results have clearly indicated that the IQRD RLS is the most suitable algorithm for adaptive beamforming due to its better throughput, hardware efficiency and latency.

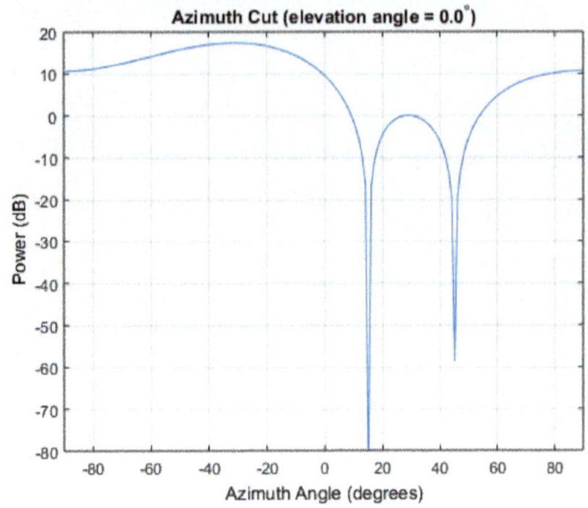

Fig. 10. Radiation pattern.

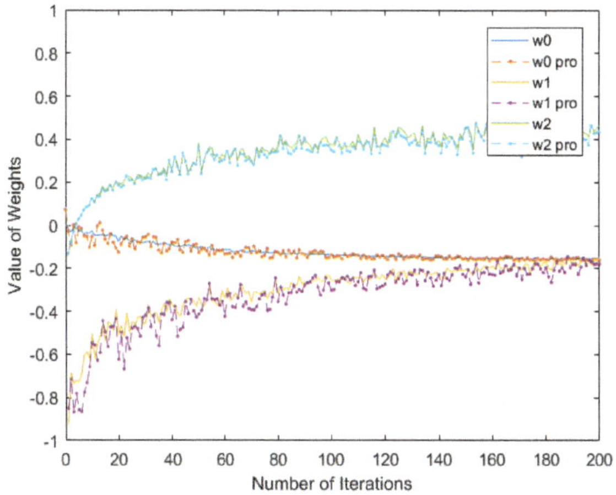

Fig. 11. Weight error.

Algorithm	IQRD RLS			QRD RLS [8]	
Modules	**MD_0**	**MD_1**	**MD_2**	**Modules**	**M_2**
BC	46	46	3	BC	5
IC	5	5	5	IC	3
FC	5	5	5	---	---
Total	455	455	52	Total	76
Throughput [MSamples/sec]	0.44	0.44	3.92	**Throughput [MSamples/sec]**	2.45

Tab. 3. Comparison between IQRD and QRD [8] latency and throughput.

Hardware Resources	MD_0	MD_1	MD_2
Slices	61%	28%	3%
4-input LUT	18%	14.4%	5%
DSP48	55%	41%	5%
ROM	0%	0%	52%
Avg. Area	38.2%	21.6%	13.92%
Area Efficiency	0.011518	0.02037	0.2292 3

Tab. 4. Hardware resources.

No.	System	Slices.	Throughput
1.	Spartan-3 (exc3s500)	5891	3.01
2.	Virtex-4 (vsx55)	5603	3.24
3.	Virtex-5 (vtx150)	5001	3.92

Tab. 5. Comparison between FPGAs.

6. Conclusion

A low latency architecture is presented that implements the IQRD algorithm for beamforming application using system level hardware tools. The proposed architecture improves the divider and square root operations via LUT approach. Moreover, it is mapped in Virtex5 FPGA by Xilinx and achieved the minimum errors between the real and ideal weight values. Hence, the proposed architecture is outperforming in terms of area efficiency, throughput, and latency, and the most suitable algorithm for adaptive beamforming. In future work, the multiplier and adder units can be implemented via LUT, for further improvement in latency and throughput.

References

[1] FENN, A. J. Adaptive antenna and phased arrays. *Lectures MIT Lincoln Laboratory*, Available at: http://www.ll.mit.edu/workshops/education/videocourses/antennas/lecture8/lecture.pdf

[2] MAILLOUX, R. J. *Phased Array Radar Handbook*. 2nd ed., rev. London (UK): Artech House, 2005. ISBN: 1580536891

[3] FLORENS, C., RAIDA, Z. Adaptive beamforming using genetic algorithms. *Radioengineering*, 1998, vol. 7, no. 3, p. 1–6.

[4] HAYKIN, S. O. *Adaptive Filter Theory*. 4th ed., Prentice Hall, 2003. (p. 436–465, 506–534) ISBN: 0130901261

[5] DINIZ, P. S. R. *Adaptive Filtering: Algorithms and Practical Implementation*. 2nd ed. USA: Kluwer Academic Press, 1997. (p. 195–389). DOI: 10.1007/978-0-387-68606-6

[6] JIAN LI, STOICA, P. *Robust Adaptive Beamforming*. Wiley Series in Telecommunication and Signal Processing, 2006, (p. 49–79). ISBN: 9780471678502

[7] MCWHIRTER, J. G. Recursive Least Square Minimization using a Systolic Array. *Proc SPIE 0431, Real-time Signal Processing VI*, 1983, vol. 431 (p. 105–112). DOI: 10.1117/12.936448

[8] ALIZADEH, M. S., BAGHERZADEH, J., SHARIFKHANI, M. A low latency QRD RLS architecture for high throughput applications. *IEEE Transactions on Circuits and Systems*, 2016, vol. 63, no. 7 p. 708–712. DOI: 10.1109/TCSII.2016.2530169

[9] DESHPANDE, A. P., MURTHY. N. S., GOVIND RAO, D., et al. Efficient filter implementation using QRD-RLS algorithm for phased array radars application. In *IEEE International Conference for Technological Advances in Electrical, Electronics and Computer Engineering (TAEECE)*. Konya (Turkey), 2013. DOI: 10.1109/TAEECE.2013.6557275

[10] MARTINEZ, M. E. I. Implementation of QRD RLS algorithm on FPGA: Application to Noise Canceller System. *IEEE Latin America Transactions*, 2011, vol. 9, no. 4, p. 458–462. DOI: 10.1109/TLA.2011.5993728 (in Spanish)

[11] APOLINARIO Jr., J. A. (Ed.) *QRD RLS Adaptive Filtering*. New York (USA): Springer Media, 2009. (p. 51–113). DOI: 10.1007/978-0-387-09734-3

[12] ALEXANDER, S. T., GHIRNIKAR, A. L. A method for recursive least squares filtering based upon an inverse QR decomposition. *IEEE Transactions on Signal Processing*, 1993, vol. 41, no. 1, p. 20–30. DOI: 10.1109/TSP.1993.193124

[13] SHOAIB, M., WERNER, S., APOLINARIO JR., J. A., et al., Equivalent output-filtering using Fast QRD-RLS algorithm for burst-type training applications. In *IEEE International Symposium on Circuits and Systems (ISCAS)*. Island of Kos (Greece), 2006, DOI: 10.1109/ISCAS.2006.1692540

[14] SAXENA, P., KOTHARI. A. G. Performance analysis of adaptive beamforming algorithms for smart antennas. *IERI Procedia* (special issue International Conference on Future Information Engineering), 2014, vol. 10, p. 131–137. DOI: 10.1016/j.ieri.2014.09.101

[15] AZIZ, A., QURESHI, M. A, JUNAID IQBAL, J., et al. Performance and quality analysis of adaptive beamforming algorithms (LMS, CMA, RLS & CGM) for smart antennas. In *IEEE 3rd International Conference on Computer and Electrical Engineering (ICCEE)*. Chengdu (China), 2010, p. V6-302–V6-306. ISBN: 978-1-4244-7224-6

[16] GENTLEMAN, W. M., KUNG, H. T, Matrix triangularization by systolic arrays. *Proc. SPIE Real-time Signal Processing*, 1981, vol. 298, p. 19–26. DOI: 10.1117/12.932507

[17] MILOVANOVIC, E. I., STOJCEV, M. K., MILOVANOVIC, I. Z., et al. Design of linear systolic arrays for matrix multiplication. *Advances in Electrical and Computer Engineering (AECE)*, 2014, vol. 14, no. 1, p. 37–42. DOI: 10.4316/AECE.2014.01006

[18] TANG, C. E. T., LIU, K. J. R., TRETTER, S. A. Optimal weight extraction for adaptive beamforming using systolic array. *IEEE Transactions on Aerospace and Electronics Systems,* 1994, vol. 30, no. 2, p. 367–385. DOI: 10.1109/7.272261

[19] FRANTZESKAKIS, E. N., LIU, K. J. R. A class of square root and division free algorithms and architectures for QRD based adaptive signal processing. *IEEE Transactions on Signal Processing*, 2002, vol. 42, no. 9, p. 2455–2469. DOI: 10.1109/78.317867

[20] KARKOOTI, M., CAVALLARO, J. R., DICK, C. FPGA implementation of matrix inversion using QRD-RLS algorithm. In *Conference Record of IEEE 39th Asilomar Conference on Signals, Systems and Computers.* Pacific Grove (CA, USA), 2005, p. 1625 to 1629. DOI: 10.1109/ACSSC.2005.1600043

[21] YHONG, T. M., MADHUKUMAR, A. S., CHIN, F. QRD-RLS adaptive equalizer and its CORDIC-based implementation for CDMA systems. *International Journal on Wireless and Optical Communications*, 2003, vol. 1, no. 1, p. 25–39. DOI: 10.1142/S0219799503000033

[22] HOLLASCH, S. *IEEE Standard 754 Floating Point Numbers*. Available at: http://steve.hollasch.net/cgindex/coding/ieeefloat.html

[23] RAMESH, A. P, TILAK, A. V. N., PARSAD, A. M. An FPGA based high speed IEEE-754 double precision floating point multiplier using Verilog. In *2013 International Conference on Emerging Trends in VLSI Embedded System, Nano Electronics and Telecommunication System (ICEVENT).* Tiruvannamalai (India), 2013. DOI: 10.1109/ICEVENT.2013.6496575

[24] KODALI, R. K, GUNDABATHULA, S. K., BOPPANA, L. FPGA implementation of IEEE 754 floating point Karatsuba multiplier. In *International Conference on Control, Instrumentation, Communication and Computational Technologies (ICCICCT).* Kanyakumari (India), 2014. DOI: 10.1109/ICCICCT.2014.6992974

[25] PALDURAI, K., HARIHARAN, K. FPGA implementation of delay optimized single precision floating point multiplier. In *IEEE International Conference on Advanced Computing and Communication System (ICACCS).* Coimbatore (India), 2015. DOI: 10.1109/ICACCS.2015.7324094

About the Authors ...

Raafia IRFAN has received her MSc degree in Electronics from Quaid-e-Azam University (QAU) in 2004 and done her MS System Engineering from the Pakistan Institute of Engineering and Applied Sciences (PIEAS) in 2006. Currently, she is working towards the Ph.D. degree in Electrical Engineering from PIEAS. For last 10 years, she is working as a senior scientist in a research organization. Her expertise and research interest are FPGA based communication system design, radar signal processing and high speed VLSI designs.

Haroon-ur-RASHEED is a senior faculty member at the Department of Electrical Engineering, Pakistan Institute of Engineering and Applied Sciences (PIEAS). He has done his Bachelors in Electrical Engineering from the University of Engineering and Technology (UET) Lahore in 1990 and completed his master from Iowa State University (ISU), USA. He completed his Ph.D. from the School of Computer Science & Technology, Beijing Institute of Technology in 2009. At PIEAS he has taught courses in "Digital Design in Verilog", "Digital Communication", "Microprocessor based Design" and "Digital Image Processing". He has supervised many final year projects at NUST as well as PIEAS during his career as a teacher. He and his students have published a number of research papers in international conferences before starting PhD studies at BIT. He is also a professional member of IEEE.

Waqas Ahmed TOOR did his Bachelors in Electrical Engineering from NUST in 2003. He received his MS System Engineering degree from PIEAS in 2006. Now he is doing his MS leading to Ph.D. in Electrical Engineering from the Capital University of Science and Technology (CUST). For last ten years, he is working in an R&D organization as a Senior Engineer. His research interests are high power microwave devices and their applications, power electronics and radar signal processing.

18

A Compact Dual Band-notched Circular Ring Printed Monopole Antenna for Super-wideband Applications

Murli MANOHAR [1], Rakhesh Singh KSHETRIMAYUM [2], Anup Kumar GOGOI [2]

[1] Dept. of Electronics and Communications Engineering, IIIT Manipur, 795002 Imphal, India
[2] Dept. of Electronics and Electrical Engineering, IIT Guwahati, 781039 Guwahati, India

murli@iiitmanipur.ac.in, krs@iitg.ernet.in, akg@iitg.ernet.in

Abstract. *In this article, a simple and compact dual band-notched (DBN) super wideband (SWB) printed monopole antenna (PMA) has been proposed. The proposed antenna composed of a circular PMA, which is connected through a 50-Ω triangular tapered microstrip fed line (TTMFL) and a round-cornered finite ground plane (RCFGP). It exhibits a very wide frequency band from 1.6–25 GHz (ratio bandwidth of 15.63:1) with a voltage standing wave ratio (VSWR) ≤ 2. By employing a U-shaped parasitic element (USPE) near the RCFGP and a T-shaped protruded stub (TSPS) inside the radiating patch, a single band-notched (SBN) characteristic in the frequency band of 3.2–4.4 GHz (WiMAX/C-band) is generated. In order to realize the second band-notched function for X-band satellite communication systems (7.2–8.4 GHz), a U-shaped slot (USS) has been inserted in the RCFGP. The overall dimension of the proposed antenna is $24 \times 30 \times 0.787$ mm^3 and occupies a relatively small space compared to the existing DBN antennas. Good agreement has been attained between predicted and measured results.*

Keywords

Dual band-notch, super wideband (SWB) antenna, triangular tapered microstrip feed line

1. Introduction

Nowadays there is an increasing demand of super wideband (SWB) radios in the modern wireless communication systems, owing to their extremely large bandwidth (BW) and very high data transmission rate. Printed monopole antenna (PMA) is an ideal candidate for SWB applications due to its several fascinating features such as small size, low cost, planar structure, operation over extremely large impedance BW and ease to accommodate with small space provided by hand held gadgets. Existing ultra-wideband (UWB) PMA (BW of 3.1 to 10.6 GHz) [1] also have features like that of SWB except that its ratio bandwidth (RB) falls short of SWB antennas and is just 3.42 : 1. SWB antennas should offer RB of more than 10 : 1 and were first supposedly developed by Rumsey et al. in the late 1950 and early 1960 and were called frequency-independent antennas [2].

Recently, few SWB antennas have been investigated by many researchers in the published literature [3–7]. Unfortunately, it is to be noted that some of the existing narrow-band systems such as WiMAX (3.3–3.6 GHz), C-band (3.7–4.2 GHz) and X-band satellite communication systems operating in 7.25–8.395 GHz (for down link: 7.25–7.745 GHz and uplink: 7.9–8.395 GHz) may cause electromagnetic interference (EMI) to the SWB system. So, to mitigate this electromagnetic interference issue, SWB antennas with band-notched characteristics is required. Several dual band-notched (DBN) UWB antennas have been investigated by many researchers across the globe [8–13].

However, the above mentioned antennas have been designed to have notch within the UWB BW and occupy a relatively large space. But to the author's knowledge, there are no DBN SWB antennas reported in the literature. In this communication, a novel and compact DBN SWB PMA is presented. The designed antenna offers a wide impedance BW from 1.6 to 25 GHz with a VSWR ≤ 2 except in the DBN of 3.2–4.4 GHz and 7.2–8.4 GHz. The proposed antenna offers wide BW with DBN characteristics and is small in size compared to the DBN UWB antennas reported in [8–13]. By tuning the dimensions of U-shaped parasitic element (USPE) and U-shaped slot (USS), two suitable notches are obtained. Good agreement has been obtained between simulation and experimental results of the proposed DBN SWB PMA. Simulation results have been carried out with the finite element method (FEM) based commercial software Ansoft high-frequency structure simulator (HFSS).

2. Antenna Configuration

The configuration of the proposed DBN SWB circular ring (CR) PMA is illustrated in Fig. 1. The CR PMA is printed on the RT/Duroid 5870 substrate of thickness $t = 0.787$ mm with dielectric constant ε_r of 2.23, which is

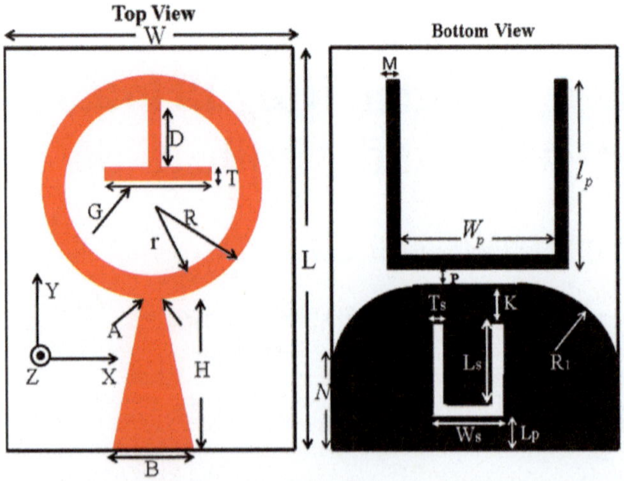

Fig. 1. Geometry of the proposed DBN SWB PMA.

excited through a 50-Ω triangular tapered microstrip fed line (TTMFL). For the signal transmission, a two-hole Sub Miniature version A (SMA)-connector has been utilized. We have specifically chosen the above mentioned substrate since FR4 substrate becomes highly lossy at higher frequencies. The proposed antenna (the overall dimensions of about $24 \times 30 \times 0.787$ mm^3) consists of a CR radiating patch with a T-shaped protruded stub (TSPS) inside CR radiating patch. The rear side of the RT/Duroid substrate is composed of a round-cornered finite ground plane (RCFGP) with embedded USS and USPE, which is placed under the CR patch. The TTMFL is used for broadband impedance matching and the reflection coefficient ($|\Gamma(\theta)|$) response of triangular taper has minima for $\beta l = 2m\pi$ where $m = 1, 2, 3, \ldots$ and corresponding frequencies. A closed form solution of the Riccati equation for triangularly tapered feed line is given by [14]

$$\Gamma(\theta) = \frac{1}{2}\exp\left(-j\beta l_k\right)\ln\left(\frac{Z_{l_k}}{Z_0}\right)\left[\frac{\sin\left(\frac{\beta l_k}{2}\right)}{\frac{\beta l_k}{2}}\right]^2 \quad (1)$$

where $\Gamma(\theta)$ is the reflection coefficient, β is the phase constant and $l_k = H$ is the triangular tapered feed line length. To get the widths at the two ends of the tapered lines, method reported in [15] is followed.

The optimal dimensions of the proposed antennas are as follows: $L = 30$ mm, $W = 24$ mm, $H = 13$ mm, $A = 1$ mm, $B = 3$ mm, $G = 5.5$ mm, $D = 4.5$ mm, $T = 0.5$ mm, $R = 8$ mm, $r = 6$ mm, $N = 6.2$ mm, $W_p = 14.5$ mm, $l_p = 16.5$ mm, $M = 0.5$ mm, $p = 1$ mm, $T_S = 0.2$ mm, $L_S = 5.2$ mm, $W_s = 1.5$ mm, $L_p = 4.5$ mm, $K = 1.6$ mm and $R_1 = 4.9$ mm. Here the optimized electrical length of the USPE (U_1) and USS (U_2) are set to approximately $0.5\,\lambda_g$ and $0.25\,\lambda_g$, where $U_1 = W_p + 2l_p + 2M$, $U_2 = W_s + 2L_S + 2T_S$. λ_g is the guided wavelength corresponding to band notch frequencies at 3.8 and 7.8 GHz respectively.

3. Results and Discussions

Figure 2 presents design evolution of the proposed SWB DBN PMA which illustrates the steps for enhancing the impedance BW as well as formation of DBN characteristics. The corresponding simulated VSWR characteristics comparison of Antennas 1-5 are shown in Fig. 3. Antenna 1 consists of a simple monopole disc connected with a rectangular feed line and rectangular finite ground plane that provides BW of 2.8–9.9 GHz (refer to Fig. 3). To improve the impedance BW of Antenna 1, the rectangular feed line has been replaced by a triangular tapered microstrip fed line (TTMFL) in Antenna 2 and it offers wide BW of 1.6–23.5 GHz without increasing the dimension as depicted in Fig. 3.

To further increase the BW of Antenna 2, we modified the rectangular ground plane of Antenna 2 into round-cornered finite ground plane (RCFGP) in Antenna 3. In Antenna 3, the ground plane also participates in radiation [16–17], additional resonance is excited at the higher frequencies and hence the impedance matching characteristics tend to improve over the entire band, resulting in an extremely large BW of 1.6–25 GHz or above. It is capable of supporting SWB radios without increasing the antenna dimensions. Now to generate the first band-stop (BS) characteristics for X-band satellite communication systems (7.2–8.4 GHz), a U-shaped slot (USS) is inserted in the RCFGP (Antenna 4). By employing a U-shaped parasitic element (USPE) on the rear side of the substrate and an inverted T-shaped protruded stub (TSPS) within the circular ring (CR) shaped radiation patch we generate the second BS characteristics for WiMAX/C-band (3.2–4.4 GHz) application (Antenna 5).

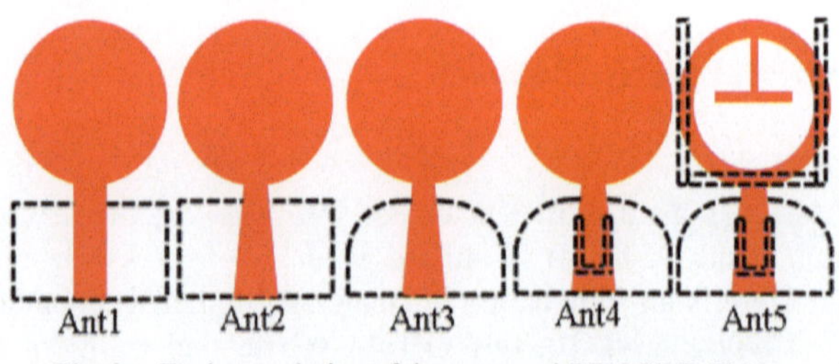

Fig. 2. Design evolution of the proposed DBN SWB PMA.

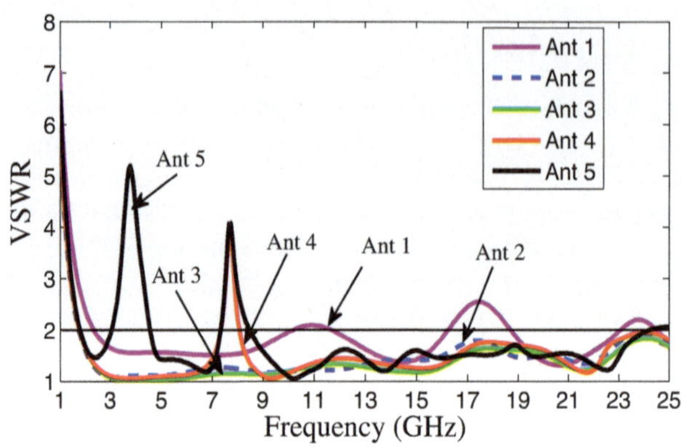

Fig. 3. Simulated VSWR of antennas 1-5.

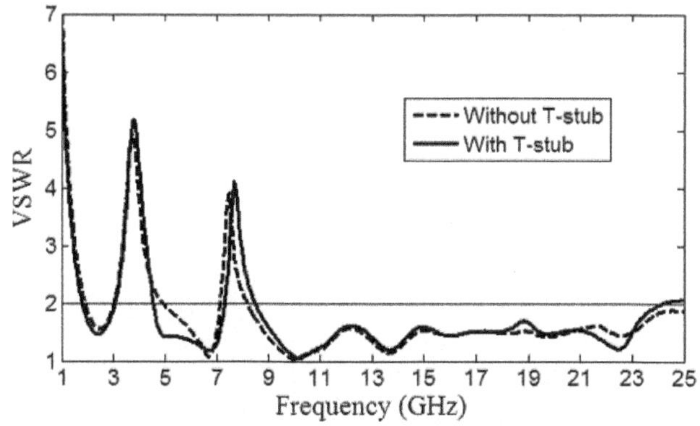

Fig. 4. Simulated VSWR versus frequency graph for the proposed compact DBN SWB PMA with and without inverted TSPS.

The effect of inverted TSPS on the WiMAX/ C-band (3.2–4.4 GHz) and X-band satellite communication systems (7.2–8.4 GHz) is shown in Fig. 4. It is found that without inverted TSPS, the upper edge frequency of WiMAX/C band increases from 4.4 GHz to 4.9 GHz, while upper edge frequency of the X-band satellite communication systems decreases from 8.4 to 8.1 GHz, which does not cover the uplink frequency (7.9–8.395 GHz) of X-band satellite communication systems. The widened frequency BW of WiMAX/C band notched characteristics (3.2–4.9 GHz) unnecessarily blocks useful frequencies from 4.4 to 4.9 GHz.

To know more details about dual band-notched functionality of the proposed SWB antenna, parametric study has been carried out. Dual band-notched characteristics of the SWB antenna can be controlled mainly by the following parameters such as R, W_p and L_S for the radiating ring patch, U-shaped parasitic element and U-shaped slot respectively. Figure 5 depicts the simulated VSWR curves for different values of outer radius R of circular ring patch. It can be observed that by increasing the value of R from 8 to 9 mm, the first notch band blocks deeper (amplitude of VSWR is about 6.5) within WiMAX/C-band, however center frequency of the second notch band (X-band) is shifted from 7.8 to 7.4 GHz and also there is an impedance mismatch at higher frequencies (12–13 GHz and 16.5–18 GHz). When the position of $R = 7$ mm, the impedance matching within SWB region is excellent, the center frequency of the first (WiMAX/C-band) and second (X-band) notched band is also appropriate. So the optimum value of $R = 7$ mm. The simulated VSWR for different values of W_p (width of U-shaped parasitic element) is plotted in Fig. 6. In this arrangement by varying the value of W_p from 13.5 to 15.5 mm, the center frequency of the second (X-band) notched band characteristics is insensitive; however the bandwidth of the first notch band (WiMAX/C-band) is decreased significantly. Therefore, the optimum value of W_p is decided to take 14.5 mm.

Figure 7 shows the effect of the length L_S of U-shaped slot on the simulated VSWR curves. From Fig. 7 it can be found that, when the length L_S of U-shaped slot changes from 4.7 mm to 5.7 mm, the center frequency of the second

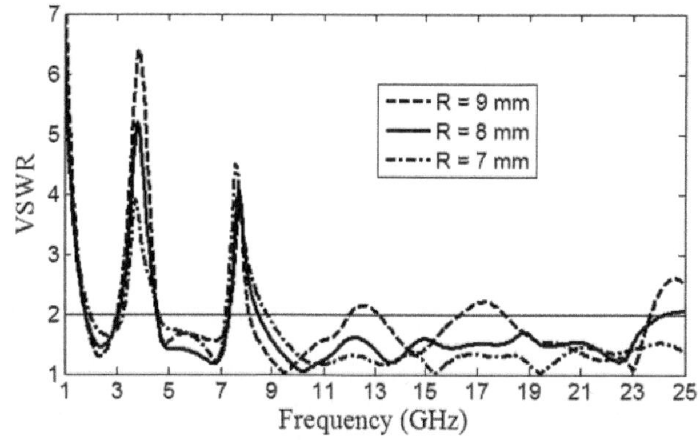

Fig. 5. Simulated VSWR versus frequency graph for the proposed compact SWB dual band-notched antenna with different value of R.

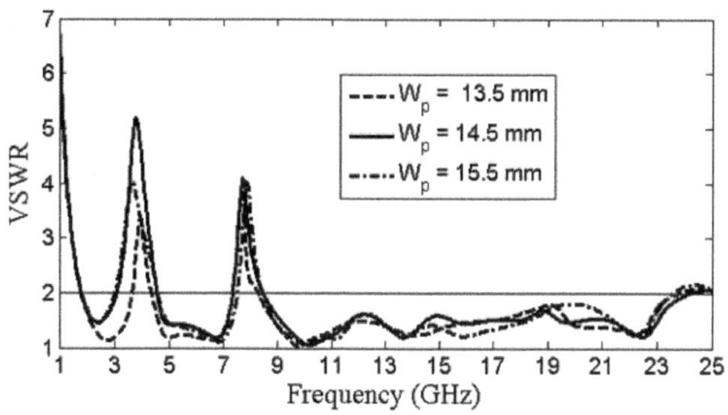

Fig. 6. Simulated VSWR versus frequency graph for the proposed compact SWB dual band-notched antenna with different value of W_p.

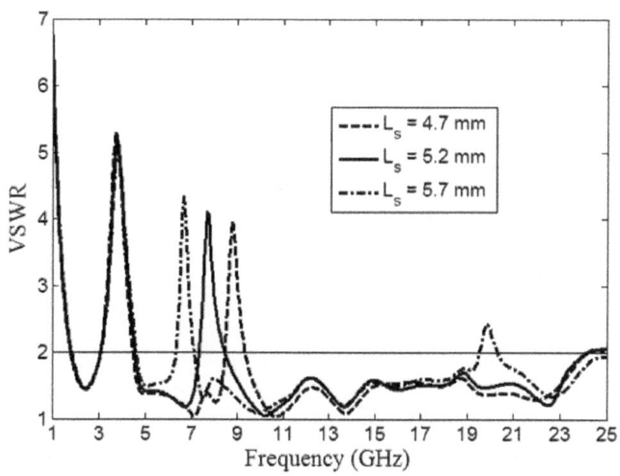

Fig. 7. Simulated VSWR versus frequency graph for the proposed compact SWB dual band-notched antenna with different value of L_S.

notched band is varied from 9 GHz to 6.5 GHz, however there is no variation within the first notched band region. Therefore the value of L_S has been selected to be 5.2 mm, which controls the variable notched band with center frequency of 3.8 GHz.

Figure 8 (a) and (b) illustrates the simulated surface current distribution for the proposed SWB PMA at band-notch center frequencies of 3.8 and 7.8 GHz. From Fig. 8(a)

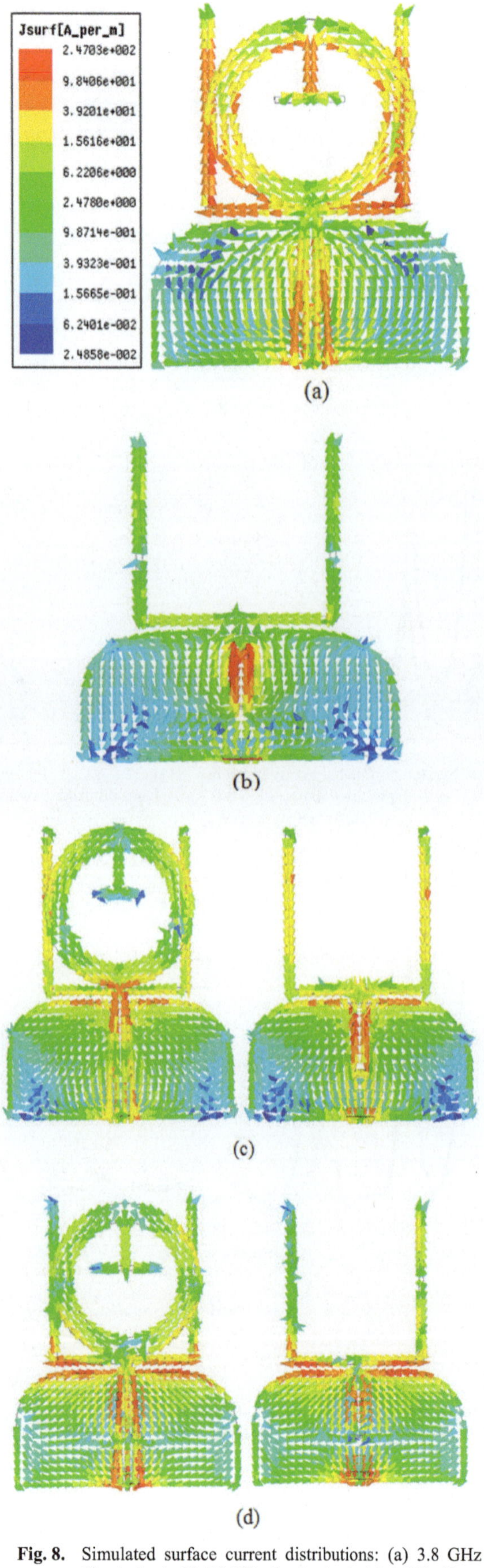

Fig. 8. Simulated surface current distributions: (a) 3.8 GHz (front view), (b) 7.8 GHz (back view), (c) 7 GHz (front and back view), (d) 10 GHz (front and back view).

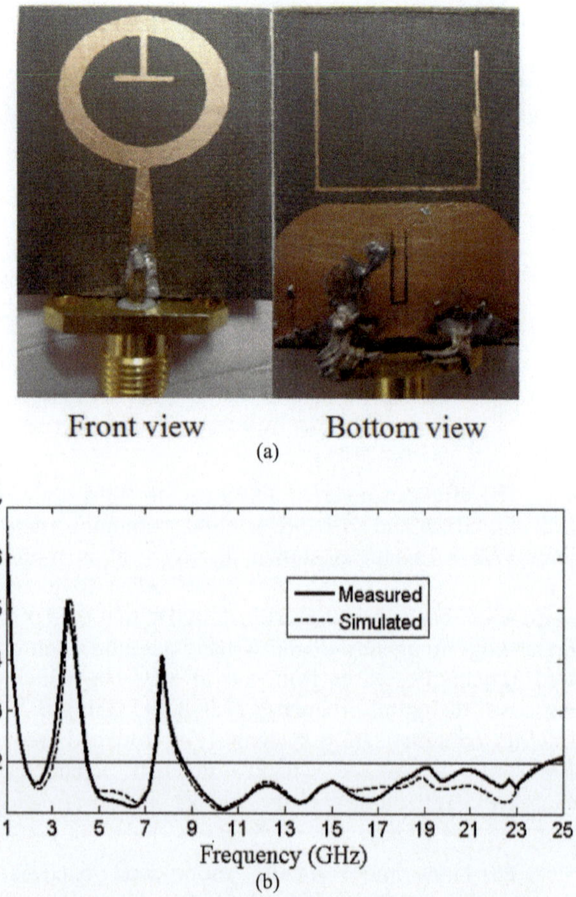

Fig. 9. (a) Image of the fabricated proposed DBN SWB PMA (b) simulated and measured VSWR versus frequency graph.

it is clearly seen that high current is gathered on the USPE around the 3.8 GHz notched frequency and it is opposite in direction (out-of-phase) to the current flowing on the CR patch as well as inverted TSPS, which cancel out the effective radiation. Similarly, at frequency 7.8 GHz, it can be observed that strong current is crowded on USS and very small current is concentrated on the ground plane, which offers strong attenuation near the notch frequency (7.8 GHz). It can be observed that small surface currents are flowing on both USPE and USS element at other frequencies such as 7 GHz and 10 GHz as depicted in Fig. 8(c) and (d) respectively.

Image of the fabricated antenna is shown in Fig. 9(a). The measurement is carried out with a Rohde and Schwarz ZVA24 vector network analyzer. Simulated and measured VSWR versus frequency graph for the proposed DBN SWB antenna is depicted in Fig. 9(b). Good agreement has been obtained up to 19 GHz between simulation and experimental VSWR results, while small deviation can be observed between simulated results and measured one if frequency is higher than 19 GHz. This may be owing to the maximum suitable frequency for the SMA-connector is up to 18 GHz.

Figure 10 shows the simulated and measured gain radiation pattern of the proposed SWB PMA. It indicates that at low frequencies (2.5 and 10 GHz), the antenna has omni-

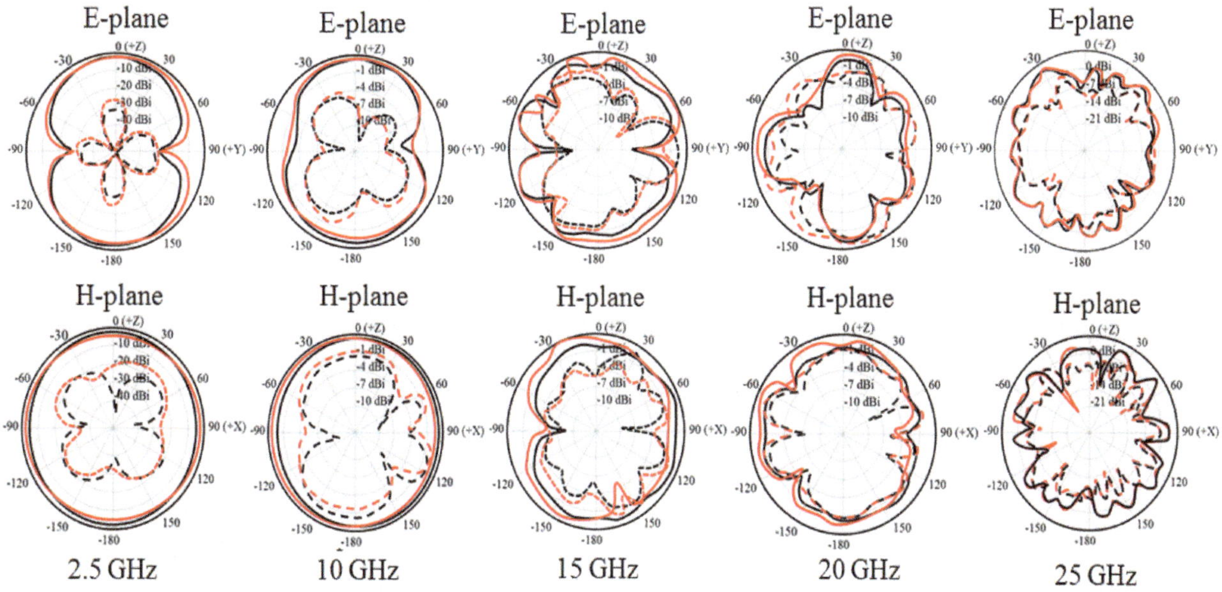

Fig. 10. Simulated (red) and measured (black) gain radiation pattern of DBN SWB PMA (solid line is co-polarization and dashed line is cross-polarization).

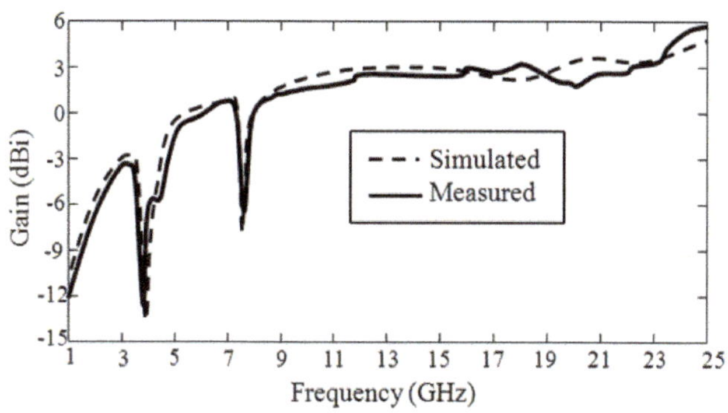

Fig. 11. Simulated and measured peak gain versus frequency graph for the proposed compact DBN SWB PMA.

Antenna of ref. no.	Dimensions ($W \times L \times h$) mm^3	RB	Substrate
[8]	46.4 × 38.5 × 1	6.25 : 1	Not specified ($\varepsilon_r = 2.65$)
[9]	26 × 32 × 0.8	3.93 : 1	Rogers4003 ($\varepsilon_r = 3.38$)
[10]	30 × 30 × 1.6	5.03 : 1	FR4 ($\varepsilon_r = 4.4$)
[11]	24 × 36 × 1.524	4.00 : 1	Rogers4003 ($\varepsilon_r = 3.38$)
[12]	12 × 18 × 0.8	4.59 : 1	FR4 ($\varepsilon_r = 4.4$)
[13]	10 × 16 × 1.6	6.17 : 1	FR4 ($\varepsilon_r = 4.4$)
Our antenna	24 × 30 × 0.787	15.63 : 1	RT/Duroid5870 ($\varepsilon_r = 2.23$)

Tab. 1. Comparison of the size and RB of the proposed DBN SWB PMA to DBN UWB antennas.

directional radiation pattern in the H-plane and bidirectional (eight shaped) patterns in the E-plane with negligible cross polarization. However, at the higher frequencies 15, 20 and 25 GHz, the cross-polarization level rises in the E and H-planes, which is probably owing to the rising horizontal component of the surface currents on the radiating patch at higher frequencies. The simulated efficiency η of the antenna varies from 82% – 95% throughout SWB frequency except at the DBN frequency (1st BS $\eta \cong 65\%$ and 2nd BS $\eta \cong 45\%$). The simulated and measured peak gain versus frequency graph for the proposed DBN SWB PMA is shown in Fig. 11. From the graph, it can be observed that the peak gain of the proposed antenna is 5.8 dBi, while gain sharply drops in the vicinity of 3.8 and 7.8 GHz BS. Table 1 depicts the dimension and RB of the proposed DBN SWB antenna with existing DBN UWB antennas. It can be observed that our proposed antenna has significantly larger RB than the antennas present in the published literature [8–13]. It also has smaller size than the DBN UWB antennas [8–11] except that of [12] and [13]. Note that DBN antennas of [12], [13] are designed on FR4 substrate which are meant for low frequency applications up to 2 GHz [18]. FR4 substrate also has higher relative permittivity of 4.4 and size also depends on relative permittivity.

4. Time Domain Performance

The time domain performance is the significant characteristics of UWB antennas in which that is required to have least distortion in transmitting and receiving scenarios. To explore the time-domain characteristics of a band-notched SWB antenna, two identical antennas were employed as transmitter and receiver. They are estranged from each other by a distance of 60 cm and placed in face-to-face and side-by-side orientations. To act as a transmitter, one antenna is excited by a time-domain transient pulse, while the other one is a receiver.

A 5th order derivative Gaussian input pulse waveform (generated in Tektronix AWG 7122B arbitrary signal generator) in the time domain with a width of 300 ps and its power spectral density in frequency domain (dBm/MHz) with reference to federal communication commission (FCC) spectral mask are displayed in Fig. 12 and Fig. 13. The UWB short pulse reported in [19] is used to excite the proposed band-notched antennas is a fifth-order derivative of the Gaussian pulse and is given by

$$G_5(t) = A\left(-\frac{t^5}{\sqrt{2\pi}\sigma^{11}} + \frac{10t^3}{\sqrt{2\pi}\sigma^9} - \frac{15t}{\sqrt{2\pi}\sigma^7}\right).\exp\left(-\frac{t^2}{2\sigma^2}\right) \quad (2)$$

where A is the amplitude parameter, t is the time, and σ is the standard deviation, whose value is 51 ps to ensure that the shape of the spectrum fit inside the FCC spectral mask. The measured results suggest that the chosen fifth derivative of Gaussian pulse fits into the FCC spectral mask reasonably well, indicating that this UWB pulse complies with the FCC spectral mask.

Figure 14 and 15 plot the received pulse for face-to-face and side-to-side configurations. It is seen that ringing distortion is observed in both the cases face-to-face and side-by-side scenarios, owing to the impedance mismatching at the notched bands. This confirms that the antenna does not response at all at the band-notched region. Moreover, it is also observed that the received signal amplitude is

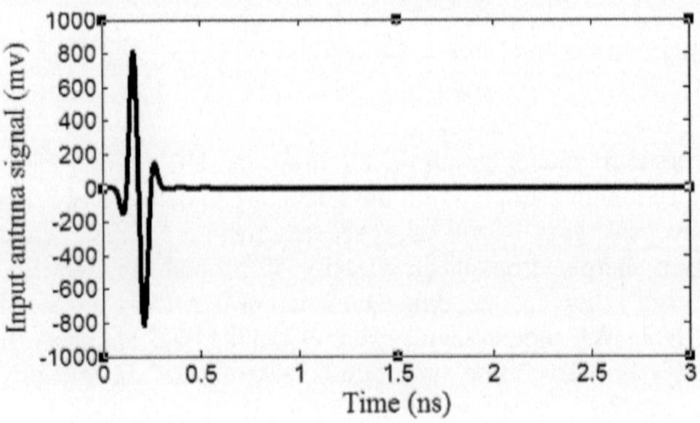

Fig. 12. Transmitted 5th order derivative Gaussian input signal.

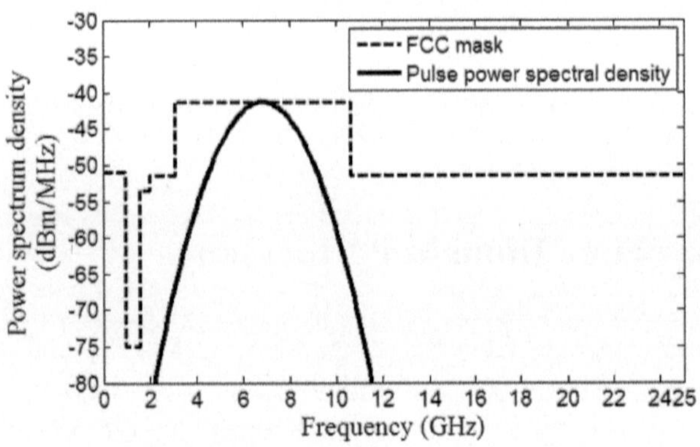

Fig. 13. Power spectral density.

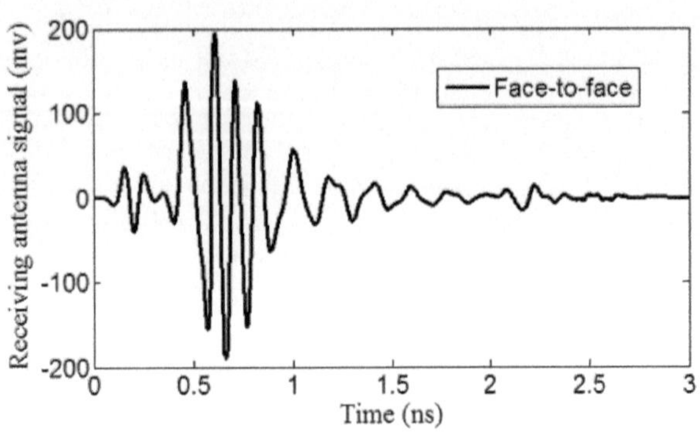

Fig. 14. Received signal in face-to-face scenario.

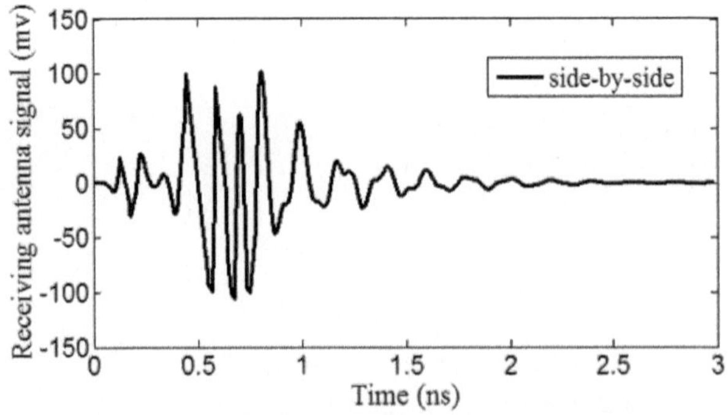

Fig. 15. Received signal in side-by-side scenario.

higher when two identical antennas are placed face-to-face (200 mV) with respect to side-by-side (100 mV) orientation. Overall ringing distortion is better than the existing UWB antennas in the literature [20]. To authenticate the similarity between received pulse $S_2(t)$ and transmitted or excited pulse $S_1(t)$ of the proposed antenna, the fidelity factor ρ is calculated which is given as follows

$$\rho = \max_{\tau}\left(\frac{\int S_1(t)S_2(t-\tau)\mathrm{d}t}{\sqrt{\int S_1^2(t)\mathrm{d}t}\sqrt{\int S_2^2(t)\mathrm{d}t}}\right) \quad (3)$$

where τ is a delay which is varied to make numerator in (3) a maximum. Using (3), the fidelity factor is calculated to be 0.8521 when the two identical antennas were placed face-to-face scenario, and the value becomes 0.8132 when they were placed side-by-side scenario. It has been reported that the fidelity factor for UWB antenna [8] is 91.2 % for face-to face-link.

5. Conclusion

A compact DBN SWB PMA has been proposed and investigated in this article. We have used TTMFL as well as RCFGP for broadband matching. The proposed SWB monopole antenna can operate from 1.6–25 GHz (RB of 15.63 : 1), except in the DBN of 3.2–4.4 GHz and 7.2 GHz to 8.4 GHz. By employing USS and USPE, DBN characteristics have been obtained. The peak gain of the proposed antenna is 6 dBi, except in the DBN of 3.8 GHz and 7.8 GHz.

References

[1] FCC, First report and order in the matter of revision of part 15 of the commission's rules regarding ultra-wideband transmission systems. Apr. 22 2002, p. 98–153. ISSN: 1937-8718.

[2] RUMSEY, V. *Frequency Independent Antennas*. New York: Academic Press, 1966. ISSN: 0018-926X

[3] CHEN, K. R., SIM, C., ROW, J. S. A compact monopole antenna for super wideband applications. *IEEE Antennas and Wireless Propagation Letters,* 2011, vol. 10, p. 488–491. ISSN: 1536-1225. DOI: 10.1109/LAWP.2011.2157071

[4] DONG, Y., HONG, W., LIU, L., ZHANG, Y., KUAI, Z. Performance analysis of a printed super-wideband antenna. *Microwave*

and Optical Technology Letters, 2009, vol. 51, no. 4, p. 949–956. ISSN: 1098-2760. DOI: 10.1002/mop.24222

[5] DOROSTKAR, M. A., ISLAM, M. T., AZIM, R. Design of a novel super wide band circular-hexagonal fractal antenna. *Progress In Electromagnetics Research*, 2013, vol. 139, no. 4, p. 229–245. ISSN: 1070-4698. DOI: 10.2528/PIER13030505

[6] MANOHAR, M., KSHETRIMAYUM, R. S., GOGOI, A. K. Printed monopole antenna with tapered feed line, feed region and patch for super wideband applications. *IET Microwaves, Antennas and Propagation*, 2014, vol. 8, no. 1, p. 39–45. ISSN: 1751-8725. DOI: 10.1049/iet-map.2013.0094

[7] WALADI, V., MOHAMMADI, N., ZEHFOROOSH, Y., HABASHI, A., NOURINIA, J. A novel modified star-triangular fractal (MSTF) monopole antenna for super-wideband applications. *IEEE Antennas and Wireless Propagation Letters*, 2013, vol. 12, p. 651 to 654. ISSN: 1536–1225. DOI: 10.1109/LAWP.2013.2262571

[8] LI, W. T., HEI, Y. Q., FENG, W., SHI, X. W. Planar antenna for 3G/Bluetooth/WiMAX and UWB applications with dual band-notched characteristics. *IEEE Antennas and Wireless Propagation Letters*, 2012, vol. 11, p. 61–64. ISSN: 1536–1225. DOI: 10.1109/LAWP.2012.2183671

[9] JIANG, W., CHE, W. A novel UWB antenna with dual notched bands for WiMAX and WLAN applications. *IEEE Antennas and Wireless Propagation Letters*, 2012, vol. 11, p. 293–296. ISSN: 1536-1225. DOI: 10.1109/LAWP.2012.2190490

[10] SHOKRI, M., SHIRZAD, H., MOVAGHARNIA, S., VIRDEE, B., AMIRI, Z., ASIABAN, S. Planar monopole antenna with dual interference suppression functionality. *IEEE Antennas and Wireless Propagation Letters*, 2013, vol. 12, p. 1554–1557. ISSN: 1536-1225. DOI: 10.1109/LAWP.2013.2292921

[11] RYU, K., KISHK, A. UWB antenna with single or dual band-notches for lower WLAN band and upper WLAN band. *IEEE Transaction on Antennas and Propagation*, 2009, vol. 57, no. 12, p. 3942–3950. DOI: 10.1109/TAP.2009.2027727

[12] OJAROUDI, M., OJAROUDI, N., GHADIMI, N. Dual band-notched small monopole antenna with novel coupled inverted U-ring strip and novel fork-shaped slit for UWB applications. *IEEE Antennas and Wireless Propagation Letters*, 2013, vol. 12, p. 182 to 185. ISSN: 1536-1225. DOI: 10.1109/LAWP.2013.2245296

[13] MEHRANPOUR, M., NOURINIA, J., GHOBADI, C., OJAROUDI, M. Dual band-notched square monopole antenna for ultrawideband applications. *IEEE Antennas and Wireless Propagation Letters*, 2012, vol. 11, p. 172–175. ISSN: 1536-1225. DIO: 10.1109/LAWP.2012.2186552

[14] POZAR, D. M. *Microwave Engineering.* John Wiley and Sons, 2005, p. 260. ISSN: 1931-7360. DOI: 10.1109/TAP.2003.816303

[15] MATHUR, S. P., SINHA, A. K. Design of microstrip exponentially tapered lines to match helical antennas to standard coaxial transmission lines. *IEE Proceedings on Microwaves, Antennas and Propagation,* 1988, vol. 135, no. 4, p. 272–274. ISSN: 0950-107X. DOI: 10.1049/ip-h-2.1988.0055.

[16] CHEN, Z. N., SEE, T. S., and QING, X. Small printed ultrawideband antenna with reduced ground plane effect. *IEEE Transactions on Antennas and Propagation, 2007,* vol. 55, p. 383–388. ISSN: 0018-926X. DOI: 10.1109/TAP.2006.889823

[17] DONG, Y., HONG, W., LIU, L., ZHANG, Y., KUAI, Z. Performance analysis of a printed super-wideband antenna. *Microwave and Optical Technology Letters*, 2009, vol. 51, no. 4, p. 949–956. ISSN: 1098–2760. DOI: 10.1002/mop.24222

[18] HUANG, Y., BOYLE, K. *Antennas from Theory to Practice*, Wiley, 2008, p. 64.

[19] KIM, H., PARK, D., JOO, Y. All-digital low-power CMOS pulse generator for UWB system. *Electronics Letter*, 2004, vol. 40, no. 24, p. 1534–1535. DOI: 10.1049/el.20046923

[20] GAO, G., HU, B., CONG, X., HE, L. Investigation of a novel dual band-notched UWB antenna by the equivalent circuit model and time domain characteristics. *Microwave and Optical Technology Letters,* 2013, vol. 55, no. 12, p. 2993–3000. ISSSN: 1096-4290. DOI: 10.1002/mop.27948.

About the Authors ...

Murli MANOHAR was born in Bhagalpur, India in 1984. He received the Ph.D. degree in Electronics and Electrical Engineering from the Indian Institute of Technology Guwahati, India in April 2015. He is presently working as an assistant professor in the Dept. of Electronics and Communication Engineering, Indian Inst. of Information Technology Manipur, India. His main research area is in antenna and RF engineering with focus on SWB printed monopole antenna.

Rakhesh Singh KSHETRIMAYUM received the Ph.D. degree from the School of Electrical and Electronic Engineering (EEE), Nanyang Technological University (NTU) Singapore in 2005. Since 2005, he has been with the Dept. of EEE, IIT Guwahati as an Associate Professor (2010-), Assistant Professor (2006-10) and Senior Lecturer (2005-2006). He worked as a Postdoctoral Scholar at the Dept. of EE, Pennsylvania State University (PSU), USA (2005), Research Associate Provisional at the Dept. of Electrical Communication Engineering (ECE), Indian Inst. of Science (IISc) Bangalore (2004-2005), Teaching Assistant at the School of EEE, NTU Singapore (2002-2003) and Trainee Software Engineer at Mphasis, India (2000-2001). Dr. Kshetrimayum is the recipient of SEFOGG Young Engineer Award (2011), Dept. of Science & Technology India (SERC) Fast Track Scheme for Young Scientists (2007-2010) and NTU Research Scholarship from 2001-2004. His current areas of research interests are in printed antennas and circuits, UWB communications and MIMO wireless communications. He has been involved in organizing several IEEE international conferences as Technical program co-chair, Publication chair, Program chair, Session chair and Technical program committee. He is the Editor-in-Chief of Inderscience journal International Journal of Ultra Wideband Communications and Systems. He is a Life Fellow of the Institution of Electronics and Telecommunication Engineers (IETE), India, Optical Society of India (OSI) and Antenna Test & Measurement Society (ATMS), India; a senior member of the International Association of Computer Science and Information Technology (IACSIT), Singapore, and the Inst. of Electrical and Electronics Engineers (IEEE), USA; a member of the Applied Computational Electromagnetics Society (ACES), USA and the European Microwave Association (EuMA), Belgium.

Anup Kumar GOGOI is a Professor of Electronics and Electrical Engineering and Dean of Outreach Education Programme at the Indian Institute of Technology Guwahati. He received his Ph.D. degree from the Dept. of Electrical Engineering, Indian Inst. of Technology Kanpur. His areas of research interests are in electromagnetics, microwave engineering, RF circuits and systems design.

Study of Horizontally Polarized Omnidirectional Microstrip Antenna Arrays

Kun WEI, Jian-Ying LI, Ling WANG, Rui XU

School of Electronics and Information, Northwestern Polytechnical University, 710072 Xi'an, Shannxi, China

weikun916@163.com, jianyingli@nwpu.edu.cn, lingwang@nwpu.edu.cn, rxuilj@163.com

Abstract. *This paper presents two microstrip antenna arrays for horizontally polarized (HP) omnidirectional application, namely rectangular patch antenna array and H-shaped patch antenna array. There are eight patch elements placed with back-to-back structure, four patch elements on each side. Antenna arrays are fed by a split eight power divider. The H-shaped patch antenna array has better omnidirectional performance than rectangular patch antenna array. The H-shaped antenna array is fabricated and measured. Both simulated and measured results show that the bandwidth of the designed H-shaped antenna array is 36 MHz with a center frequency 2.35 GHz. Horizontally polarized gains are greater than 7 dBi over the resonant band ($S_{11} < -10$ dB), while the cross-polarization level is less than −25 dB. The proposed H-shaped antenna array has high directivity (half-power beam-width is only 20 deg) and good omnidirectional performance (gain variation less than 1.5 dBi) at the center frequency.*

Keywords

Antenna array, horizontal polarization, omnidirectional radiation pattern, microstrip antenna

1. Introduction

Omnidirectional antenna array has been widely used in mobile communication, indoor radio, and wireless LAN systems because of their better performance in multiple paths environment. The need for horizontally polarized (HP) omnidirectional antenna array is growing considerably. However, it is still a challenge to design a HP omnidirectional antenna array with high gain and high directivity, which is of great importance in many applications.

Some HP omnidirectional antenna arrays have been reported. In [1], [2], HP omnidirectional antenna array with zero-phase-shift line loop elements is presented, which has a gain greater than 6 dBi and a gain variation less than ±2.5 dBi. The HP omnidirectional dipole array antennas with gain variations less than 2 dBi are presented in [3], [4]. The planar dipole array configuration is designed by using back-to-back dipole elements with omnidirectional pattern, and is constructed by printing on both sides of a dielectric substrate. In [5], a HP omnidirectional antenna utilizing stacked octagonal dipole array is designed from brass sheet. It has a gain of 1.1 dBi to 2.3 dBi along the frequency range.

Slot antenna arrays for HP omnidirectional application are also studied. In [6], [7], an omnidirectional array of broad wall slot doublets in a rectangular waveguide is presented. An HP omnidirectional slot antenna array is proposed in [8]. The antenna comprises two slot-elements that cut onto the ground plane on a substrate. The maximum gain is 5.5 dBi, and the gain variation is within ±1.5 dBi. In [9], an HP omnidirectional antenna using circular array of axial slot on cylinder is presented. The maximum gain is 5.63 dBi and the gain variation is less than 2 dBi. In [10], an HP slot array on a coaxial cylinder is achieved by arranging slots with opposite lean angles. However, slot antennas or arrays are large in size, heavy in weight, and difficult to fabricate.

An HP omnidirectional planar loop antenna that employs an artificial μ-negative transmission line (MNG-TL), which is designed by periodically loaded parallel-plate lines, is proposed in [11]. The four-element MNG-TL loop antenna array offers HP omnidirectional radiation patterns with enhanced gain of 6.5 to 7.9 dBi and the gain variation in the azimuth plane is below 2.1 dBi. A periodic leaky-wave antenna with HP omnidirectional pattern is proposed in [12]. The structure formed by 16 unit cell is developed at the 2.4 GHz band, which offers a HP omnidirectional radiation pattern with enhanced gain of 9.9 dBi to 10.6 dBi and a gain variation less than 1.8 dBi.

Though some HP omnidirectional antenna arrays have been reported, no previous works in the open literature (to the best of the present authors' knowledge) are available concerning the microstrip antenna array with HP omnidirectional radiation pattern. There are a lot works about omnidirectional planar microstrip antennas presented in [13–15], all those works focus on how to obtain omnidirectional radiation pattern, none of them are HP microstrip antenna array with high gain and high directivity. Antenna in [16] is omnidirectional horizontally polarized, but the antenna gain is only about 2 dBi.

This paper presents two new HP omnidirectional microstrip antenna arrays, namely rectangular patch antenna array and H-shaped patch antenna array. The designed antenna arrays have back-to-back structure, which consists of 8 patch elements. There are 4 patch elements on each side. The radiation model of the back-to-back antenna array is studied, and the directivity formula on main radiation plane is obtained. The H-shaped patch [17] antenna array has better omnidirectional radiation pattern than the rectangular patch antenna array. Ansoft HFSS is applied to design and simulate the antenna arrays for the purpose of optimizing the omnidirectional characteristic, drawing the better technical parameters. The proposed H-shaped antenna array is fabricated and measured. Both measured and simulated results are presented and compared.

The remaining parts of this paper are organized as follows. In Sec. 2, the study for the antenna arrays structure is presented. The ways to achieve better omnidirectional radiation are given in Sec. 3. Section 4 shows the simulation and measurement results of the H-shaped antenna array. The conclusion is given in Sec. 5.

2. Antenna Array Design

The proposed antenna arrays consist of four layers as shown in Fig. 1, two external layers are single-feedmicrostrip antenna array with back-to-back structure. There are total eight patches, four patches on each side. Two internal layers formed to be a classic stripline-type power divider, eight output ports from port 1 to port 8 have the same signal amplitude and phase. Each patch connects to the output port of the power divider by a probe, where port *i* is used to feed the patch *i*. In the top view of the designed antenna array, the physical length of the patch is *l*, the physical width of the patch is *w*, and the center-to-center distance between adjacent patches is *d*.

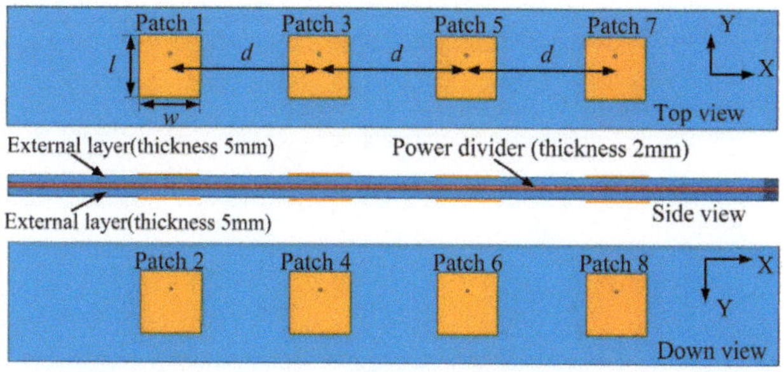

Fig. 1. Geometries of the designed antenna array.

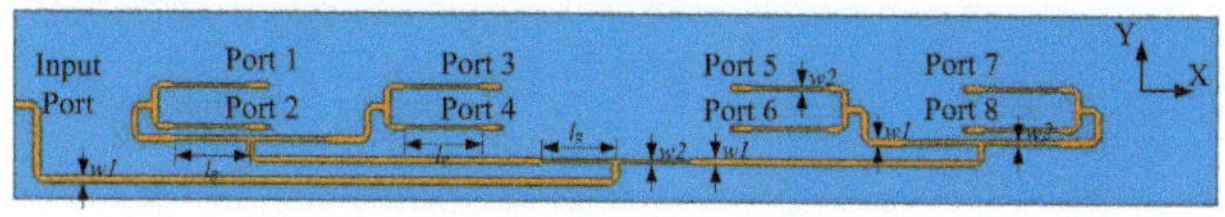

Fig. 2. Model of the designed power divider.

Figure 2 shows the model of the equal-split power divider. Width of the 50 Ω transmission lines is $w_1 = 1.42$ mm. The length of 50 Ω transmission lines is adjusted to settle outputs for feeding position, because it has little effects on divider matching. The width of the quarter-wave transmission lines, in the divider should have a characteristic impedance of 100 Ω, is $w_2 = 0.8$ mm. The length of the quarter-wave transmission line is $l_g = 19.1$ mm. The dielectric constant of the external layer is 16 and the thickness of the substrate is 5 mm, while the dielectric constant of the internal layer is 2.65 with a thickness of 1 mm. Since the designed antenna array is formed by two external layers and two internal layers, the overall substrate thickness is 12 mm. Overall dimension of the antenna is $300 \times 45 \times 12\ \text{mm}^3$.

For the main TM_{01} mode, the E-field radiation is given by (1) [18] where $u = k_0 w \sin\theta\cos\varphi$, $v = k_0 l \sin\theta\cos\varphi$, $k_0 = 2\pi/\lambda$, w is the effective width of the patch, l is the effective length of the patch. U_{01} denotes the voltage of mode TM_{01} at patch corner point.

$$\begin{cases} E_\theta = \mathrm{j}\dfrac{4k_0U_{01}}{\lambda r}\exp(-\mathrm{j}k_0 r)\exp\left(\mathrm{j}\dfrac{u+v}{2}\right)\sin(\dfrac{u}{2})\cos(\dfrac{v}{2})\left[\dfrac{w^2}{u^2}+\dfrac{l^2}{v^2-\pi^2}\right]\sin\theta\sin\varphi\cos\varphi \\ E_\varphi = \mathrm{j}\dfrac{4k_0U_{01}}{\lambda r}\exp(-\mathrm{j}k_0 r)\exp\left(\mathrm{j}\dfrac{u+v}{2}\right)\sin(\dfrac{u}{2})\cos(\dfrac{v}{2})\left[\dfrac{w^2\cos^2\varphi}{u^2}+\dfrac{l^2\sin^2\varphi}{v^2-\pi^2}\right]\sin\theta\cos\theta \end{cases} \quad (1)$$

In E-plane ($\varphi = 90°$): $u = 0$, $v = k_0 l\sin\theta$, equation (1) can be simplified as

$$\begin{cases} E_\theta = \mathrm{j}\dfrac{2wU_{01}}{\lambda r}\exp(-\mathrm{j}k_0 r)\exp\left(\mathrm{j}\dfrac{k_0 l}{2}\sin\theta\right)\cos(\dfrac{k_0 l}{2}\sin\theta) \\ E_\varphi = 0 \end{cases} \quad (2)$$

Taking positive *z*-axis side antenna array into consideration, four patches placed on *x-y* plane, formed to be array element 1, with adjacent center-to-center distance $d = \lambda/2$ as shown in Fig. 3. In E-plane ($\varphi = 90°$, *y-z* plane), $r_1 = r_2 = r_3 = r_4 = r$ in *z*-axis direction and the E-field radiation can be given by [18]:

$$E_{1\theta} = \mathrm{j}\frac{8wU_{01}}{\lambda r_1}\exp(-\mathrm{j}k_0 r_1)\exp\left(\mathrm{j}\frac{k_0 l}{2}\sin\theta\right)\cos(\frac{k_0 l}{2}\sin\theta). \quad (3)$$

Assuming that the proposed antenna array is formed by array element 1 and array element 2 on each side, thus the back-to-back structure under investigation is an array of two HP antenna array elements positioned along the *z*-axis, as shown in Fig. 4. The total field radiated by two array elements, assuming no coupling and no difference in excitation between the array elements, is equal to the sum of the two array elements. The field radiated by array element 1 in the *y-z* plane is given by

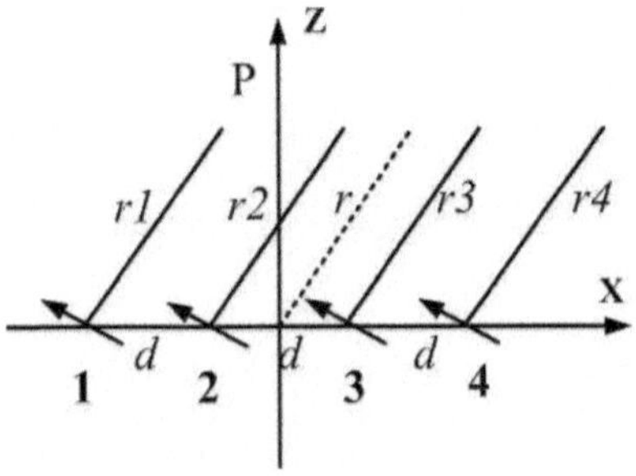

Fig. 3. Uniform four-antenna linear antenna array in x-z plane.

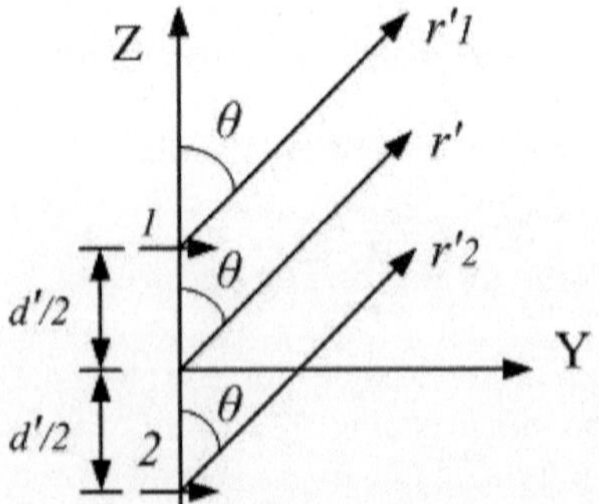

Fig. 4. Uniform linear antenna arrays in y-z plane.

$$E_0'(1) = \mathrm{j}\frac{8wU_{01}}{\lambda r_1'}\exp(-\mathrm{j}k_0 r_1')\exp\left(\mathrm{j}\frac{k_0 l}{2}\sin\theta\right)\cos(\frac{k_0 l}{2}\sin\theta) . \quad (4)$$

Since the array element 2 rotates 180° around x-axis compared to array element 1, field radiated by array element 2 in y-z plane is given by

$$E_0'(2) = \mathrm{j}\frac{8wU_{01}}{\lambda r_2'}\exp(-\mathrm{j}k_0 r_2')\exp\left(\mathrm{j}\frac{k_0 l}{2}\sin(\pi+\theta)\right)\cos(\frac{k_0 l}{2}\sin(\pi+\theta)). \quad (5)$$

As in this case, $d' = 0$, so $r_1' = r_2' = r'$. The total field radiated by the pair of array elements is given by

$$\begin{aligned} E_0' &= E_0'(1) + E_0'(2) \\ &= \mathrm{j}\frac{8wU_{01}}{\lambda r'}\exp(-jk_0 r')\cos(\frac{k_0 l}{2}\sin\theta) \\ &\quad \left[\exp\left(\mathrm{j}\frac{k_0 l}{2}\sin\theta\right) + \exp\left(-\mathrm{j}\frac{k_0 l}{2}\sin\theta\right)\right] \\ &= \mathrm{j}\frac{16wU_{01}}{\lambda r'}\exp(-jk_0 r')\cos^2(\frac{k_0 l}{2}\sin\theta). \end{aligned} \quad (6)$$

Thus, for the two array elements of constant amplitude, the total approximate formula for directivity as a function of the directional angles in y-z plane is represented by:

$$D = D_0\cos^4(\frac{k_0 l}{2}\sin\theta) \quad (7)$$

where $k_0 = 2\pi/\lambda$, D_0 is the maximum directivity. Different normalized pattern can be obtained with given different l. Suggesting that the antenna pattern roundness is $R = D_{min}/D_{max}$, and the gain variation around the E-plane is $V = |10\log R|$ dBi, so the directivity in y-z plane can be given by:

$$D = D_0\cos^4(\frac{k_0 l}{2}\sin\theta) = D_0\cos^4(\pi g\sin\theta) \quad (8)$$

where $l = g\lambda$. In general, the effective length of the patch is less than $\lambda/2$, so $0 < g < 0.5$. Since the effective length is a little bit longer than the physical length, and the effective length is in proportion to the physical length, it is tenable to replace the effective length with the physical length in (8) at the same frequency.

3. Achieving Better Omnidirectional Performance

As omnidirectivity is proportional to patch length, it is efficient to reduce the patch length for getting better omnidirectional radiation pattern when antenna arrays have the same resonant frequency. There are mainly two ways to decrease the patch length while keeping the antenna array work at the same frequency. One is increasing the substrate dielectric constant, the other way is using H-shaped patch to replace the rectangular patch.

3.1 Increasing Substrate Dielectric Constant

In order to verify the accuracy of (8) and achieve better omnidirectional performance for directivity in y-z plane of the proposed antenna array, comparisons between HFSS simulated and formula calculated gain variation around y-z plane with different patch length are made. To simplify the simulation process, the power divider is removed from the designed antenna array, the patches are fed by 8 ports with the same signal amplitude and phase.

The normalized radiation patterns of HFSS simulated and calculated by (8) in y-z plane are illustrated in Fig. 5 when tuning substrate dielectric constant. Length l_1 of patch a in Fig. 5(a) is 29.2 mm when substrate dielectric constant is 4. Length l_2 of patch b in Fig. 5(b) is 23.4 mm when substrate dielectric constant is 6. Length l_3 of patch c in Fig. 5(c) is 20.4 mm when dielectric constant is 8. Length l_4 of patch d in Fig. 5(d) is 16.3 mm when dielectric constant is 12. The antenna arrays with different patch lengths have the same resonant frequency of 2.35 GHz.

Table I summarizes the specified numeric comparison of gain variation calculated by (8) and HFSS by given different patch length. Length l of patch a is 29.2 mm, which is nearly 0.229 λ, so $g = 0.229$. Approximate equation for directivity is $D = D_0\cos^4(\pi g\sin\theta)$, thus $D_{max} = D_0$ and $D_{min} = 0.32\,D_0$, so antenna pattern roundness $R = D_{min}/D_{max} = 0.32$ and calculated gain variation is $V = |10\log R| \approx 4.9$ dBi. By using the same procedure, gain variation of antenna array with other patch physical length is calculated. Comparison of calculated gain variations (8) with HFSS simulated ones against substrate dielectric constant are plotted in Fig. 6.

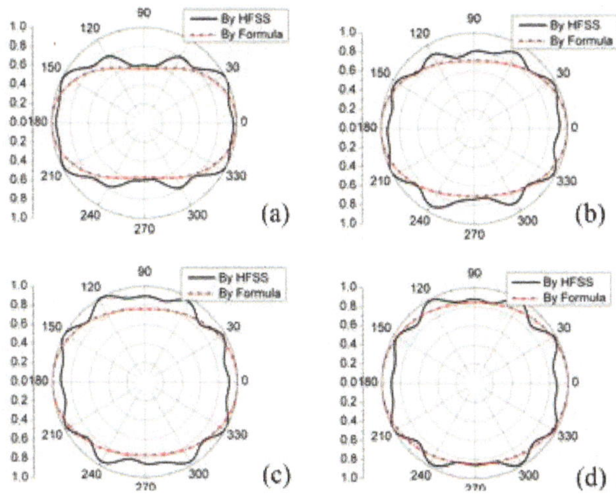

Fig. 5. Normalized formula calculated and HFSS calculated radiation pattern on E-plane when tuning dielectric constant: (a) $\varepsilon_r = 4$, (b) $\varepsilon_r = 6$, (c) $\varepsilon_r = 8$, (d) $\varepsilon_r = 12$.

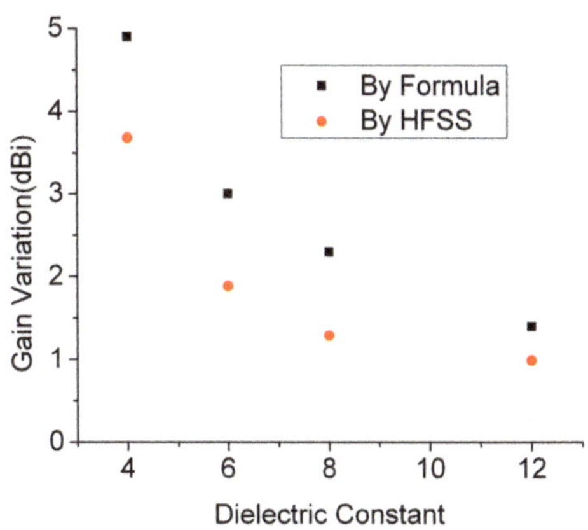

Fig. 6. Comparison between formula calculated gain variations and HFSS calculated ones when tuning substrate dielectric constant.

Dielectric constant	Length l (mm)	g	Formula Calculated (dBi)	HFSS simulated (dBi)	Optimal S11(dB)
4	29.2	0.229	4.9	3.7	−23
6	23.4	0.182	3.0	1.9	−25
8	20.4	0.161	2.3	1.3	−21
12	16.3	0.128	1.4	1.0	−24

Tab. 1. Comparison of the gain variation when tuning substrate dielectric constant.

Table 1 and Figure 6 summarize the gain variation on main radiation plane against substrate dielectric constant. From Tab. 1 and Fig. 6, the following results are clear: (a) The back-to-back structure microstrip antenna array radiates omnidirectional radiation pattern; (b) By increasing the substrate dielectric constant, the gain variation on the main radiation plane decreases. In general, the patch length can be decreased by increasing the substrate dielectric constant when the antenna arrays have the same resonant frequency. In this way, the gain variation of the omnidirectional pattern can be decreased, better omnidirectional characteristic is achieved. The optimized parameter values of the rectangular patch antenna array at 2.35 GHz are: w_1 =1.42 mm, w_2 = 0.8 mm, l_g = 19.1 mm and $l = w$ = 20.4 mm, substrate dielectric constant 8 and thickness 5 mm.

However, the increased dielectric constant often brings decreased bandwidth, increased losses (lower efficiency) and matching problems. It is important to select a proper substrate dielectric constant for achieving better antenna performances.

3.2 Using H-shaped Patch

As omnidirectivity is proportional to patch length, it is efficient to reduce the patch length for getting better omnidirectional radiation pattern. The H-shaped patch antenna has smaller dimension than that of the rectangular patch antenna when they resonate in the same frequency. Using H-shaped patch can achieve better omnidirectional performance without degrading the antenna array performance. Figure 7 shows the physical dimensions of single H-shaped patch antenna. When the H-shaped antenna resonates at 2.35 GHz, the optimized parameter values are: a = 2.5 mm, b = 9 mm, s = 13 mm, w = 17 mm, l = 20 mm, dielectric constant 10 and thickness 5 mm.

Table 2 and Figure 8 summarize the specified numeric comparison of improved gain variation by using H-shaped patch antenna. There are four kinds of substrates with different dielectric constant from No. 1 to No. 4, and each substrate is used to design rectangular patch antenna array and H-shaped patch antenna array. The antenna arrays have the same resonant frequency of 2.35 GHz. For example, taking dielectric constant of 6, the length l of H-shaped patch and rectangular patch are 20 mm and 23.3 mm. The formula calculated gain variations around y-z plane are 2.2 dBi and 3.0 dBi, respectively. Length of the H-shaped patch is 14.2 % shorter than that of the rectangular patch. The improved gain variation on main radiation plane is 1.2 dBi (formula calculated) and 0.8 dBi (HFSS simulated). The formula calculated improved gain variations of patch No. 1 to No. 4 are 1.3 dBi, 1.2 dBi, 0.8 dBi and 0.4 dBi. HFSS simulated improved gain variations from patch No. 1 to No. 4 are 1.13 dBi, 0.8 dBi, 0.55 dBi and 0.32 dBi. In Tab. 2 and Figure 8, it is clearly shown that using H-shaped patch antenna to replace rectangular patch antenna is efficient for achieving better omnidirectional radiation pattern.

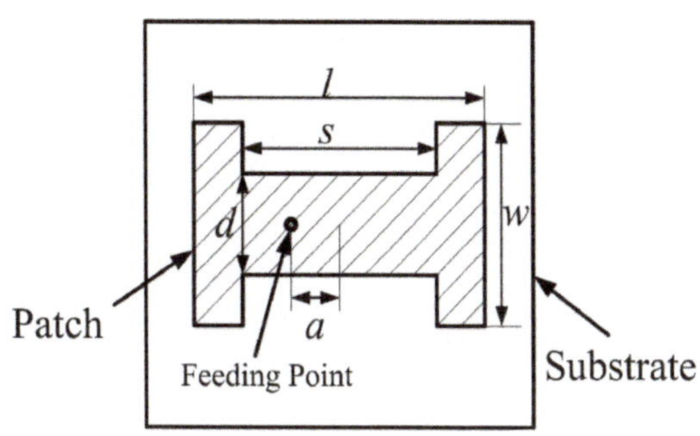

Fig. 7. Model of the H-shaped patch antenna.

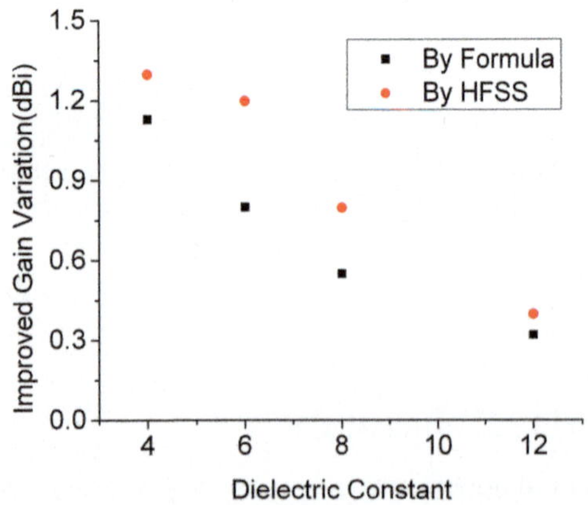

Fig. 8. The H-shaped antenna array improved gain variation comparison between formula and HFSS calculated results when tuning substrate dielectric constant.

Dielectric constant	4	6	8	12
H-shaped Length *l* (mm)	25.5	20	18.1	14.8
Rectangular patch *l* (mm)	29.2	23.3	20.5	16.3
Reduced Length	10.90%	14.20%	11.70%	9.20%
Improved gain variation by HFSS simulator (dBi)	1.13	0.8	0.55	0.32
Improved gain variation by formula (dBi)	1.3	1.2	0.8	0.4
Optimal S11 (dB) of the H-shaped antenna array	−25	−23	−26	−25

Tab. 2. Comparison of the gain variation by using H-shaped patch when tuning substrate dielectric constant.

4. Simulation and Measurement

The H-shaped antenna array is fabricated based on the optimized simulation. The optimized parameter values are: $a = 2.5$ mm, $b = 9$ mm, $s = 13$ mm, $w = 17$ mm, $l = 20$ mm, $w_1 = 1.42$ mm, $w_2 = 0.8$ mm, $l_g = 19.1$ mm.

Figure 9 shows the object of the designed H-shaped HP omnidirectional microstrip antenna array. The H-shaped antenna array uses back-to-back structure for achieving omnidirectional radiation. It has eight H-shaped patches, four patches on each side. The eight output ports of the power divider are connected with eight H-shaped patches by probes. The power divider input port is connected with a SMA connector.

The simulated and measured S_{11} (reflection coefficient) of the proposed H-shaped antenna array are shown in Fig. 10. Measured result agrees well with the simulated one. The antenna array has a 1.53 % relative bandwidth with the center frequency being 2.35 GHz, which is obtained by both simulation and measurement.

The simulated three-dimensional radiation pattern of the proposed H-shaped antenna array is shown in Fig. 11. It is obvious that the radiation pattern is omnidirectional in *y-z* plane (E-plane). Simulated and measured radiation patterns on E-plane and H-plane are indicated in Fig. 12, the maximum gain is about 7.4 dBi, and the antenna array efficiency is higher than 81 % in the resonant band. The measured and simulated results are in consistence with

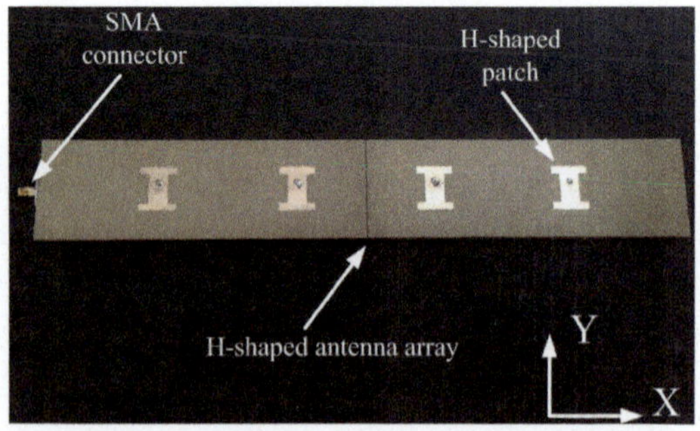

Fig. 9. Object of the designed H-shaped antenna array.

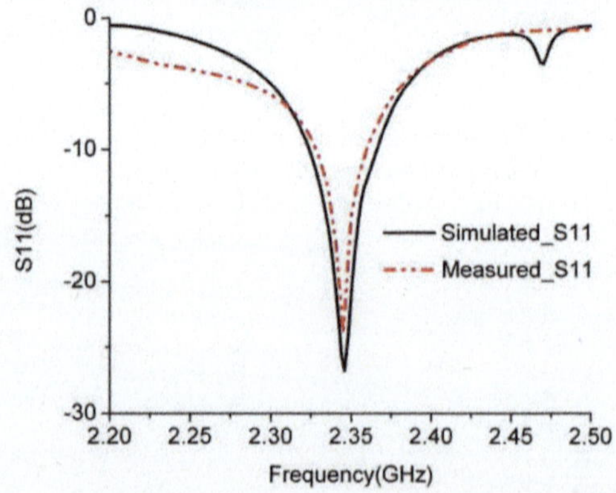

Fig. 10. Simulated and measured reflection coefficient S_{11} for the H-shaped microstrip antenna array.

each other, gain variation in the *y-z* plane is in the range of ±0.7 dBi at center frequency 2.35 GHz. This shows that the proposed H-shaped antenna array has good omnidirectional radiation pattern. The cross-polarization less than −25 dB is achieved by simulation and measurement. Furthermore, the designed antenna is with high directivity, with a half-power beam-width being only 20°.

Far fields from the two side patches of the proposed antenna array interfere constructively (add) in the *y*-axis direction and interfere destructively (cancel each other) in *z*-axis direction. When the total E-fields of two array elements in *z*-axis direction are canceled with each other, the gain of antenna array in *z*-axis direction is decreased compared to that of a single element. This is the reason why the designed omnidirectional antenna array cannot provide high gain. When total E-fields of two array elements in *y*-axis direction are added with each other, gain of the antenna array in *y*-axis direction is increased compared to that of single array element. Although the gain decreases slightly in the *z*-axis direction when achieving omnidirectional radiation pattern, the maximum gain in the *z*-axis direction is still greater than that in the *y*-axis direction.

The simulated and measured maximum gain and gain variation in the main radiation plane (*y-z* plane) are plotted in Fig. 13. The measured maximum gain is from 6.9 dBi to 7.66 dBi within 2.3 GHz to 2.4 GHz, and the measured gain variation is below 2 dBi from 2.3 GHz to 2.375 GHz. The maximum gain is higher than 7.2 dBi and gain variation is less than 1.5 dBi within operating band ($S_{11} < -10$ dB). Note that the measured gain is lower than the simulated one within the operating band. The reason is

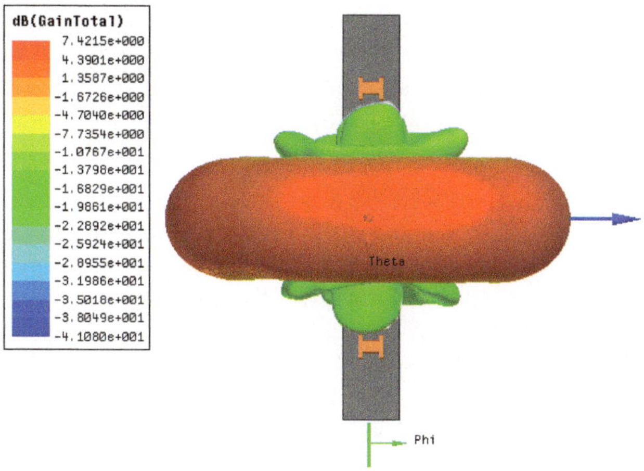

Fig. 11. Three-dimensional radiation pattern for the H-shaped microstrip antenna array.

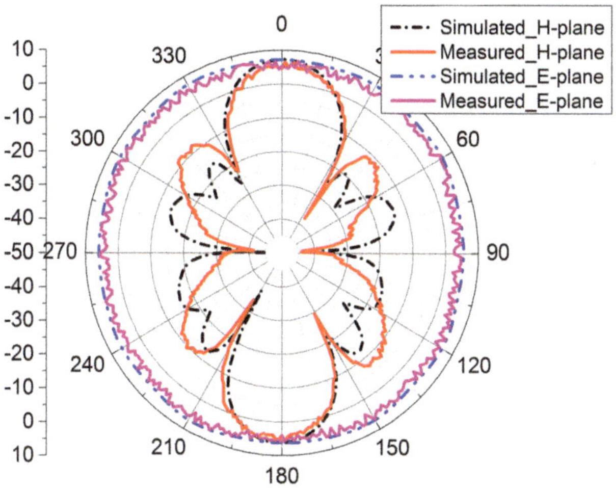

Fig. 12. Simulated and measured radiation pattern for the H-shaped microstrip antenna array at 2.35 GHz.

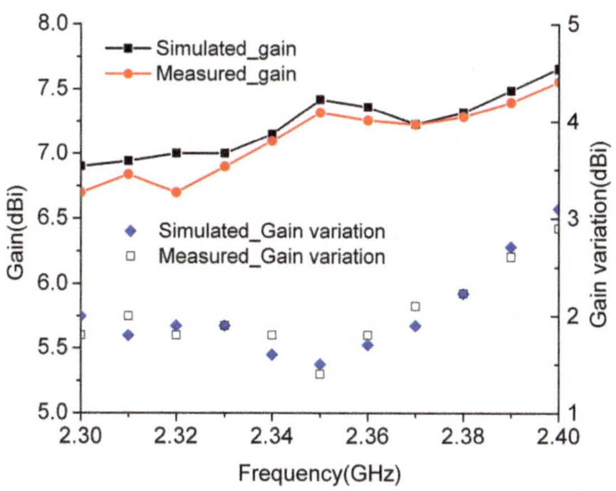

Fig. 13. Simulated and measured gain (variation) against frequency for the H-shaped microstrip antenna array.

that the dielectric loss of the fabricated antenna array is slightly greater than the value in the simulation. From the results, the proposed antenna achieves stable high gain (gain > 7.2 dBi) and good omnidirectional characteristic (gain variation less than 1.5 dBi) at the center frequency of 2.35 GHz.

5. Conclusion

Two new HP omnidirectional microstrip antenna arrays are presented in this paper. The formula of directivity in the main radiation plane with back-to-back antenna array structure is deduced. The high substrate dielectric constant and H-shaped patch are used to improve the omnidirectional performance. The H-shaped antenna array is fabricated and measured. The measured results agree well with the simulated ones. It has bandwidth 36 MHz, maximum gain 7.4 dBi at center frequency of 2.35 GHz, cross polarization lower than −25 dB, and 20° half-power beamwidth. Furthermore, the proposed H-shaped antenna array has good omnidirectional radiation pattern, and the gain variation on the main radiation plane is less than 1.5 dBi at the center frequency 2.35 GHz.

Acknowledgments

This study was supported in part by the National Natural Science Foundation of China (No. 61271416, 61601372 and 61601373), the Fundamental Research Funds for the Central Universities (No. 3102014KYJD027).

References

[1] QING, X. M., CHEN, Z. N. A compact metamaterial-based horizontally polarized omnidirectional antenna array. In *IEEE Antennas and Propagation Society International Symposium (APSURSI)*. Orlando (FL, USA), 2013, p. 1792–1793. DOI: 10.1109/APS.2013.6711555

[2] QING, X. M., CHEN, Z. N. Metamaterial-based wideband horizontally polarized omnidirectional 5-GHz WLCN antenna array. In *8th European Conference on Antennas and Propagation (EuCAP)*. The Hague (The Netherlands), 2014, p. 605–608. DOI: 10.1109/EuCAP.2014.6901831

[3] HSIAO, F. R., WONG, K. L., CHIOU, T. W. Omnidirectional planar dipole array antenna for WLCN access point. In *IEEE Antennas and Propagation Society International Symposium*. Columbus (USA), 2003, vol. 2, p. 2–5. DOI: 10.1109/APS.2003.1219165

[4] WEI, K. P., ZHANG, Z. J., CHEN, W. H., et al. A triband shunt-fed omnidirectional planar dipole array. *IEEE Antennas and Wireless Propagation Letters*, 2010, vol. 9, p. 850–853. DOI: 10.1109/LAWP.2010.2069077

[5] OSKLANG, P., LUADANG, B., PHONGCHAROENPANICH, C., et al. Horizontally polarized omnidirectional antenna using octagonal dipole array for digital television reception. In *Asia-Pacific Conference on Communications (APCC)*. Pattaya (Thailand), 2014, p. 135–138. DOI: 10.1109/APCC.2014.7091619

[6] SANGSTER, A. J., WANG, H. Moment method analysis of a horizontally polarized omnidirectional slot array antenna. *IEE Proceedings Microwaves, Antennas and Propagation,* 1995, vol. 142, no. 1, p. 1–6. DOI: 10.1049/ip-map: 19951641

[7] SANGSTER, A. J., WANG, H. Y. An entire domain analysis of a horizontally polarized omnidirectional antenna array. In *Second International Conference on Computation in Electromagnetics*. 1994, p. 343–346. DOI: 10.1049/cp: 19940087

[8] QING, X. M., CHEN, Z. N., GOH, C. K. A horizontally polarized omnidirectional slot antenna array. *IEEE Antennas and Propagation Society International Symposium (APSURSI).* Chicago (USA), 2012, p. 1–2. DOI: 10.1109/APS.2012.6348580

[9] PHONGCHAROENPANICH, C., WOUNCHOURN, P., KOSULVIT, S., et al. A horizontally polarized omnidirectional beam antenna using array of axial slot on cylindrical surface. In *3rd International Conference on Proceedings Microwave and Millimeter Wave Technology (ICMMT).* Bejing (China), 2002, p. 576–579. DOI: 10.1109/ICMMT.2002.1187765

[10] IIGUSA, K., TANAKA, M. A horizontally polarized slot-array antenna on a coaxial cylinder. In *Asia-Pacific Microwave Conference.* Sydney (Australia), 2000, p. 1444–1447. DOI: 10.1109/APMC.2000.926108

[11] WEI, K. P., ZHANG, Z. J., FENG, Z. H., et al. A MNG-TL loop antenna array with horizontally polarized omnidirectional patterns. *IEEE Transactions on Antennas and Propagation*, 2012, vol. 60, no. 6, p. 2702–2710. DOI: 10.1109/TAP.2012.2194643

[12] WEI, K. P., ZHANG, Z. J., FENG, Z. H., et al. Periodic leaky-wave antenna array with horizontally polarized omnidirectional pattern. *IEEE Transactions on Antennas and Propagation*, 2012, vol. 60, no. 7, p. 3165–3173. DOI: 10.1109/TAP.2012.2196930

[13] WANG, L., WEI, K. P., FENG, J. F., et al. A wideband omnidirectional planar microstrip antenna for WLCN applications. In *IEEE Electrical Design of Advanced Packaging and Systems Symposium (EDAPS).* Hanzhou (China), 2011, p. 1–4. DOI: 10.1109/EDAPS.2011.6213811

[14] BANCROFT, R., BATEMAN, B. An omnidirectional planar microstrip antenna. *IEEE Transactions on Antennas and Propagation*, 2004, vol. 52, no. 11, p. 3151–3154. DOI: 10.1109/TAP.2004.832338

[15] BRAS, L., BORGES CARVALHO, N., PINHO, P. Planar omnidirectional microstrip antenna array for 5 GHz ISM and UNII band. In *IEEE Antennas and Propagation Society International Symposium. (APSURSI).* Orlando (FL, USA), 2012, p. 1590–1591. DOI: 10.1109/APS.2013.6711454

[16] WEI, K., LI, J. Y., WANG, L., et al. Study on horizontally polarized omnidirectional microstrip antenna. *International Journal of Antenna and Propagation,* 2016, vol. 2016, p. 1–8. DOI: 10.1155/2016/8214153

[17] PALANISAMY, V., GARG, R. Rectangular ring and H-shaped microstrip antennas—alternatives to rectangular patch antenna. *Electronics Letters*, 1985, vol. 21, no. 19, p. 874–876. DOI: 10.1049/el: 19850617

[18] ZHONG, S. S. Slot antenna and microstrip antenna. *Antenna Theory and Techniques.* Beijing (China), 2011, sec. 1, p. 272–274. ISBN: 978-7-121-14286-4 (In Chinese)

About the Authors ...

Kun WEI received the degrees of B.Sc and M.Sc in Electronic Engineering from the Northwestern Polytechnial University in 2011 and 2014, respectively. Currently he is working on Ph.D. degree in the Northwestern Polytechnial University. His recent research interests include microstrip antennas, satellite communication antenna, circularly-polarized antenna, periodic structure, and defected ground structure (DGS).

Jian-Ying LI received the degrees of B.Sc. in Mathematics, and M.Eng.Sc. and Ph.D. both in Electromagnetic Field and Microwave Technology from Henan Normal University, Xinxiang, China, in 1986, and Xidian University, Xi'an, China, in 1992 and 1999, respectively. From 1992 to 1996, he worked at Xi'an Electronic Engineering Research Institute, Xi'an, China as a Research Engineer. From 1999 to 2004, he is first with the Dept. of Electrical and Computer Engineering at the National University of Singapore (NUS) where he is a Postdoctoral Research Fellow and then with High Performance Computation for Engineered Systems (HPCES) Programme at the Singapore-MIT Alliance (SMA) where he is a Research Fellow. From 2005 to 2010, he is with the Temasek Laboratories, in the National University of Singapore (NUS) where he is a research scientist. Since 2011 he has been with the School of Electronic and Information, Northwestern Polytechnial University. His current research interests include the fast algorithms and their applications to radar cross sections, analysis and design of phased arrays, waveguide slot antennas, and microstrip antennas, and EM Periodic Structure.

Ling WANG received the degrees of B.Sc. M.Sc. and Ph.D. from Xidian University, Xi'an, China, in 1999, 2002 and 2004, respectively, all in Electronic Engineering. From 2004 to 2007, he worked at Siemens and Nokia Siemens Networks. Since 2007, he has been with the School of Electronic and Information, Northwestern Polytechnical University and was promoted to Professor in 2012. His current research interests include array processing and smart antennas, wideband communications, cognitive radio, adaptive anti-jamming for satellite communications, satellite navigation, and date link systems.

Rui XU is born in Shaanxi, China in 1989. He received the degrees of B.Sc in Electronic Engineering in 2013 in Northwestern Polytechnical University Xi'an, China. He received the Master Degree in Electronic Engineering in Northwestern Polytechnial University in 2015. He is currently working toward the Ph.D. degree in Electronic Engineering in Northwestern Polytechnical University. His current research is ultra-wideband linear and circularly polarization antennas, waveguide slot antenna arrays, print slot antennas, microstrip antennas and EM Periodic Structure.

5G Terminals with Multi-Streaming Features for Real-Time Mobile Broadband Applications

Tomislav SHUMINOSKI, Toni JANEVSKI

Ss. Cyril and Methodius University, Faculty of Electrical Engineering and Information Technologies, Karpos 2 bb, 1000 Skopje, Republic of Macedonia

{tomish, tonij}@feit.ukim.edu.mk

Abstract. *In this paper we present a novel QoS framework on the network layer for 5G terminals with vertical multi-homing and multi-streaming capabilities by using radio networks aggregation. The proposed framework is leading to high performance utility networks with QoS provisioning for real-time multimedia services by achieving low packet delays, stochastic queuing network stability and highest mobile broadband capabilities i.e. bitrates. The proposed QoS algorithm is implemented within the mobile terminals on one side, and in dedicated proxy servers on mobile core network side. It is based on Lyapunov optimization techniques and it is targeted to handle simultaneously multiple multimedia service flows via multiple radio network interfaces in parallel.*

Keywords

5G, Lyapunov optimization, Quality of Service, Radio Access Technology, vertical multi-homing

1. Introduction

The exponential expansions and developments of mobile broadband networks are followed and supported by a novel research and development works towards the 5G era. This paper provides 5G framework that could lead to high performance delay-sensitive utility networks, spectral efficiency, network stability and high QoS provisioning for the key QoS parameters for any given multimedia service in nowadays and future Radio Access Technologies (RATs). Our framework is following the 5G key drivers and 5G enormous spectrum of advanced capabilities introduced in [1–7]. In that context our proposed algorithm with Lyapunov optimization technique implemented in both: the mobile terminal (MT) and a proxy server (placed in the core part of the mobile network) is following the 5G mobile networks' device-centric concept, with the multi-RAT or cloud-RAT paradigm using simultaneously multiple RAT interfaces. In order to exploit future 5G mobile broadband multi-RAT networks, and reaching the maximum average throughput, the proposed 5G terminal (5GT) is having vertical multi-homing and multi-streaming features [8], [9].

2. Related Work

The analysis and design methodology of our proposed 5G framework is based on the adaptive queuing Lyapunov optimization techniques [10–28], which are very powerful techniques for optimizing time average queuing networks and giving joint stability, efficiency and performance optimization. The existing solutions and applications are including maximization of the network uplink throughput subject to average power constraints on the terminal interfaces, minimizing average queue backlogs, subject to minimal queue network delay and in the same time achieving network stability and spectrum efficiency. Such Lyapunov drift-plus-penalty techniques were firstly applied to wireless networks in [12] by Tassiulas and Ephremides, where stochastic Lyapunov drift is used to develop a joint optimal routing and scheduling algorithm for the telecommunication networks. However, the Lyapunov drift has since become a powerful technique for the development of stable scheduling strategies for wireless and mobile systems and networks [15], [16], [19–24], [26–28], computer networks and switches [14], ad-hoc mobile networks [17], wireless sensor networks [18] and wireless mesh networks [25]. Moreover, example of services including maximization of the average network throughput with subject to average power constraints, minimizing the average power expenditure subject to network stability is presented in [15], [19], [22], [25] and [29]. In comparison with all above mentioned works, this paper applies a version of the Lyapunov drift-plus-penalty technique in the future 5G mobile and fixed terminals, here called with one name 5GTs, achieving high performance utility networks, spectral efficiency, network stability and high level of QoS provisioning for real-time mobile broadband services. Moreover, all changes and improvements we are implementing in the network layer of the 5GT, because the common part for all of existing and future RATs, which connect them all as a unifying technology, is IP.

Our proposed 5GT system model is device-centric,

multi-RAT, with several (m) interfaces; each for different RAT (overall M RATs) and has been previously presented in Fig. 1 [30]. However, despite the presented system model and AQUAplus optimization algorithm given in [30], the novelty of our framework presented in this paper is the development of novel adaptive QoS Module with advanced QoS user-centric network aggregation algorithm using vertical multi-homing and multi-streaming features for delay sensitive, real-time broadband services (applications). We refer to it as Massive throughput Advanced QoS-based User-centric Aggregation (MassiveAQUA) algorithm, which is defined independently from the wireless and mobile technologies and is doing Lyapunov drift-plus-penalty optimization for packet optimal scheduling in the 5GT with multiple RAT interfaces (in order to achieve massive throughput), achieving queue stability and simultaneously minimizing the queue delays, consequently having minimal overall end-to-end delay. Despite all, the AQUAplus algorithm has no mechanisms for dropping packets in the real queues, which leading to higher queue delays in comparison with the furthermore proposed MassiveAQUA. Plus, the MassiveAQUA algorithm is another extension of the AQUA algorithm (presented in related papers: [31] and [32]) placed within 5GMTs and proxy servers, because the AQUA algorithm is doing only multi-homing (selecting the best RAT for given service) with optimization problem solved by linear programming or genetic algorithms, without considering the network delay, the network stability and the queue backlogs.

The 5GT system model (Fig. 1) is almost the same as the system model in [30], with several differences: different utility functions (enriched with the delay constraints and penalties), different Lyapunov function, and existence of a packet drop decision vector for the network of queues – leading to delay sensitive and optimal multi-streaming and multi-homing processes. All novelties and differences with the AQUAplus algorithm optimization process are given in the following sections.

3. Optimization Algorithm

In Fig. 1 the 5GT system model with several (m) interfaces, each for different RAT (overall M RATs) is presented. Each source arrival process $x_i(t)$ after passing over the transport layer, is entering in the dedicated Network Layer Processor NLP_i inside the Processors part of the Upper Network layer. Despite the real M queue network in [30], here we are considering M queue network with queue vector $\boldsymbol{Q}(t) = (Q_1(t), Q_2(t), \ldots, Q_M(t))$ that evolves in slotted time $t \in \{0,1,2,3,\ldots\}$ with the following update equation:

$$Q_m(t+1) = \max[Q_m(t) + A_m(t) - \mu_m(t) - d_m(t), 0] \quad (1)$$

where $A_m(t) = \sum_{i=1}^{N} w_{i,m} x_i(t)$ is arrival rate variable for the m-th queue, and the parameter $\mu_m(t)$ is output serving rate variable of the m-th queue, where $m \in \{1,2,3,\ldots, M\}$. After

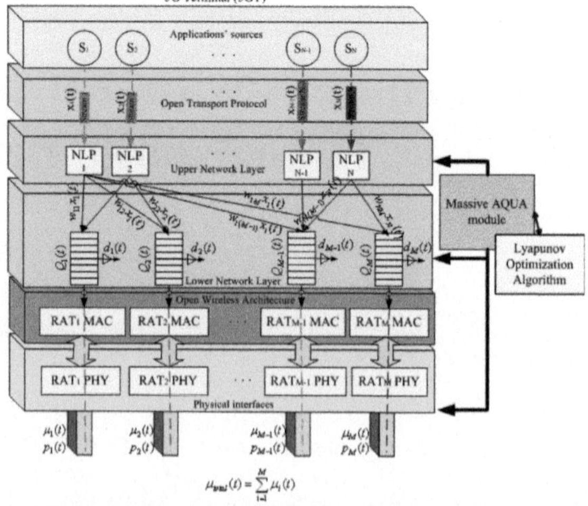

Fig. 1. The system model of the 5G terminal (fixed and mobile).

each queue we are trying to achieve maximal output serving rate $\mu_j(t)$ on each RAT interface $j = 1,2,.., M$, so the sum of all output serving rates, over the time, it will be also with maximal value. For the network of queues we are introducing a packet drop decision vector $\boldsymbol{d}(t) = (d_1(t), d_2(t), \ldots, d_M(t))$ and their adequate time average values: $\overline{\mathbf{d}} \underline{\underline{\Delta}} (\overline{d_1(t)}, \overline{d_2(t)}, \ldots, \overline{d_M(t)})$. The packet drop decision $d_m(t)$, of the m-th queue, allows packets already admitted to the network layer queue $Q_m(t)$ to be dropped if their delay is too large. Drop decisions $d_m(t)$ are chosen subject to the constraints:

$$0 \le d_m(t) \le A_{\max}, \text{ for } m \in \{1,2,3\ldots M\} \quad (2)$$

where $A_{\max}$ is a finite constant, chosen to be the maximal value of all arrival rate variables $A_m(t)$ at a given time slot t (which limits the amount of traffic we can send into the Lower Network Layer). Furthermore, let us consider a random queue network state: $\psi(t) \underline{\underline{\Delta}} [(\mu_1(t), \ldots, \mu_M(t)); (A_1(t), \ldots A_M(t))]$, and a control policy action $\alpha_{NL}(t)$ which is made every slot t with knowledge of the current state process $\psi(t)$ and is chosen within some abstract action set $A_{\psi(t)}$ (set containing all possible control actions). For simplicity we are assuming that the $\psi(t)$ process is i.i.d. over each slot.

Also, we are considering a vector with time average power values for each interface as $\overline{\mathbf{p}} \underline{\underline{\Delta}} (\overline{p_1(t)}, \overline{p_2(t)}, \ldots, \overline{p_M(t)})$ (the same as in [30]). Let the $\overline{\mathbf{x}} \underline{\underline{\Delta}} (\overline{x_1(t)}, \overline{x_2(t)}, \ldots, \overline{x_N(t)})$ be a vector of the time average arrival rates ($x_i(t)$). We are considering the following two separable utility functions of the above defined vectors:

$$\theta(\overline{\mathbf{x}}) \underline{\underline{\Delta}} \sum_{m=1}^{M} \sum_{i=1}^{N} w_{i,m} \overline{x_i(t)}, \text{ and} \quad (3)$$

$$\Gamma(\overline{\mathbf{p}}, \overline{\mathbf{d}}) \underline{\underline{\Delta}} \sum_{m=1}^{M} \left(\log\left(1 + \frac{\overline{p_m(t)}}{n_m}\right) - \beta \overline{d_m(t)} \right) \quad (4)$$

where $\log(\cdot)$ denotes the natural logarithm, and the function $\log\left(1+\frac{p_m(t)}{n_m}\right)$ has maximum derivative $1/n_m$, and is 0 when $p_m = 0$. The value n_m is finite constant and we assume $1/n_m < \infty$. Moreover, β is a constant that satisfies $1 \le \beta < \infty$. The utility function (3) is defined in [30] and equation (4) is novel utility function which incorporates the average packet drop decision vectors and the average capacity per radio interface, expressed with the logarithmic function given above. Then, for each 5GT system we are applying stochastic utility maximization framework to our streaming based system model, and the desired solution to the following optimization problem in uplink (when MT is sending information) and downlink (when the proxy server is sending information) is:

$$\text{Maximize: } \theta(\overline{\mathbf{x}}) + \Gamma(\overline{\mathbf{p}}, \overline{\mathbf{d}}) \quad (5)$$

Subject to: 1. Time average stream (flow) over the queue $Q_m(t)$ is less or equal to the time average maximal output serving rate $\mu_m^{\max}$ on the interface m, i.e.:

$$\sum_{i=1}^{N} w_{i,m}\overline{x_i(t)} \le \mu_m^{\max}, \ \forall m \in \{1,2,...,M\} \quad (6)$$

2. All queues $Q_m(t)$ are rate *stable*, i.e.:

$$\lim_{t\to\infty} \frac{\mathbf{E}\{|Q_m(t)|\}}{t} = 0, \forall m \in \{1,2,...,M\} \quad (7)$$

3. The desired time average power constraints are fulfilled [22] ($p_m(t)$ is the power incurred in interface m of the network on slot t, and p_m^{av} is a required time average power expenditure), i.e.:

$$\limsup_{t\to\infty} \overline{p_m(t)} \le p_m^{\text{av}}, \forall m \in \{1,2,...,M\} \quad (8)$$

4. For the control policy action: $\alpha_{\text{NL}}(t) \in A_{\psi(t)}$.

5. For finite parameter $\mu_{\min} > 0$, the average output serving rate of the m-th queue satisfies:

$$\overline{\mu_m(t)} \ge \mu_{\min}, \ \forall m \in \{1,2,3...M\} \quad (9)$$

The constraint (9) requires each queue $Q_m(t)$ to transmit with a time-average rate of at least $\mu_{\min}$, which ensures all queues are getting at least a minimum rate $\mu_{\min}$ of service. However, we shall not enforce this constraint, because if the input rate for the m-th queue is less than $\mu_{\min}$, then this constraint is wasteful.

For $t > 0$, the used variables as time average over the first t slots for $m \in \{1,2,3,..., M\}$ are defined with:

$$\begin{aligned} &\overline{x_m(t)} \triangleq \lim_{t\to\infty} \frac{1}{t}\sum_{\tau=0}^{t-1} \mathbf{E}\{x_m(t)\}; \ \overline{p_m(t)} \triangleq \lim_{t\to\infty} \frac{1}{t}\sum_{\tau=0}^{t-1} \mathbf{E}\{p_m(t)\}; \\ &\overline{\mu_m(t)} \triangleq \lim_{t\to\infty} \frac{1}{t}\sum_{\tau=0}^{t-1} \mathbf{E}\{\mu_m(t)\}; \ \overline{d_m(t)} \triangleq \lim_{t\to\infty} \frac{1}{t}\sum_{\tau=0}^{t-1} \mathbf{E}\{d_m(t)\}. \end{aligned} \quad (10)$$

4. MassiveAQUA Algorithm

For solving the above optimization problem we are using the Lyapunov drift-plus-penalty method (using fixed positive penalty control parameters V_1 and V_2) given in [10]. We enforce the constraints 1 and 2 of the optimization problem with the actual queue $Q_m(t)$ (1) and the constraint 3 with virtual queue $Z_m(t)$ for each m, given in [30]. It is very easy to prove that the queues are mean rate stable [10] (because they are with finite queue length).

Also, to ensure worst-case delay is bounded and to enforce the constraint (9), we define a $\mu_{\min}$-persistent service queue, being a virtual queue $U_m(t)$ for each $m \in \{1,..., M\}$ with $U_m(0) = 0$ and with update equation:

$$\begin{aligned} U_m(t+1) &= \max[U_m(t) - \mu_m(t) - d_m(t) + \mu_{\min}] \\ &\quad \text{if} \quad Q_m(t) > \mu_m(t) + d_m(t), \\ \text{and} \quad U_m(t+1) &= 0 \quad \text{if} \quad Q_m(t) \le \mu_m(t) + d_m(t). \end{aligned} \quad (11)$$

We assume average throughout that $\mu_{\min} \le A_{\max}$. The condition $Q_m(t) \le \mu_m(t) + d_m(t)$ is satisfied whenever the backlog $Q_m(t)$ is cleared (by service and/or drops) on slot t. If this constraint is not active, then $U_m(t)$ has a departure process that is the same as $Q_m(t)$, but it has an arrival of size $\mu_{\min}$ every slot t. Furthermore, we are defining a combining queue vector $\mathbf{S}(t) \triangleq [\mathbf{Q}(t), \mathbf{Z}(t), \mathbf{U}(t)]$ and the Lyapunov function (see Sec. 3.1.2 for Lyapunov functions in [10]) as:

$$L(\mathbf{S}(t)) \triangleq \frac{1}{2}\left[\sum_{i=1}^{M}\left(Q_i^2(t) + Z_i^2(t) + U_i^2(t)\right)\right]. \quad (12)$$

If we define the one-step conditional Lyapunov drift with $\Delta(\mathbf{S}(t)) \triangleq \mathbf{E}\{L(\mathbf{S}(t+1)) - L(\mathbf{S}(t)) | \mathbf{S}(t)\}$, then the Lyapunov drift-plus-penalty expression is defined as:

$$\begin{aligned} &\Delta(\mathbf{S}(t)) + V_1 \cdot \mathbf{E}\{\mathbf{u}(\mathbf{x}(t), \alpha_{NL}(t), t) | \mathbf{S}(t)\} + \\ &+ V_2 \cdot \mathbf{E}\{g(\mathbf{p}(t), \mathbf{d}(t), \alpha_{NL}(t), t) | \mathbf{S}(t)\}. \end{aligned} \quad (13)$$

So, our proposed MassiveAQUA algorithm seeks to minimize the upper bound of (13) for all t, all possible values of $\boldsymbol{S}(t)$, and all control parameters $V_1 \ge 0$ and $V_2 \ge 0$. In (13) we are introducing a "penalty" vector process with $\mathbf{u}(\mathbf{x}(t), \alpha_{NL}(t), t) = (u_0^1(t), u_0^2(t),..., u_0^M(t))$, which is already defined in more details in [30]. The novel "penalty" process $g(\mathbf{p}(\text{t}), \mathbf{d}(t), \alpha_{NL}(t), t)$, given here is defined with:

$$\begin{aligned} &g(\mathbf{p}(t), \mathbf{d}(t), \alpha_{NL}(t), t) = \\ &-\sum_{m=1}^{M} \log\left(1 + \frac{\xi_m(t) + p_m^{\text{av}}}{n_m}\right) + \sum_{i=1}^{M} \beta d_i(t). \end{aligned} \quad (14)$$

An optimal solution to (5) and also (14) has $d_m(t) = 0$ for all m. That is, the objective (14) can equivalently be replaced by the objective of minimizing $-\sum_{m=1}^{M} \log\left(1 + \frac{\xi_m(t) + p_m^{\text{av}}}{n_m}\right)$ and by adding the constraint

$d_m(t) = 0$ for all m. This is because the penalty for dropping is β. Thus, it can be shown that it is always better to restrict data at the Upper Network Layer (or even at transport layer) rather than admitting it and later dropping it. We recommend choosing β such that $1 \leq \beta \leq 2$. Because larger value of β will trade packet drops at the Lower Network Layer for packet non-admissions at the flow controller. Furthermore, assuming that: $u_{0\min}{}^m$ is deterministic low bound on $u_0{}^m(t)$ for all m in each slot t, the $g_{\min}{}^0$ is a deterministic lower bound on $g(\mathbf{p}(t),\mathbf{d}(t), \alpha_{NL}(t), t)$ for all t, plus assuming that there is a finite constant $K > 0$ such that for all choices of $\alpha_{NL}(t)$ on slot t we have:

$$\mathbf{E}\{ A_m(\alpha_{\mathrm{NL}}(t),\psi(t))^4 \} \leq K \ \forall m \in \{1,2,..,M\}, \quad (15)$$

$$\mathbf{E}\{ \mu_m(\alpha_{\mathrm{NL}}(t),\psi(t))^4 \} \leq K \ \forall m \in \{1,2,..,M\}, \quad (16)$$

$$\mathbf{E}\{ u_m(\alpha_{\mathrm{NL}}(t),\psi(t))^4 \} \leq K \ \forall m \in \{1,2,..,M\}, \quad (17)$$

$$\mathbf{E}\{ d_m(\alpha_{\mathrm{NL}}(t),\psi(t))^4 \} \leq K \ \forall m \in \{1,2,..,M\}, \quad (18)$$

$$\mathbf{E}\{ \xi_m(\alpha_{\mathrm{NL}}(t),\psi(t))^4 \} \leq K \ \forall m \in \{1,2,..,M\} \quad (19)$$

where the expectations are taken with respect to the distribution of the i.i.d. $\psi(t)$ process, and possibility randomized decisions $\alpha_{NL}(t) \in A_{\psi(t)}$. The expressions (15)–(19) are implying all major forms of stability and further are helping in the process of determining the algorithm's performance bounds. By squaring (2), (12) and (15); then summarizing according to the (16), doing the one-step conditional Lyapunov drift (given in Sec. 4.5 in [10]) and then adding the "penalties" $V_1 \cdot \mathbf{E}\{\mathbf{u}(\mathbf{x}(t), \alpha_{NL}(t), t)|\mathbf{S}(t)\}$ and $V_2 \cdot \mathbf{E}\{g(\mathbf{p}(t),\mathbf{d}(t), \alpha_{NL}(t), t)|\mathbf{S}(t)\}$ to both sides, it is not difficult to show that the expression (13) has the following upper bound for all t (for more details see the Theorem 4.2 in [10]):

$$\begin{aligned} & B - V_1 \mathbf{E}\left\{ \sum_{m=1}^{M} \sum_{i=1}^{N} w_{i,m} x_i(t) \middle| \mathbf{S}(t) \right\} - \\ & V_2 \mathbf{E}\left\{ \sum_{m=1}^{M} \left(\log\left(1 + \frac{\xi_m(t) + p_m{}^{av}}{n_m}\right) - \beta d_m(t) \right) \middle| \mathbf{S}(t) \right\} + \\ & \sum_{m=1}^{M} Q_m(t) \mathbf{E}\left\{ \sum_{i=1}^{N} w_{i,m} x_i(t) - \mu_m(t) - d_m(t) \middle| \mathbf{S}(t) \right\} + \\ & \sum_{m=1}^{M} U_m(t) \mathbf{E}\{ \mu_{\min} - \mu_m(t) - d_m(t) | \mathbf{S}(t) \} + \\ & \sum_{l=1}^{M} Z_l(t) \mathbf{E}\{ \xi_l(t) | \mathbf{S}(t) \}. \end{aligned} \quad (20)$$

Here B is a finite constant related to the worst-case second moments of $A_m(t)$, $\xi_m(t)$, $d_m(t)$ and $\mu_m(t)$ processes:

$$\begin{aligned} B \triangleq & \frac{1}{2}\left(\sum_{m=1}^{M} \mathbf{E}\left\{ \sum_{i=1}^{N} \left(\left(w_{i,m} x_i(t)\right)^2 + \mu_m{}^2(t) + d_m{}^2(t) \right) \middle| \mathbf{S}(t) \right\} \right) + \\ & \frac{1}{2}\left(\sum_{m=1}^{M} \mathbf{E}\left\{ \left(\mu_{\min}{}^2 + \mu_m{}^2(t) + d_m{}^2(t) \right) \middle| \mathbf{S}(t) \right\} \right) + \\ & \frac{1}{2}\left(\sum_{l=1}^{M} \mathbf{E}\{ \xi_m{}^2(t) | \mathbf{S}(t) \} \right). \end{aligned} \quad (21)$$

Such a constant B exists by the boundedness assumptions on the processes. Furthermore, the goal of our MassiveAQUA algorithm is to minimize the expression (20), by choosing the most appropriate control policy action $\alpha_{NL}(t) \in A_{\psi(t)}$ every slot t, observing the real and virtual queues vectors $\mathbf{Q}(t)$, $\mathbf{Z}(t)$, $\mathbf{U}(t)$ and the current state $\psi(t)$. In that way, the MassiveAQUA algorithm is solving the above discussed optimization problem, decoupling and reducing it into separate algorithms and does the following:

1) Stream control and QoS traffic routing over different RATs interfaces at the 5GTs (vertical multi-homing and multi-streaming [8],[9]): Every slot t, for each queue $m \in \{1,2,3,\ldots, M\}$ observes new arrivals $x_m(t)$, the actual queue $Q_m(t)$ (2) and the virtual queues $Z_m(t)$ (12) and $U_m(t)$ (15), and chooses $x_m(t)$ to go over that queues to:

$$\begin{aligned} & \text{Minimize: } \sum_{m=1}^{M} Q_m(t) \sum_{i=1}^{N} w_{i,m} x_i(t) - V_1 \cdot \sum_{j=1}^{M} \sum_{i=1}^{N} w_{i,j} x_i(t) \\ & \text{subject to: } 0 \leq \sum_{i=1}^{N} w_{i,m} x_i(t) \leq \mu_m{}^{\max}, \forall m \in \{1,2,\ldots,M\} \\ & \text{and } \sum_{j=1}^{M} w_{i,j} = 1, \forall i \in \{1,2,\ldots,N\}. \end{aligned} \quad (22)$$

2) Average Power Allocation: Every slot t, observe virtual queue $Z_m(t)$ for $m \in \{1,2,3,\ldots, M\}$ and allocate power vector $\mathbf{p}(t)$ on each interface (so there is power-aware uplink/downlink transmission scheduling), as the solution for the auxiliary variable $\xi_m(t)$ to:

$$\text{Minimize: } \sum_{l=1}^{M} Z_l(t)\xi_l(t) - V_2 \sum_{m=1}^{M} \log\left(1 + \frac{p_m{}^{av} + \xi_m(t)}{n_m}\right) \quad (23)$$

$$\text{subject to: } -p_m{}^{av} \leq \xi_m(t) \leq \xi_m{}^{\max}, \forall m \in \{1,2,\ldots,M\}.$$

3) Packet Dropping: For every slot t, observe virtual queue $U_m(t)$ (11) and real queue $Q_m(t)$ (1), and for $m \in \{1,2,3,\ldots, M\}$ choose $d_m(t)$ by:

$$d_m(t) = \begin{cases} A_{\max}, & \text{if } U_m(t) + Q_m(t) > \beta V_2 \\ 0, & \text{else} \end{cases} \quad (24)$$

4) Queues Updates: For $m \in \{1,2,3,\ldots, M\}$ update actual queues $Q_m(t)$, virtual queues $Z_m(t)$ and $U_m(t)$ according to (2), (12) and (15), respectively.

5) Downlink packet routing: IP packets, which are coming from all RATs interfaces and going to the application layer, are received in the incoming recipient's (another additional actual and virtual) queues, and sent from MassiveAQUA module to the Upper Network Layer. Then all packets are resent to Transport Layer where they are synthesized in one stream (flow) $x_m(t)$ and finally delivered to the peer application. As before mentioned, the arrivals, $\psi(t)$ process and channel states are assumed to be i.i.d. every slot t, and the queue network is assumed to be initially empty at time 0.

5. Performance Bounds

Our novel 5GMT is multi-RAT node, with several (n) interfaces, each for different RAT (overall n RATs). Assuming there is a finite constant $C \geq 0$ such that our MassiveAQUA algorithm chooses the most appropriate control policy action $\alpha_{NL}(t) \in A_{\psi(t)}$ every slot t, observing the real and virtual queues vectors $\mathbf{Q}(t)$, $\mathbf{Z}(t)$, $\mathbf{U}(t)$ i.e. with combining queue vector $\mathbf{S}(t)$, and minimizing the upper bound (25), then we have the following theorem:

Theorem 1 (Performance Bounds of MassiveAQUA): If we suppose that the arrivals and channel state information for each time slots is stochastic i.i.d. over slots with probabilities $\pi(\psi)$, the problem for maximizing $\sum_{m=1}^{M}\sum_{i=1}^{N} w_{i,m}\overline{x_i(t)}$ and (7-9) are feasible, and if we fixed a value for the constant $C \geq 0$, using C-additive approximation of the MassiveAQUA algorithm every slot t, and having appropriate control policy action ($\alpha_{NL}(t)$), plus setting the penalty control parameters $V_1 = V_2 = V$ then:

1) The time average expected cost satisfies:

$$\limsup_{t\to\infty}\frac{1}{t}\sum_{\tau=0}^{t-1}\left(\sum_{m=1}^{M}\mathbf{E}\{u_0^{\,m}(\tau)\}+\mathbf{E}\{g(\tau)\}\right) \geq X_\Sigma^{\,\mathrm{optim}}-\frac{B+C}{V}-g^{\mathrm{optim}} \tag{25}$$

where $X_\Sigma^{\mathrm{optim}}$ is the optimal infimum time average of the weighted sum of admitted rates into the queue network, achievable by any policy that stabilizes the queues and satisfied average power constraints and all other required constraints. The g^{optim} is the optimal infimum time average of the objective function (19), also achievable by any policy that stabilizes the queues and satisfied average power constraints and all other required constraints. Also B is defined with (21).

2) If we assume that there is a finite constraints $K > 0$, $\rho \geq 0$ and $\delta > 0$ for all (possibly random) choices of the control policy action $\alpha_{NL}(t)$ where (15)–(19) are met and the following constraints are satisfied for all m:

$$\mathbf{E}\{\,\xi_m(\alpha_{\mathrm{NL}}(t),t)\} \leq -\delta+\rho\,, \tag{26}$$

$$\mathbf{E}\{\,d_m(\alpha_{\mathrm{NL}}(t),t)\} \leq -\delta+\rho\,, \tag{27}$$

$$\mathbf{E}\{\,A_m(\alpha_{\mathrm{NL}}(t),t)-\mu_m(\alpha_{\mathrm{NL}}(t),t)-d_m(\alpha_{\mathrm{NL}}(t),t)\} \leq -\delta+\rho, \tag{28}$$

$$\mathbf{E}\left\{\sum_{m=1}^{M}\sum_{i=1}^{N} w_{i,m}x_i(\alpha_{\mathrm{NL}}(t),t)\right\} \leq X_\Sigma^{\,\mathrm{optim}}+\rho. \tag{29}$$

$$\mathbf{E}\left\{\sum_{m=1}^{M} g(\alpha_{\mathrm{NL}}(t),t)\right\} \leq g^{\mathrm{optim}}+\rho. \tag{30}$$

Then:

$$\limsup_{t\to\infty}\frac{1}{t}\sum_{\tau=0}^{t-1}\sum_{m=1}^{M}\mathbf{E}\{Q_m(\tau)+Z_m(\tau)+U_m(\tau)\} \leq \frac{B+C}{\delta}-\frac{V(X_\Sigma^{\,\mathrm{optim}}+u^{\min}-g^{\mathrm{optim}}+g_{\min}^{\;0})}{\delta} \tag{31}$$

where

$u^{\min} \leq \frac{1}{t}\sum_{\tau=0}^{t-1}\sum_{m=1}^{M}\mathbf{E}\{u_0^{\,m}(\alpha_{NL}(\tau),\tau)\} \leq u^{\max}, \forall t \in \{1,2,\ldots,t-1\}$, and

$g_{\min}^{\;0} \leq \frac{1}{t}\sum_{\tau=0}^{t-1}\mathbf{E}\{g(\alpha_{\mathrm{NL}}(\tau),\tau)\} \leq g_{\max}^{\;0}, \forall t \in \{1,2,\ldots,t-1\}$.

3) The real and virtual queue vectors $\mathbf{Q}(t)$, $\mathbf{U}(t)$ and $\mathbf{Z}(t)$ are mean rate stable, and all required constraints of the optimization problem (5–9) are satisfied.

Proof. (Theorem 1)

Due to the choice of the most appropriate control policy action $\alpha_{NL}(t) \in A_{\psi(t)}$ every slot t, by the MassiveAQUA algorithm, observing the combining queue vector $\mathbf{S}(t)$, finite constant $C \geq 0$, which choice is minimizing (25), setting the penalty control parameters $V_1 = V_2 = V$, we have for all t and all possible vectors $\mathbf{S}(t)$:

$$\begin{aligned}
&\Delta(\mathbf{S}(t))+V\cdot\mathbf{E}\{\,\mathbf{u}(\mathbf{x}(t),\alpha_{\mathrm{NL}}(t),t)|\mathbf{S}(t)\}+\\
&V\cdot\mathbf{E}\{\,g(\mathbf{p}(t),\mathbf{d}(t),\alpha_{\mathrm{NL}}(t),t)|\mathbf{S}(t)\} \leq\\
&\quad B+C-V\mathbf{E}\left\{\sum_{m=1}^{M}\sum_{i=1}^{N} w_{i,m}x_i(\alpha_L^{\,e}(t),t)|\mathbf{S}(t)\right\}-\\
&\quad V\mathbf{E}\left\{\sum_{m=1}^{M}\left(\log\left(1+\frac{\xi_m(\alpha_L^{\,e}(t),t)+p_m^{\,\mathrm{av}}}{n_m}\right)-\beta d_m(\alpha_L^{\,e}(t),t)\right)|\mathbf{S}(t)\right\}+\\
&\quad \sum_{m=1}^{M}Q_m(t)\mathbf{E}\left\{\sum_{i=1}^{N} w_{i,m}x_i(\alpha_L^{\,e}(t),t)-\mu_m(\alpha_L^{\,e}(t),t)-d_m(\alpha_L^{\,e}(t),t)|\mathbf{S}(t)\right\}+\\
&\quad \sum_{m=1}^{M}U_m(t)\mathbf{E}\{\,\mu_{\min}-\mu_m(\alpha_L^{\,e}(t),t)-d_m(\alpha_L^{\,e}(t),t)|\mathbf{S}(t)\}+\\
&\quad \sum_{l=1}^{M}Z_l(t)\mathbf{E}\{\,\xi_l(\alpha_L^{\,e}(t),t)|\mathbf{S}(t)\}.
\end{aligned} \tag{32}$$

where $\alpha_L^{\,e}(t)$ is another decision action that can be implemented on a slot t. If we fixed δ in the interval $(0, \delta_{\max}]$, and $\rho > 0$, plus using the control policy action $\alpha_L^{\,e}(t)$ designed to achieve (26)–(30) and nothing that this policy action makes decisions independent of the $\mathbf{S}(t)$ (the values $x_i(\alpha_{NL}(t),t)$, $\mu_m(\alpha_{NL}(t),t)$, $d_m(\alpha_{NL}(t),t)$, and $\xi_m(\alpha_{NL}(t),t)$ are independent of current queue backlog and state $\mathbf{S}(t)$), yields:

$$\begin{aligned}
&\Delta(\mathbf{S}(t))+V\cdot\mathbf{E}\{\,\mathbf{u}(\mathbf{x}(t),\alpha_{\mathrm{NL}}(t),t)|\mathbf{S}(t)\}+\\
&V\cdot\mathbf{E}\{\,g(\mathbf{p}(t),\mathbf{q}(t),\alpha_{\mathrm{NL}}(t),t)|\mathbf{S}(t)\} \leq\\
&\quad B+C-V\left(X_\Sigma^{\,\mathrm{optim}}+\rho\right)+V\left(g^{\mathrm{optim}}+\rho\right)+\\
&\quad (\rho-\delta)\sum_{m=1}^{M}U_m(t)+(\rho-\delta)\sum_{m=1}^{M}Q_m(t)+(\rho-\delta)\sum_{l=1}^{M}Z_l(t).
\end{aligned} \tag{33}$$

Also, the (33) holds for all $\rho > 0$. Now if we take the limit as $\rho\to 0$ yields:

$$\begin{aligned}
&\Delta(\mathbf{S}(t))+V\cdot\mathbf{E}\{\,\mathbf{u}(\mathbf{x}(t),\alpha_{NL}(t),t)|\mathbf{S}(t)\}+\\
&V\cdot\mathbf{E}\{\,g(\mathbf{p}(t),\mathbf{q}(t),\alpha_{NL}(t),t)|\mathbf{S}(t)\} \leq\\
&\quad B+C-V\left(X_\Sigma^{\,\mathrm{optim}}\right)+V\left(g^{\mathrm{optim}}\right)-\\
&\quad \delta\sum_{m=1}^{M}\left(Q_m(t)+U_m(t)+Z_m(t)\right)\,.
\end{aligned} \tag{34}$$

Furthermore, let the fourth moment bounds (15)–(19) are holding and applying Theorem 4.2 (Lyapunov Optimization) from [10], so the constraints (6)–(9) are satisfied, by using the low of telescoping sums over $\tau \in \{0,1,2,\ldots,t-1\}$ for $t > 0$, the (34) yields:

$$\mathbf{E}\{L(\mathbf{S}(t))\}-\mathbf{E}\{L(\mathbf{S}(0))\}+ V\cdot\sum_{\tau=0}^{t-1}\sum_{m=1}^{M}\mathbf{E}\{u_0^{m}(\tau)\}+V\cdot\sum_{\tau=1}^{t-1}\mathbf{E}\{g(\tau)\}\leq \left(B+C-V\cdot X_{\Sigma}^{\text{optim}}+V\cdot g^{\text{optim}}\right)\cdot t- \delta\sum_{\tau=0}^{t-1}\sum_{m=1}^{M}(Q_m(\tau)+U_m(\tau)+Z_m(\tau)). \tag{35}$$

Moreover, rearranging terms and neglecting non-negative terms when appropriate, it is easy to show that the inequality (35) directly implies the following two inequalities for all $t > 0$:

$$\frac{1}{t}\cdot\sum_{\tau=0}^{t-1}\left(\sum_{m=1}^{M}\mathbf{E}\{u_0^{m}(\tau)\}+\mathbf{E}\{g(\tau)\}\right)\leq\left(\frac{B+C}{V}-X_{\Sigma}^{\text{optim}}+g^{\text{optim}}\right)- \frac{\delta}{V\cdot t}\sum_{\tau=0}^{t-1}\sum_{m=1}^{M}(Q_m(\tau)+U_m(\tau)+Z_m(\tau))+\frac{\mathbf{E}\{L(\mathbf{S}(0))\}-\mathbf{E}\{L(\mathbf{S}(t))\}}{V\cdot t}. \tag{36}$$

and

$$\frac{1}{t}\sum_{\tau=0}^{t-1}\sum_{m=1}^{M}(Q_m(\tau)+U_m(\tau)+Z_m(\tau))\leq\frac{B+C}{\delta}- \frac{V}{\delta}\left(X_{\Sigma}^{\text{optim}}-g^{\text{optim}}+\frac{1}{t}\cdot\sum_{\tau=0}^{t-1}\sum_{m=1}^{M}\mathbf{E}\{u_0^{m}(\tau)\}+\frac{1}{t}\cdot\sum_{\tau=0}^{t-1}\mathbf{E}\{g(\tau)\}\right)+ \frac{\mathbf{E}\{L(\mathbf{S}(0))\}-\mathbf{E}\{L(\mathbf{S}(t))\}}{\delta\cdot t}. \tag{37}$$

where (36) follows by dividing (35) by V and t, and (37) follows by dividing (35) by δ and t. Now, recalling that the time average expectation for the penalty sum of all $u_0^{m}(\alpha_{\text{NL}}(t),t)$ is deterministically lower bounded by $u^{\min}$ (possibly negative value), also the time average expectation for the objective function (14) is deterministically lower bounded by $g_{\min}^{0}$ (possible negative value) and finally, taking limits of the (42) as $t\rightarrow\infty$ proves (31). Moreover, by taking limits of the (36) as $t\rightarrow\infty$ plus multiplying the result inequality by –1 proves (25).

6. Simulation Results and Analysis

In this section, we provide a simulation results and analysis of the presented MassiveAQUA algorithm for a scenario including total wireless and mobile network coverage from several RATs, for the 250 5G mobile terminals and one 5G fixed terminal (server at the core network). From the three different scenarios, we are obtaining the simulation results for average throughput, as well as the average queue backlog values for different network conditions, together with the average queue delays of the 5GTs. Our simulation scenarios are multi-cell case, plotted in the dense urban area, with random initial locations of mobile terminals uniformly distributed within the entire urban area. To emphasize that the used maximal output rates are carefully chosen to be adequate to the maximal uplink serving rates (by dividing the uplink bit rates with the average number of bits per packet) of LTE, LTE-Advanced, IEEE 802.11n, IEEE 802.16e, IEEE 802.16m, IEEE 802.11ac and IEEE 802.11b RATs, shared between 250 mobile terminals in one cell. In Tab. 1 the most important

Scenario	Parameters and values	Description
Scenario 1	v = 40 km/h	Average velocity of the mobile terminals
	$\mathbf{x}_{10}$=[68.9655; 50; 24.316; 100; 50; 50; 24.316; 50; 68.9655; 68.9655] $\mathbf{x}_9$=[68.9655; 50; 24.316; 100; 50; 50; 24.316; 50; 68.9655] $\mathbf{x}_8$=[68.9655; 50; 24.316; 100; 50; 50; 24.316; 50] $\mathbf{x}_7$=[68.9655; 50; 24.316; 100; 50; 50; 24.316] $\mathbf{x}_6$=[68.9655; 50; 24.316; 100; 50; 50] $\mathbf{x}_5$=[68.9655; 50; 24.316; 100; 50] $\mathbf{x}_4$=[68.9655; 50; 24.316; 100] $\mathbf{x}_3$=[68.9655; 50; 24.316] $\mathbf{x}_2$=[68.9655; 50] $\mathbf{x}_1$=[68.9655]	Vectors of the time average arrival rates for different number of services
	μ_1= (1/250) [44643]; μ_2= (1/250) [44643; 12500]; μ_3= (1/250) [6694.4; 37000; 12500]; μ_4= (1/250) [6694.4; 12500; 5000; 44643]; μ_5= (1/250) [6694.4; 12500; 5000; 44643; 5000]; μ_6= (1/250) [6694.4; 12500; 5000; 44643; 12500; 6694.4].	Vectors of the output serving rates for different number of RAT interfaces (M queues).
Scenario 2	μ_1= (1/250) [6694.4]; μ_2= (1/250) [6694.4; 12500]; μ_3= (1/250) [6694.4; 5000; 12500]; μ_4= (1/250) [6694.4; 12500; 5000; 6694.4]; μ_5= (1/250) [6694.4; 12500; 5000; 12053.57; 2250].	Vector of the output serving rates for different number (M) of RAT interfaces.
	$\mathbf{x}$=[68.9655; 50; 24.316]	Vector for the time average arrival rates for N=3 services
Scenario 1 & 2	$V=V_1=V_2$=10	Penalty control parameters
Scenario 3	$\mathbf{x}$=[69; 50]	Vector for the time average arrival rates for N=2 services
	v = 40 km/h	Average velocity of the mobile terminals
	For Rout=76.7776: μ_2= (1/250) [6694.4; 12500]; For Rout=93.4456: μ_2= (1/250) [6694.4; 16667]; For Rout=96.7776: μ_3= (1/250) [6694.4; 12500 5000]; For Rout=98.2160: μ_2= (1/250) [12054; 12500]; For Rout=103.5552: μ_3= (1/250) [6694.4; 12500; 6694.4].	Vectors of the output serving rates for different number of RAT interfaces (M queues) and different aggregated throughput.
	V_1=10	First penalty control parameter
Scenario 1,2,3	p_m^{av} =0.7 W	Average power for each interface (W)

Tab. 1. Simulation parameters for scenario 1, 2 and 3.

simulation parameters for all three scenarios are presented. Each simulation was run over 10^7 time slots. In Fig. 2 the average queue backlog versus different number of multimedia sources is presented. The time average arrival rates for different number of used services (applications) are given in Tab. 1. Undoubtedly, as the number of used interfaces in 5GTs is higher, the average queue backlog is becoming smaller. The case when they are 6 interfaces is achieving superior results for any number of data sources over other four cases (especially with the case when there is only one used interface (for $M = 1$ in Fig. 2)). By the Little's theorem [33], dividing the average queuing backlog with the average arrival rates yields an upper bound on average time delay. Consequently, in the case with more interfaces within 5GT with MassiveAQUA algorithm will cause minimal average time delay for any used service. Furthermore, Fig. 3 presents the average uplink throughput per user versus the average velocity of the 5G mobile terminals, when the number of sources is fixed ($N = 3$). This is Scenario 2 and the services are: VoIP-conference, VoIP and delay-sensitive data service. It is evident that as the number of RAT interfaces is higher (i.e. for $M = 5$) the throughput is higher for each velocity, even it is with diminishing trend versus velocity. Also, with smaller number of interfaces the average throughput is smaller and is supporting only those RAT interfaces belonging to the RAT with the best mobility support. A key observation about the above algorithm is that it does optimal flow control decisions and simultaneously causing network queuing stability and minimal network queue backlog. As expected, the worst case uplink throughput we have when there are mobile terminals with one interface ($M = 1$). In Scenario 3, we have only two sources of information (two services), and the capacity region is shown in Fig. 4, together with the illustration for the parameter ro (For $\lambda = (50, 69)$ (i.e., point D illustrated) different lengths for the segment OA are allowed (the parameter ro is OA/OD)). Moreover, Fig. 5 shows the average queue delays versus different values of the parameter ro, for different values of the penalty parameter V_2, when the penalty parameter for the first penalty function is $V_1 = 10$. As the parameter ro is increasing, it is near it's maximal value (which is 1, i.e. when the segment OA = OD in Fig. 4), so the average delays are becoming more stable and bounded. As can be notice from Fig. 5, when we set the penalty control parameter half of the previous value (i.e. when $V_2 = 5$), in comparison with the case when the V_2 is 10, the delays for all five cases are becoming lower with maximal values around 0.15 slots. At the same time, it tends to push the queues towards a lower congestion state, but we get large packet dropping ratio (which is proportional with the decreasing of the control parameter V_2, according to (24)). Thus, we minimize a weighted sum of drift and "penalty" for the second (novel) penalty function (14), decreasing the delay, but we are increasing the packet dropping ratio for all queues. For both values of penalty parameter V_2 (i.e. when $V_2 = 10$ and when $V_2 = 5$), the case when we have Rout = 98.2160, i.e.

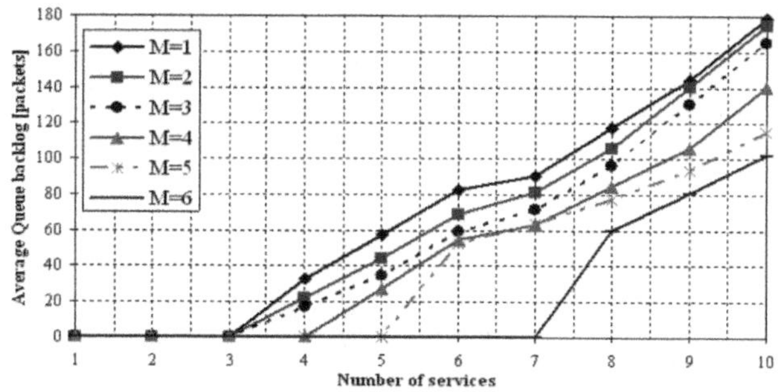

Fig. 2. The average queue backlog versus number of sources.

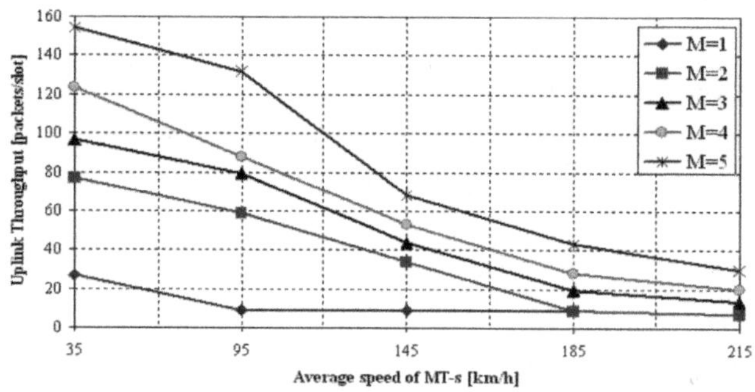

Fig. 3. The average throughput vs. average velocity of MTs.

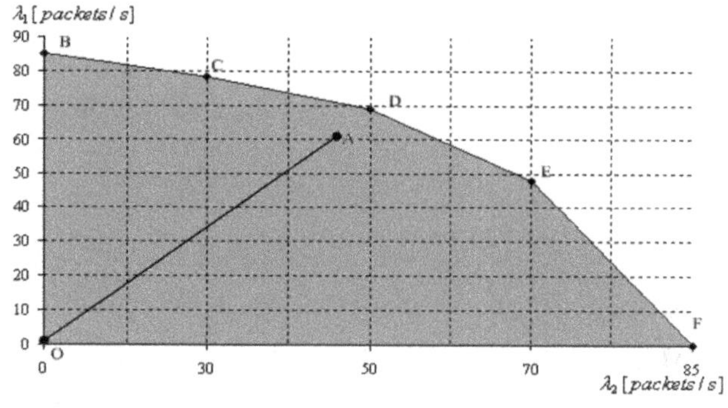

Fig. 4. The capacity region Λ.

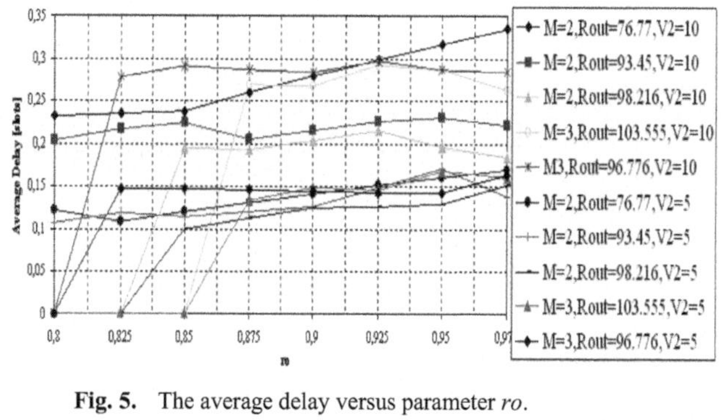

Fig. 5. The average delay versus parameter ro.

balanced queue with almost equal values for their output serving rates (i.e. when $\mu_2 = (1/250)$ [12054; 12500]), we have the lowest delay in comparison with all other cases, even when we have three RAT interfaces within 5GTs. Thus, we can conclude that the values for the output serving rates of the RAT interfaces are having very important impact in the performances of the MassiveAQUA algorithm, and should be quite balanced.

7. Conclusion

This paper presents joint stochastic Lyapunov stability and performance optimization, and develops utility optimal traffic control for the future 5G mobile terminals and 5G fixed terminals (proxy servers in the mobile core network part). It defines a novel analytical framework and performance bounds for 5GT with vertical multi-homing and multi-streaming features for 5G broadband networks, which maximizes the time average throughput and at the same time stabilizes the queuing, simultaneously providing minimal queue backlog and delay for each real-time delay-sensitive application. The 5GTs with Lyapunov drift-plus-penalty technique is implemented in user MTs and in the core network (as 5G proxy server), and it is handling simultaneously multiple multimedia real-time applications via multiple wireless/mobile network interfaces with optimal packet scheduling and high level of QoS provisioning.

References

[1] JANEVSKI, T. *Internet Technologies for Fixed and Mobile Networks*. USA: Artech House, 2015, ISBN: 978-1-60807-921-6

[2] BOCCARDI, F., et al. Five disruptive technology directions for 5G. *IEEE Communications Magazine*, 2014, vol. 52, no. 2, p. 74 to 80. DOI: 10.1109/MCOM.2014.6736746

[3] BANGERTER, B., et al. Networks and devices for the 5G era. *IEEE Communications Magazine*, 2014, vol. 52, no. 2, p. 90–96. DOI: 10.1109/MCOM.2014.6736748

[4] CHENG-XIANG WANG, et al. Cellular architecture and key technologies for 5G wireless communication networks. *IEEE Communications Magazine*. 2014, vol. 52, no. 2, p. 122–130. DOI: 10.1109/MCOM.2014.6736752

[5] JANEVSKI, T. 5G Mobile phone concept. In *IEEE Consumer Communications and Networking Conference* (*CCNC*). Las Vegas (USA), 2009. DOI: 10.1109/CCNC.2009.4784727

[6] WANG, P., et al. Multi-Gigabit millimeter wave wireless communications for 5G: From fixed access to cellular networks. *IEEE Communications Magazine*, 2015, vol. 53, no. 1, p. 168–178. DOI: 10.1109/MCOM.2015.7010531

[7] QIAO, J., et al. Enabling device-to-device communications in millimeter-wave 5G cellular networks. *IEEE Communications Magazine*, 2015, vol. 53, no. 1, p. 209–215. DOI: 10.1109/MCOM.2015.7010536

[8] Rec. ITU-T *Y.2052* (02/2008): Framework of multi-homing in IPv6-based NGN, February 2008.

[9] Rec. ITU-T *Y.2056* (08/2011): Framework of vertical multihoming in IPv6-based Next Generation Networks, August 2011.

[10] NEELY, M. J. *Stochastic Network Optimization with Application to Communication and Queueing Systems*. USA: Morgan & Claypool, 2010. DOI: 10.2200/S00271ED1V01Y201006CNT007

[11] MALISOFF, M., MAZENC, F. *Constructions of Strict Lyapunov Functions*. London: Springer, 2009. ISBN: 978-1-84882-535-2

[12] TASSIULAS, L., EPHREMIDES, A. Stability properties of constrained queueing systems and scheduling policies for maximum throughput in multihop radio networks. *IEEE Transactions on Automatic Control*, Dec. 1992, vol. 37, no. 12, p. 1936–1948. DOI: 10.1109/9.182479

[13] TASSIULAS, L., EPHREMIDES, A. Dynamic server allocation to parallel queues with randomly varying connectivity. *IEEE Transactions on Information Theory*, March 1993, vol. 39, no. 2, p. 466–478. DOI: 10.1109/18.212277

[14] KUMAR, P. R., MEYN, S. P. Stability of queueing networks and scheduling policies. *IEEE Transactions on Automatic Control*, Feb. 1995, vol.40, no. 2, p. 251–260. DOI: 10.1109/9.341782

[15] NEELY, M. J., MODIANO, E., LI, C. Fairness and optimal stochastic control for heterogeneous networks. In *Proceedings of IEEE 24th Annual Joint Conference of the IEEE Computer and Communications Societies INFOCOM*. Miami (FL, USA), March 2005. DOI: 10.1109/INFCOM.2005.1498453

[16] STOLYAR, A. Maximizing queueing network utility subject to stability: Greedy primal-dual algorithm. *Queueing Systems*, 2005, vol. 50, no. 4, p. 401–457. DOI: 10.1007/s11134-005-1450-0

[17] NEELY, M. J., MODIANO, E., ROHRS, C. E. Dynamic power allocation and routing for time varying wireless networks. *IEEE Journal on Selected Areas in Communications*, Jan. 2005, vol. 23, no. 1, p. 89–103. DOI: 10.1109/JSAC.2004.837349

[18] SRIDHARAN, A., MOELLER, S., KRISHNAMACHARI, B. Making distributed rate control using Lyapunov drifts a reality in wireless sensor networks. In *6th International Symposium on Modeling and Optimization in Mobile, Ad Hoc, and Wireless Networks and Workshops WiOPT 2008*. Berlin (Germany), 2008, p. 452–461. DOI: 10.1109/WIOPT.2008.4586106

[19] GEORGIADIS, L., NEELY, M. J., TASSIULAS, L. Resource allocation and cross-layer control in wireless networks. *Foundations and Trends in Networking*, 2006, vol. 1, no. 1, p. 1 to 149. DOI: 10.1561/1300000001

[20] NEELY, M. J. Optimal backpressure routing for wireless networks with multi-receiver diversity. In *40th Annual Conference on Information Sciences and Systems (CISS)*. Princeton (NJ, USA), 2006, p. 18–25. DOI: 10.1109/CISS.2006.286424

[21] LEE, J. W., MAZUMDAR, R. R., SHROFF, N. B. Opportunistic power scheduling for dynamic multiserver wireless systems. *IEEE Transactions on Wireless Communications*, June 2006, vol. 5, no. 6, p. 1506–1515. DOI: 10.1109/TWC.2006.1638671

[22] NEELY, M. J. Energy optimal control for time varying wireless networks. *IEEE Transactions on Information Theory*, July 2006, vol. 52, no. 7, p. 2915–2934. DOI: 10.1109/TIT.2006.876219

[23] NEELY, M. J. Super-fast delay tradeoffs for utility optimal fair scheduling in wireless networks. *IEEE Journal on Selected Areas in Communications*, Aug. 2006, vol. 24, no. 8, p. 1489–1501. DOI: 10.1109/JSAC.2006.879357

[24] ERYILMAZ, A., SRIKANT, R. Joint congestion control, routing, and mac for stability and fairness in wireless networks. *IEEE Journal on Selected Areas in Communications*, Special Issue on Nonlinear Optimization of Communication Systems, Aug. 2006, vol. 14, p. 1514–1524. DOI: 10.1109/JSAC.2006.879361

[25] NEELY, M. J., URGAONKAR, R. Cross-layer adaptive control for wireless mesh networks. *Ad Hoc Networks (Elsevier journal)*, August 2007, vol. 5, no. 6, p. 719–743. Available at: http://dx.doi.org/10.1016/j.adhoc.2007.01.004

[26] NEELY, M. J. Optimal energy and delay tradeoffs for multi-user wireless downlinks. *IEEE Transactions on Information Theory*, Sept. 2007, vol. 53, no. 9, p. 3095–3113. DOI: 10.1109/TIT.2007.903141

[27] ERYILMAZ, A., SRIKANT, R. Fair resource allocation in wireless networks using queue-length-based scheduling and congestion control. *IEEE/ACM Transactions on Networking*, 2007, vol. 15, no. 6, p. 1333–1344. DOI: 10.1109/TNET.2007.897944.

[28] LI, Q., NEGI, R. Scheduling in wireless networks under uncertainties: A greedy primal-dual approach. *Arxiv Technical Report*: arXiv:1001:2050v2, June 2010.

[29] NEELY, M. J., URGAONKAR, R. Optimal backpressure routing in wireless networks with multi-receiver diversity. *Ad Hoc Networks (Elsevier journal)*, July 2009, vol. 7, no. 5, p. 862–881. DOI: 10.1016/j.adhoc.2008.07.009

[30] SHUMINOSKI, T., JANEVSKI, T. Lyapunov optimization framework for 5G mobile nodes with multi-homing. *IEEE Communications Letters*, May 2016, vol. 20, no. 5, p. 1026–1029. DOI: 10.1109/LCOMM.2016.2540622

[31] SHUMINOSKI, T., JANEVSKI, T. 5G mobile terminals with advanced QoS-based user-centric aggregation (AQUA) for heterogeneous wireless and mobile networks. *Wireless Networks*, 2016, vol. 22, no. 5, p. 1553–1570. DOI: 10.1007/s11276-015-1047-4

[32] SHUMINOSKI, T., JANEVSKI, T. Radio network aggregation for 5G mobile terminals in heterogeneous wireless and mobile networks. *Wireless Personal Communication*. May 2014, vol. 78, no. 2, p. 1211–1229. DOI: 10.1007/s11277-014-1813-0

[33] BERTSEKAS, D. P., GALLAGER, R. *Data Networks*. New Jersey: Prentice-Hall, 1992. ISBN 0132009161

About the Authors ...

Tomislav SHUMINOSKI (corresponding author) received the B.Sc. (2008), M.Sc. (2010) and Ph.D. (2016) degrees from the Faculty of Electrical Engineering and Information Technologies, University Ss Cyril and Methodius, Skopje, Macedonia, where he is an Assistant Professor.

Toni JANEVSKI is a Full Professor at the Faculty of Electrical Engineering and Information Technologies (FEEIT), Ss. Cyril and Methodius University, Skopje, Macedonia, from which he received his Dipl. Ing., M.Sc. and Ph.D. degrees in 1996, 1999 and 2001, respectively.

A Compact Flexible and Frequency Reconfigurable Antenna for Quintuple Applications

Mubashir Ul HASSAN [1], *Farzana ARSHAD* [1], *Syeda Iffat NAQVI* [1], *Yasar AMIN* [1, 2]
Hannu TENHUNEN [2, 3]

[1]ACTSENA Research Group, University of Engineering and Technology (UET), Taxila, 47050, Pakistan
[2] iPack VINN Excellence Center, Royal Institute of Technology (KTH), SE-16440, Stockholm, Sweden
[3]TUCS, University of Turku, Turku-20520, Finland

Farzana.arshad@uettaxila.edu.pk, mmubashirulhassan@gmail.com

Abstract. *A novel, compact coplanar waveguide-fed flexible antenna is presented. The proposed design uses flexible Rogers RT/duroid 5880 (0.508 mm thickness) as a substrate with small size of 30 × 28.4 mm². Two switches are integrated on the antenna surface to change the current distribution which consequently changes the resonance frequency under different conditions of switches, thereby making it a frequency reconfigurable antenna. The antenna design is simulated on CST®MWS®. The proposed antenna exhibits VSWR < 2 and appreciable radiation patterns with positive gain over desired frequency bands. Good agreement exists between simulated and measured results. On the basis of results, the proposed antenna is envisioned to be deployed for the following applications; aeronautical radio navigation [4.3 GHz], AMT fixed services [4.5 GHz], WLAN [5.2 GHz], Unlicensed WiMAX [5.8 GHz] and X-band [7.5 GHz].*

Keywords

Flexible antennas, reconfigurable antenna, coplanar waveguide-fed, switches

1. Introduction

An antenna operating on multiple frequency bands has gained a lot of attention due to the proliferation of modern wireless technology and customer demand for multiple services in a single device. Conventionally, a frequency band is associated with a particular wireless service; therefore multi-band antenna is required to support multiple services in a single wireless device. Multi-band antennas can operate over different frequency bands exhibiting good gain and stable radiation pattern. Albeit, multiband antennas transmit electromagnetic waves simultaneously at all the supported frequencies in addition to the desired frequency. Also, electromagnetic radiations from the wireless device have an adverse effect on human health.

The reconfigurable antennas mitigate the above-mentioned problems associated with the multi-band antennas. Such type of antennas can be reconfigured at the desired frequency band, radiation pattern and polarization. Reconfiguration can be achieved by deploying switches within the radiating element of the antenna [1]. A reconfigurable antenna reduces interference from adjacent unused bands and minimizes the filter requirements of the front end circuits, thus making the design compact [2]. Frequency reconfigurable antenna with wide bandwidth is generally chosen because of its miniaturization, cost effectiveness and better tuning ability between different frequency bands without affecting the gain and stability of radiation pattern.

Frequency agility can be realized using different types of switches like varactor diodes [3], pin diodes [4], RF MEMS [5] and FET switches [6]. In [3], varactor diodes are used for reconfiguration, but varactor diodes are nonlinear and their continuous tuning range is narrow in nature [4]. Switching between multiple bands requires a large number of pin diodes that increases the insertion loss and complicates the biasing circuitry [7]. RF MEMS has a low loss, but its deployment is expensive [8]. In [9], three pin diodes are employed in U-shaped and L-shaped slots for LTE, AMT Fixed Services, and WLAN applications. However, it uses antenna element on both sides of the substrate. Microstrip based frequency and pattern reconfigurable antenna is reported in [10] that utilizes five pin diodes. It has three operating modes; omnidirectional at 2.4 GHz, unidirectional at 5.4 GHz and both omnidirectional and unidirectional operating concurrently. In [11], frequency reconfigurable antenna using a thick substrate (3.3 mm thickness) is presented in which resonance is controlled by shorting strips; moreover, the conical radiation pattern is maintained even at higher frequencies. A compact frequency reconfigurable antenna proposed in [12] utilizes a simple square shaped radiating patch for Bluetooth, WLAN, and WiMAX applications. Three pin diodes are inserted in ground plane that controls switching bands. A novel frequency reconfigurable antenna using the FR4 substrate that switches between an ultra wide band, narrow band, and dual band mode is proposed in [13].

Switching is achieved by four pin diodes along with the slotted structure that is created on the ground. Microstrip based frequency reconfigurable antenna is proposed in [14]. Using five pin diodes it achieves six switchable bands from 2.2 GHz to 4.75 GHz. However, the above-mentioned designs suffer from three main drawbacks; the first is their larger dimension, the second is limited impedance bandwidth, the third is design complexity in terms of numerous switches and intricate structure.

Nowadays, flexible antennas have gained much importance because of their low profile, light weight, and robustness [15]. Different flexible substrates have been reported in [15–18]. In [15], the crescent-shaped antenna is presented using flexible RO4003 Rogers with the impedance bandwidth of 7.1 GHz. Kapton® polyamide- based multi-band antenna is proposed in [16]. Paper-based antenna for 2.4 GHz WLAN application is introduced in [17]. Dual frequency rejection at 5.25 GHz and 5.775 GHz is successfully achieved in [18] using flexible Liquid Crystal Polymer. Above-mentioned flexible antennas have non-reconfigurable functions. Various feeding techniques have been used in flexible antennas, but coplanar waveguide feeding is preferable as it reduces complication by placing an antenna element and a patch on the same side of the substrate. One pin diode is employed in the T-shaped antenna for WLAN and WiMAX applications. However, its gain is comparatively low and its fabrication is expensive [19].

In this paper, a novel compact, flexible and frequency reconfigurable antenna is proposed. Both features; flexibility, and reconfiguration are added in this design which makes it attractive for conformal and many other applications. Switches are employed to change the electrical length of the radiator which subsequently changes the resonant frequency. Thus, by applying switches at an appropriate location in the proposed design, frequency reconfiguration is possible for five different applications.

The main contributions of this paper are as follows:

- A novel, compact, flexible and frequency reconfigurable antenna is proposed for Aeronautical Radio Navigation (ARN), AMT fixed Services, WLAN, WiMAX and X-band applications.
- Useful frequency bands are achieved for every On/Off state of the switches.
- Gain and bandwidth enhancement using the flexible substrate.

2. Antenna Design and Reconfiguration

The proposed design is shown in Fig. 1. The proposed antenna uses Flexible Rogers RT/Duroid 5880 as a substrate. The dielectric constant of the substrate is 2.2 and loss tangent is 0.0009 with the thickness of 0.508 mm. The

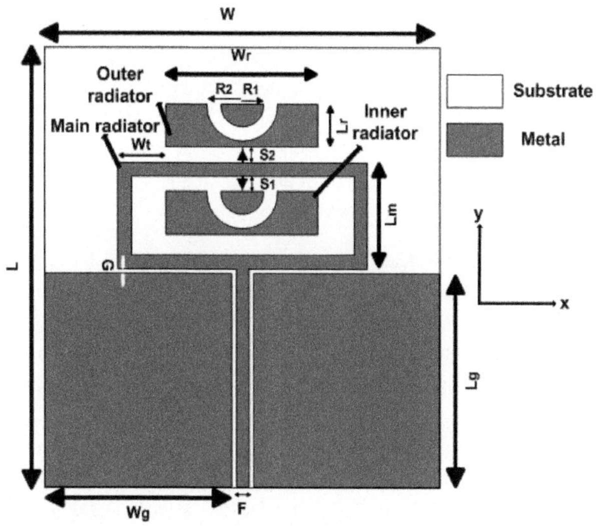

Fig. 1. Antenna configuration (top view).

Parameter	Value (mm)	Parameter	Value (mm)
L	30	Lm	7.3
W	28.4	Lr	3
Lg	14.642	Wt	5.2
Wg	13.45	Wr	11
F	1	R1	1.5
G	0.358	R2	2.5

Tab. 1. Optimized parameters of the proposed design.

proposed antenna has a compact size of 30 mm × 28.4 mm. The antenna is fed by a 50 Ω microstrip line.

The CPW feed line having the width of 1 mm is connected to the main radiator. The inner and outer radiator is connected to the main radiator via switch S1 and S2. First of all, CPW fed rectangular antenna which has a single band at 5.8 GHz is designed. The rectangle is introduced inside and outside the main radiator to acquire more resonance frequencies. The arc-shaped slots are introduced in appropriate locations in the inner and outer rectangle to obtain the desired band. The width of slot controls the current intensity and minimizes the return loss. The lumped element boundary condition is used to implement the switches in CST® MWS®. With the four states of switches five resonance modes at 4.2 GHz, 4.3 GHz, 5.1 GHz, 5.8 GHz and 7.5 GHz are excited with good impedance matching. Different parameters are described in Tab. 1.

3. Results and Discussion

The prototype of the proposed antenna is fabricated and tested to validate the performance of the design. Measurements are taken using the Vector Network Analyzer (VNA) R&S ZVL13.

The description of states (1 to 4) in terms of the position of the two switches is described in Tab. 2. The status of the switch, i.e. whether the switch is On/off actually defines the electrical length of the antenna structure that contributes for radiating a particular frequency band. S1 and S2 are the switches that are implemented using conductor/conducting wires between two conductors to provide the path. Although diodes can be used to provide the path, but this conductor is used because of limitations. When both S1 & S2 are shorted simultaneously, the current circulates in the main radiator as well as in the inner and outer radiator. When both S1 and S2 are open, the current circulates only in the main radiator.

The simulated surface current distribution of the proposed antenna at various frequencies under different states of switches is shown in Fig. 2 (a-e). Figure 2(a) suggests that inner radiator radiates due to coupling with surrounding walls of the main radiator. The main radiator and the outer radiator also radiate. In this case, the current follows the longer path, hence antenna resonates at a low frequency of 4.2 GHz with an impedance bandwidth of 630 MHz (3.9–4.53 GHz) that covers 4.3 GHz Aeronautical Radio Navigation. Figure 2(b) indicates that due to strong current intensity around S1 and S2 another frequency band from 7.2 GHz to 7.8 GHz with an impedance bandwidth of 600 MHz is also observed in state 1. It is noticed that in state 2 the antenna covers the frequency range of (4.7–5.4 GHz) with an impedance bandwidth of 700 MHz that covers 5.2 GHz WLAN. The impedance bandwidth in state 3 is 700 MHz (3.9–4.6 GHz) that is sufficient for the standard of 4.5 GHz AMT Fixed Services. In state 4, the only main radiator is contributing to the radiation, hence, current follows the shortest path so, resonance at 5.5 GHz with a very wide bandwidth of 1 GHz (5–6 GHz) is achieved. It covers 5.5 GHz WLAN and 5.8 GHz unlicensed WiMAX. As mentioned earlier that single rectangular antenna operates at 5.8 GHz. Therefore, the shift in frequency from 5.8 GHz to 5.5 GHz is due to the presence of strong coupling that exists in the gap between the main radiator, arc-shaped radiators as well as around the switches.

The prototype of the proposed antenna is illustrated in Fig. 3. Input reflection coefficient, VSWR, gain and radiation patterns are discussed in this section. Figure 4 shows the simulated and measured reflection coefficient of the proposed antenna for different states of switches. The measured S_{11} response is in accordance with the simulated results. However, the slight shift between simulated and measured frequencies is due to fabrication inaccuracy, but it still covers the desired frequency bands. It has been observed that bandwidth for 4.2 GHz, 7.5 GHz, 4.3 GHz, 5.1 GHz and 5.5 GHz band is 18%, 8%, 16.2%, 13.7% and 18.18%, respectively. Figures 5 and 6 show that $VSWR < 2$ has been successfully achieved at required resonance bands.

States	S1	S2	Frequency (GHz)	Bandwidth (MHz)
State 1	ON	ON	4.2 & 7.5	630 & 600
State 2	ON	OFF	5.1	700
State 3	OFF	ON	4.3	700
State 4	OFF	OFF	5.5	1000

Tab. 2. Configuration of switches.

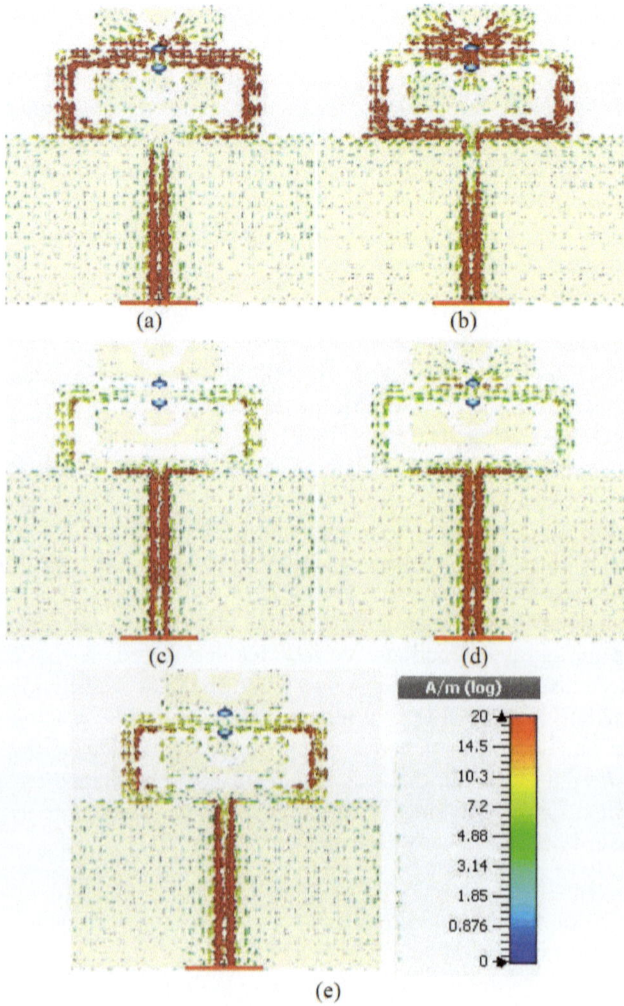

Fig. 2. Current distribution: (a) 4.2 GHz at state 1, (b) 7.5 GHz at state 1, (c) 5.1 GHz at state 2, (d) 4.3 GHz at state 3, (e) 5.5 GHz at state 4.

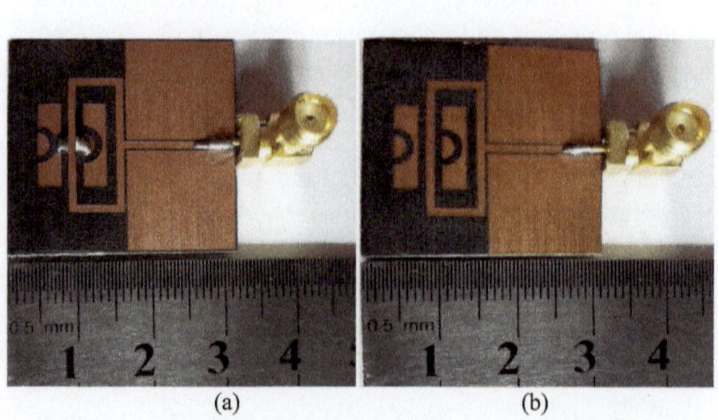

Fig. 3. The photograph of the fabricated antenna: (a) without switches (b) with switches.

The computed and measured gain of the proposed antenna in different states is shown in Fig. 7. The gain is computed at discrete frequencies that are of particular interest. The gain requirement is fulfilled in our proposed antenna as compared to [2], [19]. In [2], pin diode diminishes the gain and efficiency of the bow-tie antenna because of insertion loss introduced by the resistance diodes.

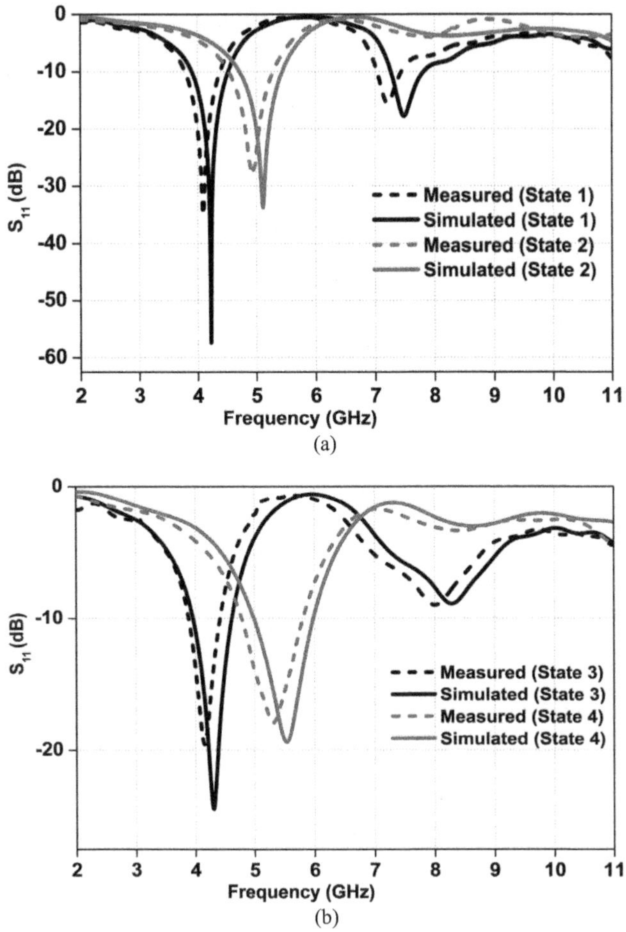

Fig. 4. Simulated and measured reflection coefficient for states (1-4).

Table 3 summarizes the gain comparison of the proposed antenna. Mostly literature does not mention radiation efficiency, although it is an important parameter for the reconfigurable antennas [20]. Due to reconfiguration techniques, radiation efficiency is emulated for such type of antennas. In the proposed work, radiation efficiency, greater than 90% is successfully achieved at required bands except for 7.5 GHz where radiation efficiency is 74%. (The radiation efficiency at 4.2 GHz, 4.3 GHz, 5.1 GHz, 5.5 GHz and 7.5 GHz is 94.8%, 95.3%, 98.83%, 97.4% and 74.6%, respectively.) This is comparable to the radiation efficiency of a conventional dipole.

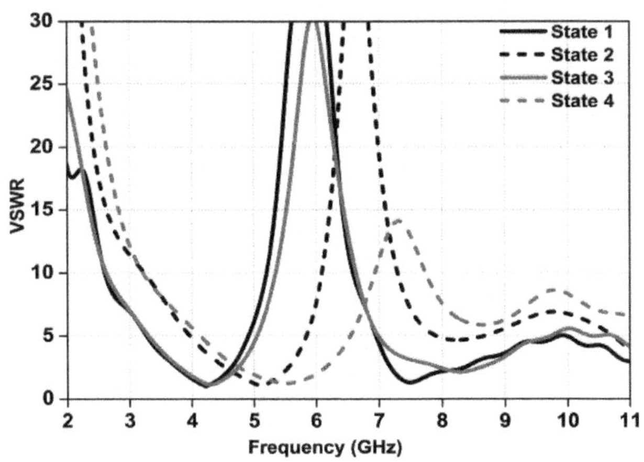

Fig. 5. Simulated VSWR.

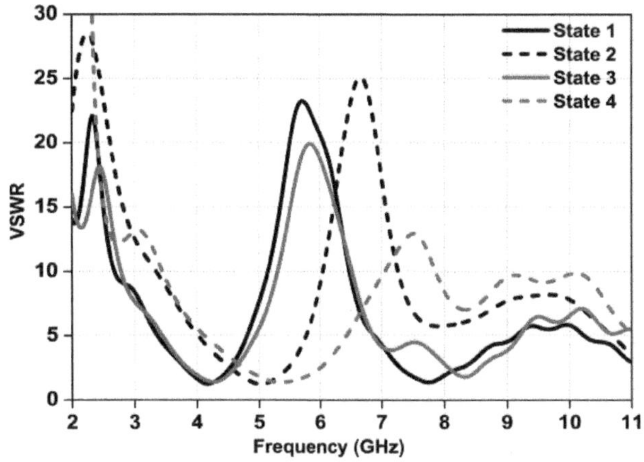

Fig. 6. Measured VSWR.

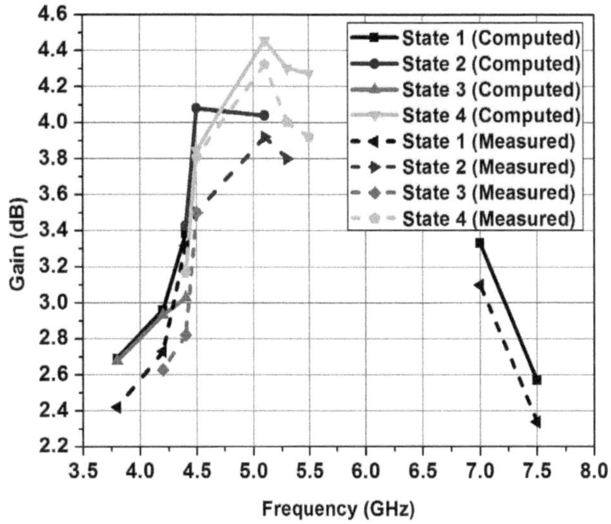

Fig. 7. Computed and measured gain (1-4).

Frequency	4.2 GHz	4.3 GHz	5.1 GHz	5.5 GHz	7.5 GHz
Simulated Gain (dB)	2.96	2.98	4.04	4.27	2.57
Measured Gain (dB)	2.73	2.72	3.92	3.92	2.34

Tab. 3. Gain comparison of the proposed antenna.

Figure 8 exhibits the simulated and measured radiation patterns in E and H-planes at 4.2 GHz, 4.3 GHz, 5.1 GHz, 5.5 GHz and 7.5 GHz. From Fig. 8, it can be concluded that the proposed antenna exhibits the good omnidirectional pattern in the H-plane and bidirectional pattern in the E-Plane. This shows that the antenna is apposite for integration with portable devices.

The proposed antenna is compared with few recently published works; the brief comparison is presented in Tab. 4. Table 4 shows that [9], [12], [14] uses relatively thick substrates. Also, their bandwidth is less compared to the proposed work. Although [10] has the bandwidth greater than 500 MHz at three bands, but it employs five switches and achieves just four resonance bands. In [19], a very thin flexible substrate is used, but the bandwidth is even less than 300 MHz in all the three resonance bands. It can be concluded that the proposed antenna has small size and shows better performance in terms of impedance bandwidth.

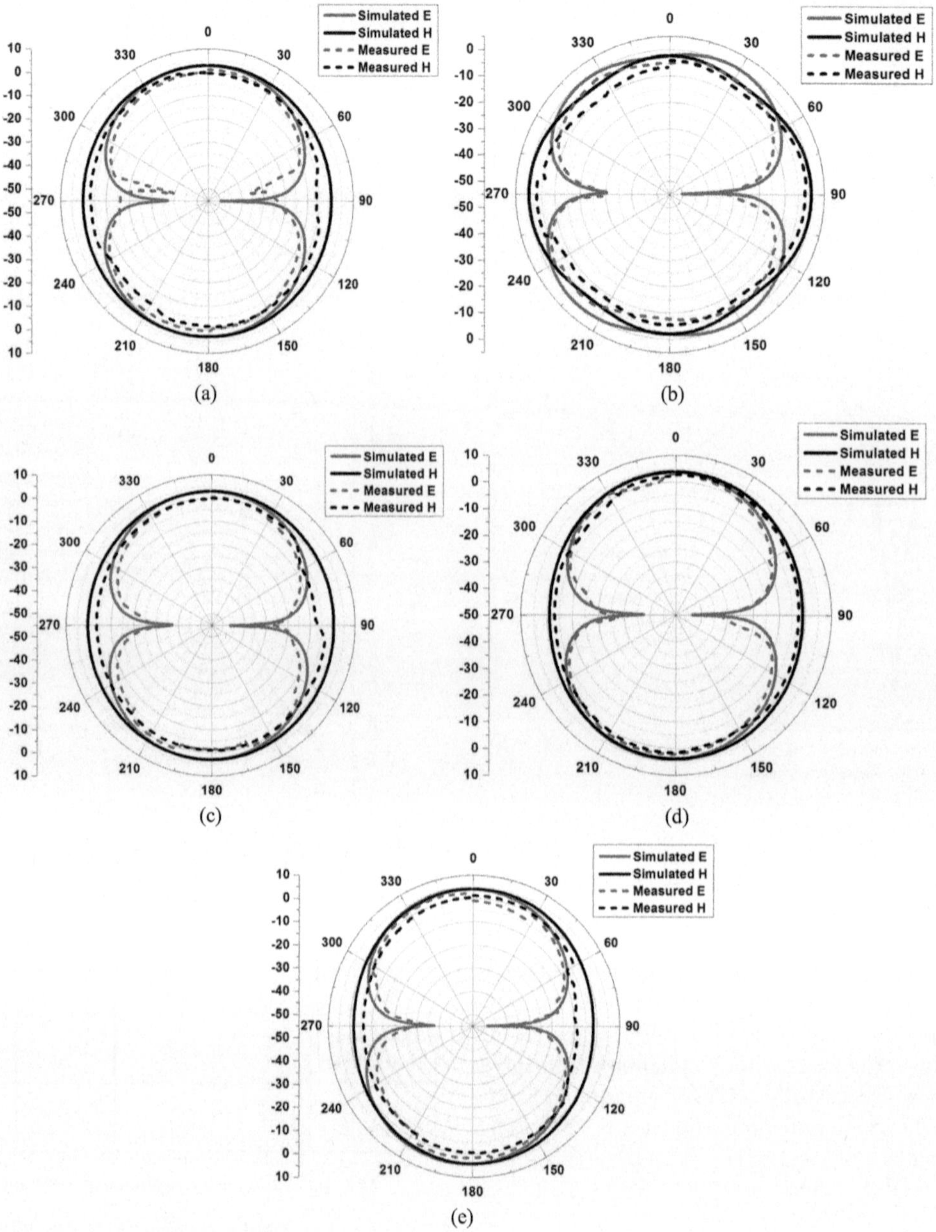

Fig. 8. Simulated and measured radiation patterns: (a) 4.2 GHz, (b) 7.5 GHz, (c) 5.1 GHz, (d) 4.3 GHz, (e) 5.5 GHz.

Characteristics	[9]	[10]	[12]	[14]	[19]	This work
Area (mm²)	675	1852.3	400	2300	1829	852
Thickness(mm)	0.8	1.5	0.8	1.52	0.1	0.508
Substrate	RO4350B	RO4350	FR4	Taconic RF35	PET	RT 5880
No. of switches	3	5	3	5	1	2
No. of resonances	6	4	3	6	3	5
Bandwidth at different resonance bands (MHz)	100; 120; 280; 220;100;320	690; 300; 740; 620	210; 400; 580	250; 310; 300, 300; 260; 210	160; 180; 270	630; 600; 700; 1000; 700

Tab. 4. Comparison with previous work.

4. Conclusion

A frequency reconfigurable antenna for Aeronautical Radio Navigation, AMT Fixed services, WLAN, unlicensed WiMAX and X-band applications is proposed. The effective electrical length of the antenna is changed by employing switches that provide the wide tunability of the operating bands. The simulated and measured results are in good agreement. The antenna operates efficaciously at desired bands and has good radiation patterns. Simplicity, compactness, reconfigurability and flexibility are some features that make it a promising candidate for wireless applications.

Acknowledgments

This work was financially supported by Vinnova (The Swedish Governmental Agency for Innovation Systems) and University of Engineering and Technology Taxila, Pakistan through the Vinn Excellence Centers program and ACTSENA research group funding, respectively.

References

[1] SHAH, S. S. A., KHAN, M. F., ULLAH, S., et al. Design and measurement of planar monopole antenna for multi-band wireless applications. *IETE Journal of Research,* 2017, vol. 63, no. 2, p. 194–204. DOI: 10.1080/03772063.2016.1261049

[2] LI, T., ZHAI, H., WANG, X., et al. Frequency-reconfigurable bow-tie antenna for Bluetooth, WiMAX, and WLAN applications. *IEEE Antennas and Wireless Propagation Letters*, 2015, vol. 14, p. 171–174. DOI: 10.1109/LAWP.2014.2359199

[3] GE, L., LUK, K. M. Frequency-reconfigurable low-profile circular monopolar patch antenna. *IEEE Transactions on Antennas and Propagation,* 2014, vol. 62, no. 7, p. 3443–3449. DOI: 10.1109/TAP.2014.2318077

[4] LI, T., ZHAI, H., LI, L. Frequency-reconfigurable bow-tie antenna with a wide tuning range. *IEEE Antennas and Wireless Propagation Letters*, 2014, vol. 13, p. 1549–1552. DOI: 10.1109/LAWP.2014.2344676

[5] CETINER, B. A., CRUSATS, G. R., JOFRE, L., et al. RF MEMS integrated frequency reconfigurable annular slot antenna. *IEEE Transactions on Antennas and Propagation,* 2010, vol. 58, no. 3, p. 626–632. DOI: 10.1109/TAP.2009.2039300

[6] YANG, X. L., LIN, J. C., CHEN, G., et al. Frequency reconfigurable antenna for wireless communications using GaAs FET switch. *IEEE Antennas and Wireless Propagation Letters*, 2015, vol. 14, p. 807–810. DOI: 10.1109/LAWP.2014.2380436

[7] BHELLAR, B., TAHIR, F. A. Frequency reconfigurable antenna for handheld wireless devices. *IET Microwaves, Antennas & Propagation,* 2015, vol. 9, no. 13, p. 1412–1417. DOI: 10.1049/iet-map.2015.0199

[8] CHEN, G., YANG, X. L., WANG, Y. Dual-band frequency-reconfigurable folded slot antenna for wireless communications. *IEEE Antennas and Wireless Propagation Letters*, 2012, vol. 11, p. 1386–1389. DOI: 10.1109/LAWP.2012.2227293

[9] HAN, L., WANG, C., CHEN, X., et al. Compact frequency-reconfigurable slot antenna for wireless applications. *IEEE Antennas and Wireless Propagation Letters*, 2016, vol. 15, p. 1795–1798. DOI: 10.1109/LAWP.2016.2536778

[10] LI, P. K., SHAO, Z. H., WANG, Q., et al. Frequency- and pattern-reconfigurable antenna for multi standard wireless applications. *IEEE Antennas and Wireless Propagation Letters*, 2015, vol. 14, p. 333–336. DOI: 10.1109/LAWP.2014.2359196

[11] ROW, J. S., LIN, T. Y. Frequency-reconfigurable coplanar patch antenna with conical radiation. *IEEE Antennas and Wireless Propagation Letters,* 2010, vol. 9, p. 1088–1091. DOI: 10.1109/LAWP.2010.2093118

[12] BORHANI, M., REZAEI, P., VALIZADE, A. Design of a reconfigurable miniaturized microstrip antenna for switchable multiband systems. *IEEE Antennas and Wireless Propagation Letters,* 2016, vol. 15, p. 822–825. DOI: 10.1109/LAWP.2015.2476363

[13] BOUDAGHI, H., AZARMANESH, M., MEHRANPOUR, M. A frequency-reconfigurable monopole antenna using switchable slotted ground structure. *IEEE Antennas and Wireless Propagation Letters*, 2012, vol. 11, p. 655–658. DOI: 10.1109/LAWP.2012.2204030

[14] MAJID, H. A., RAHIM, M. K. A., HAMID, M. R., et al. A compact frequency-reconfigurable narrowband microstrip slot antenna. *IEEE Antennas and Wireless Propagation Letters*, 2012, vol. 11, p. 616–619. DOI: 10.1109/LAWP.2012.2202869

[15] SALLAM, M. O., KANDIL, S. M., VOLSKI, V., et al. 2.4/5 GHz WLAN crescent antenna on flexible substrate. In *10th European Conference on Antennas and Propagation (EuCAP).* Davos (Switzerland), 2016, p. 1-3. DOI: 10.1109/EuCAP.2016.7481498

[16] AHMED, S., TAHIR, F. A., SHAMIM, A., et al. A compact Kapton-based inkjet-printed multiband antenna for flexible wireless devices. *IEEE Antennas and Wireless Propagation Letters,* 2015, vol. 14, p. 1802–1805. DOI: 10.1109/LAWP.2015.2424681

[17] ANAGNOSTOU, D. E., GHEETHAN, A. A., AMERT., et al. A direct-write printed antenna on paper-based organic substrate for flexible displays and WLAN applications. *Journal of Display Technology*, 2010, vol. 6, no. 11, p. 558–564. DOI: 10.1109/JDT.2010.2045474

[18] GHEETHAN, A. A., ANAGNOSTOU, D. E. Dual band-reject UWB antenna with sharp rejection of narrow and closely-spaced bands. *IEEE Transactions on Antennas and Propagation*, 2012, vol. 60, no. 4, p. 2071–2076. DOI: 10.1109/TAP.2012.2186221

[19] SAEED, S. M., BALANIS, C. A., BIRTCHER, C. R. Inkjet-printed flexible reconfigurable antenna for conformal WLAN/WiMAX wireless devices. *IEEE Antennas and Wireless Propagation Letters*, 2016, vol. 15, p. 1979–1982. DOI: 10.1109/LAWP.2016.2547338

[20] DEL BARRIO, S. C., FOROOZANFARD, E., MORRIS, A., et al. Tunable handset antenna: Enhancing efficiency on TV white spaces. *IEEE Transactions on Antennas and Propagation*, 2017, vol. 65, no. 4, p. 2106–2111. DOI: 10.1109/TAP.2017.2662221

About the Authors...

Mubashir UL HASSAN received his Bachelor's degree in Telecommunication Engineering from University of Engineering and Technology, Taxila in 2014. Currently, in the same University, he is enrolled as an MS Full time scholar. His research interests include planar antennas, multi-band and reconfigurable antennas.

Farzana ARSHAD has received her M.Sc. degree in Telecommunication Engineering from University of Engineering and Technology, Taxila in 2010. Currently, she is an Assistant Professor at the Telecommunication Engineering Department and part time Ph.D. researcher of ACTSENA group, UET Taxila. Her research focuses on low profile multi-band and reconfigurable antenna design.

Syeda Iffat NAQVI received her M.Sc. degree in Telecommunication Engineering from University of Engineering and Technology Taxila, in 2011. Currently, she is an Assistant Professor at the Telecommunication Engineering Department and part time Ph.D. researcher of ACTSENA group, UET Taxila. Her research focuses on RF and microwave antenna designing for cutting edge wireless technologies.

Yasar AMIN is Chairman and Associate Professor of the Telecommunication Engineering Department, University of Engineering and Technology Taxila, Pakistan. He is founder of the ACTSENA Research Group at UET Taxila, Pakistan. He has done his B.Sc. in Electrical Engineering in 2001 with specialization in Telecommunication, and M.Sc. in Electrical Engineering in 2003 with specialization in System-on Chip Design from Royal Institute of Technology (KTH), Sweden. His Ph.D. is in Electronic and Computer Systems from Royal Institute of Technology (KTH), Sweden, with research focus on printable green RFID antennas for embedded sensors, while having an MBA in Innovation and Growth from Turku School of Economics, University of Turku, Finland. He has done several specialized courses from Stanford University, California, USA and Massachusetts Institute of Technology (MIT), USA. He has supervised over 15 M.Sc. thesis, and presently supervising 8 doctoral thesis. He is presently serving as leading Guest Editor at two international journals and an active reviewer of more than a dozen well reputed international journals. He has contributed to over 20 journal papers, over 30 reviewed international conference papers. Dr. Yasar is a member of IEEE, IET, ACM and ACES.

Hannu TENHUNEN is Chair Professor of Electronic Systems at Royal Institute of Technology (KTH), Stockholm, Sweden. Prof. Tenhunen has held professor position as full professor, invited professor or visiting honorary professor in Finland (TUT, UTU), Sweden (KTH), USA (Cornel U), France (INPG), China (Fudan and Beijing Jiatong Universities), and Hong Kong (Chinese University of Hong Kong), and has an honorary doctorate from Tallinn Technical University. He has been director of multiple national large scale research programs or being an initiator and director of national or European graduate schools. He has actively contributed in VLSI and SoC design in Finland and Sweden via creating new educational programs and research directions, most lately at European level as being the EU-level Education Director of the new European flagship initiative European Institute of Technology and Innovations (EIT), and its Knowledge and Innovation Community EIT ICT Labs. Prof. Tenhunen has been active in promoting the innovation system and innovation support mechanism in research and education both at national and European level. Prof. Tenhunen has been a board member in science parks, startup companies, and has served as advisor or expert for high technology companies and venture capitalists, as well as evaluator for EU and national programs and research institutes. He has supervised over 70 M.Sc. thesis, 39 doctoral theses, and 8 post-doc. From his doctoral students and post-docs, as of today, 21 are currently professors and associate professors. Prof. Tenhunen has served on the Technical Program Committees of all major conferences in his area, have been general Chairman or vice-chairman or member of Steering Committee of multiple conferences in his core competence areas. He has been one of the founding editorial board members of 3 scientific journals, has been questing editor for multiple special issues of scientific journals or books, and has contributed numerous invited papers to journals. He has contributed to over 110 journal papers, over 625 reviewed international conference papers, over 170 non-reviewed papers, local conference papers, or other publications, and 9 international patents granted in multiple countries. Prof. Tenhunen is a member of the Academy of Engineering Science of Finland.

Promotion of GM-PHD Filtering Approach for Single-Target Tracking in Raw Data of Synthetic Aperture Radar in Spotlight Imaging Mode

Navid DARYASAFAR[1], *Ramazanali SADEGHZADEH*[2], *Mohammad NASER-MOGHADASI*[1]

[1] Faculty of Engineering, Science and Research Branch, Islamic Azad University, Tehran, Iran
[2] Faculty of Electrical Engineering, K. N. Toosi University of Technology, Tehran, Iran

navid_daryasafar@yahoo.com, sadeghz@eetd.kntu.ac.ir, mn.moghaddasi@srbiau.ac.ir

Abstract. *So far multi-antenna techniques have been used in Synthetic Aperture Radar (SAR) to track moving targets. These techniques carry out the tracking of moving targets in an imaging area, using a combination of the data received by two or several antennas. The aim of this paper is single-target tracking in SAR Spotlight imaging mode based on the promoted PHD filter. In most applications, target tracking in densely cluttered environment using radar system demands robust filtering so as to increase the tracking efficiency. Therefore, tracking of moving targets in the presence of high density clutters in environment, as the particular capability of the PHD filter, has turned it into a robust approach in SAR to track moving targets. Also as the simulation results show, using Range Cell Migration Compensation (RCMC) on SAR raw data before tracking, makes it possible to track a moving target with high quality.*

Keywords

Synthetic Aperture Radar (SAR), Probability Hypothesis Density (PHD) filter, Linear Frequency Modulation (LFM), Range Cell Migration Compensation (RCMC), Random Finite Set (RFS)

1. Introduction

Among different target tracking methods, the concentration of this paper is on Probability Hypothesis Density (PHD) filter which spreads the first order moment of targets state Random Finite Set (RFS) or the intensity of targets state RFS in time. The RFS method considers targets as a set-valued state and also observation values as a set-valued observation; so by the viewpoint of RFS, the problem is related to the framework of Bayesian filtering. This efficient method for multi-target tracking is in fact a delicate generalization of single-target Bayesian filtering. As a matter of fact, the multi-target Bayes filter, PHD filter and also different implementation methods of it are desirable versions of filters based on RFS method [1].

In order to explain the PHD, the target state and the sensor observation models are assumed as known. In these models, γ_k is the generation rate of the RFS Γ_k at instance k and Γ_k represents RFS of spontaneous birth at time k; $\beta_{k|k-1}$ is the proliferation rate (ramification rate) of the set $\beta_{k|k-1}(\zeta)$ at instance k for a target with a previous state of ζ; $p_{s,k}(\zeta)$ is the probability that a target with a previous state of ζ, i.e. at instance $k-1$, also exists at instance k (target survival probability); $p_{D,k}(x)$ is the probability that a target with state x at instance k is detected by the sensor at the same instance, and $k_k(\cdot)$ denotes the clutter severity of the set k_k at instance k.

Using the following recurrence time equations, the PHD function is diffusible [2]:

$$\begin{aligned} v_{k|k-1}(x) &= \int p_{s,k}(\zeta) f_{k|k-1}(x\,|\,\zeta) v_{k-1}(\zeta)\mathrm{d}\zeta \\ &+ \int \beta_{k|k-1}(x\,|\,\zeta) v_{k-1}(\zeta)\mathrm{d}\zeta + \gamma_k(x) \end{aligned} \tag{1}$$

where $f_{k|k-1}$ is the state transition probability density function and

$$\begin{aligned} v_k(x) &= [1 - p_{D,K}(x)] v_{k|k-1}(x) \\ &+ \sum_{z \in Z_k} \frac{p_{D,K}(x) g_k(z\,|\,x) v_{k|k-1}(x)}{K_k(z) + \int p_{D,K}(\zeta) g_k(z\,|\,\zeta) v_{k|k-1}(\zeta)\mathrm{d}\zeta}. \end{aligned} \tag{2}$$

Observations vector defined as random finite set is $\mathbf{Z}_k = \{z_{k,1}, \ldots, z_{k,M_k}\}$, where M_k denotes the measurement number at time k. $v_{k|k-1}(x)$ and $v_k(x)$ are predicted and updated (modified) values of the PHD function respectively.

The Sequential Monte Carlo (SMC) [3], [4] and the Gaussian Mixture (GM) [5] are the two main approaches in implementing the PHD recursion. As a large number of particles is used to approximate the multi-dimensional integrals in the SMC-PHD filter, the main drawback is its high computational burden. Moreover, to extract the target state estimates some clustering techniques are required, which might be often unreliable. To overcome these disadvantages, the GM-PHD filter was developed for linear Gaussian target dynamics and Gaussian birth model, in which the weights, means and covariance matrices are propagated analytically by the Kalman Filter (KF). To deal with nonlinear target dynamics and measurement models, the nonlinear Kalman filter counterparts can be directly employed. The convergence properties of two implementations were analyzed in [4], [6]. As shown in [6], the true PHD filter to any desired degree of accuracy can be approximated by the GM-PHD filter under the linear Gaussian assumption of the dynamic model. For tracking multiple maneuvering targets, similar results have been extended to handle jump Markov models [7–10]. To derive PHD smoothers, the particle and Gaussian mixture techniques have also been used [11–18]. In [19], the GM-PHD filter is extended to multi-sensor tracking system and the target state estimates are obtained sequentially at each sensor. However, the sensor registration errors are neglected.

Multi-antenna methods (hardware methods) are referred to as Interferometric Synthetic Aperture Radar (IFSAR). In these methods two or more images (complex matrix of SAR) are used to extract more information regarding target region [20], [21]. It can be noted that IFSAR system is actually an applied generalization and a different implementation of SAR systems in which independent images of a particular region are extracted using two or more imaging SAR radars. In this study, single-target tracking implementation in the received raw data by SAR is fulfilled using PHD filtering in software mode (in the MATLAB software).

2. The Proposed Algorithm

SAR uses an actual small scale antenna to propagate waves along the trajectory of the platform and receive the echoes of the area surface. Having received the returned echoes from the ground, a special processing begins aiming at image acquisitions from the area. The main characteristic of the hardware performance of SAR is therefore to move on the area surface, radiating and receiving waves. Having received a sufficient number of echoes, they undergo signal processing; thereby the small radar antenna behaves as a large antenna [22–26].

The impulse response of SAR, i.e., the echoes received from a spot target [27], are as follows:

$$h(\tau,\eta) \approx A w_{\mathrm{r}}\left(\tau - \frac{2R(\eta)}{c}\right) w_{\mathrm{a}}(\eta - \eta_c) \exp\left\{-j\frac{4\pi f_0 R(\eta)}{c}\right\} \exp\left\{j\pi K_{\mathrm{r}}\left(\tau - \frac{2R(\eta)}{c}\right)^2\right\}. \quad (3)$$

η is the azimuth time, A is the amplitude fluctuation, $w_{\mathrm{r}}(\tau - 2R(\eta)/c)$, is a range window centered on a delay of $2R(\eta)/c$, $w_{\mathrm{a}}(\eta - \eta_{\mathrm{c}})$ is an azimuth window centered on the Doppler centroid time. $R(\eta)$, K_{r} and c are ranges corresponding to the azimuth time η, Chirp rate of the signal transmitted by the radar and light velocity, respectively. The principle of superposition, which is the convolution of the reflection function of the area with the SAR impulse response, can be used for the calculation of the signal received from an extensive area by SAR because of the linearity of the SAR system. After the receiver oscillator and after being transferred to base band, the signal received by SAR from an area with the reflection function $g(\tau,\eta)$ is as follows [27]:

$$s_{\mathrm{bb}}(\tau,\eta) = g(\tau,\eta) \otimes h(\tau,\eta) + n(\tau,\eta). \quad (4)$$

$n(\tau,\eta)$ and $\otimes$ represent the additive noise in the receiver and convolution operator, respectively.

On the range-Doppler domain, applying Fourier transform on the signal received by the SAR receiver along the azimuth, it will be as follows:

$$S_{\mathrm{rd}}(\tau, f_\eta) \approx w_{\mathrm{r}}\left(\tau - \frac{2R_{\mathrm{rd}}(f_\eta)}{c}\right) w_{\mathrm{a}}(f_\eta - f_{\eta_c}) \exp\{j\theta_{\mathrm{rd}}\}. \quad (5)$$

$w_{\mathrm{r}}(\tau - 2R_{\mathrm{rd}}(f_\eta)/c)$ is a range window centered on a delay of $2R_{\mathrm{rd}}(f_\eta)/c)$. θ_{rd}, R_{rd}, f_η and f_{η_c} denote azimuth phase in the range Doppler domain, Range Cell Migration (RCM), azimuth frequency, and the Doppler centroid frequency, respectively. Moreover, $w_{\mathrm{a}}(f_\eta - f_{\eta c})$ stands for azimuth window centered on the Doppler centroid frequency.

The RCM which is present in the data received by SAR for the azimuth frequency of f_η is as follows [27]:

$$R_{\mathrm{rd}}(f_\eta) \approx R_0 + \frac{\lambda^2 R_0}{8V_{\mathrm{r}}^2} f_\eta^2. \quad (6)$$

R_0 and V_{r} show the range of closest approach and the radar velocity, respectively. Combining (5) and (6), on the range Doppler domain, the signal received by the SAR receiver is approximately as follows:

$$S_{\mathrm{rd}}(\tau, f_\eta) \approx w_{\mathrm{r}}\left(\tau - \frac{2R_0}{c} - \frac{\lambda^2 R_0}{4cV_{\mathrm{r}}^2} f_\eta^2\right) w_{\mathrm{a}}(f_\eta - f_{\eta_c}) \exp\{j\theta_{\mathrm{rd}}\}. \quad (7)$$

The mentioned observation array, after sampling, is delivered to the SAR processor as an input signal, and as it can be seen in Fig. 1, it is processed to form radar output image on one hand and as the input to the target tracking processor, on the other hand (the proposed plan).

Now, in the tracking section of the proposed receiver, the Range Cell Migration Compensation (RCMC) process is applied to the received observation array and then the output is given to the GM-PHD algorithm to track moving targets in the area surface. Therefore it is only needed to add the RCMC approach to the PHD filtering equations

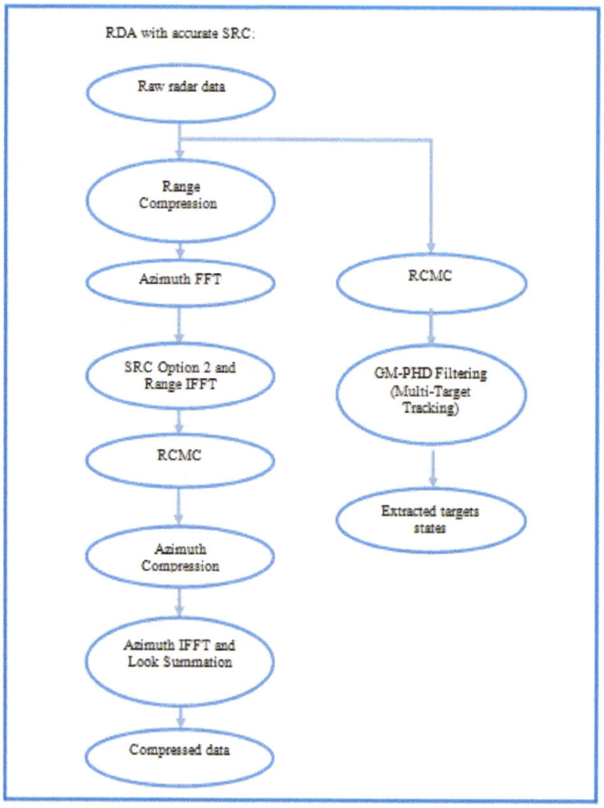

Fig. 1. The complete block diagram of the SAR processor containing the block proposed to track moving targets as well as the image formation algorithm of Range-Doppler Algorithm (RDA).

and using an additional processing of RCMC to update the sensor observation array. To this end, the $R(\cdot)$ operator is used with the following function to illustrate the RCMC operation on the observation vector received by the SAR sensor:

$$\boldsymbol{Z}^{\mathrm{RCMC}} = R(\boldsymbol{Z}). \tag{8}$$

The RCM value available in the received observation vector, which is approximately as follows, should be modified from (6) as

$$\Delta R(f_\eta) \approx \frac{\lambda^2 R_0}{8V_{\mathrm{r}}^2} f_\eta^2. \tag{9}$$

ΔR indicates the proportion of RCM which needs to corrected and it is a function of the azimuth frequency f_η, range R_0, radar wavelength λ, and the radar platform velocity V_{r}. After implementing the RCMC interpolation, the signal received by SAR will be as follows:

$$S^{\mathrm{RCMC}}(\tau, f_\eta) \approx A p_{\mathrm{r}}\left(\tau - \frac{2R_0}{c}\right) w_{\mathrm{a}}\left(f_\eta - f_{\eta_c}\right) \exp\left\{-\mathrm{j}\frac{4\pi f_0 R_0}{c}\right\} \exp\left\{\mathrm{j}\frac{f_\eta^2}{K_{\mathrm{a}}}\right\}. \tag{10}$$

Sampling $S^{\mathrm{RCMC}}(\tau, f_\eta)$ on the time-range domain, the received observation vector of SAR after RCMC is obtained and denoted by z^{RCMC}. Given the observation vector z^{RCMC}, beginning the iteration process of the Bayesian filtering with an initial density function such as $\rho_0(\cdot)$, the posterior density function at the moment k is tracked by the two following equations:

$$p_{k|k-1}(x_k \mid \mathbf{z}^{\mathrm{RCMC}}{}_{1:k-1}) = \int f_{k|k-1}(x_k \mid x) p_{k-1}(x \mid \mathbf{z}^{\mathrm{RCMC}}{}_{1:k-1}) \mathrm{d}x, \tag{11}$$

$$p_k(x_k \mid \mathbf{z}^{\mathrm{RCMC}}{}_{1:k}) = \frac{g_k(\mathbf{z}^{\mathrm{RCMC}}{}_k \mid x_k) p_{k|k-1}(x_k \mid \mathbf{z}^{\mathrm{RCMC}}{}_{1:k-1})}{\int g_k(\mathbf{z}^{\mathrm{RCMC}}{}_k \mid x) p_{k|k-1}(x \mid \mathbf{z}^{\mathrm{RCMC}}{}_{1:k-1}) \mathrm{d}x}. \tag{12}$$

Knowing the state x_k at time k, probability density of receiving observation $z_k \in \boldsymbol{Z}$ is $g_k(z_k \mid x_k)$ and, knowing all the observations $z_{1:k} = (z_1, \ldots, z_k)$ up to time k, the probability density of state x_k at time k is given by $p_k(x_k \mid z_{1:k})$.

The above filtering can be implemented approximately with an acceptable accuracy utilizing GM-PHD filter equations. It is therefore enough to update the GM-PHD equations obtained in [5] with $\boldsymbol{Z}^{\mathrm{RCMC}}$ instead of $\boldsymbol{Z}$.

After RCMC approach on received observation by receiver sensor SAR, filter prediction and update relations of GM-PHD in tracking process of SAR is followed as below:

Assume posterior intensity function at instant $k-1$ is in the form of Gaussian mixture as:

$$v_{k|k-1}(x) = \sum_{i=1}^{J_{k-1}} \omega_{k-1}^{(i)} \mathrm{N}(x; m_{k-1}^{(i)}, P_{k-1}^{(i)}) \tag{13}$$

where $\mathrm{N}(.;m,P)$ and ω denote a Gaussian density (with mean m and covariance P) and weight coefficients, respectively. J_{k-1} is the number of Gaussian components.

Thus, predicted intensity function at instant k is in the form of Gaussian mixture as:

$$v_{k|k-1}(x) = v_{S,k|k-1}(x) + v_{\beta,k|k-1}(x) + \gamma_k(x) \tag{14}$$

where $\gamma_k(x)$ is also represented by (15):

$$\gamma_k(x) = \sum_{i=1}^{J_{\gamma,k}} \omega_{\gamma,k}^{(i)} \mathrm{N}(x; m_{\gamma,k}^{(i)}, P_{\gamma,k}^{(i)}), \tag{15}$$

$$v_{S,k|k-1}(x) = p_{S,k} \sum_{j=1}^{J_{k-1}} \omega_{k-1}^{(j)} \mathrm{N}(x; m_{S,K|k-1}^{(j)}, P_{S,k|k-1}^{(j)}), \tag{16}$$

$$m_{S,k|k-1}^{(j)} = \boldsymbol{F}_{k-1} m_{k-1}^{(j)}, \tag{17}$$

$$P_{S,k|k-1}^{(j)} = Q_{k-1} + \boldsymbol{F}_{k-1} P_{k-1}^{(j)} \boldsymbol{F}_{k-1}^{\mathrm{T}}, \tag{18}$$

in which $\boldsymbol{F}_{k-1}$ and Q_{k-1} are the state transition matrix and the process noise covariance, respectively.

$$v_{\beta,k|k-1}(x) = \sum_{j=1}^{J_{k-1}} \sum_{\ell=1}^{J_{\beta,k}} \omega_{k-1}^{(j)} \omega_{\beta,k}^{(\ell)} \mathrm{N}(x; m_{\beta,k|k-1}^{(j,\ell)}, P_{\beta,k|k-1}^{(j,\ell)}), \tag{19}$$

$$m_{\beta,k|k-1}^{(j,\ell)} = F_{\beta,k-1}^{(\ell)} m_{k-1}^{(j)} + d_{\beta,k-1}^{(\ell)}, \quad (20)$$

$$P_{\beta,k|k-1}^{(j,\ell)} = Q_{\beta,k-1}^{(\ell)} + F_{\beta,k-1}^{(\ell)} P_{\beta,k-1}^{(j)} (F_{\beta,k-1}^{(\ell)})^{\mathrm{T}}. \quad (21)$$

Assume that the predicted intensity is a Gaussian mixture of the following form at time k given time $k-1$:

$$v_{K|K-1}(x) = \sum_{i=1}^{J_{k|k-1}} \omega_{k|k-1}^{(i)} N(x; m_{k|k-1}^{(i)}, P_{k|k-1}^{(i)}). \quad (22)$$

Then, the posterior intensity is also a Gaussian mixture at time k, and is given by:

$$v_K(x) = (1 - p_{D,k}) v_{K|K-1}(x) + \sum_{z^{\mathrm{RCMC}} \in Z_k^{\mathrm{RCMC}}} v_{D,k}(x; z^{\mathrm{RCMC}}) \quad (23)$$

where

$$v_{D,k}(x; z^{\mathrm{RCMC}}) = \sum_{j=1}^{J_{k|k-1}} \omega_k^{(j)}(z^{\mathrm{RCMC}}) N(x; m_{k|k}^{(j)}(z^{\mathrm{RCMC}}), P_{k|k}^{(j)}), \quad (24)$$

$$\omega_k^{(j)}(z^{\mathrm{RCMC}}) = \frac{p_{D,k} \omega_{k|k-1}^{(j)} q_k^{(j)}(z^{\mathrm{RCMC}})}{K_k(z^{\mathrm{RCMC}}) + p_{D,k} \sum_{\ell=1}^{J_{k|k-1}} \omega_{k|k-1}^{(\ell)} q_k^{(\ell)}(z^{\mathrm{RCMC}})}, \quad (25)$$

$$q_k^{(j)}(z^{\mathrm{RCMC}}) = N(z^{\mathrm{RCMC}}; H_k m_{k|k-1}^{(j)}, R_k + H_k P_{k|k-1}^{(j)} H_k^{\mathrm{T}}), \quad (26)$$

$$m_{k|k}^{(j)}(z^{\mathrm{RCMC}}) = m_{k|k-1}^{(j)} + K_k^{(j)}(z^{\mathrm{RCMC}} - H_k m_{k|k-1}^{(j)}), \quad (27)$$

$$P_{k|k}^{(j)} = [I - K_k^{(j)} H_k] P_{k|k-1}^{(j)}, \quad (28)$$

$$K_k^{(j)} = P_{k|k-1}^{(j)} H_k^T (H_k P_{k|k-1}^{(j)} H_k^{\mathrm{T}} + R_k)^{-1} \quad (29)$$

in which H_k and R_k are the observation matrix and the observation noise covariance, respectively.

In the next section, simulation results of the aforementioned cases will be presented.

3. Simulation of the Single-Target Tracking

Owing to the SAR platform movements, the velocity data of moving targets across the region are changed and transformed in the received observation array. As a result, it should be modified and the SAR movement effect should be compensated before the implementation of the tracking process by the PHD filter. This compensation process of the received observation array (algorithm input) which is added to the PHD filtering algorithm is the same as the RCMC approach in all the SAR image formation algorithms except that it is applied to different data.

The outcome of the convolution of the signal sent by the radar and the impulse response of the SAR is equivalent to the echo received by the SAR receiver (without the noise and clutter though). Clutter which has always a Poisson distribution in the time domain and a uniform distribution in the spatial domain [28], [29] (at any point in time) is generated accordingly and added to the echo received by the radar (the convolution of the chirp signal sent and the impulse response). After that, the observation array received by the sensor (the tracking algorithm input) is formed.

Care must be taken that fixed targets play the role of clutters and therefore considered undesirable targets. Furthermore, the main feature of Kalman and PHD filters is their ability to eliminate clutters, which was the impetus for applying the PHD filter in this paper to track moving targets in the SAR data. It should be noted that given the idea of moving targets detection and tracking, the SAR data are considered a type of radar data with high levels of clutters. The use of these filters is therefore regarded as a desirable approach for the detection and tracking of moving targets. It should however be noted that in the presence of multiple targets, the use of the Kalman filter necessitates applying techniques dedicated to data association of various targets and applying each set of data on a separate filter. The PHD filter does not face this restriction, but its only problem is the processing of actual high volume SAR data, which needs very powerful processors. The main problem in multi-target tracking is the association of data to targets, that means which observation is related to which target and this will lead to an increase in calculation load and the processing of tracking process. PHD filter operates on a single-target space and it has not the data association complexities. In fact, PHD filter transforms multi-target tracking into a single-target space and implements tracking process. Data association algorithms are often divided into Bayesian and non-Bayesian approaches. Non-Bayesian approaches utilize an optimization algorithm to find a solution for data association. On the other hand, Bayesian algorithms employ statistical theory for data association and conclude techniques such as Maximum A Posteriori (MAP).

In these simulations, the space borne Synthetic Aperture Radar (SAR) is in the spotlight imaging mode. The values of the radar's main parameters are presented in Tab. 1.

Parameter	Description	Value
R_0	radar near range (km)	825
V_r	velocity of platform (m/s)	7619
ρ_r	radar range resolution (m)	0.6
V_t	velocity of targets (m/s)	20-60
PRF	pulse repetition frequency (kHz)	3
f_S	sampling rate (MHz)	187.5
f_c	radar center frequency (GHz)	9.6
Chirp_BW	transmitted chirp bandwidth (Hz)	2.5415e+08
T_r	pulse width (sec)	8.0000e-05

Tab. 1. Simulation parameters.

A moving target with the trajectory shown in Fig. 2 is considered as an example.

For better illustration of the tracking performance over time, the target movement and also the sensor observations are shown along the perpendicular axes X and Y separately over time. Figure 3 shows the observation values of the targets and clutters along the perpendicular axes separately over time. The parameter N_Clutter represents the mean number of the Poisson process of generating clutter at every instance of time, which is equal to 20 in Fig. 3.

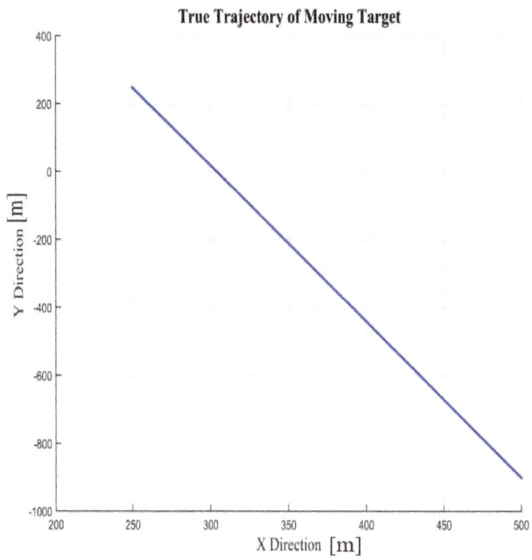

Fig. 2. A sample trajectory of a moving target in the single target scenario.

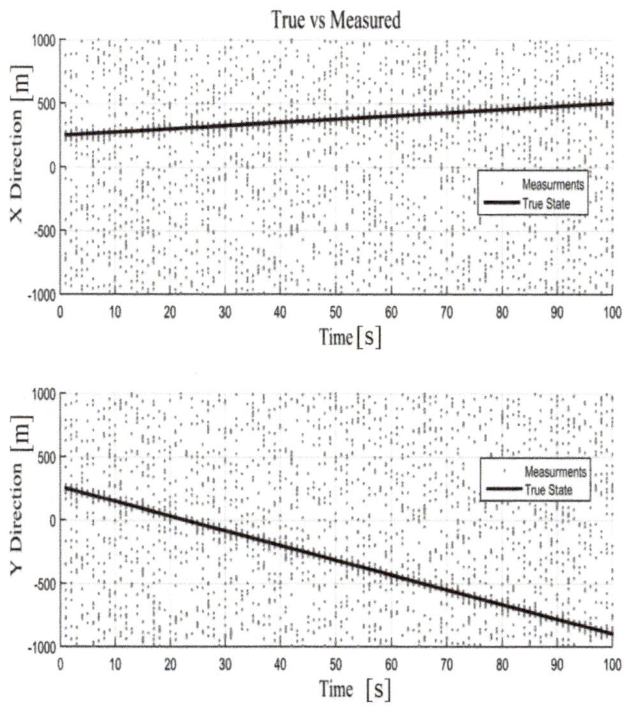

Fig. 3. Observation values of the target and clutter by the sensor along the perpendicular axes X and Y separately over time for the single target shown in Fig. 2, N_Clutter = 20 and after RCMC.

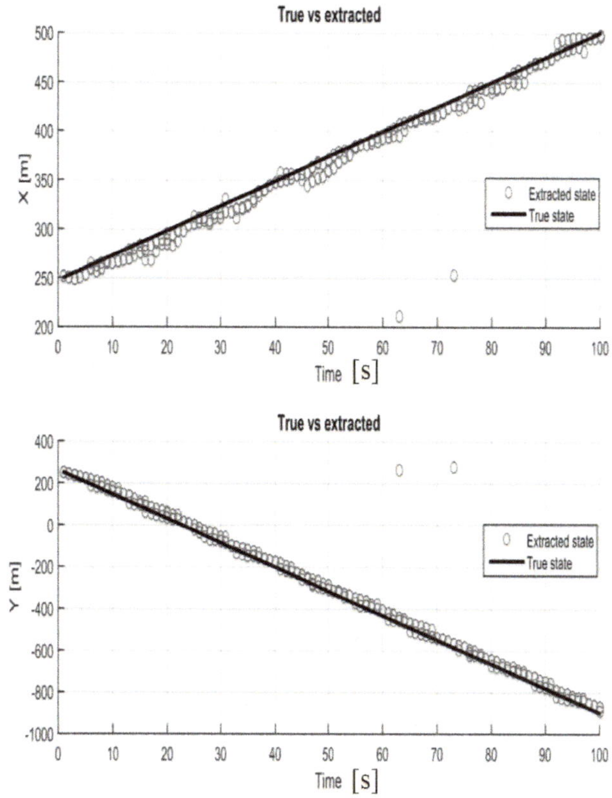

Fig. 4. State estimation of the target using the promoted GM-PHD algorithm along the perpendicular axes X and Y separately over time for the single target shown in Fig. 2 and N_Clutter = 20.

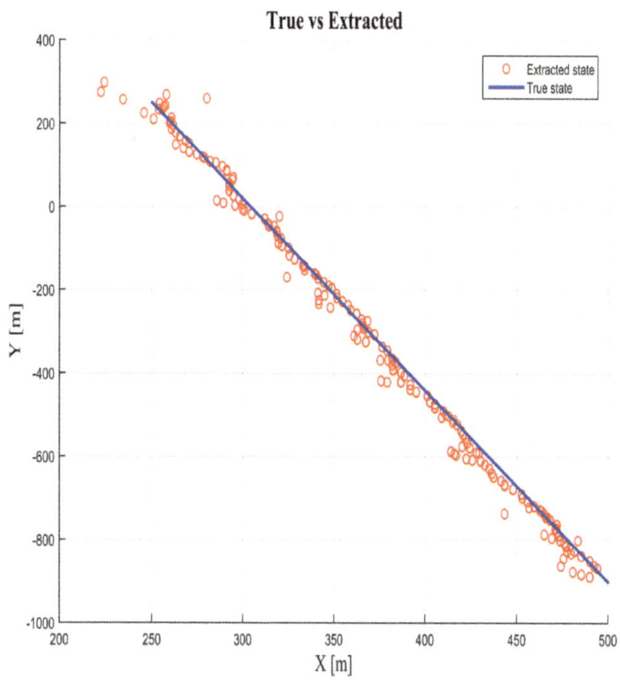

Fig. 5. The 2D target trajectory and the output of the tracking approach by the use of the promoted GM-PHD algorithm for the single target shown in Fig. 2 and N_Clutter = 20.

An important parameter for the quality assessment of the tracking process suggested is N_Clutter. It means that it is expected that the tracking approach performs well for different and particularly high values of the parameter. The output of the proposed approach (the target state at every moment) for the same target shown in Fig. 2 is exhibited in Fig. 4. The continuous curve in Fig. 4 represents the true state (position) of the target, while circles denote the target state at any instance extracted by the promoted GM-PHD algorithm.

The 2D target trajectory as well as the tracking result obtained by the promoted GM-PHD algorithm is shown in Fig. 5.

Moreover, in order to ensure that the proposed tracking plan is robust in the presence of highly dense clutter, the recent simulations have been repeated for N_Clutter = 200, which are all shown in Figs. 6–8.

The 2D target trajectory as well as the tracking result obtained by the promoted GM-PHD algorithm is shown in Fig. 8 (N_Clutter = 200).

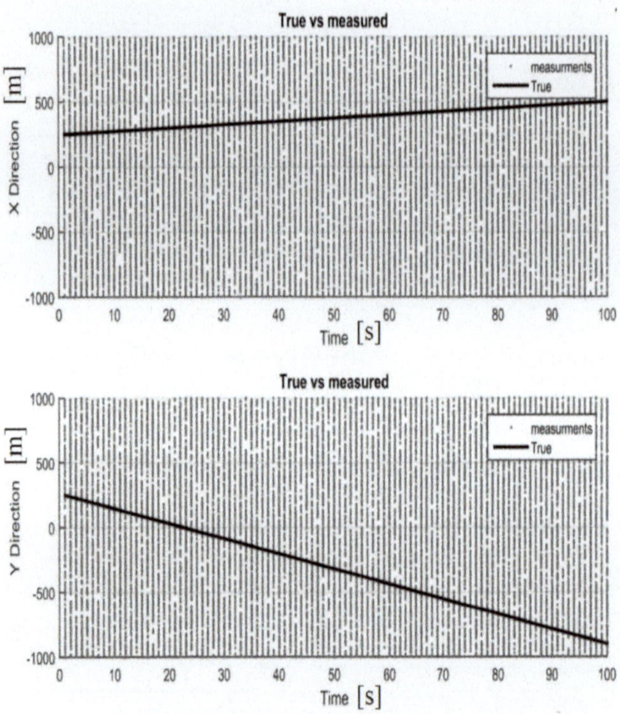

Fig. 6. Observation values of the target and clutter by the sensor along the perpendicular axes X and Y separately over time for the single target shown in Fig. 2, N_Clutter = 200 and after RCMC.

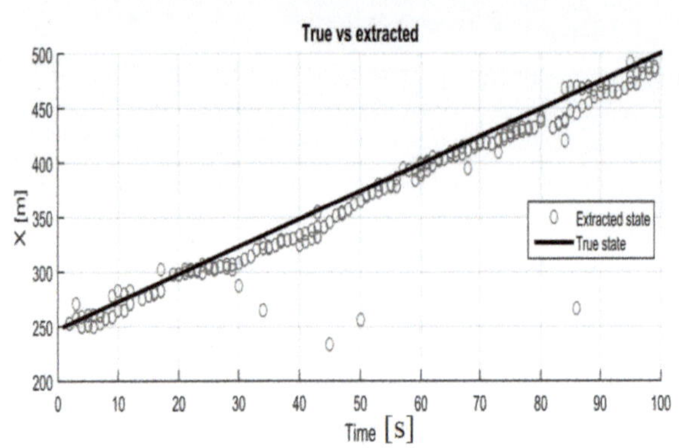

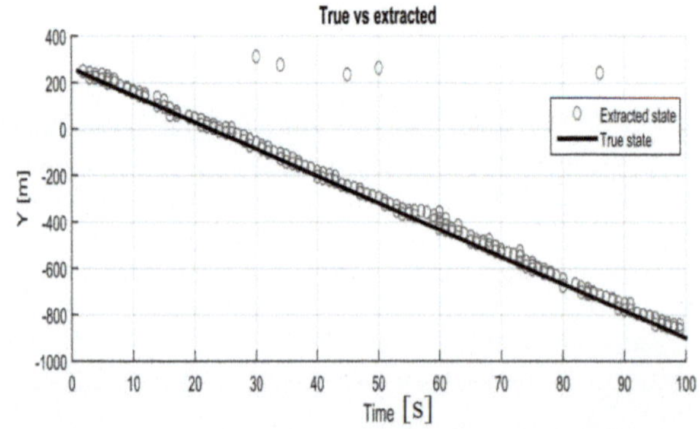

Fig. 7. State estimation of the target using the promoted GM-PHD algorithm along the perpendicular axes X and Y separately over time for the single target shown in Fig. 2 and N_Clutter = 200.

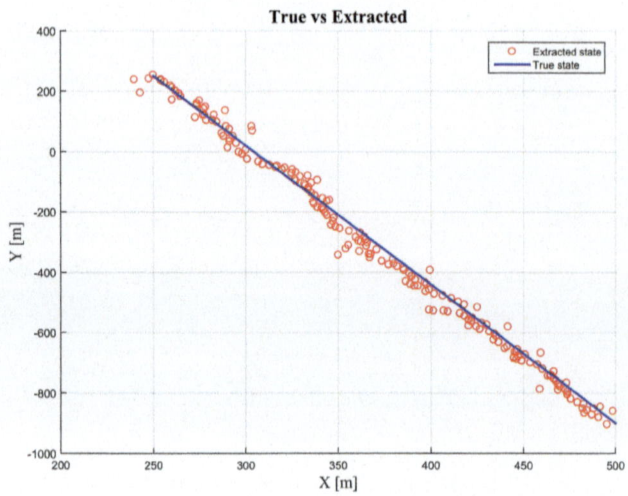

Fig. 8. The 2D target trajectory and the output of the tracking approach by the use of the promoted GM-PHD algorithm for the single target shown in Fig. 2 and N_Clutter = 200.

In order to compare the performance of the promoted GM-PHD algorithm with ordinary GM-PHD algorithm for tracking of a single moving target in the received raw data by SAR, the tracking results of GM-PHD algorithm are also considered without implementing RCMC. In this case, the observations obtained from sensors (of the target and clutter) are shown in Fig. 9.

The sensor observations of the target with greater density than the clutters with shift-invariance along X axis (due to radar squint angle) and a curve deviation along Y axis (due to RCM caused by SAR imaging geometry and motion) are shown in Fig. 9. Due to the faulty observations of sensors regarding the true value in both X and Y axis, it is necessary to implement the RCMC process in sensor observations (a radar receiver), before PHD filtering tracking process (or Kalman for single target only) initiates, so as to obtain a true tracking of the moving target in the received raw data by SAR. In fact, Figure 3 is the result of implementation of RCMC process on the observations shown in Fig. 9. Moreover, the estimation results of the state of the moving target shown in previous figures can be obtained and evaluated based on the observations shown in Fig. 9. These results are illustrated in Fig. 10.

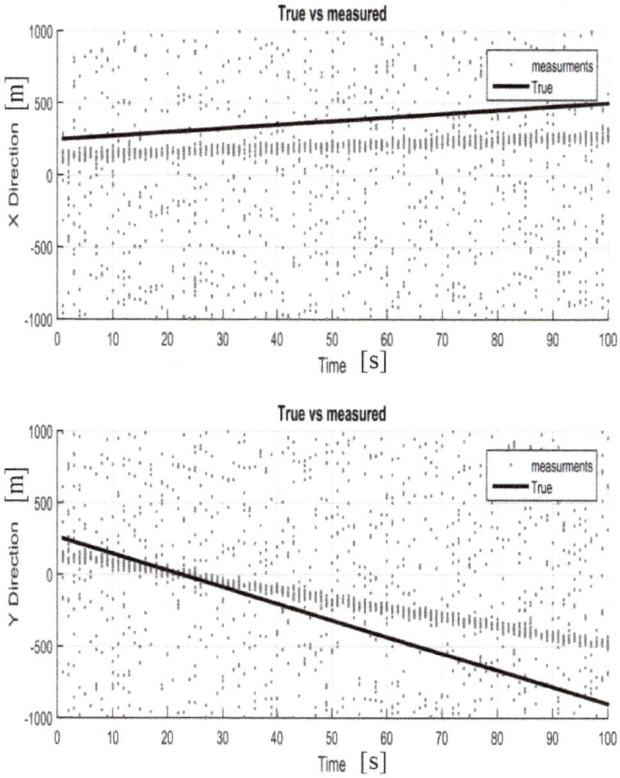

Fig. 9. Observation values of the target and clutter by the sensor along the perpendicular axes X and Y separately over time for the single target shown in Fig. 2, N_Clutter = 10 and without RCMC.

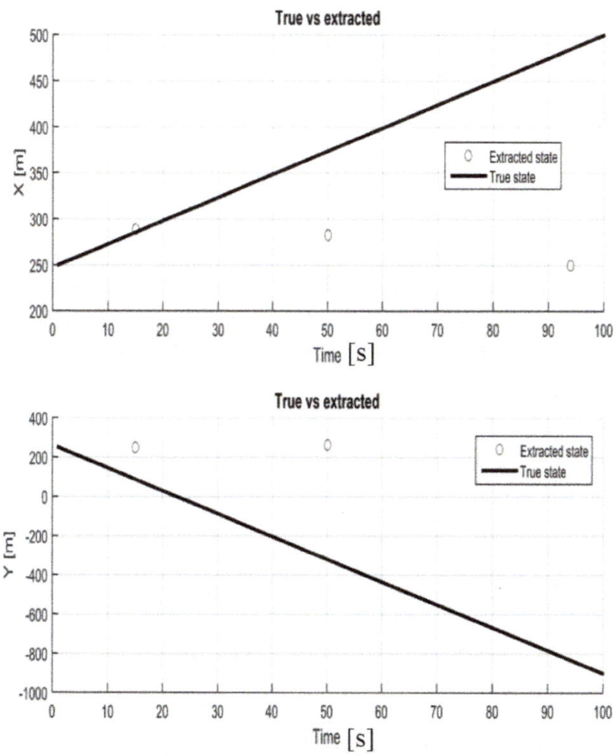

Fig. 10. State estimation of the target using the ordinary GM-PHD algorithm along the perpendicular axes X and Y separately over time for the single target shown in Fig. 2 and N_Clutter = 10 (without RCMC).

According to Fig. 10, if RCMC compensation process is not implemented in the sensor observations, the estimation of the state of target will be erroneous and the moving target tracking process in the received raw data by SAR material virtually fails. 2-dimensional target trajectory and the result of tracking by ordinary GM-PHD algorithm for this case (if N_Clutter = 10) is shown in Fig. 11.

As shown in Fig. 12, the increase in surface density of the clutters (10 times in this sample) results in an increase in the erroneous estimations which shows that the erroneous estimations are caused by clutters not the target(s).

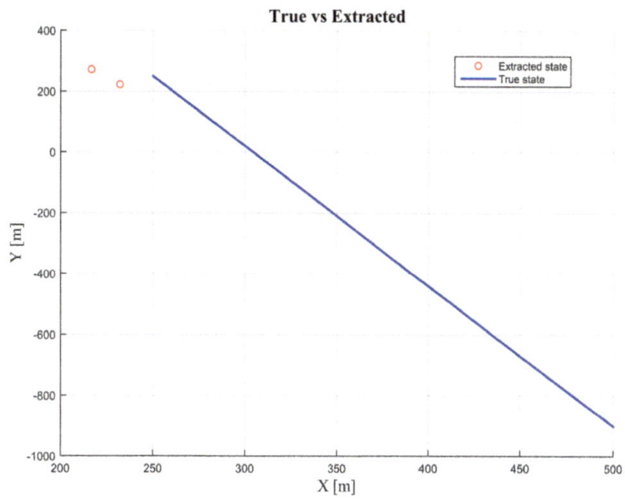

Fig. 11. The 2D target trajectory and the output of the tracking approach by the use of the ordinary GM-PHD algorithm for the single target shown in Fig. 2 and N_Clutter = 10 (without RCMC).

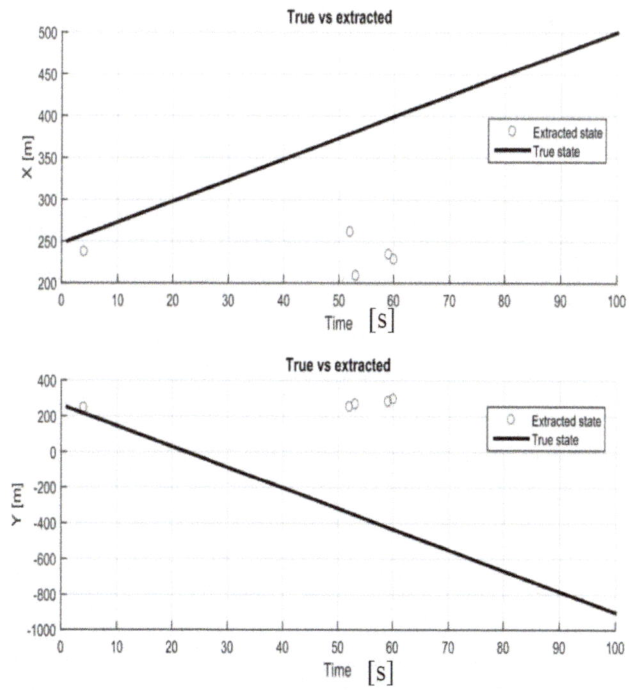

Fig. 12. State estimation of the target using the ordinary GM-PHD algorithm along the perpendicular axes X and Y separately over time for the single target shown in Fig. 2 and N_Clutter = 100 (without RCMC).

4. Conclusion

This paper proposed a novel technique based on the GM-PHD filtering for the simultaneous tracking of single moving target in the data received by the SAR in the Spotlight imaging mode. Moreover, the prediction and update equations of the PHD function are in the single-target state space. The implementation of the PHD filter has therefore less computational load than the ideal prediction and update equation in the Bayesian filter. Applying range cell migration compensation on the raw data received by SAR before tracking, made it possible to track single moving target with a good quality.

References

[1] MAHLER, R. Multi-target Bayes filtering via first-order multi-target moments. *IEEE Transactions on Aerospace and Electronic Systems*, 2003, vol. 39, no. 4, p. 1152–1178. DOI: 10.1109/TAES.2003.1261119

[2] VO, B.N., SINGH, S. Technical aspects of the probability hypothesis density recursion. *Technical Report TR05-006 EEE Department the University of Melbourne, Australia*, 2005.

[3] VO, B. N., SINGH, S., DOUCET, A. Sequential Monte Carlo implementation of the PHD filter for multi-target tracking. In *Proceedings of the 6th International Conference on Information Fusion*. Cairns (Australia), 2003, p. 792–799. DOI: 10.1109/ICIF.2003.177320

[4] VO, B. N., SINGH, S., DOUCET, A. Sequential Monte Carlo methods for Bayesian multi-target filtering with random finite sets. *IEEE Transactions on Aerospace and Electronic Systems*, 2005, vol. 41, no. 4, p. 1224–1245. DOI: 10.1109/TAES.2005.1561884

[5] VO, B. N., MA, W. K. The Gaussian mixture probability hypothesis density filter. *IEEE Transactions on Signal Processing*, 2006, vol. 54, no. 11, p. 4091–4104. DOI: 10.1109/TSP.2006.881190

[6] CLARK, D. E., VO, B. N. Convergence analysis of the Gaussian mixture PHD filter. *IEEE Transactions on Signal Processing*, 2007, vol. 55, no. 4, p. 1204–1211. DOI: 10.1109/TSP.2006.888886

[7] VO, B. N., PASHA, A., TUAN, H. D. A Gaussian mixture PHD filter for nonlinear jump Markov models. In *Proceedings of the 45th IEEE Conference on Decision and Control*. San Diego (CA, USA), December 13–15, 2006, p. 3162–3167. DOI: 10.1109/CDC.2006.377103

[8] PUNITHAKUMAR, K., KIRUBARAJAN, T., SINHA, A. Multiple model probability hypothesis density filter for tracking maneuvering targets. *IEEE Transactions on Aerospace and Electronic Systems*, 2008, vol. 44, no. 1, p. 87–98. DOI: 10.1109/TAES.2008.4516991

[9] PASHA, A., VO, B. N., TUAN, H. D, et al. A Gaussian mixture PHD filter for jump Markov system models. *IEEE Transactions on Aerospace and Electronic Systems*, 2009, vol. 45, no. 3, p. 919 to 936. DOI: 10.1109/TAES.2009.5259174

[10] LI, W. JIA, Y. Gaussian mixture PHD filter for jump Markov models based on best-fitting Gaussian approximation. *Signal Processing*, 2011, vol. 91, no. 4, p. 1036–1042. DOI: 10.1016/j.sigpro.2010.08.004

[11] NANDAKUMARAN, N., PUNITHAKUMAR, K., KIRUBARAJAN, T. Improved multi-target tracking using probability hypothesis density smoothing. In *Proceedings of the SPIE Conference on Signal and Data Processing of Small Targets*, August 2007, vol. 6699, p. 1–6. ISBN: 9780819468475

[12] NANDAKUMARAN, N., THARMARASA, R., LANG, T., et al. Gaussian mixture probability hypothesis density smoothing with multistatic sonar. In *Proceedings of the SPIE Conference on Signal Processing, Sensor Fusion and Target Recognition*. March 2008, vol. 6968, p. 1–6. ISBN: 9780819471598

[13] NANDAKUMARAN, N., KIRUBARAJAN, T. Maneuvering target tracking using probability hypothesis density smoothing. In *Proceedings of the SPIE Conference on Signal Processing, Sensor Fusion and Target Recognition*. March 2009, vol. 7336, p. 1–6. ISBN: 9780763776473

[14] LI, W., JIA, Y., DU, J., et al. Gaussian mixture PHD smoother for jump Markov models in multiple maneuvering targets tracking. In *Proceedings of the 2011 American Control Conference*. San Francisco (CA, USA), June 29–July 1, 2011, p. 3024–3029. DOI: 10.1109/ACC.2011.5991161

[15] MAHLER, R., VO, B. T., VO, B. N. The forward–backward probability hypothesis density smoother. In *Proceedings of the13th International Conference on Information Fusion*. Edinburgh (UK), 2010, p. 1–8. DOI: 10.1109/ICIF.2010.5711920

[16] MAHLER, R., VO, B. T., VO, B. N. Multi-target forward–backward smoothing with the probability hypothesis density. *IEEE Transactions on Aerospace and Electronic Systems*, 2012, vol. 48, no. 1, p. 707–728. DOI: 10.1109/TAES.2012.6129665

[17] VO, B. N., VO, B. T., MAHLER, R. A closed form solution to the probability hypothesis density smoother. In *Proceedings of the 13th International Conference on Information Fusion*. Edinburgh (UK), 2010, p. 1–8. DOI: 10.1109/ICIF.2010.5711983

[18] VO, B. N., VO, B. T., MAHLER, R. Closed form solutions to forward–backward smoothing. *IEEE Transactions on Signal Processing*, 2012, vol. 60, no. 1, p. 2–17. DOI: 10.1109/TSP.2011.2168519

[19] VO, B. N., SINGH, S., MA, W. K. Tracking multiple speakers with random sets. In *Proceedings of the International Conference on Acoustics, Speech and Signal Processing*. Montreal (Canada), 2004, vol. II, p. 357–360. DOI: 10.1109/ICASSP.2004.1326268

[20] CHEN, F.L., LIN, H., ZHOU, W., et al. Surface deformation detected by ALOS PALSAR small baseline SAR interferometry over permafrost environment of Beiluhe section, Tibet Plateau, China. *Remote Sensing of Environment*, 2013, vol. 138, p. 10–18. DOI: 10.1016/j.rse.2013.07.006

[21] YOSHIDA, T., RHEEM, C. K. Time-domain simulation of along-track interferometric SAR for moving ocean surfaces. *Sensors*, 2015, vol. 15, p. 13644–13659. DOI: 10.3390/s150613644

[22] JIN, M. J., WU, C. A SAR correlation algorithm which accommodates large range migration. *IEEE Transactions on Geoscience and Remote Sensing*, November 1984, vol. 22, no. 6, p. 592–597. DOI: 10.1109/TGRS.1984.6499176

[23] CURLANDER, J., MCDONOUGH, R. *Synthetic Aperture Radar*. New York: John Wiley & Sons, 1991. ISBN: 978-0-471-85770-9

[24] OLIVER, C. J. Synthetic-aperture radar imaging. *Journal of Physics D: Applied Physics*, 1989, vol. 25, p. 871–890. DOI: 10.1088/0022-3727/22/7/001

[25] RIEDL, M., POTTER, L., BRYANT, C., et al. Joint synthetic aperture radar and space-time adaptive processing on a single receive channel. *IEEE Transactions on Aerospace and Electronic Systems*, January 2015, vol. 51, p. 331–341. DOI: 10.1109/TAES.2014.130596

[26] XIA, Y. Synthetic aperture radar interferometry. In *Sciences of Geodesy - I*. Ed. G. Xu, Springer Berlin Heidelberg, 2010, p. 415 to 474. DOI: 10.1007/978-3-642-11741-1

[27] CUMMING, I. G., WONG, F. H. *Digital Processing of Synthetic Aperture Radar Data*. Norwood: Artech House, 2005. ISBN: 978-1580530583

[28] BLACKMAN, S. S. *Multiple-target Tracking with Radar Applications*. Artech House, 1986. ISBN: 978-0890061794

[29] SMITH, F. W., MALIN, J. A. Models for radar scatterer density in terrain images. *IEEE Transactions on Aerospace and Electronic Systems*, 1986, vol. 22, no. 5, p. 642–647. DOI: 10.1109/TAES.1986.310730

About the Authors ...

Navid DARYASAFAR received his B.Sc. and M.Sc. degree in Electrical Telecommunication from Islamic Azad University of Bushehr in 2008 and 2011. Now he is a PhD student of Science & Research Branch, Islamic Azad University, Tehran. He has been as a senior researcher and lecturer in Islamic Azad University of Bushehr from 2012. His research interests are signal processing, filter design, image processing, acoustic signal, radar recognition, MIMO-OFDM systems.

Ramazanali SADEGHZADEH received the B.Sc. degree in Telecommunication Engineering from K. N. Toosi University of Technology, Tehran, Iran, in 1984, the M.Sc. in Digital Communication Engineering from the University of Bradford, Bradford, UK, and University of Manchester, Institute of Science and Technology (UMIST), Manchester, UK, as a joint program in 1987, and the Ph.D. degree in Electromagnetic and Antenna from the University of Bradford in 1991. During 1992–1997, he worked as a Postdoctoral Research Assistant in the field of propagation, electromagnetic, antenna, biomedical, and wireless communication at the University of Bradford. From 1984 to 1985, he was with the Iran Telecommunication Company, Tehran, Iran, working on networking. Since 1997, he has been with the Faculty of Electrical and Computer Engineering, K. N. Toosi University of Technology. He has published more than 120 referable papers in international journals and conferences. His research interests include numerical techniques in electromagnetics, antenna, propagation, radio networks, wireless communications, nano antennas, and radar systems.

Mohammad NASER-MOGHADASI was born in Saveh, Iran, in 1959. He received the B.Sc. degree in Communication Engineering in 1985 from Leeds Metropolitan University (formerly Leeds Polytechnic), UK. Between 1985 and 1987 he worked as an RF design engineer for the Gigatech Company in Newcastle Upon Tyne, UK. From 1987 to 1989, he was awarded a full scholarship by the Leeds Educational Authority to pursue an M. Phil. studying in CAD of microwave circuits. He received his Ph.D. in 1993, from the University of Bradford, UK. He was offered then a 2years Post Doc. to pursue research on Microwave Cooking of Materials at the University of Nottingham, UK. From 1995, he joined Islamic Azad University, at Science & Research Branch, Iran, where currently he is an Associate Professor and head of Telecommunications group. His main areas of interest in research are microstrip antenna, microwave passive and active circuits, RF MEMS. He is a member of the Institution of Engineering and Technology, MIET, and the Institute of Electronics, Information and Communication Engineers (IEICE). He has so far published over 160 papers in different journals and conferences.

Dual Band Square Planar Dipole Antennas with Interdigital Capacitors for Base Stations

Pongsathorn CHOMTONG [1], *Prayoot AKKARAEKTHALIN* [2]

[1]Dept. of Teacher Training in Electrical Engineering, Faculty of Technical Education
[2]Dept. of Electrical and Computer Engineering, Faculty of Engineering
King Mongkut's University of Technology North Bangkok, 1518 Pracharat 1 Road, Bangsue, Bangkok, Thailand

pongsathornc@kmutnb.ac.th, prayoot@kmutnb.ac.th

Abstract. *A dual band antenna based on the square planar dipole structure with interdigital capacitors is proposed. With this technique, the harmonic frequency band can be controlled as desired. The dipole element consists of two square copper plates with overall size of* 21.6×13.8 *cm*^{2}*. The bandwidths of the proposed antenna are 170 MHz (1.71-1.88 GHz) and 200 MHz (5.15-5.35 GHz). The antenna gain of each band is obtained to be 10 dB and 8 dB, respectively. To increase the antenna gain, the array antenna composing of the two designed dipoles is studied. The size of array antenna is 21.6 cm × 24.4 cm. The improved gains of 13.1 dB for the first band and 10.1 dB for the second band are obtained. The array antenna with four dipoles is also investigated. The high gains of 16.8 dB for the first band and 12.6 dB for the second band are obtained. The proposed array antennas have higher gains compared with the conventional array antennas with the same element numbers. Therefore, the proposed antennas can be suitably used for LTE and WLAN base stations.*

Keywords

Square dipole antenna, array antenna, interdigital technique

1. Introduction

Development of wireless communication systems has been continuously increased to support user demands in daily life. Antenna is an important part of any wireless system used for receiving and sending information. Improvement of antenna characteristics including return loss and gain is needed to serve long distance communications. Also, nowadays some modern antennas with multiband operation are required. However, to design antennas, appropriate size, ease of design, and affordable cost must be carefully considered. Also, higher power handling capacity is necessarily required for base station antenna design. Currently, the base station antennas with multiband operation have been developed continuously for both indoor and outdoor applications. There are different types of base station antennas, especially, using microstrip and metal plate structures. Many of them are designed based on the dipole and monopole antenna structures, whose radiation patterns are commonly omnidirectional. The advantage is that the signals can be received or transmitted around the antenna, but there are low gains. For multiband antenna design, various techniques were proposed such as fractal antennas [1–3]. However, the antenna sizes were large about half wavelength and antenna gains were still low. Also, the fractal antenna radiation patterns are very difficult to be controlled. Another technique is to design the wideband antennas with built-in notches creating multiband responses [4–6]. Nevertheless, it is still difficult to design antennas with higher gain. To improve antenna gains, some reflectors were applied [7–9]. Using the reflectors, interferences between back lobes and side lobes are significantly reduced. Another technique to increase antenna gain was to use genetic algorithm optimization for suitable array antenna geometry and feed position, but complicated computation was required [10]. Also, antenna array could be applied for higher gain, but large number of antenna elements may be required, resulting in large size.

For base station antenna design, high power handling capacity is a main concern. Therefore, dielectric loss or breakdown of material used to build up the base station antenna must be carefully determined. The indoor base station antennas with low power handling capacity were designed to spread omnidirectional or directional patterns but the antenna gains were low [11], [12]. Some outdoor base station antennas in [13–15] were developed by using metal materials for high power handling, resulting in high gain as 17 dB. These antenna structures are very complicated to design and their sizes are also too large. However, these proposed base stations could not support multiband operation. Up to now, several multiband base station antennas were proposed. A dual band base station antenna using U-shaped planar slots was proposed for WLAN and WiMAX applications. However, implementation of both radiation elements and the feed network of this antenna was somehow difficult [16]. Also, the antenna gain and power handling capacity are obviously low. The dual-band base-station antenna whose high-frequency elements are embedded in low frequency elements to reduce volume was also proposed in [17]. This antenna has a low profile but there is a limitation in impedance matching for both bands. It is found that not only its structure is very complicated but also the gain and power handling capacity are low.

This paper presents dual band antennas to be used as indoor and outdoor base station antennas. The proposed antennas will be designed using a simple structure on copper plates for improved characteristics and power handling capacity. The antenna structure is based on the dipole with built-in interdigital capacitors and a reflector. The advantage of the proposed antennas is that the second harmonic frequency band can be freely tuned as desired. This paper is divided into two main parts; the single dipole element with interdigital capacitors and the array antenna design for increasing gain. The details of antenna design will be given in Sec. 2. In Sec. 3, the antenna implementation and measurement will be presented. Finally, conclusion will be provided in the last section.

2. Antenna Design

This section presents the design of the dipole antenna with embedded interdigital lines [18–20], as the structure shown in Fig. 1. The antenna is based on a planar dipole structure with a reflector. The copper plate with thickness of 1.5 mm was employed. Both a single dipole antenna and array antennas were designed and investigated using the CST Studio Suit. All details of the antenna design will be shown in the following subsections.

2.1 Design of a Dipole Antenna with Interdigital Lines and a Reflector

The rectangular patch dipole antenna with a length of λ/2 was designed and simulated at resonant frequency of 1.8 GHz (LTE band: 1.72–1.83 GHz) as shown in Fig. 2(a). Size of each patch is 70.9 mm × 66 mm. This conventional dipole is compared to the proposed interdigital structure in Fig. 2(b). As a result of $|S_{11}|$ shown in Fig. 3, the harmonic frequency band of the conventional antenna without interdigital structure does not cover WLAN application band (5.15–5.35 GHz). However, with the proposed interdigital technique, the harmonic frequency band at 5.2 GHz could be adjusted to cover the required WLAN band caused by changing capacitive value of the dipole patch. This capacitive value depends on sizes of fingers and gaps of the interdigital structure. Therefore, some interdigital parameters of the proposed antenna in Fig. 1 including w_1 and w_3 were investigated. The parameter w_1 was then varied from 2 to

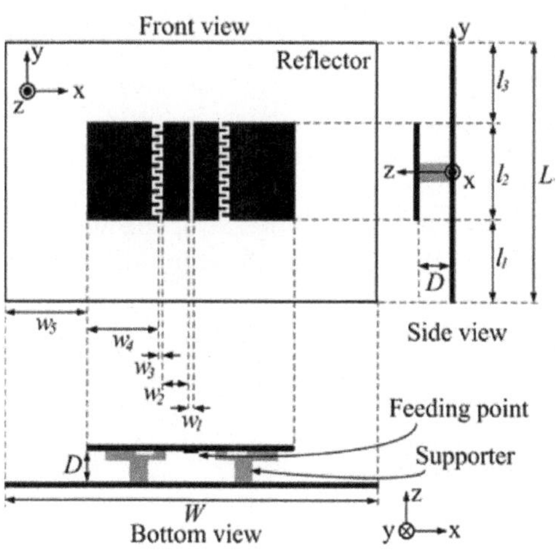

Fig. 1. Symmetrical configuration element of the antenna.

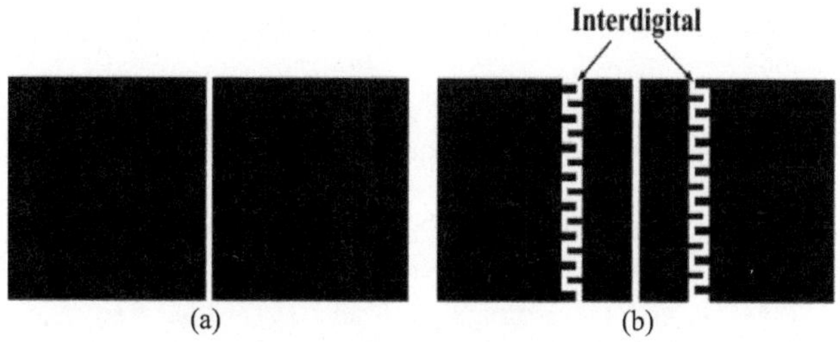

Fig. 2. The square elements of dipole antenna without interdigital and (b) with interdigital.

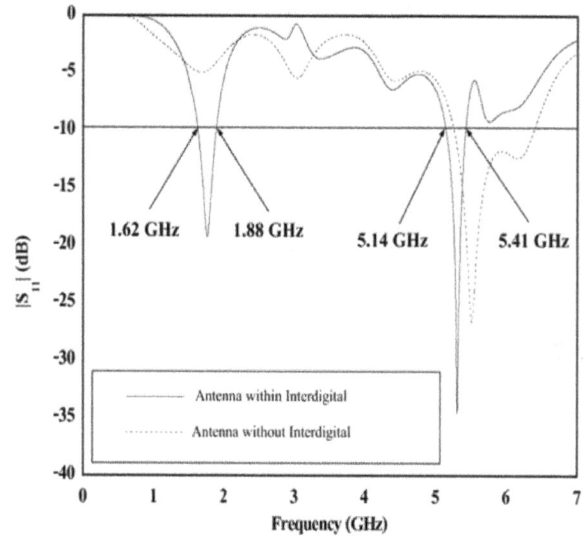

Fig. 3. Comparison of $|S_{11}|$ for the dipole antennas with and without interdigital structure.

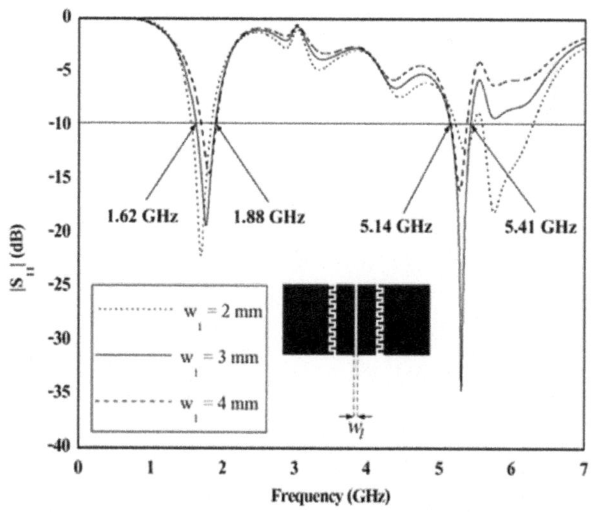

Fig. 4. Comparison of $|S_{11}|$ for the effect of dipole gap when varying W_1 from 2 mm to 4 mm.

4 mm when other parameters were fixed as initial values of $w_2 = 18$ mm, $w_3 = 2.5$ mm, $w_4 = 50.4$ mm, and $D = 14$ mm. It is clearly seen in Fig. 4 that the parameter w_1 slightly affects on both frequency bands. However, the proper distance of gap w_1 is 3 mm that results to better matching with the $|S_{11}|$ lesser than –10 dB. The bandwidths are obtained to be 1.62–1.88 GHz (260 MHz) and 5.14–5.41 GHz (270 MHz) for the fundamental and harmonic frequency bands, respectively. Figure 5 shows simulated results when varying the interdigital gap w_3 between 1.5 and 3.5 mm. It is clearly seen that the fundamental frequency has a little effect. At minimum gap of 1.5 mm, the capacitive value is higher, resulting in shifting the harmonic frequency close to the fundamental frequency. At maximum gap of 3.5 mm, the harmonic frequency is shifted away from the fundamental frequency due to lower capacitance. Therefore, the proper distance of gap to increase capacitive value on the patch antenna for resonance at the second harmonic frequency is 2.5 mm, which results to the resonance frequency at 5.2 GHz as desired to cover the bandwidth between 5.14 and 5.41 GHz (270 MHz).

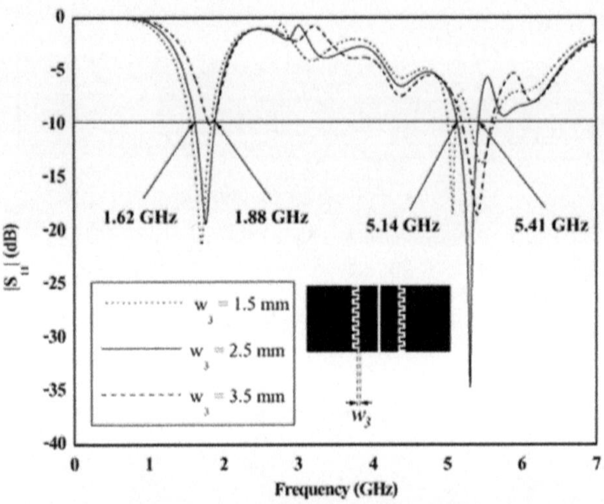

Fig. 5. Comparison of $|S_{11}|$ for the effect of interdigital gap when varying w_3 from 1.5 mm to 3.5 mm.

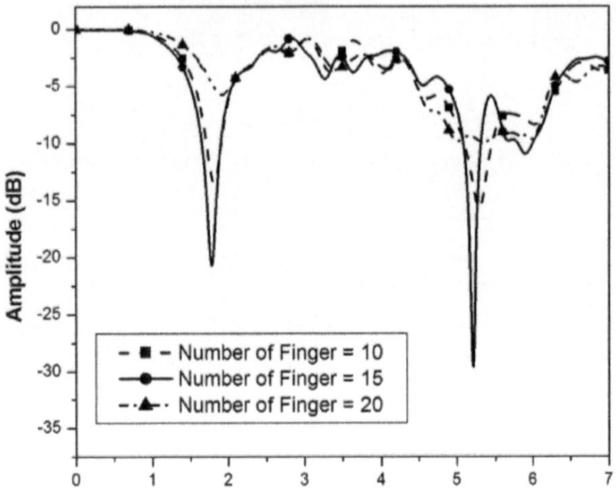

Fig. 6. Comparison $|S_{11}|$ for the effect number of finger when varying number of fingers from 10 to 20.

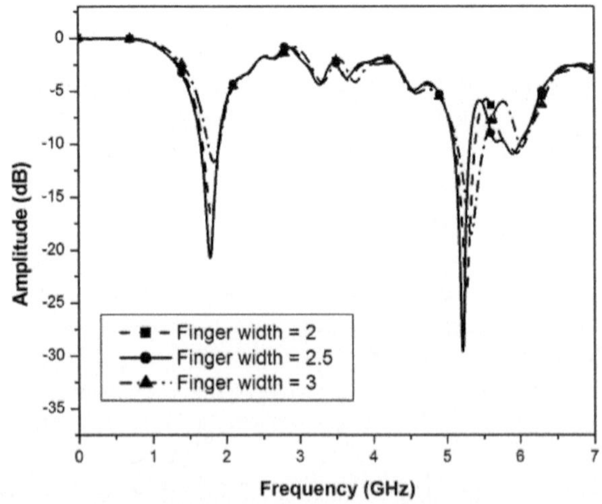

Fig. 7. Comparison $|S_{11}|$ for the effect of finger width when varying from 2 mm to 3 mm.

In addition to control harmonic frequency, it can be seen that the good impedance matching is obtained with the $|S_{11}|$ lower than –10 dB at both frequency bands. The final $|S_{11}|$ of the proposed antenna when $w_1 = 3$ mm and $w_2 = 2.5$ mm is already shown in Fig. 3. The number of fingers and finger width were also investigated. Figure 6 shows the results when varying the number of fingers to be 10, 15 and 20. It is found that the number of fingers slightly affects to the frequency band. However, 15 fingers were chosen to give a suitable matching for both bands. Then, the finger width was varied between 2 and 3 mm, resulting in the frequency response as shown in Fig. 7. It is clearly seen that the finger width has the effect on shifting the second harmonic band. Nevertheless, the finger width of 2.5 mm was chosen for covering the application bands of LTE and WLAN. It is clearly seen that the dipole with interdigital lines can control the second harmonic as desired without affecting to the fundamental frequency. All important parameter dimensions of the proposed interdigital capacitors are shown in Tab. 1.

Next, a reflector was applied to the dipole antenna in order to increase its gain. Figure 8 shows the structure of the antenna with a reflector. The reflector size is 232 mm × 153.5 mm to cover the dipole. Between the antenna and the reflector, a support material with low effect on antenna radiation, lightweight and high power handling was needed. The material of polyoxymethylene (POM) was then chosen, as the structure shown in Fig. 1. Normally, the distance between the antenna and reflector D is approximately about $\lambda/4$. However, the distance was slightly changed when using interdigital structure with the dipole antenna. To find the appropriate value of D, it was then varied from 14 and 41.6 mm.

Parameter	Description	Size (cm)
w_1	feeding gap	0.3
w_2	width of smaller side of each element	1.8
w_3	width of incision gap	0.25
w_4	width of larger side of each element	5.04

Tab. 1. Optimal parameters of the proposed dipole antenna.

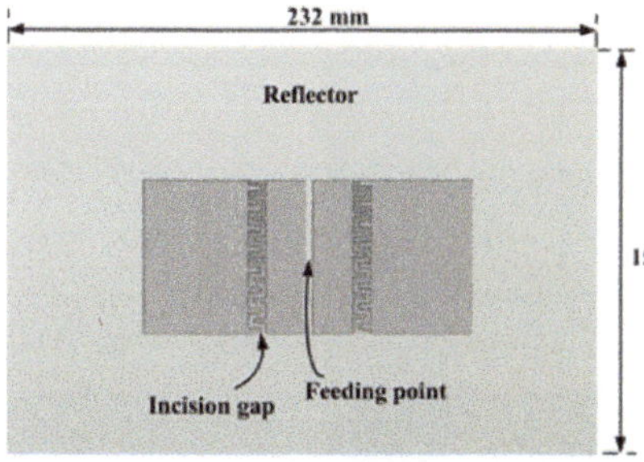

Fig. 8. Structure of the proposed antenna with a reflector.

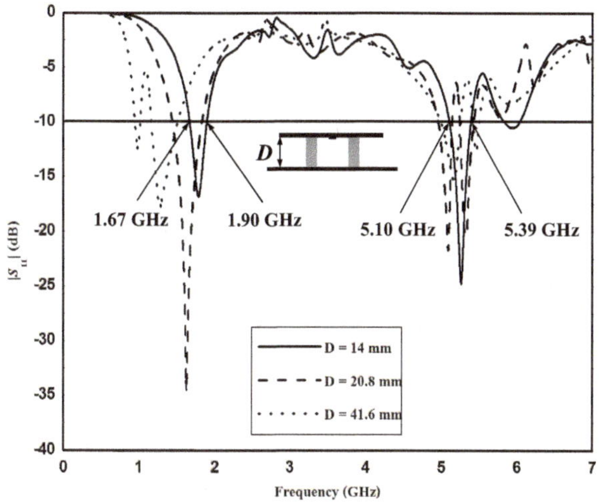

Fig. 9. Optimal distance *D* between the antenna element and the reflector.

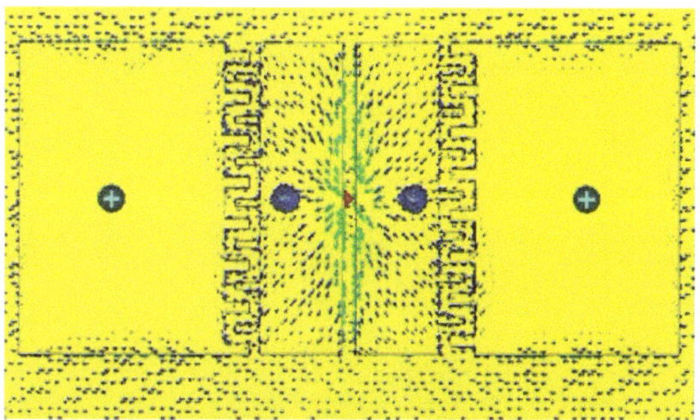

(a)

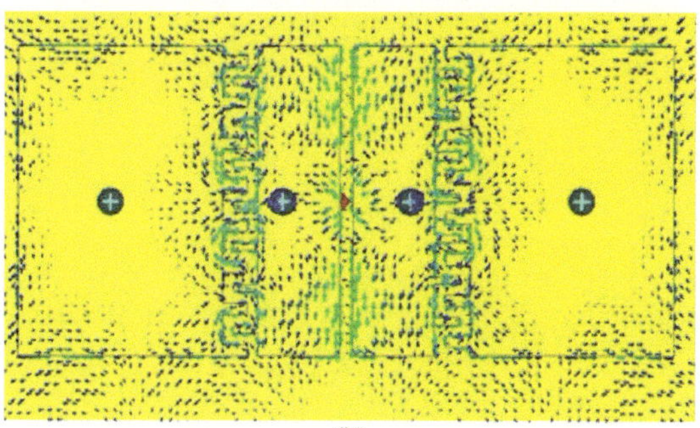

(b)

Fig. 10. Simulated surface currents on the dipole antenna (a) 1.8 GHz and (b) 5.2 GHz.

It is found that the distance *D* affects to the frequency shifting at both fundamental and harmonic frequency bands. The most appropriate distance is 14 mm, that results to good impedance matching with |S11| lesser than –10 dB. The bandwidths of 1.67–1.9 GHz (230 MHz) and between 5.1 and 5.39 GHz (290 MHz) are obtained for the fundamental band of 1.8 GHz and the second harmonic band of 5.2 GHz, respectively. Simulated surface currents of the proposed dipole antenna are shown in Fig. 10. Figure 11 also shows the current distributions at the fundamental and second harmonic frequencies. It is clearly seen that the higher band at 5.2 GHz has more surface current and current distribution at the fingers of interdigital structure than the lower band at 1.8 GHz. This means that the dipole with interdigital capacitor has a strong effect at harmonic frequency.

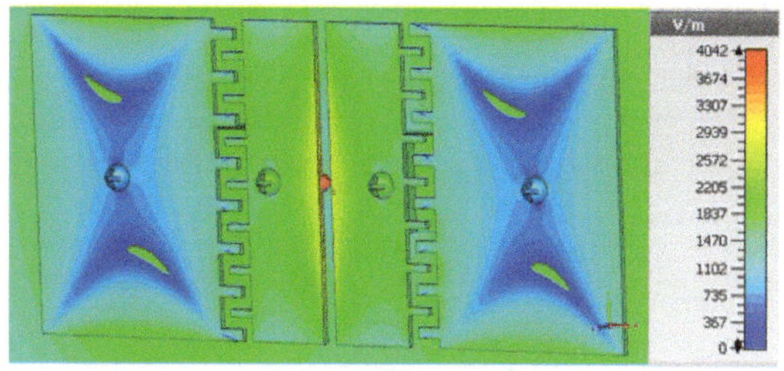

(a)

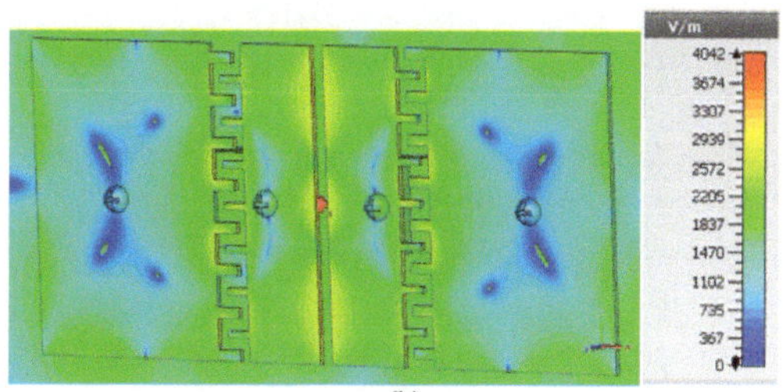

(b)

Fig. 11. Simulated current distribution of the dipole antenna: (a) 1.8 GHz and (b) 5.2 GHz.

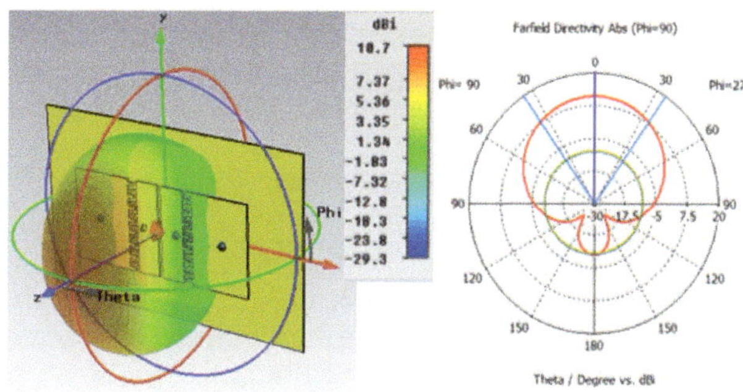

(a)

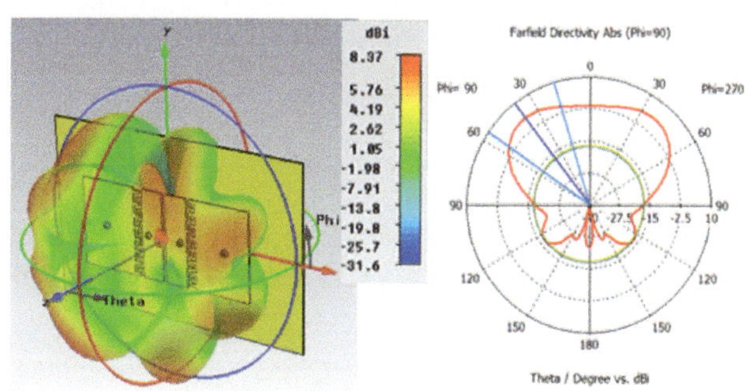

(b)

Fig. 12. Simulated radiation patterns of the dipole antenna: (a) 1.8 GHz and (b) 5.2 GHz.

Parameter	Description	Size(cm)
L	length of reflector	15.3
W	width of reflector	21.5
D	distance between array antenna and reflector	1.4
w_1	feeding gap	0.3
w_2	width of smaller side of each element	1.8
w_3	width of incision gap	0.25
w_4	width of larger side of each element	5.04
w_5	distance from reflector to antenna element	3.56
l_1	distance between of each antenna element	3.5
l_2	length of each element	6.65
l_3	distance from upper reflector to upper antenna element	3.57

Tab. 2. Optimal parameters of the dipole antenna with the reflector.

Property	1.8 GHz	5.2 GHz
Angular beamwidth (degree)	42.1	22.9
Main lobe direction (degree)	0	29
Back lobe (dB)	−10.1	−23.29
Side lobe (dB)	−20.8	−13.3
Gain (dBi)	10.71	8.37

Tab. 3. Radiation properties of the antenna with the reflector.

Figure 12 shows the simulated radiation patterns which are directional at both frequency bands. The simulated gain at 1.8 GHz is 10.7 dB at direction 0 degree and size of main lobe is 42.1 degree, while at 5.2 GHz the gain is 8.37 dB at the best direction of 22.9 degree. All key parameters of the dipole antenna with the reflector in Fig. 1 are shown in Tab. 2. Also the important characteristics of the antenna are included in Tab. 3. This antenna could be used for indoor applications.

2.2 Design of Array Antennas

From the previous subsection, it is found that the low antenna gain is obtained. To develop for higher gain, the array structure must be employed. Therefore, two dipole antennas were connected to be an array as shown in Fig. 13. The distance l_1 between each antenna affects the radiation patterns and gains. It is found that the appropriate distance is 3.5 cm, resulting in highest gain at both resonance frequencies. Table 4 shows all key parameters of the array antenna. Simulated $|S_{11}|$ of the proposed array antenna is shown in Fig. 14, which has dual resonance frequencies as desired at 1.8 GHz and 5.2 GHz. There are bandwidths of 1.67–1.89 GHz (220 MHz) and between 5.1 and 5.39 GHz (290 MHz), respectively, which are equal to the single dipole antenna. The simulated radiation patterns are also shown in Fig. 15. It is clearly seen that the array antenna radiates the directional patterns, the same as the proposed dipole antenna patterns but higher gain. The gains of array antenna are 12.6 dB at 1.8 GHz and 9.84 dB at 5.2 GHz. With these higher gains, the proposed array antenna may be suitable for outdoor applications.

The radiation pattern characteristics of array antenna with two elements are concluded in Tab. 5. From this table, it is shown that the directions of main lobes are at the same for both bands, but it is split into two maximums at 23 degree for the second band.

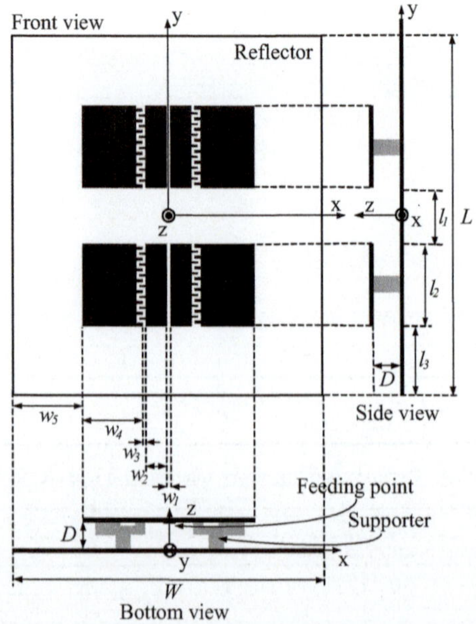

Fig. 13. Layout of the array antenna with two elements.

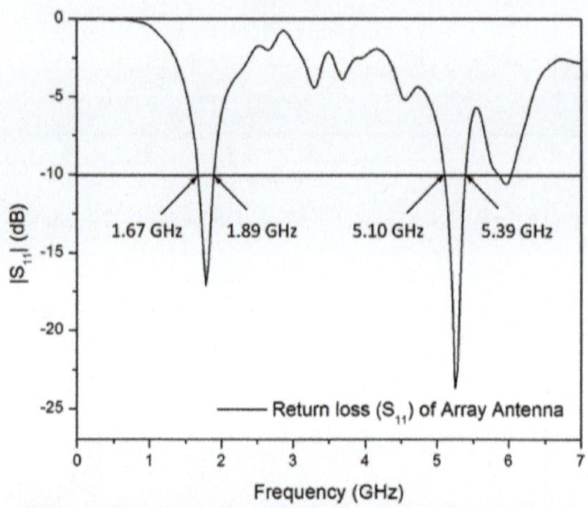

Fig. 14. Simulated result of array antenna with two elements.

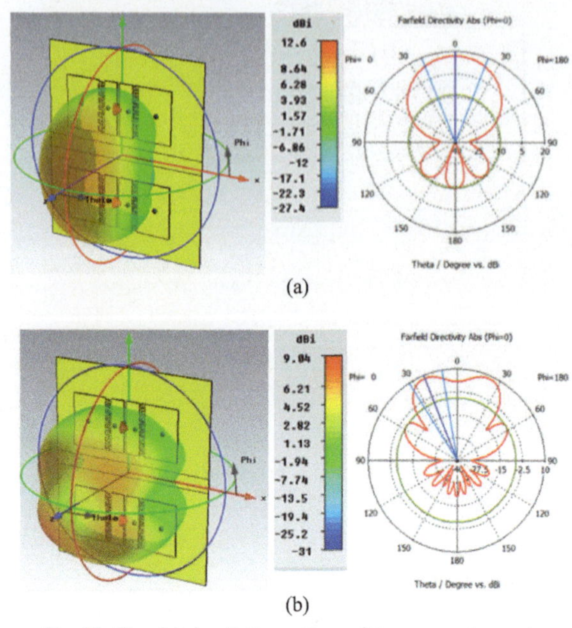

Fig. 15. Simulated radiation pattern of the array antenna at: (a) 1.8 GHz and (b) 5.2 GHz.

Parameter	Description	Size(cm)
L	length of reflector	24.5
W	width of reflector	21.5
D	distance between array antenna and reflector	1.4
w_1	feeding gap	0.3
w_2	width of smaller side of each element	1.8
w_3	width of incision gap	0.25
w_4	width of larger side of each element	5.04
w_5	distance from reflector to antenna element	3.56
l_1	distance between of each antenna element	3.5
l_2	length of each element	6.65
l_3	distance from upper reflector to upper antenna element	3.57

Tab. 4. Parameters of the array antenna with two elements.

Property	1.8 GHz	5.2 GHz
Angular beamwidth (degree)	48	33.25
Main lobe direction (degree)	0	23
Back lobe (dB)	–13	–19
Side lobe (dB)	–19.5	–12.3
Gain (dBi)	12.6	9.84

Tab. 5. Properties of radiation pattern for the array antenna with two elements.

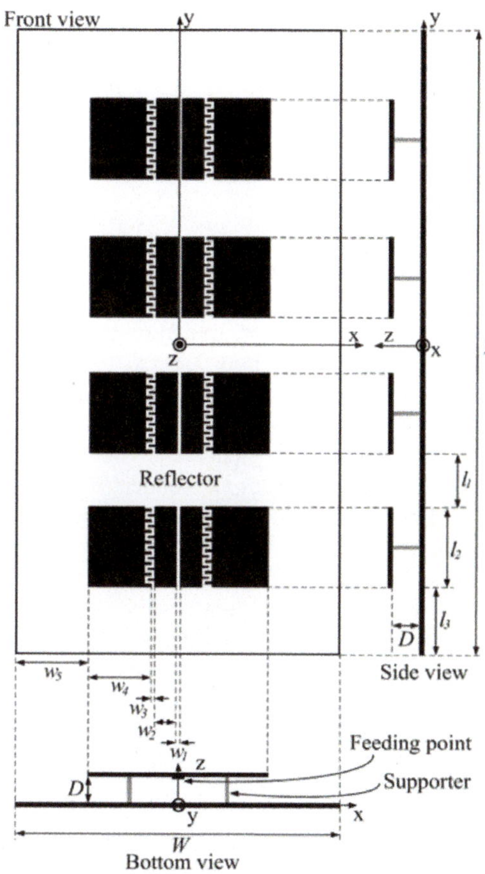

Fig. 16. Layout of the array antenna with four elements.

To further increase the antenna gain suitable for a standard base station, the array antenna with 4 elements was presented as a layout shown in Fig. 16. The simulated result of the array antenna with four elements is shown in Fig. 17. The resonance frequencies are still at 1.8 GHz and 5.2 GHz and the bandwidths are 1.68–1.88 GHz (260 MHz) and 5.14–5.41 GHz (270 MHz), respectively.

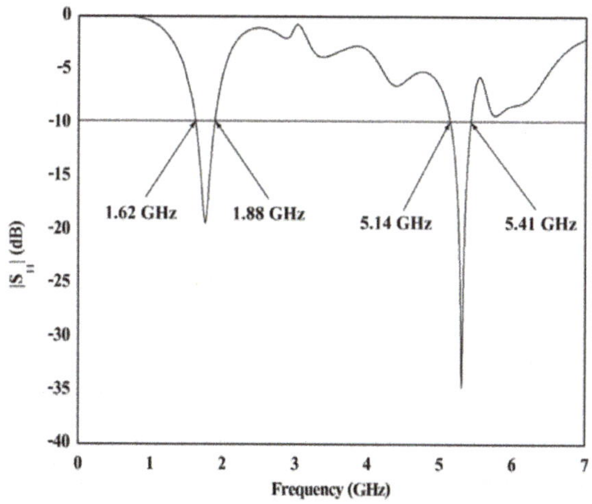

Fig. 17. Simulation result of the array antenna with 4 elements.

Parameter	Description	Size(cm)
L	length of reflector	55.85
W	width of reflector	21.5
D	distance between array antenna and reflector	1.4
w_1	feeding gap	0.3
w_2	width of smaller side of each element	1.8
w_3	width of incision gap	0.25
w_4	width of larger side of each element	5.1
w_5	distance from reflector to antenna element	4.3
l_1	distance between of each antenna element	6.75
l_2	length of each element	6.75
l_3	distance from upper reflector to upper antenna element	4.3

Tab. 6. Parameters of the array antenna with 4 elements.

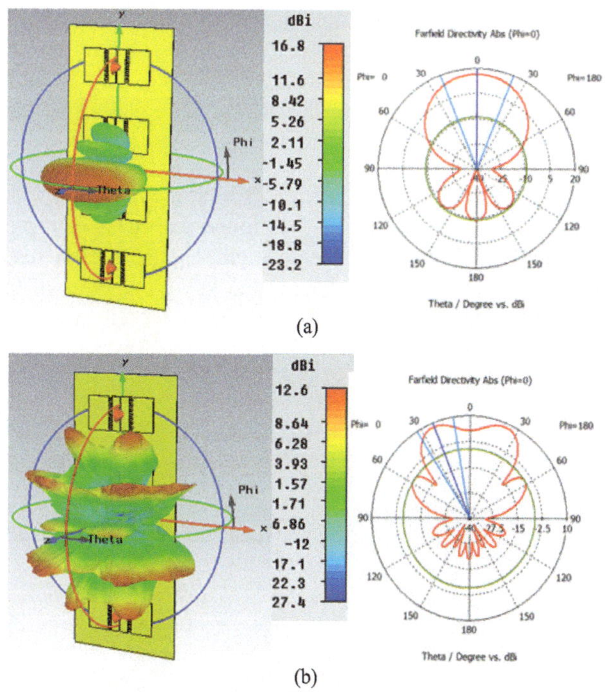

Fig. 18. Simulated radiation patterns of the array antenna with four elements at (a) 1.8 GHz and (b) 5.2 GHz.

Property	1.8 GHz	5.2 GHz
Angular beamwidth (degree)	43.5	22.3
Main lobe direction (degree)	0	22
Back lobe (dB)	–8.87	–20.25
Side lobe (dB)	–25.7	–14.2
Gain (dBi)	16.84	12.6

Tab. 7. Properties of radiation pattern for the array antenna with four elements.

The simulated radiation patterns are also shown in Fig. 18. It is found that at 1.8 GHz, the antenna radiates directionally at 0 degree with gain of 16.8 dB as shown in Fig. 18(a). In Fig. 18(b), the main lobe is split into two maximums in the direction of 22 degree at 5.2 GHz with gain of 12.6 dB. The size of the proposed array antenna is 23.2 cm × 55.8 cm. Table 6 concludes all key parameters of the antenna structure in Fig. 16. The properties of radiation pattern are also shown in Tab. 7. This proposed antenna can be potentially used for outdoor applications as a dual band base station antenna because of its good performances, especially high gain and high power handling capacity.

The proposed array antenna is smaller than the conventional base station antenna. Also, it is easy to design and control harmonic frequency or the second band frequency as needed. Next, antenna fabrication and measurement results will be presented.

3. Implementation and Results

After designing the proposed antenna as mentioned in previous sections, this section presents the antenna fabrication. The interdigital dipole antenna with the reflector was firstly constructed and then the array antenna using interdigital dipole antenna with two elements was implemented in order to verify the concept of gain improvement. The proposed dipole antenna with the reflector fabricated on copper plates with thickness of 1.5 mm was fed by the coaxial RG-142 cable with 50 Ω characteristic impedance. Another end of the coaxial cable was connected with SMA-51 as shown in Fig. 19. Figure 20 shows the picture of the completed antenna prototype. The proposed interdigital dipole antenna as shown in Fig. 21 was then measured by using Agilent 8791ES network analyzer, resulting in the same resonant frequencies as simulated.

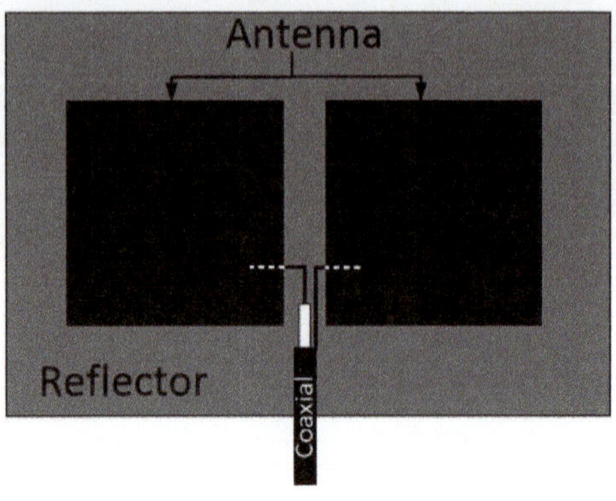

Fig. 19. Schematic of the proposed dipole antenna.

Fig. 20. Photograph of the fabricated dipole antenna.

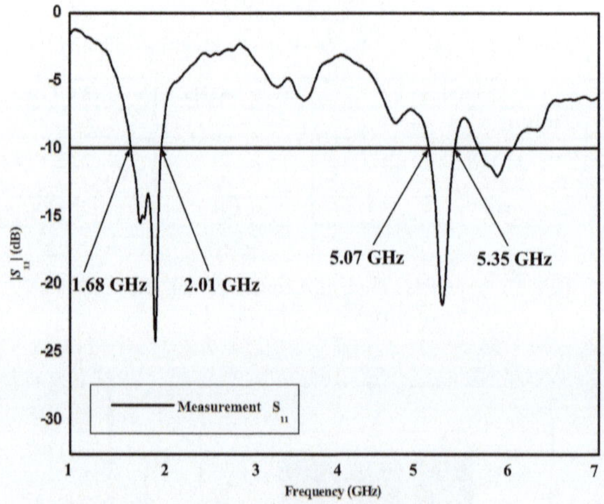

Fig. 21. Measured result of the dipole antenna.

From the measurement results, the values of return losses S_{11} at the fundamental and the second harmonic frequencies are –24 dB and –23 dB, respectively. The bandwidths are 300 MHz (1.68–2.01 GHz) and 280 MHz (5.07–5.35 GHz), respectively. Radiation patterns as shown in Fig. 22 are directional in both XZ and YZ planes for co-polarization and non-directional in both XZ and YZ planes for cross-polarization. The array antenna with two elements was then implemented as schematic shown in Fig. 23, which two interdigital dipoles are connected using RG-142 coaxial cable with 50 Ω. A power splitter was employed to combine those two dipole elements and then the proposed array antenna was mounted above the reflector using POM. The fabricated array antenna was measured by Agilent 8791ES network analyzer, resulting in the same resonant frequencies as simulated. From the measurement results as shown in Fig. 25, the values of return losses S_{11} are –17 dB and –24 dB for the fundamental and the second resonant frequencies, respectively.

The bandwidths of both bands are 390 MHz (1.62 to 2.01 GHz) and 620 MHz (4.93–5.55 GHz), respectively. For radiation patterns at 1.8 GHz, there are directional patterns on both XZ and YZ planes for co-polarizations with small back lobe and lower cross-polarization as shown in Fig. 26(a). At 5.2 GHz, the radiation patterns are also directional with lower cross-polarization on both XZ and YZ planes. That means the proposed array antenna still has directional patterns with high gain and works well for co-polarizations.

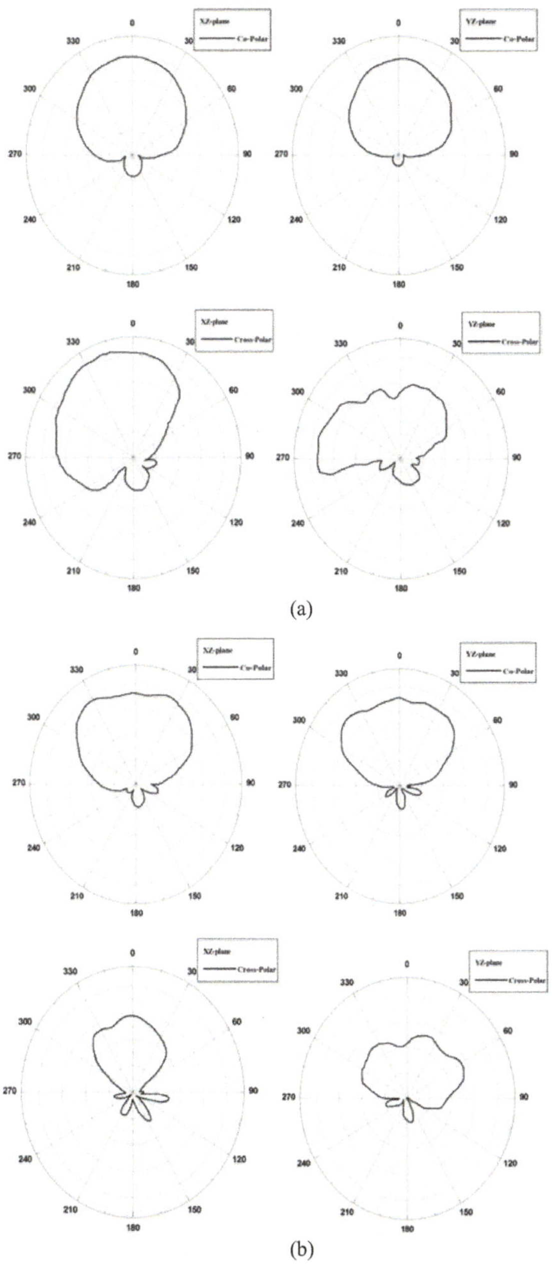

Fig. 22. Measured radiation patterns of the dipole antenna at (a)1.8 GHz and (b) 5.2 GHz.

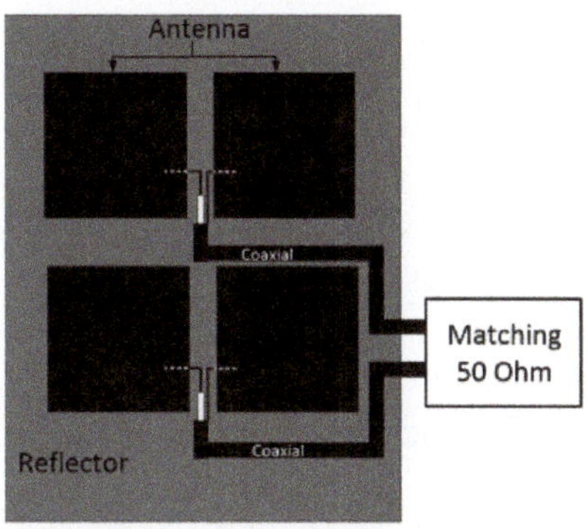

Fig. 23. Schematic of the proposed array antenna.

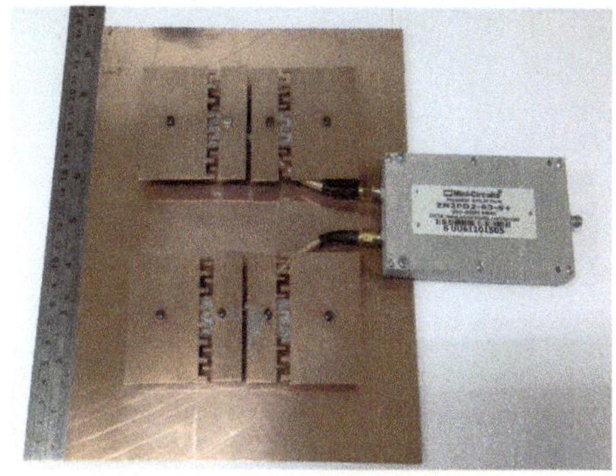

Fig. 24. Photograph of the fabricated array antenna.

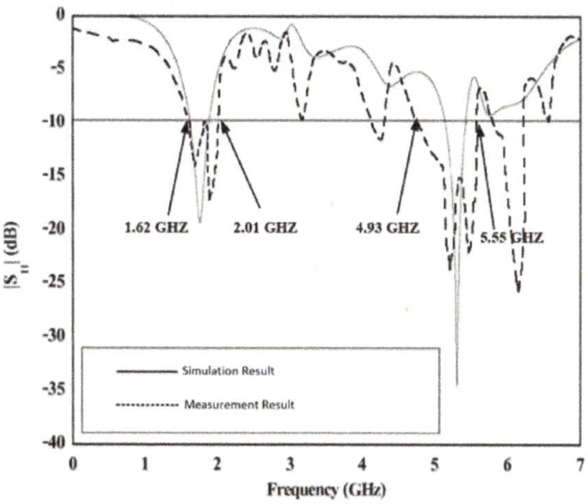

Fig. 25. Comparison of $|S_{11}|$ between simulation and measured result of the array antenna.

4. Conclusions

The interdigital dipole antenna has been designed and it is found that by using the proposed technique the second harmonic band can be freely tuned as desired by changing interdigital capacitive value. The proposed antennas can be operated at 1.8 GHz and 5.2 GHz bands. The results show good agreement between simulation and measurement with good impedance matching (return loss below –10 dB), good impedance bandwidth, high gain and high radiation efficiency. The proposed antennas also demonstrate good performances with high power handling capacity because of its copper plate structure and they can support LTE and WLAN systems as indoor and outdoor base station antennas.

Acknowledgments

This work has been supported by the Thailand Research Fund through the TRF Grant for New Researcher No. TRG5780231 and the TRF Senior Research Scholar Program Grant No. RTA5780010.

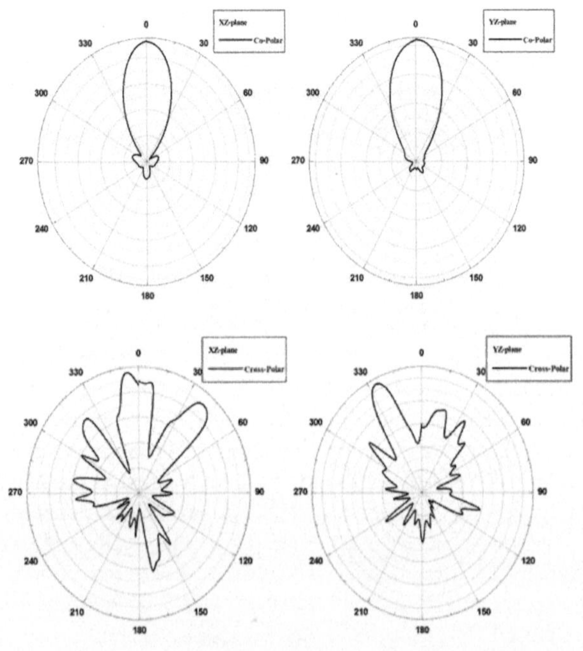

(a)

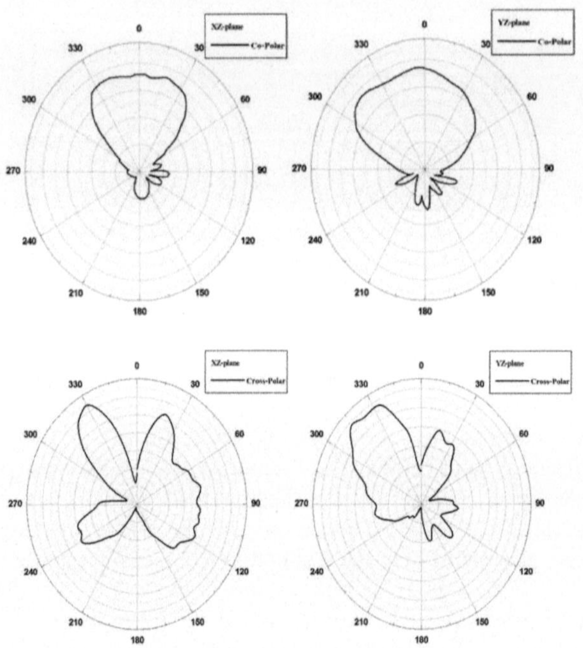

(b)

Fig. 26. Measured radiation patterns of the array antenna: (a) 1.8 GHz and (b) 5.2 GHz.

References

[1] KINGSLEY, N., ANAGNOSTOU, D. E., TENTZERIS, M., PAPAPOLYMEROU, J. RF MEMS sequentially reconfigurable Sierpinski antenna on a flexible organic substrate with novel DC-biasing technique. *Journal of Microelectromechanical Systems*, 2007, vol. 16, no. 5, p. 1185–1192. DOI: 10.1109/JMEMS.2007.902462

[2] WERNER, D. H., GANGULY, S. An overview of fractal antenna engineering research. *IEEE Antennas and Propagations Magazine*, 2003, vol. 45, no. 1, p. 38–57. DOI: 10.1109/MAP.2003.1189650

[3] MARAGHECHI, P., ELEZZABI, A. Y. Fractal structures for THz radiation emitters. *IEEE Journal of Selected Topics in Quantum Electronics*, 2013, vol. 19, no. 1, p. 8400310 (10 p.). DOI: 10.1109/JSTQE.2012.2198196

[4] GUNDETI, P., GEORGE, T. K. Performance analysis of a compact directional monopole antenna. In *Applied Electromagnetics Conference (AEMC)*. Kolkata (India), 2011, 4 p. DOI: 10.1109/AEMC.2011.6256794

[5] HONG-YAN LIANG, HONG-CHUN YANG, JING ZHANG. A cylindrical conformal directional monopole antenna for borehole radar application. *IEEE Antennas and Wireless Propagation Letters*, 2012, vol. 11, p. 1525–1528. DOI: 10.1109/LAWP.2012.2231852

[6] MOODY, R. A., SHARMA, S. K. Ultrawide bandwidth (UWB) planar monopole antenna backed by novel pyramidal-shaped cavity providing directional radiation patterns. *IEEE Antennas and Wireless Propagation Letters*, 2011, vol. 10, p. 1469–1472. DOI: 10.1109/LAWP.2011.2179513

[7] LI WEI, QIU JINGHUI, SUO YING. Design and simulation of novel ultra wideband planar reflector antenna. In *International Conference on Microwave and Millimeter Wave Technology (ICMMT)*. Builin (China), Apr 2007, 4 p. DOI: 10.1109/ICMMT.2007.381265

[8] AGARWAL, K., NASIMUDDIN, ALPHONES, A. Wideband circularly polarized AMC reflector backed aperture antenna. *IEEE Transactions on Antenna and Propagation*, 2013, vol. 61, no. 3, p. 1456–1461. DOI: 10.1109/TAP.2012.2227446

[9] DONG-ZO KIM, HYUN-SUNG TAE, KYOUNG-SUB OH, MOON-QUE LEE, JONG-WON YU. Helical reflector antenna with a wideband CP for RFID reader. In *Asia Pacific Microwave Conference*. Singapore, Dec 2009. DOI: 10.1109/APMC.2009.5384359

[10] JAYASINGHE, J. M. J. W., ANGUERA, J., UDUWAWALA, D. N. Genetic algorithm optimization of a high-directivity microstrip patch antenna having a rectangular profile. *Radioengineering*, 2013, vol. 22, no. 3, p. 700-707.

[11] GAO FENG, LEI ZHOU, ZHAO YAN, SHAN RUNHONG, QI YIHONG, HU YANG. Influence of cable routing for indoor multi-system omnidirectional antenna horizontal pattern ripple measurement. In *2014 IEEE International Symposium on Electromagnetic Compatibility (EMC)*. Raleigh (NC, USA), 2014, p. 418–422. DOI: 10.1109/ISEMC.2014.6899008

[12] LEI ZHOU, YONGCHANG JIAO, YIHONG QI, ZIBIN WENG, LIANG LU. Wideband ceiling-mount omnidirectional antenna for indoor distributed antenna systems. *IEEE Antennas and Wireless Propagation Letters*, 2014, vol. 13, p. 836–839. DOI: 10.1109/LAWP.2014.2319087

[13] QI WU, XIAOJING DING, AIXIN CHEN. A broadband dipole antenna for multiservice indoor distribution antenna system (MS-IDAS). *IEEE Antennas and Wireless Propagation Letters*, 2015, vol. 14, p. 839–842. DOI: 10.1109/LAWP.2014.2381876

[14] NAYAN, M.K., JAMLOS, M. F., JAMLOS, M. A., LAGO, H. MIMO 2x2 RHCP array antenna for point-to-point communication. In *IEEE Symposium on Wireless Technology and Applications (ISWTA)*. Kota Kinabalu (Malaysia), 2014, p. 121 to 124. DOI: 10.1109/ISWTA.2014.6981169

[15] DONG-FANG GUAN, CAN DING, ZU-PING QIAN, YING-SONG ZHANG, WEN-QUAN CAO, DUTKIEWICZ, E. An SIW-based large-scale corporate-feed array antenna. *IEEE Transactions on Antennas and Propagation*, 2015, vol.63, no.7, p. 2969–2976. DOI: 10.1109/TAP.2015.2430369

[16] ZUBING WU, JIN PAN, SHA HU, SHUBO WEN, DEQIANG YANG. Design of a novel dual polarization wideband base station antenna. In *IEEE Symposium on Intelligent Signal Processing and Communication Systems (ISPACS)*. Chengdu (China), 2010. DOI: 10.1109/ISPACS.2010.5704705

[17] HUIQING ZHAI, QIQIANG GAO, CHANGHONG LIANG, RONGDAO YU, SHENG LIU. A dual-band high-gain base-station antenna for WLAN and WiMAX applications. *IEEE Antennas and Wireless Propagation Letters,* 2014, vol. 13, p. 876 to 879. DOI: 10.1109/LAWP.2014.2321503

[18] YEJUN HE, ZHENGZHENG PAN, XUDONG CHENG, YUAN HE, JIAN QIAO, TENTZERIS, M. M. A novel dual-band, dual-polarized, miniaturized and low-profile base station antenna. *IEEE Transaction on Antennas and Propagation*, 2015, vol. 63, no. 12, p. 5399–5408. DOI: 10.1109/TAP.2015.2481488

[19] FENGLIU XU, XUEGUAN LIU, HUIPING GUO, YING WANG, LINGFENG MAO. A compact dual mode BPF base on interdigital structure. In *Proceeding of the IEEE International Conference on Microwave and Millimeter Wave Technology (ICMMT)*. Chengdu (China), 2010, p. 1595–1597. DOI: 10.1109/ICMMT.2010.5524850

[20] CHOMTONG, P., AKKARAEKTHALIN, P. A MIMO antenna using interdigital technique for LTE and Wi-MAX on mobile applications. In *International Symposium on Antennas and Propagation (ISAP*). Hobart (Australia), Nov 2015, 4 p.

[21] MEESOMKLIN, S., CHOMTONG, P., AKKARAEKTHALIN, P. A compact mutiband BPF using step-impedance resonators with interdigital capacitors. *Radioengineering*, 2016, vol. 25, no. 2, p. 258–267. DOI: 10.13164/re.2016.0258

About the Authors ...

Pongsathorn CHOMTONG was born in Nonthaburi, Thailand. He received his M.Eng. and Ph.D. degrees in Electrical Engineering from King Mongkut's University of Technology North Bangkok (KMUTNB), Thailand, in 2006 and 2011, respectively. In 2012, he joined the Department of Teacher Training in Electrical Engineering, KMUTNB, as an instructor. His current research interests include passive and active microwave circuits, wideband and multiband antennas, and telecommunication systems. He is a member of IEEE.

Prayoot AKKARAEKTHALIN received the B.Eng. and M.Eng. degrees in Electrical Engineering from King Mongkut's University of Technology North Bangkok (KMUTNB), Thailand, in 1986 and 1990, respectively, and the Ph.D. degree from the University of Delaware, Newark, USA, in 1998. From 1986 to 1988, he worked in the Micro-tek Laboratory, Thailand. In 1988, he joined the Department of Electrical Engineering, KMUTNB. His current research interests include passive and active microwave circuits, wideband and multiband antennas, and telecommunication systems. Dr. Prayoot is a members of IEEE, IEICE Japan, and ECTI Thailand. He was the Chairman for the IEEE MTT/AP/ED Thailand Joint Chapter during 2007 and 2008 and the President of ECTI Association from 2014 to 2015. He is now the head for Senior Research Scholar Project of Thailand Research Fund (TRF).

A Modified Unequally Spaced Array Antenna Synthesis Method for Side Lobe Reduction

Soroush VEISEE, Shahrooz ASADI

Dept. of Electrical Engineering, University of Shahid Beheshti, Tehran, Iran

veisee@aut.ac.ir, sh_asadi@sbu.ac.ir

Abstract. *The aim of this paper is to demonstrate the application of Unequally Spaced Arrays (USAs) in decreasing side lobe level (SLL) in linear arrays. As well known, solving of a nonlinear equation is needed in USA antenna pattern synthesis. In this paper, an improved algorithm for USA antenna pattern synthesis is presented. This method is based on converting the array factor into a triangular system of equations capable to be solved using a recursive algorithm. This method has more accuracy and speed than reported similar analytical methods based on simulation results, which leads to lower SLL and simulation time. In addition, an improvement of 3dB beamwidth in comparison with uniform spaced array can be observed.*

Keywords

Antenna arrays, unequal spacing, pattern synthesis, side lobe level

1. Introduction

Uniformly spaced array antenna synthesis techniques have been deeply investigated in past decades. Traditional synthesis techniques such as Dolph-Chebyshev [1], Taylor [2], and Fourier transform, can lead to design of a desired pattern. These synthesis procedures change the phase and magnitude of elements current to reduce the side lobe level (SLL). However, the amplitude/phase distribution can complicate the designing of the array antenna feed network. In addition, it cannot optimize the use of all elements in terms of maximum power. Non-uniformly spaced array antenna can solve this issue. This kind of array antenna has also other significant advantages such as smaller size and reduced number of elements. As a result, non-uniformly spaced array antennas are widely used in radar, sonar, and wireless communications.

Variety of numerical and analytical methods [7–20] have been introduced for unequally spaced array (USA) pattern synthesis. The first work on USA pattern synthesis was performed by Unz [3] by using a matrix formulation to obtain the distribution of elements current of an USA to form a desired pattern. Then, Harrington [4] presented a method for reducing SLL using non-uniform element spacing and showed that SLL can be reduced to the level of $2/N^{\text{th}}$ of the main lobe level. In [5], Ishimaru used Poisson's formula to reduce SLL in comparison with a linear array with uniform amplitude distribution. Miller and Goodman [6] proposed a simple and non-iterative method for USA pattern synthesis using Prony's method. In [7], [8], Kumar proposed a recursive algorithm for USA pattern synthesis based on Legendre conversion.

On the other side, stochastic methods such as Genetic Algorithm (GA) [11], Particle Swarm Optimization (PSO) and Differential Evolution Algorithm (DEA) [10] have been recently used in pattern synthesis. However, these optimization methods require large CPU time, which further increases with the increment of array element numbers.

In USA pattern synthesis, analytical methods are affordable in time because of the non-iterative property. In this paper, an analytical method for pattern synthesis of unequally spaced arrays is proposed. It is an enhanced version of the method described in [7], in terms of higher precision and speed. Compared with the method in [7], which is used for odd number of elements, the method presented in this paper with some modification, can be used also for even number of elements.

2. Theory

The geometry of a $(2N+1)$ element non-uniform linear array is shown in Fig. 1, where I_i and d_i state respectively for the element position and current.

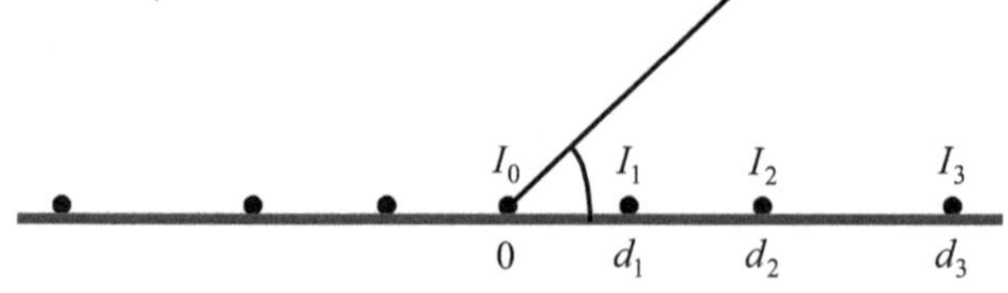

Fig. 1. Geometry of a symmetrical non-uniform linear array.

This paper proposes a new synthesis pattern approach that adjusts the element position in a linear array with a uniform current distribution, in order to quickly and accurately improve the radiation characteristics such as lower peak side lobe level (PSLL). This method can indeed determine the element positions of the array in order to fit the array pattern to a pre-specified array pattern. The desired array pattern is defined as where, $-1 \le u \le 1$ and $u = \cos\theta$, $0 \le \theta \le \pi$ radians. As known, the array pattern of a symmetric array with $(2N + 1)$ elements is defined as:

$$E(u) = \sum_{n=0}^{N} I_n \cos(kd_n u)\,. \tag{1}$$

By sampling the above relation with a sampling space $\Delta u = 1/(M - 1)$, we obtain:

$$E(u_m) = \sum_{n=0}^{N} I_n \cos(m\beta_n) \tag{2}$$

where $u_m = m\,\Delta u$ and $\beta_n = kd_n\,\Delta u$.

Because of the symmetry of the pattern in the u space this sampling is performed in $0 \le u \le 1$ interval or $m = 0,1,2,3,\ldots, N$.The element position must be chosen in such a way where:

$$E(u_m) = E_{\mathrm{d}}(u_m) \tag{3}$$

or

$$E_{\mathrm{d}}(u_m) = \sum_{n=0}^{N} I_n \cos(m\beta_n), \qquad m = 0,1,2,3,\ldots,M-1 \tag{4}$$

where $E_{\mathrm{d}}(u)$ is the desired pattern.

Relation (1) is a nonlinear equation. To solve it, both sides of (4) are multiplied by a matrix **A** defined as:

$$\mathbf{A}(\alpha) = [a_0(\alpha)\ a_1(\alpha)\ a_2(\alpha)\ \ldots\ldots a_{M-1}(\alpha)] \tag{5}$$

leading to

$$\sum_{m=0}^{M-1} a_m(\alpha) E_{\mathrm{d}}(u_m) = \sum_{n=o}^{N-1} I_n \sum_{m=0}^{M-1} a_m(\alpha)\cos(m\beta_n d) \tag{6}$$

The values of $a_m(\alpha)$ coefficients are selected as:

$$f(\alpha,\beta) = \sum_{m=0}^{M-1} a_m(\alpha)\cos(m\beta) \tag{7}$$

where [7]

$$f(\alpha,\beta) = \begin{cases} \dfrac{2}{(\cos(\beta) - \cos(\alpha))^{1/2}}, & 0 \le \beta < \alpha \\ 0 & \alpha \le \beta \le \pi \end{cases} \tag{8}$$

Also, by expressing the left side of (6) as:

$$F(\alpha_p) = \sum_{m=0}^{M-1} a_m(\alpha_p) E_{\mathrm{d}}(u_m)\,. \tag{9}$$

This later can be simplified to:

$$F(\alpha_p) = \sum_{n=0}^{N} I_n\, f(\alpha_p,\beta_n)\,. \tag{10}$$

In [7], Legendre coefficients $p_{m-0.5}(\cos\alpha)$ are used instead of $a_m(\alpha)$ in (7), which are the Fourier series coefficients of $f(\alpha,\beta)$. But in this paper the values of a_m coefficients are obtained by minimizing the error term stated below:

$$error = \sum_{i=0}^{M-1} [f(\alpha,\beta_i) - \sum_{m=0}^{M-1} a_m(\alpha)\cos(m\beta_i)]^2, \qquad \beta_i = i\,\Delta\beta, \Delta\beta = \frac{\pi}{M-1} \tag{11}$$

which leads to

$$f(\alpha,\beta_i) - \sum_{m=0}^{M-1} a_m(\alpha)\cos(m\beta_i) = 0, \quad i = 0,1,\ldots,M-1\,. \tag{12}$$

In fact, the left hand side of (7) is considered to be equal to the right hand side in β_i points, which $i = 0,1,\ldots,M - 1$.

After some manipulations, equation (12) can be stated as a matrix equation as below:

$$\mathbf{B} \times \mathbf{A}(\alpha) = \mathbf{F}\,. \tag{13}$$

Therefore, matrix $\mathbf{A}(\alpha)$ can be obtained as:

$$\mathbf{A}(\alpha) = \mathbf{B}^{-1} \times \mathbf{F} \tag{14}$$

which leads to obtain the values of $a_m(\alpha)$. In order to avoid misunderstanding, in continue, instead of $a_m(\alpha)$ which derived from (14), we use $a'_m(\alpha)$.

In (14), the matrices **B** and **F** are defined as below:

$$\mathbf{B} = \begin{bmatrix} 1 & 1 & 1\ldots\ldots\ldots\ldots\ldots\ldots1 \\ 1 & \cos(\beta_1) & \cos(2\beta_1)\ldots\ldots\cos((M-1)\beta_1) \\ 1 & \cos(\beta_2) & \cos(2\beta_2)\ldots\ldots\cos((M-1)\beta_2) \\ . & & \\ . & & \\ . & & \\ 1 & \cos(\beta_{M-1}) & \cos(2\beta_{M-1})\ldots\cos((M-1)\beta_{M-1}) \end{bmatrix}$$

and

$$\mathbf{F} = \begin{bmatrix} f(\alpha,\beta_0) \\ f(\alpha,\beta_1) \\ f(\alpha,\beta_2) \\ . \\ . \\ . \\ f(\alpha,\beta_{M-1}) \end{bmatrix}.$$

The value of α is defined as:

$$\alpha_p = \beta_p + V_p \tag{15}$$

in which the value of V_p is defined such that:

$$\alpha_p < \beta_{p+1}. \quad (16)$$

From (9) and (14), we obtain:

$$F(\alpha_p) = \sum_{n=0}^{p} I_n f(\alpha_p, \beta_n) \quad (17)$$

which can be reformatted as:

$$\begin{aligned} &F(\alpha_0) = I_0 f(\alpha_0, \beta_0) \\ &F(\alpha_1) = I_0 f(\alpha_1, \beta_0) + I_1 f(\alpha_1, \beta_1) \\ &\ldots\ldots \\ &F(\alpha_p) = \sum_{n=0}^{p-1} I_n f(\alpha_p, \beta_n) + I_p f(\alpha_p, \beta_p) \end{aligned} \quad (18)$$

The locations of the array elements for a defined current distribution can be then obtained from the above equation using a recursive algorithm. In the proposed algorithm, the distance between elements was selected as:

$$0.5\lambda \le d_p - d_{p-1} \le 0.5\lambda + \Delta(p), \quad p = 1,2,\ldots,N \quad (19)$$

where $\Delta(p)$ is the variable space broadening factor for the pth element. The lower limit of (17) is able to prevent mutual coupling effects and the upper limit is determined in order to prevent from grating lobes. It is noted that in comparison with [7] which used a constant value for $\Delta(p)$, in this paper we obtained an optimized value for each value of $\Delta(p)$. Let us define some parameters such as:

d_0, the place of the first element, assumed to be zero, which leads to $\beta_0 = 0$,

I_0, the normalized value of the first element current as:

$$I_0 = \frac{F(\alpha_0)}{f(\alpha_0, \beta_0)}, \quad (20)$$

α such as:

$$\alpha_0 = k(d_0 + \frac{\lambda}{2})\Delta u \quad (21)$$

and

$$\alpha_p = k(d_{p-1} + \frac{\lambda}{2} + \Delta(p))\Delta u. \quad (22)$$

Before starting the algorithm, the initial values of all $\Delta(p)$ are considered to be equal with Δ, where Δ is a constant space broadening factor to be specified by the user. Then the values of $\Delta(p)$ should be optimized in order to minimize PSLL. This algorithm includes N steps. In the nth step, the optimized value of $\Delta(n)$ is obtained and the other values of $\Delta(p)$, $p \neq n$ are considered as below:

For $p > n$, $\Delta(p) = \Delta$.

For $p < n$, $\Delta(p)$ is equal to the optimized value obtained in previous steps.

In order to get the optimized value of $\Delta(n)$, the value of $\Delta(n)$ varies from 0 to 0.5λ with 0.01λ step size. For each value of $\Delta(n)$, by utilizing the following procedure, a set of element positions ($d_0, d_1, \ldots, d_N$) can be obtained.

Assuming $\beta_0 = 0$, the values of β_p for $p = 1,2,3,\ldots,N$, are calculated as [7]:

$$\beta_p = \cos^{-1}\left(g(\alpha_p)\right),$$
$$g(\alpha_p) = \frac{2I_p^{\,2}}{\left[F(\alpha_p) - \sum_{n=0}^{p-1} I_n f(\alpha_p, \beta_n)\right]^2} + \cos(\alpha_p). \quad (23)$$

Thus,

$$d_p = \beta_p / (k\Delta u).$$

For each value of d_p from (23), the conditions below should be applied.

- If d_p is a complex number, $d_p = d_{p-1} + 0.5\lambda$.
- If d_p is larger than α_p, $d_p = d_{p-1} + 0.5\lambda$.
- If the element spacing between d_p and d_{p-1} is less than 0.5λ, $d_p = d_{p-1} + 0.5\lambda$.

Then, utilizing the obtained set of element positions from above procedure and using (1), the array pattern and PSLL of the related set of positions can be acquired. Therefore, in the nth step, for each value of $\Delta(n)$, there is a related value of PSLL. The optimized value of $\Delta(n)$ is the one which has the minimum related value of PSLL. This selected value of $\Delta(n)$ should be used for the $(n+1)^{th}$ step.

In order to apply the mentioned method to an array of even number of elements ($2N$), α_0 and d_0 should be selected as:

$$\alpha_0 = k(\Delta_0 + \frac{\lambda}{2})\Delta u, \; d_0 = 0.25\lambda. \quad (24)$$

In the above equation, Δ_0 is called the initial space broadening factor and its value is changed from 0 to λ. In this way, Δ_0 and Δ should be initially defined by users.

3. Numerical Results

In this section, the presented method is demonstrated through examples. In all these examples, it is assumed that the current distribution amplitude is uniform and unequally spacing is used in order to decrease SLL. The desired pattern to design an unequally spaced array with q elements is defined as:

$$E_d(u) = \begin{cases} 1 & 0 \le u \le u_m \\ E_{min} & u_m \le u \le 1 \end{cases} \quad (25)$$

The value of E_{min} is typically selected as 1×10^{-3}. It is shown that this value should be decreased with the element number increment. u_m is equal to the place of the first pattern null of an equally spaced array with the same number of elements q.

As mentioned before, the proposed method in this paper is more accurate vs. similar existing methods [7]. It is noted that the right hand side of (7) is an estimation of $f(\alpha,\beta)$ function defined in (8). Using the $a'_m(\alpha)$ coefficients and also the Legendre coefficients $p_{m-0.5}(\cos\alpha)$ [7], instead of $a_m(\alpha)$ in the right hand side of (7), two approximate function can be obtained to estimate the function $f(\alpha,\beta)$ defined in (8). In Fig. 2, these two approximate functions and $f(\alpha,\beta)$ in (8) are plotted for $M = 107$ and $\alpha = 1$. As can be seen, the $a'_m(\alpha)$ coefficients can better estimate $f(\alpha,\beta)$ than the Legendre coefficients. Another advantage of this method is its speediness since the $a'_m(\alpha)$ coefficients can be determined faster than $p_{m-0.5}(\cos\alpha)$. The reason is that in [7], it is needed to calculate the integral operation for M times to get the values of the Legendre coefficients:

$$p_{m-0.5}(\cos\alpha) = \frac{1}{\pi}\int_0^{\alpha} \frac{2}{(\cos(\beta)-\cos(\alpha))^{1/2}} d\beta \tag{26}$$

but in our proposed algorithm, it is needed just to obtain the reverse of matrix **B** which is an $M \times M$ matrix. This speediness makes the proposed approach more suitable for arrays with large number of elements. For example, the comparison of measured simulation time in both cases, shows that $a'_m(\alpha)$ coefficients can be determined 37 times faster than $p_{m-0.5}(\cos\alpha)$ for $M = 107$ and $\alpha = 1$.

To validate the proposed synthesis method, a linear unequally spaced array of 39 elements was evaluated. In its pattern, shown in Fig. 3, the SLL has been decreased by 8.15 dB in comparison with a uniform array with the same number of elements. In addition, this value was improved by 1.75 dB while compared to the one obtained by the method described in [7]. Also as can be seen, the beamwidth is narrower than the case with uniform space. The 3-dB beamwidth of the uniform and non-uniform arrays are 2.52° and 2.25° respectively. In this algorithm, the values of Δ and M were set to 0.33λ and 107, respectively. The obtained elements positions are reported in Tab. 1. As can be inferred from (23), in order to obtain the value of element positions ($d_0, d_1,\ldots, d_N$), we need to calculate the value of $F(\alpha_p)$ function with using (9) and then calculating the values of $a_m(\alpha)$. As mentioned previously, since the $a'_m(\alpha)$ coefficients can be determined faster than $p_{m-0.5}(\cos\alpha)$, therefore, using $a'_m(\alpha)$ instead of $p_{m-0.5}(\cos\alpha)$ makes our proposed method faster than the Legendre method. For the linear unequally spaced array of 39 elements discussed above, with $\Delta = 0.33\lambda$ and $M = 107$, the measured simulation time shows that for this case, the simulation speed of our proposed method is 79 time faster in comparison with the Legendre method. Table 2 provided information about the simulation time of the proposed method and Legendre method, respectively, for different number of elements. In this case, the values of Δ and M were considered to be 0.4λ and 120. According to the data shown in Tab. 2, it can be deduced, with increasing the number of elements the simulation time of the proposed method increases with a slower rate in comparison with Legendre method. As a result, the proposed method is an appropriate analytical method for synthesis of large unequally spaced array.

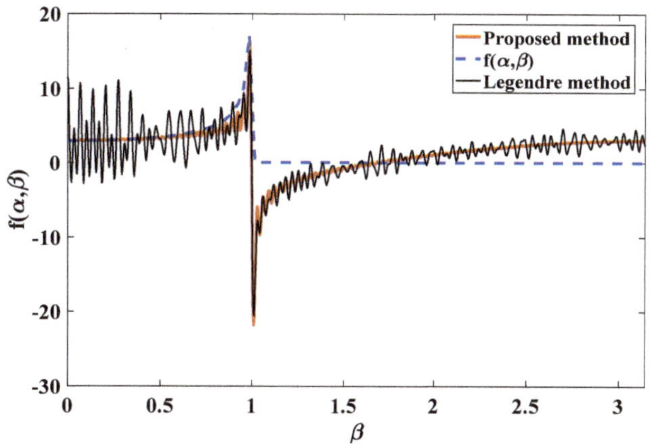

Fig. 2. Estimation of $f(\alpha,\beta)$ function using the proposed method and Legendre method [7].

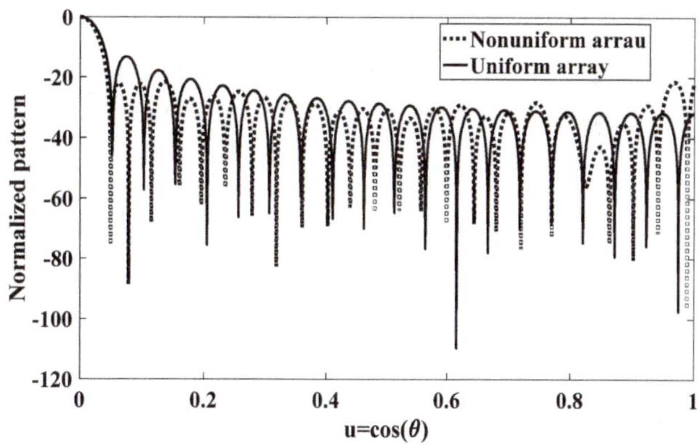

Fig. 3. Synthesized pattern of a 39 elements nonuniform array.

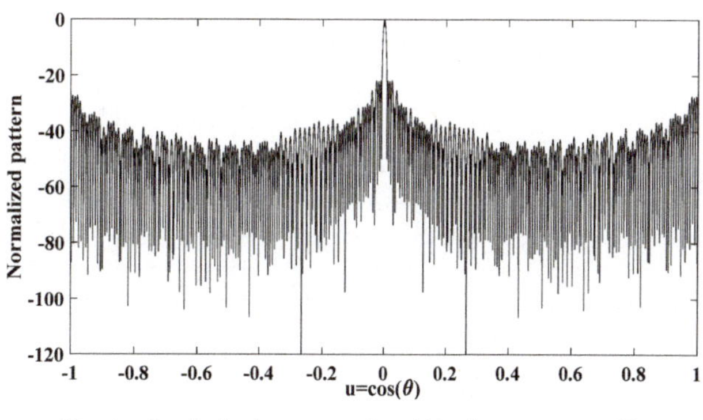

Fig. 4. Synthesized pattern of a 200 elements nonuniform array.

Element number	Element position	Element number	Element position	Element number	Element position
p	d_p/λ	p	d_p/λ	p	d_p/λ
1	0.5	8	4.00	15	8.63
2	1.0	9	4.52	16	9.46
3	1.5	10	5.12	17	10.41
4	2.0	11	5.71	18	11.32
5	2.5	12	6.40	19	12.12
6	3.0	13	7.12		
7	3.5	14	7.87		

Tab. 1. Element position for 39 elements array.

2*N*+1 element array	Simulation time (seconds)	
N	Proposed Method	Legendre Method
9	19.5	1052
14	50.2	3295.7
19	86.5	6644.5
24	139.0	11837

Tab. 2. Simulation time of the proposed method and Legendre method for different number of array elements.

Element number	Element position	Element number	Element position	Element number	Element position
p	d_p/λ	p	d_p/λ	p	d_p/λ
1	0.2500	35	17.6657	69	39.2361
2	0.7500	36	18.1657	70	39.9993
3	1.3000	37	18.6657	71	40.7753
4	1.9516	38	19.1657	72	41.5557
5	2.5571	39	19.7145	73	42.3460
6	3.1529	40	20.2219	74	43.1430
7	3.6657	41	20.7825	75	43.9469
8	4.1657	42	21.3521	76	44.7617
9	4.6657	43	21.9279	77	45.5825
10	5.1657	44	22.5179	78	46.4118
11	5.6657	45	23.1082	79	47.2480
12	6.1657	46	23.7168	80	48.0910
13	6.6657	47	24.2958	81	48.8706
14	7.1657	48	24.9018	82	49.7688
15	7.6657	49	25.5218	83	50.6699
16	8.1657	50	26.1451	84	51.5754
17	8.6657	51	26.7853	85	52.4859
18	9.1657	52	27.4052	86	53.4023
19	9.6657	53	28.0433	87	54.3240
20	10.1657	54	28.6978	88	55.2507
21	10.6657	55	29.3386	89	56.1823
22	11.1657	56	29.9970	90	57.1186
23	11.6657	57	30.6702	91	58.0594
24	12.1657	58	31.3342	92	59.0043
25	12.6657	59	32.0142	93	59.9532
26	13.1657	60	32.7085	94	60.9062
27	13.6657	61	33.3974	95	61.8625
28	14.1657	62	34.1000	96	62.8219
29	14.6657	63	34.8174	97	63.7842
30	15.1657	64	35.5323	98	64.7491
31	15.6657	65	36.2583	99	65.7165
32	16.1657	66	36.9881	100	66.6862
33	16.6657	67	37.7323		
34	17.1657	68	38.4785		

Tab. 3. Element position for 200 elements array.

To further demonstrate the proposed approach, a larger array with 200 elements (even number) was studied. The values of Δ_0, Δ, and M are 0.06λ, 0.46λ, and 237, respectively. As shown in Fig. 4, the PSLL is equal to –21.9 dB, i.e., an improvement of 8.64 dB in comparison with a uniform array with the same number of elements. In addition, the 3-dB beamwidth is 0.4140° which has decreased in comparison with 0.5040° beamwidth of the uniform array. The synthesized elements positions are presented in Tab. 3. Also in this case, as can be expected, the simulation time has been decreased to 1/89 of its value in comparison with the method using $p_{m-0.5}(\cos\alpha)$ instead of $a_m(\alpha)$ coefficients.

4. Conclusion

In this paper, a fast and accurate analytic method for synthesizing of unequally spaced arrays with uniform amplitude distribution was proposed. It can be implemented in arrays with both even and odd number of elements. This approach allows obtaining a radiation pattern with a narrow beamwidth and lowest PSLL value. The simulation results show a significant reduction of simulation time. In fact, the obtained results demonstrated its strong ability to design low side lobe level phased array antennas with unequally spaced elements.

References

[1] ELLIOT, R. S. *Antenna Theory and Design.* Englewood Cliffs (NJ): Prentice-Hall, 1981.

[2] BALANIS, C. A. *Antenna Theory.* New York: Wiley, 1997.

[3] UNZ, H. Linear arrays with arbitrarily distributed elements. *IRE Transactions on Antennas and Propagation,* 1960, vol. AP-8, no. 2, p. 222–223. DOI: 10.1109/TAP.1960.1144829

[4] HARRINGTON, R. F. Sidelobe reduction by nonuniform element spacing. *IRE Transactions on Antennas and Propagation,* 1961, vol. 9, no. 2, p. 187–192. DOI: 10.1109/TAP.1961.1144961

[5] ISHIMARU, A. Theory of unequally-spaced arrays. *IRE Transactions on Antennas and Propagation,* 1962, vol. AP-11, no. 6, p. 691–702. DOI: 10.1109/TAP.1962.1137952

[6] KUMAR, B. P., BRANNER, G. R. Generalized analytical technique for the synthesis of unequally spaced arrays with linear, planar, cylindrical or spherical geometry. *IEEE Transactions on Antennas and Propagation,* 2005, vol. 53, no. 2, p. 621–634. DOI: 10.1109/TAP.2004.841324

[7] KUMAR, B. P., BRANNER, G. R. Design of unequally spaced arrays for performance improvement. *IEEE Transactions on Antennas and Propagation,* 1999, vol. 47, no. 3, p. 511–523. DOI: 10.1109/8.768787

[8] JIN, N., RAHMAT-SAMII, Y. Advances in particle swarm optimization for antenna designs: Real-number, binary, single-objective and multi-objective implementations. *IEEE Transactions on Antennas and Propagation,* 2007, vol. 55, no. 3, p. 556–567. DOI: 10.1109/TAP.2007.891552

[9] KURUP, D. G., HIMDI, M., RYDBERG, A. Synthesis of uniform amplitude unequally spaced antenna arrays using the differential evolution algorithm. *IEEE Transactions on Antennas and Propagation,* 2003, vol. 51, no. 9, p. 2210–2217. DOI: 10.1109/TAP.2003.816361

[10] CHEN, K., HE, Z., HAN, C. A modified real GA for the sparse linear array synthesis with multiple constraints. *IEEE Transactions on Antennas and Propagation,* 2006, vol. 54, no. 7, p. 2169–2173. DOI: 10.1109/TAP.2006.877211

[11] DE LUCCIA, C. S., WERNER, D. H. Nature-based design of aperiodic linear arrays with broadband elements using a combination of rapid neural-network estimation techniques and genetic algorithms. *IEEE Antennas and Propagation Magazine,* 2007, vol. 49, no. 5, p. 13–23. DOI: 10.1109/MAP.2007.4395292

[12] DONELLI, M., CAORSI, S., DE NATALE, F., et al. Linear antenna synthesis with a hybrid genetic algorithm. *Progress in Electromagnetic Research PIER,* 2004, vol. 49, p. 1–22. DOI: 10.2528/PIER03121301

[13] MURINO, V., TRUCCO, A., REGAZZONI, C. S. Synthesis of unequally spaced arrays by simulated annealing. *IEEE Transactions on Signal Processing*, 2006, vol. 44, no. 1, p. 119–122. DOI: 10.1109/78.482017

[14] CAORSI, S., LOMMI, A., MASSA, A., et al. Peak side lobe level reduction with a hybrid approach based on GAs and difference sets. *IEEE Transactions on Antennas and Propagation,* 2004, vol. 52, no. 4, p. 1116–1121. DOI: 10.1109/TAP.2004.825689

[15] OLIVERI, G., MASSA, A. Bayesian compressive sampling for pattern synthesis with maximally sparse non-uniform linear arrays. *IEEE Transactions on Antennas and Propagation,* 2011, vol. 59, no. 2, p. 467–481. DOI: 10.1109/TAP.2010.2096400

[16] WANG, W. B., FENG, Q., LIU, D. Application of chaotic particle swarm optimization algorithm to pattern synthesis of antenna arrays. *Progress in Electromagnetic Research PIER*, 2011, vol. 115, p. 173–189. DOI: 10.2528/PIER11012305

[17] RUPCIC, S., MANDRIC, V., ZAGAR, D. Reduction of side lobes by non-uniform elements spacing of a spherical antenna array. *Radioengineering,* 2011, vol. 20, no. 1, p. 299–306.

[18] LIU, Y., NIE, Z. P., LIU, Q. H. A new method for the synthesis of non-uniform linear arrays with shaped power patterns. *Progress in Electromagnetic Research PIER*, 2010, vol. 107, p. 349–363. DOI: 10.2528/PIER10060912

[19] YANG, K., ZHAO, Z., LIU, Q. H. Fast pencil beam pattern synthesis of large unequally spaced antenna arrays. *IEEE Transactions on Antennas and Propagation,* 2013, vol. 61, no. 2, p. 627–634. DOI: 10.1109/TAP.2012.2220319

[20] GOUDOS, S. K., MOYSIADOU, V., SAMARAS, T., et al. Application of a comprehensive learning particle swarm optimizer to unequally spaced linear array synthesis with sidelobe level suppression and null control. *IEEE Antennas and Wireless Propagation Letters*, 2010, vol. 9, p. 125–129. DOI: 10.1109/LAWP.2010.2044552

Miniaturized Balanced Antenna with Integrated Balun for Practical LTE Applications

Issa ELFERGANI[1], Jonathan RODRIGUEZ[1,2], Fathi ABDUSSALAM[3], Chan H SEE[4], Read ABD-ALHAMEED[3]

[1]Instituto de Telecomunicações, Aveiro, Portugal
[2]University of South Wales, Pontypridd, CF37 1DL , UK
[3]Antennas & Applied Electromagnetics Research Group, School of Engineering and Informatics, Univ. of Bradford, Bradford, BD7 1DP, UK
[4]School of Engineering, University of Bolton, Bolton, BL3 5AB, UK

i.t.e.elfergani@av.it.pt, jonathan@av.it.pt, {r.a.a.abd, f.m.a.abdussalam}@bradford.ac.uk, C.See@bolton.ac.uk

Abstract. *A design of dual-band balanced antenna structure operating in the 700 and 2600 MHz LTE bands is studied and investigated. The overall dimensions of the radiator are 50 × 18 × 7 mm^3 allowing it to be easily concealed within mobile handsets. A broad-band balun is designed and integrated with the antenna handset in order to provide the feeding network and perform the measurements of the antenna radiation performance. Prototypes of the proposed antenna with and without balun are fabricated and verified. The simulated and practical results with and without the handheld effects in terms of reflection coefficient, power gain and radiation pattern, are studied and show reasonable agreement.*

Keywords

Balanced antenna, printed dipole, dual-band, LTE, balun

1. Introduction

The demand for researching and developing multi-system handset has increased rapidly in recent years. The design of a wireless transceiver in a smart phone or a portable device must support multi-system operations, since the forthcoming mobile networking ecosystem is expected to constitute legacy and future 5G technology. This raises numerous challenges and requirements for mobile phone antenna designers [1]. To combat these problems, antennas are required to be thin, light, compact design, reduced volume size, and low energy for better incorporating into a portable device. As a result, several works have recently been dedicated to internal multiband antennas in mobile handset applications. The planar Inverted-F antennas (PIFA) [2], [3], microstrip patch antennas [4], [5] and monopole printed structures [6], [7] have been considered as the most common internal mobile phone antennas. In particular, with the advancement of the LTE technology, smart phones are now broadly used, thus designing new antennas for legacy and future release of the LTE standard, catering for carrier aggregation and multiple wideband frequency bands are now of much interest in the research community. Numerous miniaturized LTE handset antennas have been reported [8–13].

By examining the work in [8–13], one can notice that, these antennas have only a single terminal and driven against a local ground plane. Having an antenna along with the system ground plane may help in improving the bandwidth and gain performance. These unbalanced antennas have been exploited in commercial handsets, especially as the mobile device is quite small. However, these unbalanced structures are sensitive to user hand held effect, in that the user´s hand covers a large area of the antenna ground plane in which may change the impedance matching requirements of the antenna and it may lead in performance degradation. This is because the ground plane of such antennas is being used as part of the radiator structure in which a large amount of current would be induced on the radiator as well as on the antenna ground. As a result and while the device is being held by the users' hand, the coupling with the human hand/body, could degrade the antenna performance [14, 15, 16].

To avoid such degradation performance phenomena, a balanced antenna is deemed as a promising candidate for the mobile phone since the current induced on the ground is small or approximately neglected, which leads to minor influences on the performances of the antenna in the scenario of device being held by the user's hand.

To exploit this beneficial property, a number of mobile phone balanced antenna structures operating in dual-band, multiband and wide band have been recently studied [16–27]. Tab. 1 shows the differences between these antennas in terms of operating frequency band, antenna size, power gain and efficiency.

By investigating Tab. 1, one can observe these available balanced antennas can only cover either the major existing mobile bands or/and UWB spectrum, for example: antenna design in [17] can only cover GSM900, the antenna geometry in [18] is capable to work for GSM1800, an antenna operating over the GSM900/1800 was proposed in [19]; whereas [16], [20] proposed antennas to operate in WLAN, authors in [22] proposed balanced antenna structure covering GSM and UMTS, while [23] designed a balanced antenna to operate in GSM and WLAN, and authors in [24] offered balanced antennas that work in the full operation of three mobile radio bands of GSM900/1800, PCS/1900 and UMTS/2200.

On the other hand, several works have been reported to operate over the UWB spectrum, e.g. authors in [25] have proposed an antenna design that operates in both lower bands of UWB from 2.36 to 2.56 GHz and higher bands of UWB spectrum from 5.13 to 12 GHz. Furthermore, in [26], [27] Vivaldi balanced antennas covering the whole range of UWB from 3.1 to 10.6 GHz have been designed and tested.

In contrast, due to the big demand for higher data rate as well as larger bandwidth in recent network of mobile communication, the new technology of the 4th generation namely long-term evolution (LTE) has been developed and newly released. However, none of these balanced antenna designs in [16], [27] have the capability to operate in the range of the lower LTE bands and in particular the lower band of 700 MHz. To address this, we propose a miniaturized printed folded dipole balanced antenna, which operates at dual-band frequency bands of LTE, i.e. 698 to 748 MHz and 2500 to 2690 MHz, for a mobile communication device.

By compromising the bandwidth, antenna size and frequency bands, some approaches were proposed by previous author's works [18], [20–24] to enable wide-band and dual-band functional operation, as shown in Tab. 1. In comparison to [16], [27], this version of the proposed balanced antenna has come up with such advantage of covering the dual-band of LTE namely 700/2600MHz as well as achieving a size reduction compared to previous work in [16–18], [24, 25]. Moreover, this present antenna has achieved a better efficiency in contrast to [16], [23]. Furthermore, it accomplishes an improved gain compared with works in [18, 22, 23, 25, 26].

Ref	Operating Frequency Band (GHz)	Size mm^3 including ground plane	Peak Gain Range (dBi)	Radiation efficiency (%)
16	2.4	118×62.5×0.8	NaN	75.3
17	0.9	100×50×10	NaN	NaN
18	1.8	120×50×12	4	NaN
19	0.9 and 1.8	100×50×6.6	NaN	
20	2.48,5,4 and 6.5	90×40×7	3.5-5.2	NaN
22	0.9, 1.8 and 2.2	120×50×9.5	2.5 -3.5	NaN
23	1.8-2.4	120×50×9.5	2.7-4.2	70-94
25	2.36 - 2.56 and 5.13 - 12	87×35×1	0.7- 5	NaN
26	3.1-10.6	32 × 35 ×1.6	-3-5.25	NaN
27	3.1-10.6	123.5×96.7×1.6	NaN	NaN
proposed	0.7-2.6	100×50×7	0.95-1.7 and 3.8 -4.9	79-95

Tab. 1. Comparison of the performance of the published balanced antennas.

2. Proposed Antenna Structure

The full configuration of the present dual-band LTE mobile balanced design is shown in Fig. 1. The present design has a simple structure. The antenna is printed over a FR4 material permittivity of 4.4, tangent loss of 0.025 and with a thickness of 1.6 mm. The overall size of the antenna and handheld device is 100 × 50 × 7 mm^3, where the antenna size is 50 × 18 × 1.6 mm^3. A balanced voltage source was used to feed the proposed design. The folded printed arms have a uniform width of 1 mm.

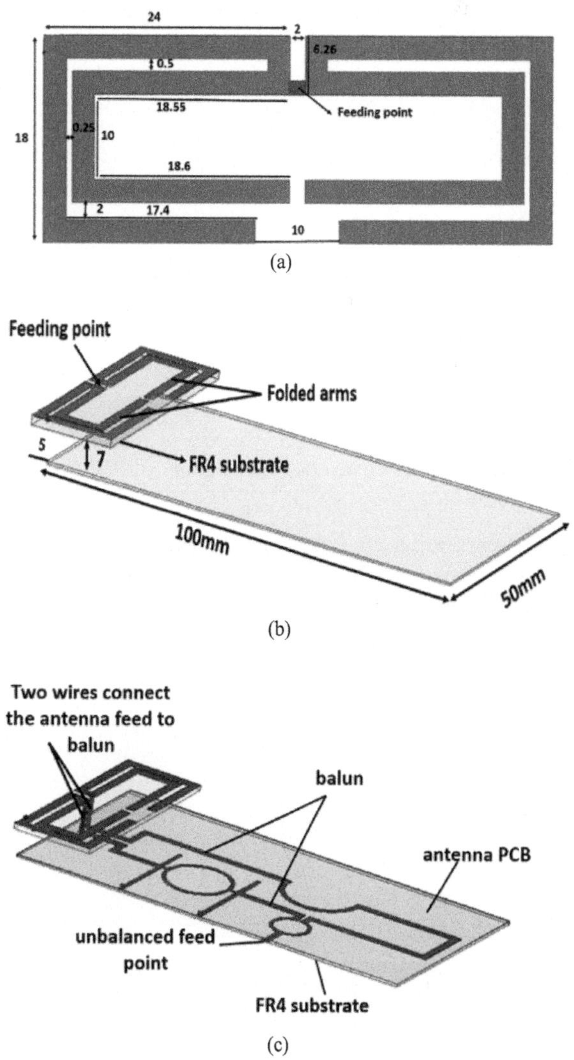

Fig. 1. Antenna structure: (a) Top view, (b) antenna without balun, (c) antenna with balun. Unit in mm.

The ground plane was shifted backward by 5 mm. This has created a defected area under the radiator which has contributed to enhance and improve the bandwidth of the LTE lower frequency of 700 MHz. The proposed antenna comprises of two printed dipoles arms with separation slot of 2 mm width. Each dipole arm is patterned in a way to create two joint U shapes. The formation of such printed dipole shapes has not only contributed towards the antenna miniaturization, but has also effectively achieved the dual-band LTE frequencies defined within this work in particular the lower band of 700 MHz. It should be noted that different set/shapes of folded arms were attempted on the top of the substrate in which it can pave the path towards the targeted dual-band frequencies of 700/2600 MHz.

To further understand the contribution of the printed folded arms technique in size miniaturization, different antenna designs with several printed folded arms shapes were modelled. In this analysis, four standard different printed arms including L-shaped, U-shaped, L-U shaped and 2U shaped were studied. The dimensions of L, U and L-U shapes are depicted in Fig. 2, while the proposed structure of 2U shapes are already shown in Fig. 1. The variation of the printed folded arms against the response of S_{11} was investigated within this study as depicted in Fig. 2. The simulated S_{11} of L-shaped, U-shaped, L-U shaped and 2U shaped arms of the proposed antenna are shown in Fig. 2. As can be seen, by implementing L-shaped radiator on each arm, the proposed antenna can only operate at 4200 MHz, but, when the U-shape was applied, the current path length of the antenna will be changed in which it will force the antenna to operate at 3500 MHz. On the other hand, by joining L and U shapes together on each side of the radiator, the dual-band paradigm started taking a place for which the proposed antenna is being able to cover the dual-band of 1000/3400 MHz. However, by employing 2U shapes configuration on each arm, the present antenna was tuned to meet the targeted dual-band of 700/2600 MHz proposed within this study.

Planar balun in [28] is used and integrated on the handset of the proposed antenna to support the balanced feeding network, as illustrated in Fig. 1c. One can notice that the ground plane of the antenna was placed on one side of the FR4 dielectric with a thickness of 0.8 mm, permittivity of 4.4, and tangent loss of 0.025, while the planar balun was located on the opposite side as depicted in Fig. 1c. The proposed antennas were modeled using HFSS software package [29]. The locations of two balanced ports of the balun were wisely designated to be exactly in direct position underneath the antenna feeding point on the upper sheet of the substrate. Dual thin cables were exploited in order to join the wide band balun to the antenna feeding point via holes. In this manner, the integration of both the antenna and its balanced feeding system were successfully accomplished. The proposed balun operates over a wider frequency range from 700 MHz to 3200 MHz in which the targeted frequency bands of 700 and 2600 MHz proposed in this work can be easily met.

To further investigate the physical behavior of the antenna, the input impedance of the proposed antennas with and without balun in free space and hand held are studied and investigated as shown in Fig. 3. The values of the input impedance of the proposed antenna for the mentioned five cases were summarized in Tab. 2.

As can be observed from Tab. 2, the proposed antenna for both free space and handheld scenarios exhibits a re-

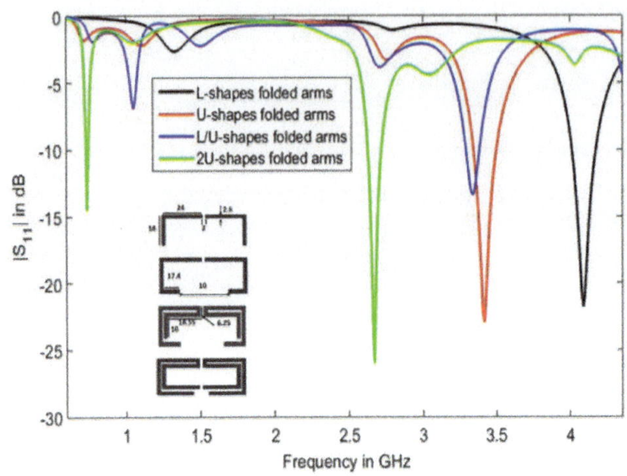

Fig. 2. The variation of the printed folded shapes against the response of S_{11}.

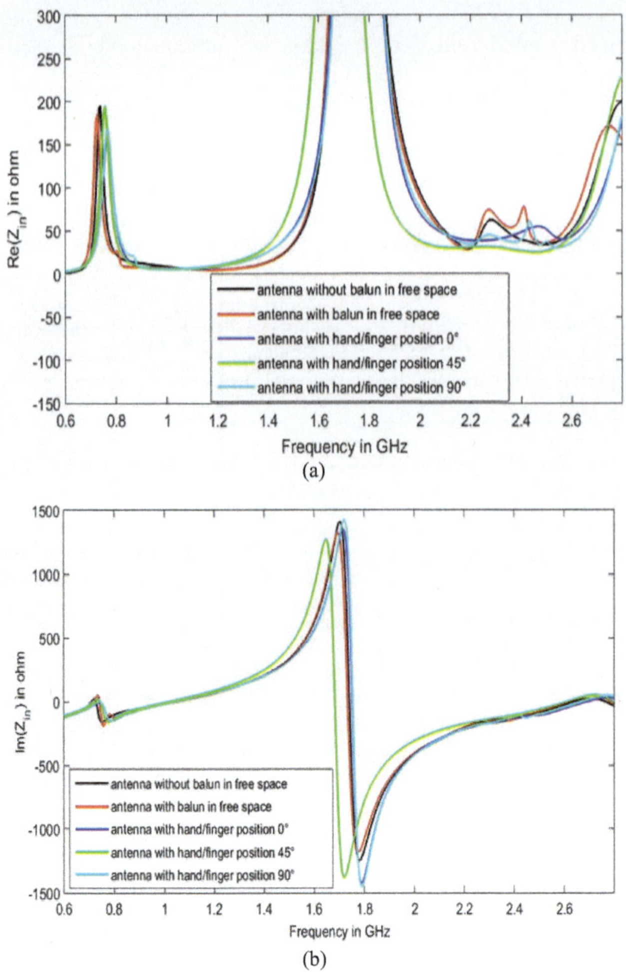

Fig. 3. Input impedance of the proposed antenna.

Impedance (Ω)	Antenna without balun in free space	Antenna with balun in free space	Antenna with finger position at 0°	Antenna with finger position at 45°	Antenna with finger position at 90°
Resistance 700 MHz	50	50	46	47	49
Reactance 700 MHz	0	0	–3	–3	–2
Resistance 2600 MHz	46	49	47	46	48
Reactance 2600 MHz	–2	–3	–3	–3	–3

Tab. 2. Input impedance of the proposed antenna in free space and handheld scenarios.

sistance of around 50 Ω (fluctuated between 46 and 50 Ω) at 700 MHz and 2600 MHz. The correspond reactance values of the five cases for the both scenarios at the dual targeted frequencies of 700/2600 MHz were varied between –3 and 0 Ω. In summary, the antenna response for both free space and hand held scenarios has shown a good impedance matching condition to a 50 Ω load.

3. Measurement and Simulation Results

The simulated reflection coefficients of the proposed designs were studied and investigated in the free space and close vicinity to human hand scenarios. The dimension of the hand model surrounding the proposed antenna is assumed as 50 × 80 × 110 mm^3. The antenna and hand model are illustrated in Fig. 4. For simplicity, the proposed hand model is considered to be a muscle tissue of only a single layer, having a relative permittivity material of 54 and a conductivity of 1.45 S·m^{-1} [22], [23].

As depicted in Fig. 4, the hand model takes three typical different configurations of holding the handset, while taking the finger positions into account, i.e., 0° (Left), 45° (middle), and 90° (Right) which are the most common talk positions.

Figure 5 depicts the computed S_{11} for the balanced antenna (i.e., the antenna with/without balun) in free space scenario and including the human hand effect. Observing Fig. 5, it is obviously seen that the $|S_{11}|$ remains below –10 dB over the targeted operational dual-band of 700/2600 MHz in free space scenarios.

On the other hand, in Fig. 5, the S_{11} of the antenna in hand effect paradigm of three positions shows approximately a stable performance in term of S_{11} and in good agreement with the free space scenario. This proves the balanced antenna is ground plane independent and can be a good candidate for practical mobile applications.

For validation purposes of the simulated S_{11} results of the antenna system without balun, a prototype antenna without the inclusion of balun is shown in Fig. 6(a) and (b). It was initially fabricated based upon on the structure and dimensions as clarified in Fig. 1(b), and then tested. Fig. 7

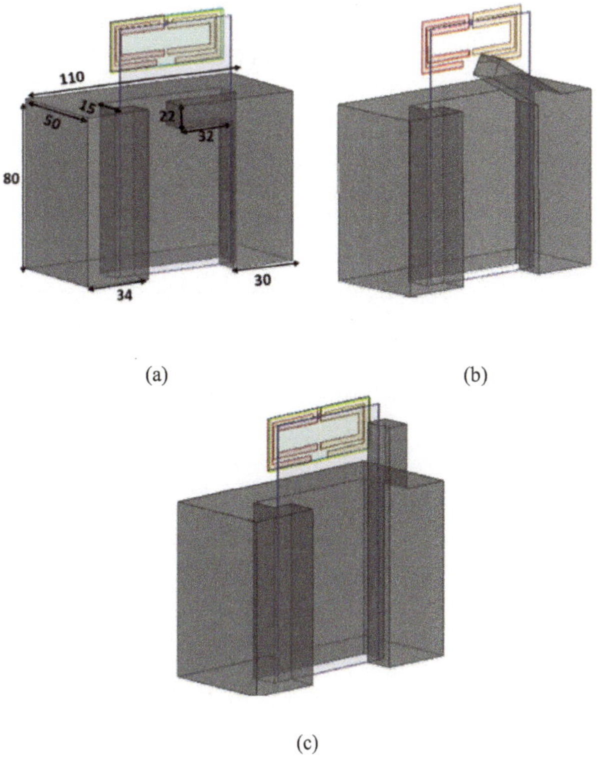

Fig. 4. Simulated hand model, with finger positions, 0° (a), 45° (b), and 90° (c).

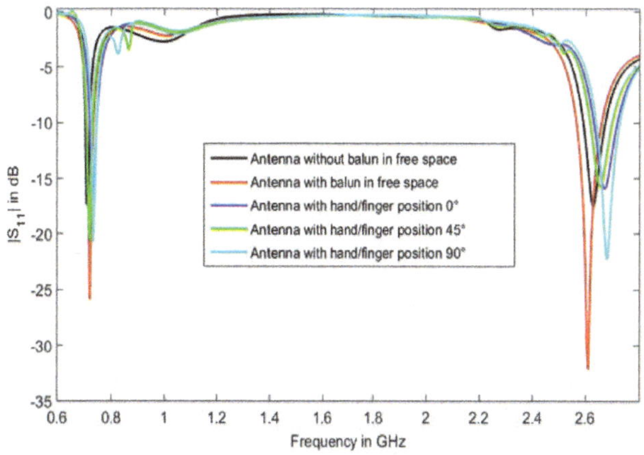

Fig. 5. Simulated reflection coefficients $|S_{11}|$ of the proposed antennas.

shows the measured S_{11} of the present balanced antennas. The S_{11} of the antenna without balun was achieved by utilizing the method of two port network analyzer whereby the integrated balun was not required. Practically, this was accomplished by direct connection of the balanced antenna two ports into the two inputs ports of a calibrated vector network analyzer. One can clearly observe that the measured results of S_{11} are said to be in fair agreement with the computed results shown in Fig. 5, where the targeted LTE dual-band, 700/2600 MHz was accomplished.

For verification purposes, the full antenna prototype assembly (antenna and balun) has been manufactured and measured as depicted in Fig. 6(c), (d). The measured S_{11} of the antenna with the integrated balun is indicated in Fig. 7.

(a)

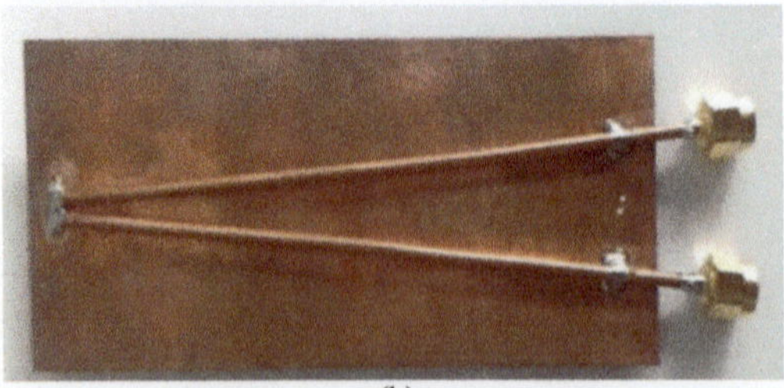

(b)

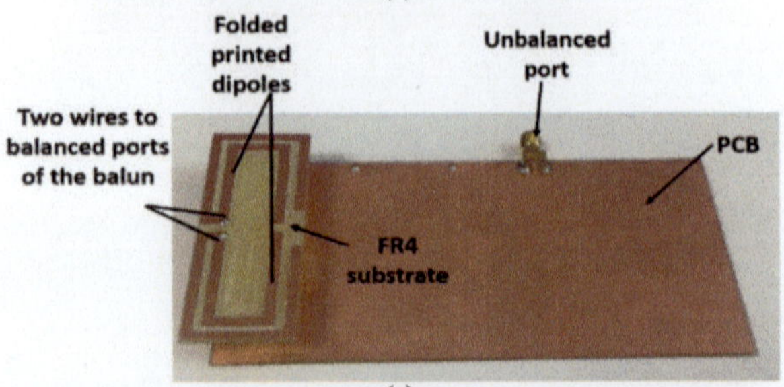

(c)

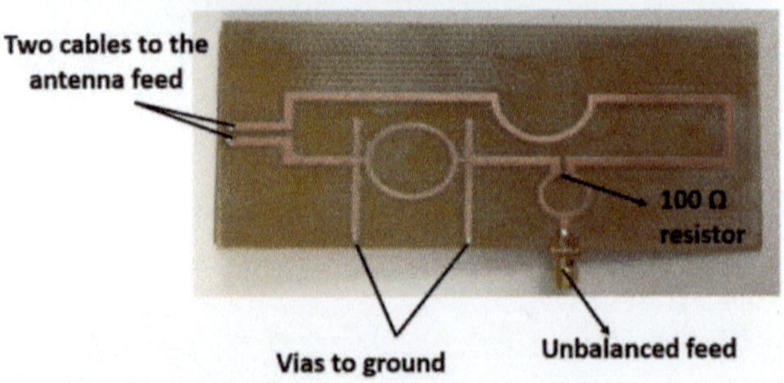

(d)

Fig. 6. The antenna prototypes, (a) 3D without balun, (b) bottom without balun, (c) 3D view with balun, (d) bottom view with balun.

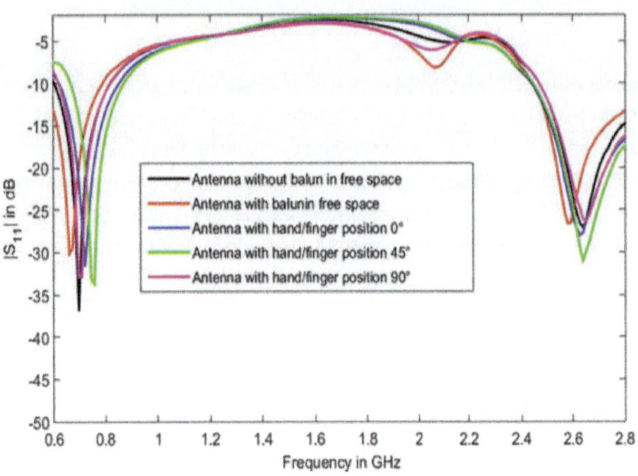

Fig. 7. Measured reflection coefficients $|S_{11}|$ of the proposed antennas.

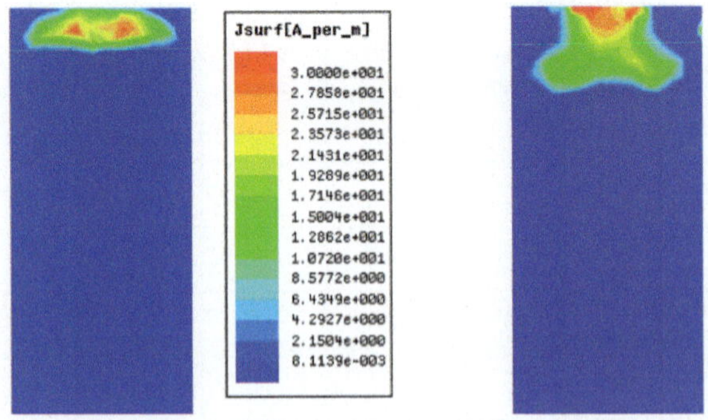

700 MHz (a) 2600 MHz

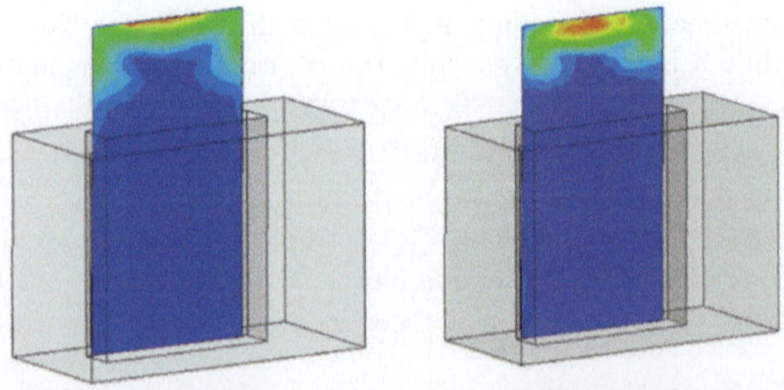

0° finger position @700 MHz (b) 0° finger position @2600 MHz

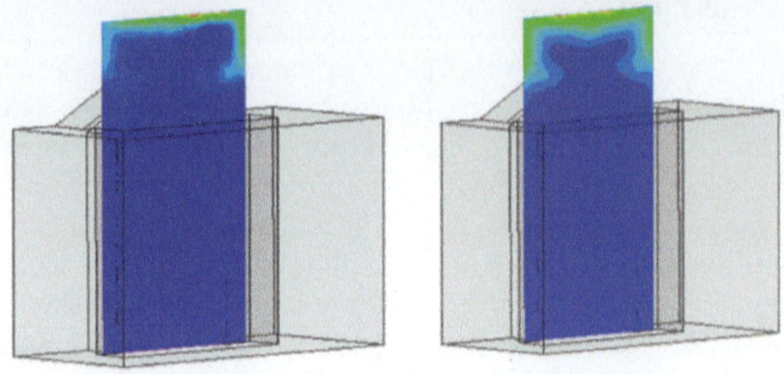

45° position @700 MHz (c) 45°position @2600 MHz

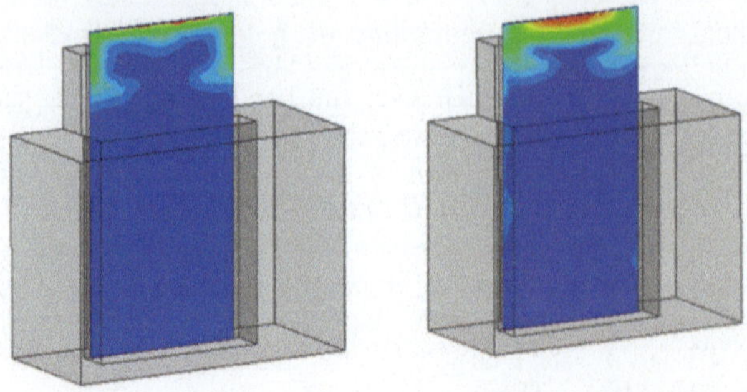

90° position @700 MHz (d) 90°position @2600 MHz

Fig. 8. Current surface for (a) antenna in free space, (b) antenna with hand finger 0° position, (c) antenna with hand finger 45° position, and (d) antenna with hand finger 90° position.

As can be observed, the measured S_{11} of the full antenna assembly shows a good result covering the suggested dual-band frequency spectrum. The results agree well with the computed results as demonstrated in Fig. 5.

The effect of hand holding scenario on the S_{11} performance of the prototype antennas was also studied and investigated in which the ground plane was considered as being held in a hand and positioned in the above-mentioned "talk" positions, shown in Fig. 4. It was noticeable that slight discrepancies occurred between the simulated and measured values of S_{11} in free space within the envisioned operating bands as shown in Fig. 7.

In order to explain how the balanced antenna is ground plane independent, a study of the current intensity of the present design in both free space and hand held scenarios are shown in Fig. 8. The surface current distributions in free space were illustrated over the dual-band of 700/2600 MHz. It is shown that the surface current induced on the antenna ground plane is strong in the area exactly underneath the feeding point for both frequency bands, while it is neglected over the rest of ground plane as indicated in Fig. 8(a), which is comparable to the results obtained from [16]. It also shows some advantages comparable to the induced current on the ground of the unbalanced antenna in [30].

In the hand model scenario for all finger/talk positions, 0°, 45° and 90°, the major current appears around more or less the similar area in the example of free space, in which the current only exists in the area below the antenna and gradually tapers as we head further away over the whole ground plane as shown in Fig. 8(b), (c) and (d). From the above-mentioned scenarios, this antenna proves the fact that the balanced antenna is ground plane independent. This also suggests that the antenna design has an advantage of being insensitive when it is held by the user's hand.

Figure 9(a) illustrates both the computed and measured gain of full assembly antenna for the 700/2600 MHz of LTE frequencies. The computed antenna gain varies from 0.9 dBi and 1.62 dBi over the lower band of 700 MHz and between 3.5 dBi and 4.4 dBi over 2600 MHz band. On the other hand, the measured gain fluctuated between 0.85 dBi and 1.4 dBi over the bandwidth of 700 MHz, and from 3.45 dBi to 4.3 dBi over the band of 2600 MHz as detected in Fig. 9(a). These minor variations may be attributed to the introduction of the physical integration of the balun with ground of the device, and the possible existence of errors of the connector pins antenna fabrication.

Figure 9(b) shows the radiation efficiency of antenna assembly. The Wheeler Cap method was adopted here as an easier and practical method to measure the radiation efficiency [31–33]; that is based on the measured antenna input resistances in free space and in the cap. It is observed that the simulated and measured results are in reasonable agreement and consistent with the variations of the gain results achieved in Fig. 9(a). It can also be observed that the simulated efficiency varies from 76 to 81 % over the lower band of 700 MHz and from 81 to 91 % over the 2600 MHz; on the other hand, the measured efficiency over the 700 MHz band fluctuates from 80 to 83.6 %, while it varies from 85 to 95 % over the higher band of 2600 MHz.

The simulated and measured far-field radiation patterns of the proposed antenna assembly (antenna and balun) were presented in Fig. 10. Two pattern planes cuts at xz and yz were considered at 700/2600 MHz, respectively. From Fig. 10, a reasonable agreement was observed be-

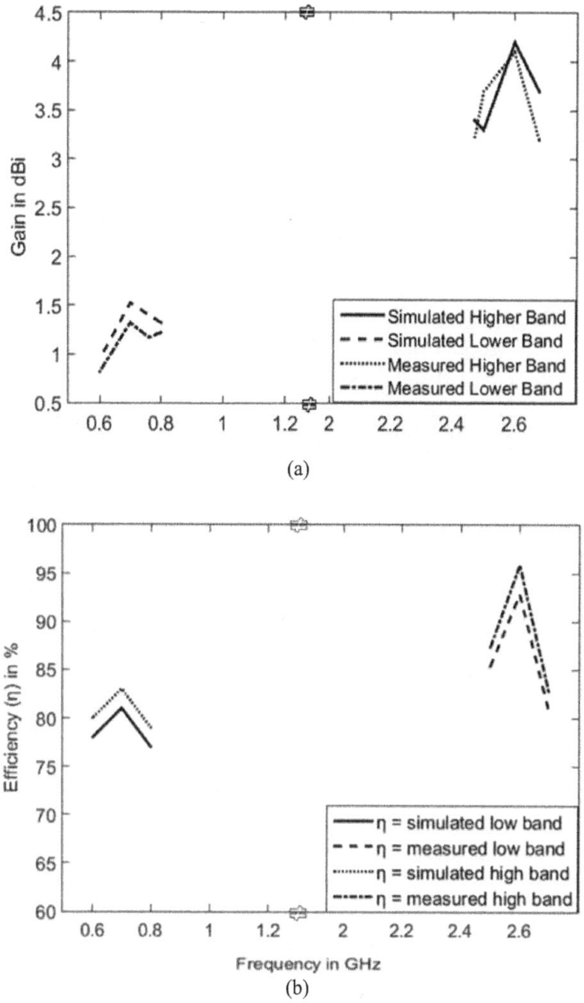

Fig. 9. (a) Simulated and measured gain of the proposed antenna. (b) Simulated and measured radiation efficiency of the proposed antenna.

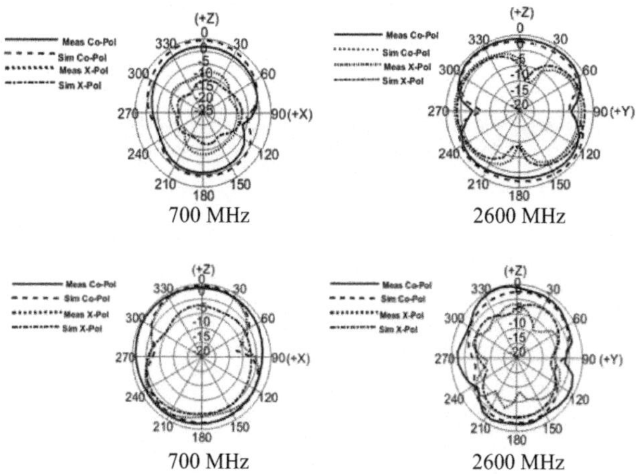

Fig. 10. Normalized antenna radiation patterns for two planes (left: *xz*, right: *yz*) at 700 MHz, 2600 MHz, '____' measured co-polarization, '– – –' simulated co-polarization, '------' measured cross-polarization, '.......' simulated cross-polarization.

tween computed and measured ones. The accomplished results indicate that the radiation patterns are nearly omnidirectional.

4. Conclusion

A balanced antenna covering dual-bands of the LTE standards 698–748 MHz and 2500–2690 MHz has been presented. Computed and measured results of reflection coefficients (S_{11}) showed sufficient impedance matching of $S_{11} \leq -10$ dB including good agreement for free space and hand held with/without balun scenarios. The antenna was demonstrated to exhibit near-omnidirectional radiation over the two operating bands. The surface current of the proposed antenna proves that the currents are diminished over the entire ground plane, except underneath the feeding point where there is improved immunity to the hand-held. This also enhanced the stability of the present mobile antenna to operate at real environment configurations. The antenna could be considered as an attractive candidate for practical applications in mobile phones.

Acknowledgments

This work is carried out under the grant of the Fundação para a Ciência e a Tecnologia (FCT - Portugal), with the reference number: SFRH / BPD / 95110 / 2013 and funded by national funds through FCT/MEC (PEst-OE/EEI/LA0008/2013 - UID/EEA/50008/2013). The research leading to these results has received funding from the Fundação para a Ciência e Tecnologia and the ENIAC JU (THINGS2DO–GA n. 621221); also, financially supported by the Engineering and Physical Sciences Research Council (EPSRC) in the UK under grant EP/E022936/1.

References

[1] ANGUERA, J., ANDÚJAR, A., HUYNH, M.C., ORLENIUS, C., PICHER, C., PUENTE, C. Advances in antenna technology for wireless handheld devices. *International Journal on Antennas and Propagation*, 2013, article ID 838364, 25 p. DOI: 10.1155/2013/838364

[2] ABRI, M., BELGACEM, N., BELGACEM, W. New GSM, DCS and GSM/DCS PIFA antenna for wireless networks applications. *International Journal of Information Network Security*, 2013, vol. 2, no. 4, p. 305–310. ISSN: 2089-3299

[3] CABEDO, A., ANGUERA, J., PICHER, C., RIBO, M., PUENTE, C. Multi-band handset antenna combining PIFA, slots, and ground plane modes. *IEEE Transactions on Antennas and Propagation*, 2009, vol. 57, no. 9, p. 2526–2533. DOI: 10.1109/TAP.2009.2027039

[4] GANDARA, T., PEIXEIRO, C. Compact double U-slotted microstrip patch antenna element for GSM1800, UMTS and HiperLAN2. In *IEEE Antennas and Propagation Society Int. Symp.* Monterey (California), 2004, p. 1459–1462. DOI: 10.1109/APS.2004.1330463

[5] BRISSOS J., PEIXEIRO, C. Triple-band microstrip patch antenna element for GSM1800, UMTS and HiperLAN2. In *Proc. of 2003 IEEE Int. Antennas and Prop. Symp.*, Columbus (OH, USA), 2003, p. 130–133. DOI: 10.1109/APS.2003.1220138

[6] FAN, S., T., YIN, Y., Z., HU, W., SONG, K., LI, B. Novel CPW-fed printed monopole antenna with an n-shaped slot for dual-band operations. *Microwave and Optical Technology Letters*, 2012, vol. 54, no. 1, p. 240–242. DOI: 10.1002/mop.26475

[7] MANOUARE, A. Z., IBNYAICH, S., EL IDRISSI, A., ABDELILAH GHAMMAZ, A. Miniaturized triple wideband CPW-fed patch antenna with a defected ground structure for WLAN/WiMAX applications. *Journal of Microwaves, Optoelectronics and Electromagnetic Applications*, 2016, vol. 15, no. 3, p. 157–169. DOI: 10.1590/2179-10742016v15i3497

[8] ANG, I., GUO, Y., X., CHIA, Y. W. Compact internal quad-band antenna for mobile phones. *Microwave and Optical Technology Letters*, 2003, vol. 38, no. 3, p. 217–223. DOI: 10.1002/mop.11019

[9] LEE, C., WONG, K. Planar monopole with a coupling feed and an inductive shorting strip for LTE/GSM/UMTS operation in the mobile phone. *IEEE Transactions on Antennas and Propagation*, 2010, vol. 58, no. 7, p. 2479–2483. DOI: 10.1109/TAP.2010.2048878

[10] DING-BING LIN, JUI-HUNG CHOU, SON-ON FU, HSUEH-JYH LI A compact dual-band printed antenna design for LTE operation in handheld device applications. *International Journal of Antennas and Propagation*, 2014, 9 p. DOI: 10.1155/2014/897328

[11] ELFERGANI, I., HUSSAINI, S.A., RODRIGUEZ, J., SEE, C. H., ALHAMEED, R. A. Wideband tunable PIFA antenna with loaded slot structure for mobile handset and LTE applications. *Radioengineering*, 2014, vol. 23, no. 1, p. 345–355.

[12] LIZZI, L., MASSA, A. Dual-band printed fractal monopole antenna for LTE applications. *IEEE Antennas and Wireless Propagation Letters*, 2011, vol. 10, p. 760–763. DOI: 10.1109/LAWP.2011.2163051

[13] WONG, K. L., LYU, C. A., CHOU, L. C. Small-size multiband planar antenna for LTE700/2300/2500 operation in the tablet computer. *Microwave and Optical Technology Letters*, 2012, vol. 54, no. 1, p. 81–86. DOI: 10.1002/mop.26507

[14] ABD-ALHAMEED, R. A., EXCELL, P. S., KHALIL, K., ALIAS, R., MUSTAFA, J. SAR and radiation performance of balanced and un-balanced mobile antennas using a hybrid computational electromagnetics formulation. *IEE Proceedings - Science, Measurement and Technology*, 2004, vol. 151, no. 6, p. 440 – 444. DOI: 10.1049/ip-smt:20040855

[15] VAINIKAINEN, P., OLLIKAINEN, J., KIVEKAS, O., KELANDER, K. Resonator-based analysis of the combination of mobile handset antenna and chassis. *IEEE Transactions on Antennas and Propagation*, 2002, vol. 50, no. 10, p. 1433–1444. DOI: 10.1109/TAP.2002.802085

[16] ARENAS, J. J., ANGUERA, J., PUENTE, C. Balanced and single-ended handset antennas: free space and human loading comparison. *Microwave and Optical Technology Letters*, 2009, vol. 51, no. 9, p. 2248–2254. DOI: 10.1002/mop.24525

[17] ILVONEN, J., KIVEKÄS, O., AZREMI, A. A. H., VALKONEN, R., HOLOPAINEN, J., VAINIKAINEN, P. Isolation improvement method for mobile terminal antennas at lower UHF band. In *Proceedings of the 5th European Conference on Antennas and Propagation*. Rome (Italy), 2010, p. 1238–1242.

[18] ZHOU, D., ABD-ALHAMEED, R. A., EXCELL, P. S. Bandwidth enhancement of balanced folded loop antenna design for mobile handsets using genetic algorithms. *PIERS Online*, 2008, vol. 4, no. 1, p. 136–139.

[19] ILVONEN, J., HOLOPAINEN, J., KIVEKÄS, O., VALKONEN, R., ICHELN, VAINIKAINEN, P. Balanced antenna structures of

mobile terminals. In *Proceedings of the 4th European Conference on Antennas and Propagation*. Barcelona (Spain), 2010, 5 p.

[20] ELFERGANI, I. T. E., ABD-ALHAMEED, R. A., ALI, N. T. et al. A novel dual band frequency tunable balanced handset antenna for WLAN application. In *18th IEEE International Conference on Electronics, Circuits and Systems*, Beirut (Lebanon), 2011, p. 516 to 519. DOI: 10.1109/ICECS.2011.6122326

[21] BEHERA, S. K., PODDAR. D. R., MISHRA, R. K. Design of sequentially fed balanced amplifying antenna for circular polarization. ICTACT *Journal on Communication Technology*, 2010, vol. 1, no. 4, p. 213–217.

[22] ZHOU, D., ABD-ALHAMEED, R. A., SEE, C. H., EXCELL, P. S. Design of wideband balanced folded-arms dipole antenna for mobile handsets. *Electromagnetics*, 2009, vol. 29, p. 641–651. DOI: 10.1080/02726340903287406

[23] ZHOU, D., ABD-ALHAMEED, R. A., SEE, C. H., ALHADDAD, A. G., EXCELL, P. S. Compact wideband balanced antenna for mobile handsets. *IET Microwaves, Antennas and Propagation*, 2010, vol. 4, no. 5, p. 600–608. DOI: 10.1049/iet-map.2009.0153

[24] COLLINS, B. S., KINGSLEY, S. P., IDE, J. M., SAARIO, S. A., SCHLUB, R. W., KEEFE, S. G. O. A multi-band hybrid balanced antenna. In *IWAT 2006 IEEE International Workshop on Antenna Technology: Small Antenna and Novel Materials*, White Plains (New York), 2006, p. 100–103.

[25] WEN-JUN LU, YA-MING BO, HONG-BO ZHU. Novel planar dual-band balanced antipodal slot-dipole composite antenna with reduced ground plane effect. *International Journal of RF and Microwave Computer-Aided Engineering*, 2012, vol. 22, no. 3, p. 3319–328. DOI: 10.1002/mmce.20600

[26] JOLANI, F., DADASHZADEH, G. R., NASER-MOGHADASI, M., DADGARPOUR, A. Design and optimization of compact balanced antipodal Vivaldi antenna. *Progress In Electromagnetics Research C*, 2009, vol. 9, p. 183–192. DOI: 10.2528/PIERC09071510

[27] VIGNESH, N., SATHISH KUMAR, G. A., BRINDHA, R. Design and development of a tapered slot Vivaldi antenna for ultra-wide band application. *International Journal of Advanced Research in Computer Science and Software Engineering*, 2014, vol. 4, no. 5, p. 174–178. ISSN: 2277 128X

[28] ELFERGANI, I. T. E., HUSSAINI, A. S., RODRIGUEZ, J., ALI, A. H., SEE, C. H., ABD-ALHAMEED, R. A. A compact and broadband balun design for LTE applications. *Progress In Electromagnetics Research C*, 2016, vol. 67, p. 85–95.

[29] HFSS version 14, Ansys Inc, USA, 2013, Available at: http://www.ansys.com/. Accessed on October 10, 2014.

[30] YUN-WEN CHI, KIN-LU WONG. Very-small-size printed loop antenna for GSM/DCS/PCS/UMTS operation in the mobile phone. *Microwave and Optical Technology Letters*, 2009, vol. 51, no. 1, p. 184–192. DOI: 10.1002/mop.24008

[31] JOHNSTON, R. W., MCRORY, J. G. An improved small antenna radiation-efficiency measurement. *IEEE Antenna and Propagation Society Magazine*, 2008, vol. 40, p. 40–48. DOI: 10.1109/74.735964

[32] AGAHI, D., DOMINO, W. Efficiency measurements of portable-handset antennas using the Wheeler cap. *Applied Microwave and Wireless*, June 2000. Available at: http://www.rfcafe.com/references/articles/Efficiency-Measurement-Antenna-Wheeler-Cap.htm

[33] HOSUNG CHOO, ROGERS, R., HAO LING. On the Wheeler cap measurement of the efficiency of microstrip antennas. *IEEE Transactions on Antennas and Propagation*, 2005, vol. 53, no. 7, p. 2328–2332. DOI: 10.1109/TAP.2005.850758

About the Authors ...

Issa ELFERGANI received his M.Sc. and Ph.D. in Electrical Engineering with Power Electronics (EEPE) from University of Bradford (UK) in 2008 and 2012, with a specialization in tunable antenna design for mobile handset and UWB applications. He is now a Senior Researcher at the Instituto de Telecomunicações - Aveiro (Portugal), working with European research funded projects, whilst serving as technical manager for ENIAC ARTEMOS and EUREKA BENEFIC. He is an IEEE member and member of American Association for Science and Technology (AASCIT). He is a TPC member and reviewer for several IEEE international conferences. He is the author of several journal and conference publications, including one book editorial. His research interests are multi-disciplinary and have a number of cross cutting themes that include research in various antenna designs including MIMO, UWB, balanced and unbalanced mobile phone antennas with the application of theoretical, computational analytical approaches, RF MEMS filter technologies and power amplifiers.

Jonathan RODRIGUEZ received his M.Sc. degree in Electronic and Electrical Engineering and Ph.D from the University of Surrey (UK), in 1998 and 2004, respectively. In 2005, he became a researcher at the Instituto de Telecomunicacoes (IT) - Portugal where he was a member of the Wireless Communications Scientific Area. In 2008, he became a Senior Researcher where he established the 4TELL Research Group (http://www.av.it.pt/4TELL/) targeting next generation mobile networks with key interests on green communications, cooperation, and electronic circuit design. He has served as project coordinator for major international research projects, which includes Eureka LOOP and FP7 C2POWER, whilst serving as technical manager for FP7 COGEU and FP7 SALUS. Since 2009, he became an Invited Assistant Professor at the University of Aveiro (Portugal), and Associate in 2015. He is author of more than 300 scientific works that includes 6 book editorials. His professional affiliations include: Senior Member of the IEEE and Chartered Engineer (CEng) since 2013, and Fellow of the IET (2015).

Fathi ABDUSSALAM was born in Bin-Waleed, Libya. He received the B.Sc. from Al Fatah University, Tripoli, Libya in 1992. After graduation, he had worked at Ministry of Education from 1992 to 2002, and then he joined University of West of England, Bristol, UK to do his M.Sc. degree from 2003 to 2004. He has appointed as lecturer at Higher Institute of Electronics, Tripoli, Libya from 2004 to 2009 and then appointed as a senior lecturer at Azzaytuna University from 2009 to 2014. Since October 2014, he joined the School of Engineering and Informatics at University of Bradford to pursue his Ph.D. research degree. The main objective of his Ph.D. research is to model and analyze complex electromagnetic problems by means of a new hybridized computational technique combining the frequency domain Method of Moments (MoM) and Finite-

Difference Time-Domain (FDTD) with the optimization process of GA, PSO and FF methods. Fathi has published three conference papers and also contributed to other three since his registration to Bradford University.

Chan H. SEE received the first-class B.Eng. Honors degree in Electronic, Telecommunication and Computer Engineering and a Ph.D. degree from the University of Bradford, U.K., in 2002 and 2007, respectively. Currently, he is a Lecturer in Electrical and Electronic Engineering at Engineering, Sport, and Sciences (ESS) Academic Group, University of Bolton, U.K. He is also appointed as an Honorary Visiting Research Fellow in the Radio Frequency, Antennas, Propagation and Computational Electromagnetics Research Group within the School of Electrical Engineering and Computer Science, University of Bradford. He has published over 100 journal articles and conference papers. He is a co-author for one book and one book chapter. Dr. See was a recipient of two Young Scientist Awards from the International Union of Radio Science (URSI) and Asia-Pacific Radio Science Conference (AP-RASC) in 2008 and 2010, respectively. He is a Chartered Engineer and Member of the Institution of Engineering and Technology (MIET). He has an NVQ level 4 in Management from the Chartered Management Institute.

Raed ABD-ALHAMEED received the B.Sc. and M.Sc. degrees from Basrah University, Basrah, Iraq, in 1982 and 1985, respectively, and the Ph.D. degree from the University of Bradford, West Yorkshire, U.K., in 1997, all in Electrical Engineering. He is a Professor of Electromagnetic and Radio Frequency Engineering at the University of Bradford. He is the senior academic responsible for radio frequency and electromagnetics research in the University of Bradford, for which new antenna design configurations and computational techniques have been developed including several patents. The expertise includes in particular the highly realistic analysis of antenna design process in the presence of large multilayer scatterers, using the Bradford-developed hybrid modeling techniques including MoM, FEM, FDTD and subgridding FDTD methods. He has published over 400 academic journal and conference papers and is co-author of two books and several book chapters. Currently, he is the leader of the Communication Research Group and head of RF and electromagnetics research in the School of Engineering and Informatics, Bradford University. Dr. Abd-Alhameed is a Fellow of the Institution of Engineering and Technology, Fellow of Higher Education Academy, and a Chartered Engineer in the U.K.

Asymmetric Turbo Code for Coded-Cooperative Wireless Communication Based on Matched Interleaver with Channel Estimation and Multi-Receive Antennas at the Destination

Shoaib MUGHAL, Fengfan YANG, Saqib EJAZ, Syed Hassan AHMED, Rahim UMAR

College of Electronic and Information Engg., Nanjing Univ. of Aeronautics and Astronautics, Nanjing 210016, China

mughalshoaib@hotmail.com, yffee@nuaa.edu.cn, saqib.ejaz@foxmail.com, syedhassan_ahmad@hotmail.com, rahim_nuaa@hotmail.com

Abstract. *This paper investigates the multiple relay coded-cooperation scheme based on asymmetric turbo code (ATC) with multiple receive antennas over Rayleigh block fading channels. An encoding scheme based on ATC is proposed for coded-cooperation, i.e. distributed asymmetric turbo code (DATC). The code matched interleaver (CMI) is selected by a rigorous comparison with a uniform-random interleaver (URI). This optimum choice of interleaver at the relay nodes provides maximum benefit from DATC coded-cooperation scheme. Practically in any wireless communication system, the channel side information (CSI) is usually unknown at the receiver. Therefore, spatial normalized least mean square (NLMS) adaptive transversal filters are employed to estimate the CSI at the destination node. Moreover, in coded-cooperation scheme, the effectiveness and validation of spatial NLMS adaptive transversal filters is also verified by simulation results. Quadrature phase shift keying (QPSK) is used in coded-cooperation scheme and corresponding soft-demodulators are employed along with joint iterative soft-input soft-output (SISO) decoder at the destination node. Monte Carlo simulations show that the proposed scheme incorporates coding gain, diversity gain and cooperation gain successfully, which eventually results in net gain of 2.7 to 3.5 dBs over non-cooperation ATC counterpart.*

Keywords

Joint iterative SISO decoder, coded-cooperation, distributed asymmetric turbo code, CMI, multiple-input multiple-output (MIMO), channel side information

1. Introduction

Recently, there has been a rapid increase in the demand of high speed data rates in wireless communication. Fading is one of the main adverse factors that affect the speed of high data rate in wireless communication. This problem of fading can be solved by employing diversity techniques such as time, frequency, and spatial diversities [1]. Spatial diversity is one of the key diversity techniques extensively used in literature [1], [2] to mitigate the harsh effects of fading channels, thereby enhancing the performance of wireless communication system. The simplest way to get the spatial diversity is to employ multiple antennas both at transmitter and receiver ends [1], [2]. However, in typical mobile equipment, there has been an always inevitable constraint due to limitations in weight, size and area etc. Thus, deploying multiple antennas in mobile equipment is not a feasible option in today's present world. The problem in this situation can be compensated by using three-terminal or relay assisted cooperative system commonly known as virtual MIMO [3]. In a virtual MIMO, each of the mobile equipment contains a single antenna, forms a group and communicates with each other in a cooperative manner. Thus, it takes the benefits of MIMO technique by creating a virtual antenna array.

The idea of relay assisted cooperative system was an outcome of the pioneered efforts made by Van der Meulen [4] to overcome the existing problem of single antenna deployment in a typical mobile device. Relay assisted cooperative communication is fundamentally based on communication protocols such as amplify-and-forward (AF) [5], compress-and-forward (CF) [6] and decode-and-forward (DF) [7]. The performance efficient user cooperation based on ad-hoc wireless commination is detailed in [5]. The implementation issues of user cooperation for wireless networks are discussed in [8]. Recently, there has been an evolution in the ground breaking idea of relay assisted cooperative system called coded-cooperation. The coded-cooperation scheme [9] is a combination of cooperative scheme with existing channel coding like Turbo codes [10], Low Density Parity Check (LDPC) codes [11], Convolution codes [12], Reed-Muller code [13] and Polar codes [14, 15]. The basic idea of coded-cooperation is to distribute the channel codes over source and relay terminals, i.e. overall code-word reaches the destination by independent fading paths. Thus, the information bits are recovered by using joint decoding scheme [16] deployed at the destination node. Recently, the author in [16] described the concept of coded-cooperation in detail by considering turbo code as distributive channel code.

In literature, coded-cooperative schemes have been mostly developed for Binary Phase Shift Keying (BPSK) modulation scheme but these schemes are bandwidth inefficient. However, bandwidth efficiency can be improved by using higher order modulation schemes like QPSK, 8-PSK, 16-PSK or other higher order modulation schemes [17]. Thus, QPSK has been considered as a modulation scheme in this paper. Moreover, we have considered ATC in coded-cooperation as to the best of our knowledge all the previous research has been done on Symmetric Turbo codes (STC) in coded-cooperation. The proposed scheme is capable of accomplishing coding gain and diversity gain with increased bandwidth efficiency. Diversity gain is achieved from multi-receive antennas while transmission links from source to destination and relays to destination provide cooperation diversity.

Turbo code performance is based on distance spectrum of the code [18]. Distance spectrum of the code can be improved by using component encoder with large minimum free distance d_{free}. Interleaver design is a crucial part of turbo code which plays an important role in improving the effective free distance of the code. The basic purpose of interleaver is to remove the bad weight code-words. The bad weight code-words are usually produced by low weight information sequences such as weight-2, 3 and 4. Different interleavers have been reported in literature [19–22]. For ATC, it has been suggested in the literature [23] that CMI outperforms uniform-random interleaver (URI) in additive white Gaussian noise (AWGN) and Rayleigh block fading channels for non-cooperative communication. In order to enhance the error performance of coded-cooperative communication system, CMI is selected as an optimum choice of interleaver in this paper. CSI is usually not known to the receiver so CSI estimation is required. Therefore, we propose a training sequence based spatial normalized least mean square (NLMS) adaptive transversal algorithm for the channel estimation at the destination [24–26]. The performance of DATC for CMI with multi-receive antenna with perfect and imperfect CSI have been analyzed.

The major contributions of this manuscript can be summarized as below:

- The ATC is used for the first time in multi-relay coded cooperative communication to the best of our knowledge.
- The ATC with CMI is used in multi-relay coded cooperative communication.
- The channel estimation is performed by spatial NLMS adaptive filter in coded cooperative communication.

The remaining manuscript is organized as follows. Section 2 presents the fundamentals of general coded-cooperative scheme with multiple relays, and multiple receive antennas at the destination node. Section 3 mainly deals with the fundamentals of ATCs and describes the proposed multi-relay coded-cooperative scheme. Moreover, it discusses the CMI design for ATC schemes. Section 4 proposes a spatial NLMS adaptive filter for channel estimation. Soft demodulation and joint iterative SISO decoder for DATC scheme are explained in Sec. 5 and Sec. 6, respectively. In Sec. 7, simulation results of coded-cooperative schemes for ATC in different scenarios are presented and discussed. Finally, the paper is concluded with the conclusion drawn in Sec. 8.

2. Model of Coded-Cooperation Communication with Multi-Relays and Multi-Receive Antennas

A basic half-duplex coded-cooperation model with multi-relays and multi-receive antennas is shown in Fig. 1. The source S node and relay (R_ω, $\omega = 1$ to L_R) nodes are composed of single antenna as they are considered to be user equipment (UE), where L_R is representing the total number of relay nodes. The destination node D can be viewed as a base station as it contains multiple receive antennas. One complete transmission from source to destination node requires two time slots. During the first time slot, the source node broadcasts information sequence to relay nodes and a destination node. During the second time slot, all relay nodes transmit their information sequence to the destination node. Moreover, the relay nodes are placed far apart from each other in such a way that they transmit their corresponding information sequences over non overlapping channels.

At the source node, code C_1 encodes the binary sequence of bits $\mathbf{u}_1$ to code-word $\mathbf{c}_S$. The modulation is performed on codeword $\mathbf{c}_S$ to get the modulated sequence of symbols which is denoted by $\mathbf{x}_S = \left[x_S^1, x_S^2, \ldots, x_S^{N_1^S}\right]$, where x_S^j ($j = 1, 2, \ldots, N_1^S$) is the modulated symbol and N_1^S is the length of this sequence. Modulated sequence of symbols is broadcasted to the relay nodes and destination node. During the first time slot, the received coded symbol vector $\mathbf{r}_{SD_\varrho}$ at the destination's ϱth ($\varrho = 1$ to L) antenna is modelled as follows

$$\mathbf{r}_{SD_\varrho} = \mathbf{h}_{SD_\varrho}\mathbf{x}_S + \mathbf{n}_{SD_\varrho} \tag{1}$$

where $\mathbf{h}_{SD_\varrho} = \left[h_{SD_\varrho}(1), h_{SD_\varrho}(2), \ldots, h_{SD_\varrho}\left(N_1^s\right)\right]$ is a Rayleigh fading channel vector, $h_{SD_\varrho}(j) \sim R(0,1), \left(j = 1, 2, \ldots, N_1^s\right)$

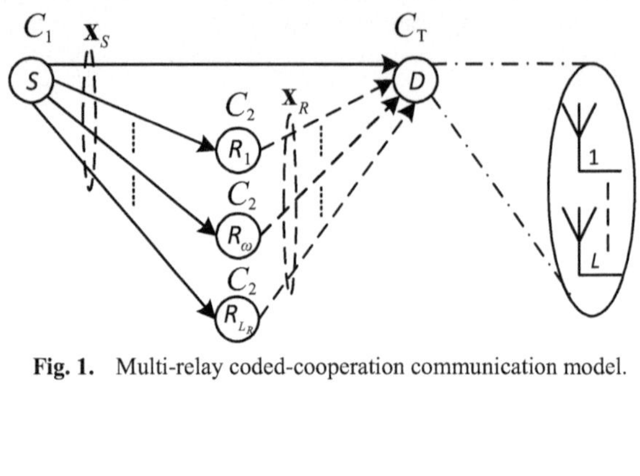

Fig. 1. Multi-relay coded-cooperation communication model.

is zero mean and unit variance complex Gaussian random variable. Since Rayleigh block fading is assumed therefore $\mathbf{h}_{SD\varrho}$ becomes $\mathbf{h}_{SD_\varrho} = \left[h_{SD_\varrho}(1) = h_{SD_\varrho}(2) =,\ldots,= h_{SD_\varrho}(N_1^s)\right]$. The vector $\mathbf{n}_{SD_\varrho} = \left[n_{SD_\varrho}(1), n_{SD_\varrho}(2),\ldots,n_{SD_\varrho}(N_1^s)\right]$ is an AWGN, $n_{SD_\varrho}(j) \sim N(0,\sigma^2)$, $(j = 1,2,\ldots,N_1^s)$ is zero mean and $\sigma^2/2$ variance per dimension. The sequences received at each antenna at the destination node are joined using maximum ratio combining (MRC) technique to get $\mathbf{r}_{SD}$ sequence.

The ω-th relay node receives the coded symbol vector $\mathbf{r}_{SR_\omega}$ during the first time slot that is modelled as below

$$\mathbf{r}_{SR_\omega} = \mathbf{h}_{SR_\omega}\mathbf{x}_S + \mathbf{n}_{SR_\omega} \tag{2}$$

where $\mathbf{h}_{SR_\omega}$ is a Rayleigh fading channel vector and $\mathbf{n}_{SR_\omega}$ is a vector of AWGN and their definitions are similar as $\mathbf{h}_{SD_\varrho}$and $\mathbf{n}_{SD_\varrho}$, respectively. The detection process at ω-th relay node is divided into two steps. The first step consists of soft demodulation of received coded symbol vector $\mathbf{r}_{SR_\omega}$ into soft coded bits. In the second step, SISO decoder decodes soft coded bits. The source to ω-th relay $(S - R_\omega)$ channel is usually considered as ideal ($\text{SNR}_{SR_\omega} = \infty$) to have a perfect decoding at relay node. In case of non-ideal ($\text{SNR}_{SR_\omega} \neq \infty$) $S - R_\omega$ channel, different cooperation protocols have been proposed in literature. Cyclic redundancy check (CRC) [9] is mostly applied to ensure error free decoding at the relay node. In this paper, ideal $S - R_\omega$ channel is assumed for proposed coded-cooperation schemes.

At the ω-th relay node, code C_2 encodes the decoded bits again. These encoded bits are transmitted to the destination node after modulation in the second time slot. The modulated sequence of symbols is denoted by $\mathbf{x}_R = \left[x_R^1, x_R^2,\ldots,x_R^{N_1^R}\right]$ where $x_R^j, (j = 1,2,\ldots N_1^R)$ is the modulated symbol and N_1^R is the length of this sequence. The received sequence at the destination's ϱ-th antenna during second time slot is modelled as

$$\mathbf{r}_{RD_\varrho} = \mathbf{h}_{RD_\varrho}\mathbf{x}_R + \mathbf{n}_{RD_\varrho} \tag{3}$$

where $\mathbf{h}_{RD_\varrho}$ is a Rayleigh fading channel vector and $\mathbf{n}_{RD_\varrho}$ is a vector of AWGN and their definition are similar as $\mathbf{h}_{SD_\varrho}$ and $\mathbf{n}_{SD_\varrho}$, respectively.

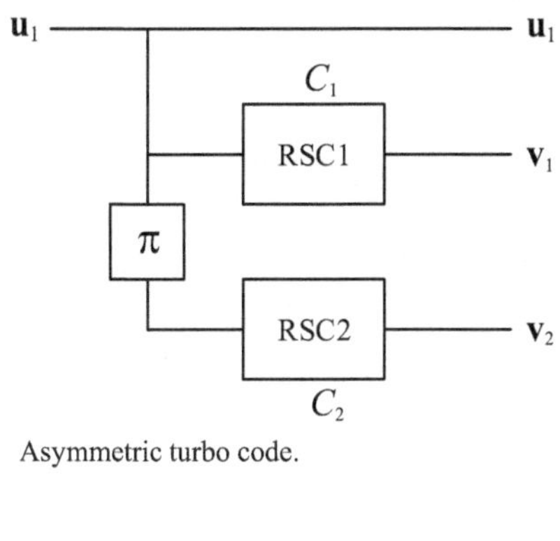

Fig. 2. Asymmetric turbo code.

At the destination node, the antenna array receives the signal that was transmitted by the relay node. The destination node combines them by using MRC technique to get the $\mathbf{r}_{RD}$ sequence. The channels are assumed to be Rayleigh block fading channels. Moreover, the channels remain static for a complete code-word. Furthermore, the transmission power at each node is assumed to be equal.

3. ATC and DATC with Code Matched Interleaver

This section discusses the basic structure of ATC. The two different RSCs in ATC are distributed in such a way that one RSC is placed at source node while other RSC is deployed at the relay node. Thus, this parallel encoding structure of ATC makes it suitable for coded-cooperation therefore a DATC scheme with CMI is discussed.

3.1 Asymmetric Turbo Code (ATC)

The turbo code is mostly a parallel concatenation of two RSC encoders with an interleaver π connected in a manner as shown in Fig. 2. Unlike conventional STC, ATC is a combination of two different constituent RSC encoders and decoders.

A parallel concatenated turbo code is denoted by $C_T = [C_1, C_2]$, where C_1 and C_2 are RSC codes with only the trellis of C_1 is terminated. For conventional STC C_1 and C_2 codes are identical but for ATC C_1 and C_2 codes are different. An RSC is denoted by generator matrix $\mathbf{G}(g_b(D)/g_a(D))$ where $g_a(D)$ is feedback polynomial while $g_b(D)$ is feed-forward polynomial in octal form. The turbo code with primitive feedback polynomial in component codes is called as primitive turbo code. The turbo codes discovered by Berrou et al. [27] are an example of non-primitive turbo codes.

The BER performance of turbo code is basically evaluated on two distinct regions, namely error-floor and waterfall region. The conventional STC either performs well in the error-floor region or in the waterfall region but not in both regions simultaneously [28]. The non-primitive STC performs well in the waterfall region while the primitive STC shows good performance in the error-floor region.

The performance of a turbo code is based on various parameters such as block length, component codes, interleaver design and the distance spectrum properties. The ATC has a capability to provide good distance spectrum properties that enables it to perform well in both regions, i.e. waterfall and error-floor [23]. At the same time choosing suitable component codes depends upon the generator matrix as it plays a vital role in determining the overall weight of code-words [23].

In this paper, ATC code having different RSCs for constraint length $K_L = 4$ is being considered. The ATC with generator matrix $\mathbf{G}(13/15, 11/15)_8$ is selected based on its better BER performance [23]. This ATC is used in the proposed coded-cooperation scheme.

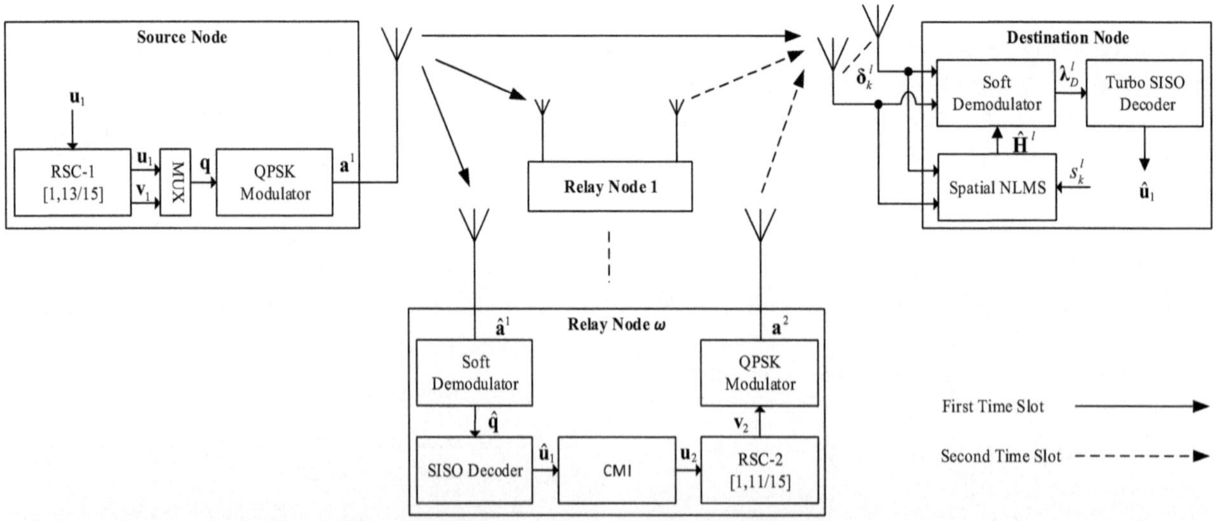

Fig. 3. Distributed multi-relay asymmetric turbo code.

3.2 Distributed Asymmetric Turbo Code (DATC)

The structure of turbo code is perfectly suitable to be effectively used in coded-cooperative scenarios. Turbo codes which are formed as the result of cooperation between source and relay nodes are usually called as Distributed Turbo Codes (DTC) in literature [29]. If the RSCs at the source and relay nodes are different then this DTC is termed as DATC. The generator polynomial use for our proposed DATC scheme is $\mathbf{G}(13/15, 11/15)_8$. This scheme takes two time slots for transmission of a complete turbo code frame. The source node consists of 1/2 rate RSC-1 encoder and QPSK modulator. The source node converts information sequence $\mathbf{u}_1$ of length K into coded sequence $\mathbf{q}$ of length N_1. These coded bits are mapped onto QPSK constellation $\mathcal{M}(\mathbf{q})$ to form coded symbol sequence $\mathbf{a}^1$ that contains N_1^S symbols. The source node simultaneously broadcasts these N_1^S symbols $\mathbf{a}^1$ to all relay nodes and the destination node in the first time slot. The destination node consists of multi-receive antennas. The relay nodes are usually placed closer to the destination node in a typical coded-cooperation scenario [30]. Therefore, we also have placed all the relay nodes near to the destination node.

The ω-th relay node demodulates N_1^S received coded symbol sequence $\hat{\mathbf{a}}^1$ to N_1 soft information sequence $\hat{\mathbf{q}}$ and this soft information sequence is decoded by SISO decoder to get the estimated information sequence $\hat{\mathbf{u}}_1$. If the estimated information sequence $\hat{\mathbf{u}}_1$ is error free then it is called ideal coded-cooperation otherwise it is called non-ideal coded-cooperation. Estimated information sequence $\hat{\mathbf{u}}_1$ is interleaved using CMI. The details of designing CMI are given in the subsequent subsection. This interleaved information sequence $\mathbf{u}_2 = \pi(\hat{\mathbf{u}}_1)$ is given as an input to 1/2 rate RSC-2 encoder at relay node. Output of RSC-2 encoder contains systematic and parity bit sequences. Systematic bit sequence of RSC-2 encoder is suppressed and parity bit sequence $\mathbf{v}_2$ of length N_2 is further passed through QPSK modulator. The modulated sequence $\mathbf{a}^2$ of length N_2^R is send to the destination node in the second time slot. Similarly all the relay nodes transmit their modulated sequences in the second time slot.

The channels are estimated by spatial NLMS adaptive filters using training sequence for each time slot. With the help of estimated channels, the received symbol sequence from each receive antenna are converted into log likelihood ratios (LLRs). These LLRs are combined using maximum ratio combining (MRC) technique. Combined LLRs are given as input to soft demodulator that results in soft bit LLRs. These soft bit LLRs for the first and second time slots are given as input to the joint iterative SISO turbo decoder to estimate the information sequence. The whole process of joint iterative SISO decoding is explained in the subsequent section. Figure 3 shows the DATC scheme. The overall code rate for DATC is $R_c = 1/3$.

3.3 Code Matched Interleaver (CMI)

The performance of a code is decided by its distance properties. The distributive construction of a code with good distance spectrum plays a crucial role in coded-cooperation. The distance spectrum of turbo code depends on interleaver; therefore interleaver design is the vital part of turbo code.

CMI is a better variant of S-random interleaver [21]. It improves the distance spectrum of ATC by removing low-weight code-words which are usually generated by low-weight information sequences such as weight-2, 3 and 4. Although all the low weight (weight-2, 3 and 4) information sequences are not bad so CMI only considers those low weight information sequences which produces low weight code-words. The performance of turbo code is based on its free distance in high SNR regime but it is dependent on the size of interleaver in low to medium SNR regime [22].

CMI mainly improves the performance of turbo code by removing low-weight information sequences which generate the low-weight code-words [23]. The design steps of CMIs for STC and ATC are discussed in literature [22],

[23]. However, we briefly discuss the construction of CMI for ATC which removes bad weight code-words and improves the performance of ATC.

Suppose that the RSC-1 encoder encodes $\mathbf{u}_1$ information sequence to $(\mathbf{u}_1, \mathbf{v}_1)$ sequence pair. Similarly RSC-2 encoder generates $(\mathbf{u}_2, \mathbf{v}_2)$ sequence pair after encoding the $\mathbf{u}_2$ information sequence where $\mathbf{u}_2$ is interleave form (URI) of $\mathbf{u}_1$ information sequence. $\mathbf{v}_1$ and $\mathbf{v}_2$ are parity sequences of RSC-1 and RSC-2, respectively. Assume that the weight of information sequences $\mathbf{u}_1$ and $\mathbf{u}_2$ is denoted by α. The weight of parity sequences of RSC-1 and RSC-2 are denoted by $\alpha(\mathbf{v}_1)$ and $\alpha(\mathbf{v}_2)$, respectively. The weight of an ATC code-word is given as follows

$$d = \alpha + \alpha(\mathbf{v}_1) + \alpha(\mathbf{v}_2). \tag{4}$$

The polynomial representation of weight-2 information sequence $\mathbf{u}_1^2$ which results in finite weight code-word is given as

$$\mathbf{u}_1^2(D) = \left(1 + D^{\varphi_1 k_1}\right) D^{\zeta_1} \tag{5}$$

where φ_1 is the minimum spacing between pair of "1"s of weight-2 information sequence $\mathbf{u}_1^2$, $k_1 = 1,2,3,4,\ldots$ and $\zeta_1 = 1,2,3,4,\ldots$ is time delay. The polynomial representation of interleave information sequence $\mathbf{u}_2^2$ is given as

$$\mathbf{u}_2^2(D) = \left(1 + D^{\varphi_2 k_2}\right) D^{\zeta_2} \tag{6}$$

where φ_2 is the minimum spacing between pair of "1"s of permuted weight-2 information sequence $\mathbf{u}_2^2$, $k_2 = 1,2,3,4,\ldots$ and $\zeta_2 = 1,2,3,4,\ldots$ is time delay.

Let the lowest weight of the i-th parity sequence is denoted by $z^i_{\min}$, which is generated by its corresponding weight-2 information sequence. For any weight-2 information sequence the weight of the i-th parity sequence is given as

$$k_j\left(z^i_{\min} - 2\right) + 2 \tag{7}$$

where $k_j = 1,2,3,4,\ldots$, $j = 1,2$ and $i = 1,2$. The weight of an ATC code-word generated by weight-2 information sequence is given as follows

$$\begin{aligned} d &= 2 + k_1\left(z^1_{\min} - 2\right) + 2 + k_2\left(z^2_{\min} - 2\right) + 2 \\ &= 6 + k_1\left(z^1_{\min} - 2\right) + k_2\left(z^2_{\min} - 2\right). \end{aligned} \tag{8}$$

Assume that the positions of "1"s are denoted by f_1, f_2 for weight-2 information sequence $\mathbf{u}_1^2$. In interleaved information sequence $\mathbf{u}_2^2$ positions of pair of "1"s are assumed to be $\pi(f_1)$, $\pi(f_2)$. Let us consider a worst scenario for weight-2 information sequence such as

$$|f_1 - f_2| \bmod \varphi_1 = 0 \text{ and } |\pi(f_1) - \pi(f_2)| \bmod \varphi_2 = 0. \tag{9}$$

In such a situation the separation between the positions of "1"s in the information sequence after interleaver mapping is still a multiple of φ_2 as shown in Fig. 4. So for such cases finite number of "1"s are generated by RSC-2 encoder for parity sequence. It results in low weight ATC code-word and degrades the performance for ATC code.

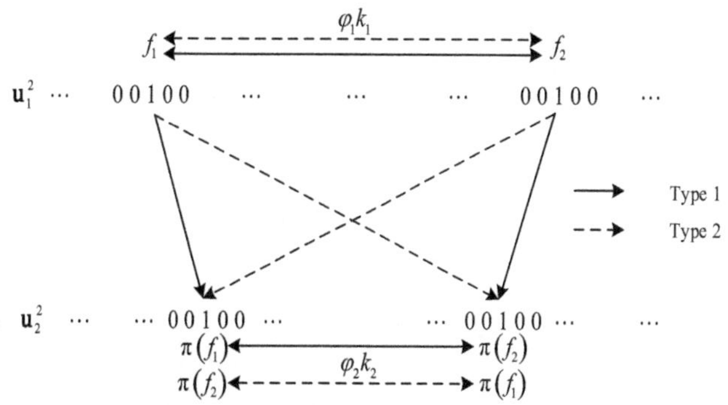

Fig. 4. Weight-2 information sequence is mapped to weight-2 sequence of similar pattern.

To prevent such outcomes, the interleaver must fulfil the mapping conditions given as

$$\text{If } |f_1 - f_2| \bmod \varphi_1 = 0 \text{ then } |\pi(f_1) - \pi(f_2)| \bmod \varphi_2 \neq 0. \tag{10}$$

However, not every weight-2 sequence is bad weight sequence. In order to remove only bad weight-2 sequences usually a parameter $d^{\alpha}_{\max}$ is specified which acts as a threshold, i.e. if the weight of code-word falls below this threshold than that weight-2 information sequence is removed by interleaving. The relative contribution integral that is used to find a suitable value of $d^{\alpha}_{\max}$ for a particular turbo code is given in [23]. Applying the parameter $d^{\alpha=2}_{\max}$ to (8) results in the following inequality

$$6 + k_1\left(z^1_{\min} - 2\right) + k_2\left(z^2_{\min} - 2\right) \leq d^{\alpha=2}_{\max}, \tag{11}$$

which can be written as

$$k_1\left(z^1_{\min} - 2\right) + k_2\left(z^2_{\min} - 2\right) \leq d^{\alpha=2}_{\max} - 6. \tag{12}$$

If $z^1_{\min} = z^2_{\min} = z_{\min}$ then (12) becomes

$$(k_1 + k_2) \leq \frac{d^{\alpha=2}_{\max} - 6}{(z_{\min} - 2)}. \tag{13}$$

If (9) and (12) are satisfied then the weight-2 information sequence $\mathbf{u}_1^2$ will always produce bad weight ATC code-word. The weight-3 information sequences which results in low weight ATC code-word can be removed by using S-random constraint. The design details of S-random interleaver are well documented in [21].

Now we consider weight-4 information sequence which is composed of compound error events, i.e. concatenation of two weight-2 information sequences as depicted in Fig. 5.

The polynomial representation of weight-4 information sequence $\mathbf{u}_1^4$ is given as

$$\mathbf{u}_1^4(D) = \left(1 + D^{\varphi_1 k'_1}\right) D^{\zeta_1} + \left(1 + D^{\varphi_1 k'_2}\right) D^{\zeta_2} \tag{14}$$

where $\zeta_2 > \zeta_1 + \varphi_1 k'_1$, $k'_1 = 1,2,3,\ldots$ and $k'_2 = 1,2,3,\ldots$. The polynomial representation of interleaved weight-4 information sequence $\mathbf{u}_1^4$ is given as

$$\mathbf{u}_2^4(D) = \left(1 + D^{\varphi_2 k'_3}\right) D^{\zeta_3} + \left(1 + D^{\varphi_2 k'_4}\right) D^{\zeta_4} \tag{15}$$

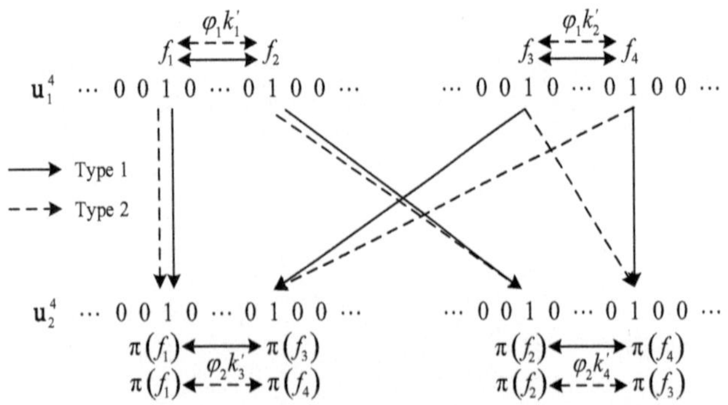

Fig. 5. Weight-4 information sequence is mapped to another weight-4 information sequence after passing through interleaver.

where $\zeta_4 > \zeta_3 + \varphi_2 k'_3$, $k'_3 = 1,2,3,\ldots$ and $k'_4 = 1,2,3,\ldots$. The weight of an ATC code-word generated by weight-4 information sequence is given as follows

$$d = 4 + \alpha(\mathbf{v}_1) + \alpha(\mathbf{v}_2), \qquad (16)$$

$$d = 12 + \left(k'_1 + k'_2\right)\left(z^1_{\min} - 2\right) + \left(k'_3 + k'_4\right)\left(z^2_{\min} - 2\right). \qquad (17)$$

Assume that the positions of "1"s are denoted by f_1, f_2, f_3 and f_4 for weight-4 information sequence $\mathbf{u}_1^4$, where $f_1 < f_2 < f_3 < f_4$. In interleaved weight-4 information sequence $\mathbf{u}_2^4$ positions of "1"s are assumed to be $\pi(f_1)$, $\pi(f_2)$, $\pi(f_3)$, and $\pi(f_4)$. The detrimental scenario for weight-4 information sequence is given as

$$\begin{cases} \text{If } |f_1 - f_2| \bmod \varphi_1 = 0 \text{ and } |f_3 - f_4| \bmod \varphi_1 = 0 \\ \text{then } |\pi(f_1) - \pi(f_3)| \bmod \varphi_2 = 0 \\ \text{and } |\pi(f_2) - \pi(f_4)| \bmod \varphi_2 = 0 \end{cases}$$

or

$$\begin{cases} \text{If } |f_1 - f_2| \bmod \varphi_1 = 0 \text{ and } |f_3 - f_4| \bmod \varphi_1 = 0 \\ \text{then } |\pi(f_1) - \pi(f_4)| \bmod \varphi_2 = 0 \\ \text{and } |\pi(f_2) - \pi(f_3)| \bmod \varphi_2 = 0. \end{cases} \qquad (18)$$

The URI again maps the weight-4 information sequence as a concatenation of two weight-2 error events as shown in Fig. 5, which results in bad weight parity sequence and generates low weight ATC code-word. Such concatenated events can be avoided if the interleaver satisfies the following mapping conditions

$$\begin{cases} \text{If } |f_1 - f_2| \bmod \varphi_1 = 0 \text{ and } |f_3 - f_4| \bmod \varphi_1 = 0 \\ \text{then } |\pi(f_1) - \pi(f_3)| \bmod \varphi_2 \neq 0 \\ \text{and } |\pi(f_2) - \pi(f_4)| \bmod \varphi_2 \neq 0 \end{cases}$$

or

$$\begin{cases} \text{If } |f_1 - f_2| \bmod \varphi_1 = 0 \text{ and } |f_3 - f_4| \bmod \varphi_1 = 0 \\ \text{then } |\pi(f_1) - \pi(f_4)| \bmod \varphi_2 \neq 0 \\ \text{and } |\pi(f_2) - \pi(f_3)| \bmod \varphi_2 \neq 0. \end{cases} \qquad (19)$$

All weight-2 information sequences that produce finite weight are not bad sequences in general. Similarly, all weight-4 information sequences are not bad sequences either. In order to remove only bad weight-4 sequences a parameter $d_{\max}^{\alpha=4}$ is specified. This parameter acts like a threshold and its functionality is similar to $d_{\max}^{\alpha=2}$ parameter. Applying this parameter $d_{\max}^{\alpha=4}$ to weight-4 sequences will result in the following inequality

$$\left(12 + \left(k'_1 + k'_2\right)\left(z^1_{\min} - 2\right) + \left(k'_3 + k'_4\right)\left(z^2_{\min} - 2\right)\right) \leq d_{\max}^{\alpha=4} \qquad (20)$$

which can be written as

$$\left(\left(k'_1 + k'_2\right)\left(z^1_{\min} - 2\right) + \left(k'_3 + k'_4\right)\left(z^2_{\min} - 2\right)\right) \leq d_{\max}^{\alpha=4} - 12. \qquad (21)$$

If $z^1_{\min} = z^2_{\min} = z_{\min}$ then (21) becomes

$$\left(k'_1 + k'_2 + k'_3 + k'_4\right) \leq \frac{d_{\max}^{\alpha=4} - 12}{\left(z_{\min} - 2\right)}. \qquad (22)$$

To check whether the weight-4 information sequence $\mathbf{u}_1^4$ is producing bad weight ATC code-word, the conditions in (18) and (21) should be satisfied.

The *S*-random constraint is sufficient for breaking the information sequences having $\alpha > 4$ [31]. Therefore, it is not required to consider each of them separately in the design of interleaver. Thus deploying CMI is considered as an optimum choice of interleaver as the mapping conditions for weight-2, 3 and 4 are satisfied in its design.

4. Spatial NLMS Adaptive Filter

The multiple receive antennas are usually deployed at the destination node that not only mitigates the harsh effects of fading channel but also achieves receive diversity. The source and relay nodes can be thought of mobile terminals while the destination node behaves like a base terminal. As the base terminals usually do not have power constraint therefore multiple receive antennas at the base terminal seems to be a quite reasonable assumption in our system model.

We have assumed an ideal coded-cooperation scenario, i.e. $S - R_\omega$ channel is ideal. Furthermore, the destination node does not have any prior knowledge of source to destination ($S - D$) and relays to destination ($R - D$) channels. Therefore, in order to estimate $S - D$ and $R - D$ channels, adaptive spatial NLMS [24] filters must be employed at the destination node. In this paper, two spatial NLMS filters are deployed to estimate the unknown aforementioned channels. As in the second time slot, all relay nodes transmit their signals to their corresponding receive antennas at the destination node. Since no co-channel interference is considered at the destination node during the second time slot, therefore employing a spatial NLMS filters is a reasonable option to estimate the aforementioned channels.

The *k*-th received sequence $\boldsymbol{\delta}^l_k$ during the *l*-th time slot for *L* receiving antennas are given by

$$\boldsymbol{\delta}_k^l = \mathbf{H}^l a_k^l + \mathbf{n}_k^l \quad (23)$$

where

$$\boldsymbol{\delta}_k^l = \left[\delta_{k,1}^l \delta_{k,2}^l \ldots \delta_{k,L}^l\right]^{\mathrm{T}}, \quad (24)$$

$$\mathbf{n}_k^l = \left[n_{k,1}^l n_{k,2}^l \ldots n_{k,L}^l\right]^{\mathrm{T}}, \quad (25)$$

and

$$\mathbf{H}^l = \left[h_1^l h_2^l \ldots h_L^l\right]^{\mathrm{T}} \quad (26)$$

where $\mathbf{H}^l$ is the memory-less Rayleigh block fading channel vector with zero mean and unit variance, a_k^l is the QPSK modulated symbol, $\mathbf{n}_k^l$ is the independent AWGN with zero mean and N_0 variance for the l-th time slot, $l = 1, 2$ for source and relay nodes, respectively.

The spatial NLMS transversal filter for the l-th time slot is shown in Fig. 6. The difference between the transmitted bit s_k^l of the training sequence and the filtered output y_k^l is termed as an error signal and is given as follows

$$\xi_k^l = s_k^l - \left(\mathbf{b}^l\right)^{\mathrm{T}} \boldsymbol{\delta}_k^l \quad (27)$$

where $\mathbf{b}^l = [b_1^l b_2^l \ldots b_L^l]^{\mathrm{T}}$ is the transversal filter vector.

The absolute square of error signal is given as follows

$$\left|\xi_k^l\right|^2 = \left|s_k^l\right|^2 - 2\,\mathrm{Re}\left[s_k^l\left\{\left(\mathbf{b}^l\right)^* \bar{\boldsymbol{\delta}}_k^l\right\}\right] + \left(\mathbf{b}^l\right)^*\left[\bar{\boldsymbol{\delta}}_k^l\left(\boldsymbol{\delta}_k^l\right)^{\mathrm{T}}\right]\mathbf{b}^l . \quad (28)$$

The gradient of $\left|\xi_k^l\right|^2$ w.r.t $\mathbf{b}$ is given by

$$\Delta_b\left(\left|\xi_k^l\right|^2\right) = 2\left[\bar{\boldsymbol{\delta}}_k^l\left(\boldsymbol{\delta}_k^l\right)^{\mathrm{T}}\right]\mathbf{b}^l - 2 s_k^l \bar{\boldsymbol{\delta}}_k^l . \quad (29)$$

The adaptive weights of filter are

$$\mathbf{b}_{k+1}^l = \mathbf{b}_k^l - \frac{\mu_k^l}{2}\Delta_b\left(\left|\xi_k^l\right|^2\right)\Big|_{\mathbf{b}^l=\mathbf{b}_k^l} . \quad (30)$$

To compute the updated weight vector of the filter, the step-size becomes

$$\mu_k^l = \frac{\rho}{c + \left\|\boldsymbol{\delta}_k^l\right\|^2} \quad (31)$$

where μ_k^l is the step-size parameter at the k-th sample, ρ is the normalized step-size ($0 < \rho < 2$) and c defines the safety factor.

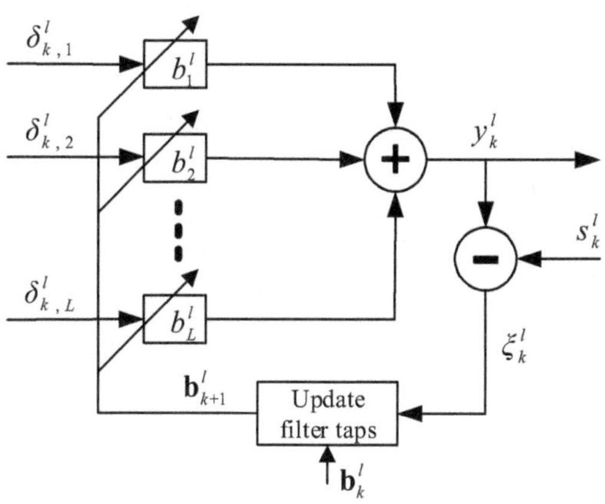

Fig. 6. Spatial NLMS adaptive filter for the l-th time slot.

The step-size is an important parameter of NLMS filter as it affects the convergence rate and the learning curve of a filter. The power of input signal changes with time, which causes the step-size to fluctuate. The small power signals will make the convergence rate slow, however, high power signals will cause convergence rate to speed up and eventually result in more error. Therefore, in order to cope in such a situation, the step-size parameter is adjusted according to the power of input signal and is usually taken as a normalized value. Sometimes $\boldsymbol{\delta}_k^l$ is very small and updating weight vector becomes difficult, so a small positive constant c known as a safety factor ($0 < c < 1$) is applied to overcome this uncertainty.

The number of iterations for estimating the channel depends on the length of the training sequence. Let K_{T} be the length of training sequence and the channel will be estimated after K_{T} iterations. The estimated channel is given by

$$\hat{\mathbf{H}}^l = \mathbf{b}_{K_{\mathrm{T}}}^l = \left[b_1^l\, b_2^l \ldots b_L^l\right]^{\mathrm{T}} \quad (32)$$

where $l = 1, 2$ for S–D and R–D channels, respectively.

5. Soft Demodulation

Soft-demodulation for MIMO is proposed in [32]. The soft-demodulator with multi-receive antennas and joint iterative SISO decoder are connected in serial concatenation. The soft-demodulator takes a-priori LLRs of coded bits and channel observations $\boldsymbol{\delta}_k^l$. Then it generates a-posteriori LLRs λ_{Di}^l for each coded bit under the assumption that the channel observations are independent of each other. For the i-th element of the k-th transmitted bit-vector $q_{k,i}$, the soft-demodulator calculates

$$\lambda_{D,k,i}^l = \ln\frac{P\left\{q_{k,i} = +1 \,|\, \boldsymbol{\delta}_k^l\right\}}{P\left\{q_{k,i} = -1 \,|\, \boldsymbol{\delta}_k^l\right\}} . \quad (33)$$

For simplicity, we remove k from the subscript and l from the superscript. Using Bayes' rule, (33) can be written as

$$\lambda_{D,i} = \ln\frac{\sum_{\Psi_{i,+1}} p(\boldsymbol{\delta} \,|\, \mathbf{q})\, p(\mathbf{q})}{\sum_{\Psi_{i,-1}} p(\boldsymbol{\delta} \,|\, \mathbf{q})\, p(\mathbf{q})} \quad (34)$$

where $\boldsymbol{\Psi} = \{\mathbf{q} \in \{-1, +1\}^2\}$ represents a set of all possible combinations of transmitted bit-vectors and $\Psi_{i,\mp 1} = \{\mathbf{q} \in \Psi \,|\, q_i = \pm 1\}$ denotes the set of bit vectors $\mathbf{q}$ having $q_i = \pm 1$. The likelihood function $p(\boldsymbol{\delta} \,|\, \mathbf{q})$ in (34) can be expressed as

$$p(\boldsymbol{\delta}|\mathbf{q}) \propto \exp\left(-\left\|\boldsymbol{\delta} - \mathbf{H}\mathcal{M}(\mathbf{q})\right\|^2 / 2\sigma^2\right) . \quad (35)$$

Since the elements of $\mathbf{q}$ are considered as independent, the a-priori probability $p(\mathbf{q})$ can be calculated as follows

$$p(\mathbf{q}) \approx \prod_i p(q_i) .$$

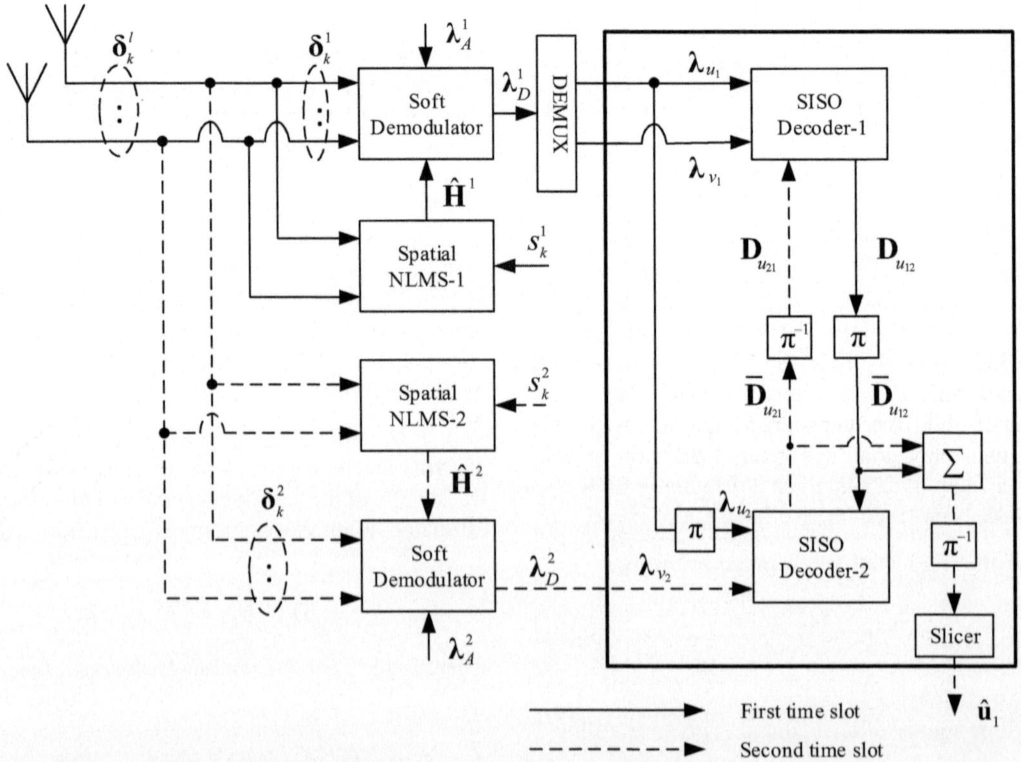

Fig. 7. Joint iterative SISO decoding for DATC scheme.

The a-priori LLR is defined as

$$\lambda_{A,i} = \ln\left(\frac{p(q_i = +1)}{p(q_i = -1)}\right). \quad (37)$$

If $\boldsymbol{\lambda}_A$ represents the vector of a-priori LLR, then $p(\mathbf{q})$ can be represented as

$$\prod_i p(q_i) \propto \exp(\boldsymbol{\lambda}_A^T \mathbf{q}/2). \quad (38)$$

Using (35) and (38), the expression in (34) can be written as

$$\lambda_{D,i} \cong \ln \frac{\sum_{\Psi_{i,+1}} \exp(-W(\mathbf{q})/(2\sigma^2))}{\sum_{\Psi_{i,-1}} \exp(-W(\mathbf{q})/(2\sigma^2))} \quad (39)$$

where

$$W(\mathbf{q}) \triangleq \|\boldsymbol{\delta} - \mathbf{H}\mathcal{M}(\mathbf{q})\|_2^2 - \sigma^2 \boldsymbol{\lambda}_A^T \mathbf{q}. \quad (40)$$

The demodulator uses (39) to generate LLR for coded bits. The complexity of the demodulator increases with increase in number of elements in the set $\Psi_{i,\mp 1}$.

6. Joint Iterative SISO Decoding for DATC Scheme

The joint iterative SISO decoding for DATC scheme is shown in Fig. 7. The SISO module for iterative decoding of serial and parallel concatenated turbo codes has been discussed in literature [33]. The joint decoder consists of two different SISO decoders. The SISO decoder consists of three inputs and one output. A-priori probabilities of systematic bits and parity bits are given as input to SISO decoder which results in extrinsic information as an output.

During the first time slot, the soft demodulator generates LLRs $\boldsymbol{\lambda}_D^1$, these LLRs are soft-bits corresponding to the systematic bits and parity bits which were transmitted from the source node. The de-multiplexer breaks the LLRs $\boldsymbol{\lambda}_D^1$ into two separate streams of systematic LLRs $\boldsymbol{\lambda}_{u_1}$ and parity LLRs $\boldsymbol{\lambda}_{v_1}$. These LLRs are received by SISO-decoder-1 which outputs the extrinsic information $\mathbf{D}_{u_{12}}$. The extrinsic information of systematic bits $\mathbf{D}_{u_{12}}$ is interleaved and given as a-priori input to SISO-decoder-2.

In the second time slot, the LLRs $\boldsymbol{\lambda}_D^2$ are generated by the soft demodulator. These LLRs are soft-bits corresponding to the parity bits of RSC-2 encoder which were transmitted from relay node. The extrinsic information $\bar{\mathbf{D}}_{u_{21}}$ is deinterleaved to form extrinsic information $\mathbf{D}_{u_{21}}$. This extrinsic information $\mathbf{D}_{u_{21}}$ is given back to SISO-decoder-1. The extrinsic information is interchanged between SISO-decoders for a specific number of iterations. In order to get the estimated information sequence $\hat{\mathbf{u}}_1$, the extrinsic information $\bar{\mathbf{D}}_{u_{21}}$ and $\bar{\mathbf{D}}_{u_{12}}$ are summed up and passed through a slicer after de-interleaving.

7. Simulated Results and Discussion

This section discusses the simulation results of ATC and the proposed DATC coded-cooperative scheme with

CMI and URI. The ATC scheme is simulated over AWGN and Rayleigh fading channels while the DATC scheme is simulated over Rayleigh fading channels with multi-receive antenna and unknown CSI at the destination node. In order to have fair comparison between the coded-cooperative scheme and non-cooperative scheme with CMI, overall code rate at the destination node is kept the same, i.e. $R_c = 1/3$. The generator matrix for ATC and DATC is **G**$(13/15, 11/15)_8$. The length of CMI and URI is $K = 1024$. The channels are assumed to be slow Rayleigh fading channels. For all coded-cooperative simulations, the S-R_ω channel is assumed to be ideal ($\mathrm{SNR}_{SR_\omega} = \infty$). The relay nodes are placed near to the destination node in order to get the advantage of SNR over the source node. In case of two relay nodes, one relay node has given 1 dB gain in SNR while the other relay node has given 2 dB gain in SNR over the source node. Furthermore, in the case of three relay nodes, two relay nodes have given 2 dB gain in SNR while one relay node has given 1 dB gain in SNR over the source node. Moreover, spatial NLMS adaptive filters, soft demodulators and joint SISO decoder are employed at the destination node. The SNR per bit is used for all simulations.

7.1 Performance Comparison between Conventional STC and ATC over AWGN Channel

The BER performance comparison between conventional STC and proposed ATC having generator matrices **G**$(11/15, 11/15)_8$ and **G**$(13/15, 11/15)_8$, respectively over AWGN channel with QPSK modulation is shown in Fig. 8. The information block length is considered to be $K = 400$ bits. The BER curve of ATC with URI clearly shows that it outperforms conventional STC with URI over the whole SNR range. The performance gain of ATC with URI at BER $\approx 3\times10^{-6}$ over conventional STC with URI is about 0.5 dB.

It can be seen that by eliminating low weight codewords from ATC using CMI, the performance in terms of BER improves significantly. ATC with CMI achieves approx. 0.4 dB gain as compared to ATC with URI at BER $\approx 3\times10^{-6}$. The performance gain of ATC is also improved by using CMI in comparison to conventional STC with URI. At BER $\approx 3\times10^{-6}$, ATC with CMI has the performance gain of about 0.9 dB over conventional STC with URI.

7.2 Performance Comparison between CMI and URI for ATC Non-Cooperation with Multi-Receive Antennas

In this subsection, we consider the ATC non-cooperation with CMI and URI under perfect and imperfect CSIs and different number of receive antennas. The spatial NLMS adaptive filter is adopted for channel estimation at the destination node. The length of training sequence is 40

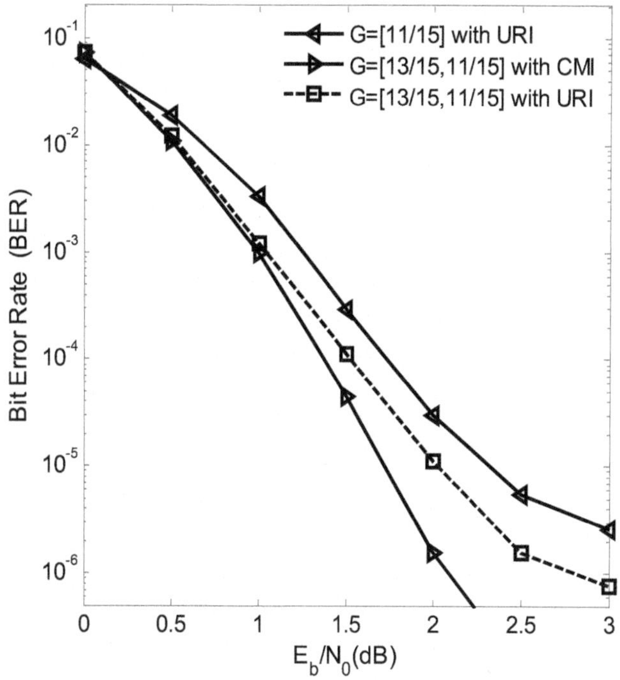

Fig. 8. BER performance of conventional STC and ATC over AWGN channel, $K = 400$ and 8 decoding iterations.

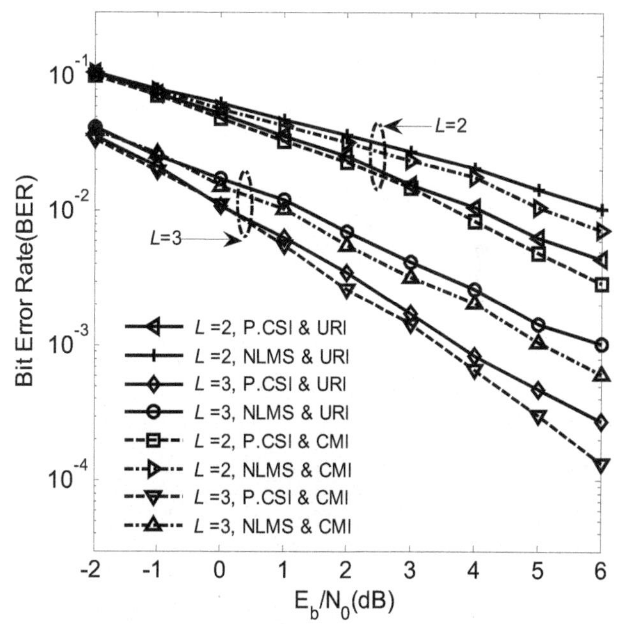

Fig. 9. BER performance of ATC non-cooperation with CMI and URI over perfect / imperfect CSI (Rayleigh fading), 8 decoding iterations, $K = 1024$ and different number of receive antennas at the destination.

for spatial NLMS filter. Safety parameter has constant value 0.05 and step-size parameter μ_k at the k-th sample is $0.1/(0.05 + \|\boldsymbol{\delta}_k\|^2)$.

The BER performance comparison between CMI and URI for ATC non-cooperation with multiple receive antennas is shown in Fig. 9. It can be observed that there is a narrow gap between BER performances of CMI and URI for ATC non-cooperation under perfect CSI. At SNR of 5 dB, the BER performance under CMI is 5×10^{-3} and under

URI is 6×10^{-3} for two receive antennas. Similarly at SNR of 5 dB, the BER performances are 3×10^{-4} and 4.5×10^{-4} for CMI and URI, respectively in case of three receive antennas.

It can be seen that there is also a small gap between BER performances of CMI and URI for ATC non-cooperation under estimated CSI by spatial NLMS filter. For instance, the performance gain of 0.4 dB is achieved at BER $\approx 1.5\times10^{-2}$ for two receive antennas. Similarly at BER $\approx 1.5\times10^{-3}$, the performance gain of 0.5 dB is observed for the case of three receive antennas. The BER performance gap between perfect CSI and estimated CSI for various receive antennas are under acceptable range. This gap can further be minimized by increasing the length of training sequence but it adds some latency to the system.

The receive diversity is increased by increasing the receive antennas at the destination node. The gain ranges from 4 to 4.5 dB between two and three receive antennas over perfect CSI at BER $\approx 4\times10^{-3}$. Similarly the gain ranges from 4 to 5 dB for estimated CSI between two and three receive antennas at BER $\approx 10^{-2}$.

7.3 Performance Comparison of DATC Coded-Cooperation ($SNR_{SR_\omega}=\infty$) and ATC Non-Cooperation with CMI and Multi-Receive Antennas over Perfect and Imperfect CSI

In this subsection, the simulated results of DATC in coded-cooperation scenario incorporating CMI are presented. As evident from the Monte Carlo simulation results shown in Fig. 10, the proposed scheme outperforms its corresponding non-cooperation scheme under identical conditions.

For the case of two relay nodes, DATC coded-cooperation scheme gives 2.7 dB performance gain over ATC non-cooperation scheme with two receive antennas at BER $\approx 3\times10^{-3}$ over perfect CSI. Similarly the BER performance gain of 3.5 dB is observed for three relay nodes at BER $\approx 0.5\times10^{-4}$ over perfect CSI.

In case of estimated CSI for multi-relay DATC coded cooperation scheme, two spatial NLMS adaptive filters are adopted for channel estimation at the destination node. The length of training sequence, safety and step-size parameters are all kept the same as for coded non-cooperation scheme. At BER $\approx 7\times10^{-3}$, under imperfect CSIs and two receive antennas, the two relays DATC coded cooperation scheme achieves the performance gain of 3 dB over ATC non-cooperation scheme. Moreover, the gain of 3.3 dB is observed for three relay nodes at BER $\approx 6\times10^{-4}$.

The BER performance of three relays DATC scheme is better than two relays DATC scheme under identical conditions. Increasing the number of relay nodes improves the information reliability by providing diversity and hence the BER performance is improved. At SNR of 6 dB, the

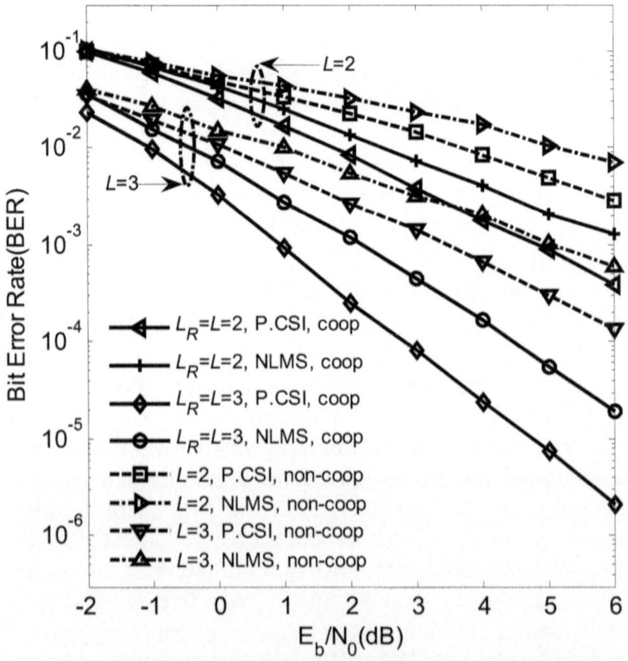

Fig. 10. BER performance of multi-relay DATC ideal coded-cooperation and non-cooperation with CMI over perfect / imperfect CSI (Rayleigh fading), eight decoding iterations, $K = 1024$ and different number of receive antennas L at the destination.

BER performance of multi-relay DATC scheme with two and three relay nodes are 4×10^{-4} and 2×10^{-6}, respectively over perfect CSI. Similarly in case of imperfect CSI, the BER performance 0.5×10^{-3} and 2×10^{-5} for two and three relay nodes, respectively is observed at 6 dB.

The better BER performance in the case of multi-relays is based on the facts that there is an increase in SNR due to the placement of additional relays near to destination. Moreover, deploying multi-relays causes additional spatial diversity gain. However, it is worth mentioning that the proposed coded-cooperation scheme introduces extra latency in the whole communication system as compared to non-cooperation scheme. In addition it also requires extra spatial NLMS filter which results in increased hardware complexity at the destination.

8. Conclusion

In this paper, we have analyzed the BER performance of ATC $\mathbf{G}(13/15, 11/15)_8$ on the basis of CMI and URI over AWGN channel. It is concluded that the performance of ATC with CMI outperforms ATC with URI under identical conditions. The proposed ATC with CMI is also compared with its counterpart STC $\mathbf{G}(11/15, 11/15)_8$. The simulated results have shown that the performance of ATC is quite superior over conventional turbo code. Moreover, the proposed ATC is also examined under perfect and imperfect CSI with multi-receive antennas over the Rayleigh block fading channel on the basis of CMI and URI. Our proposed schemes incorporate both coding and diversity gains. In addition, it is also reinforced by Monte Carlo simulations that CMI is always an optimum choice of interleaver in the design of turbo code. The CMI is further deployed in coded-cooperation scenario with multiple relays under the proposed encoding schemes. Finally, the BER performance of the coded-cooperation under the proposed encoding scheme is analyzed, and compared with its corresponding non-cooperation counterpart scheme. The coded-cooperation scheme outperforms the non-cooperation scheme under identical conditions by a significant gain of 2.7 to 3.5 dB.

Acknowledgments

The authors would like to thank the Editor and anonymous reviewers for their precious suggestions, comments and reviews that help in improving the quality of this paper.

References

[1] BARRY, J. R., LEE, E. A., MESSERSCHMITT, D. G. *Digital Communication.* 3rd ed. Springer US, 2004. ISBN: 978-1-4615-0227-2

[2] FOSCHINI, G. J., GANS, M. J. On limits of wireless communications in a fading environment when using multiple antennas. *Wireless Personal Communications,* 1998, vol. 6, no. 3, p. 311–335. DOI: 10.1023/A: 1008889222784

[3] NOSRATINIA, A., HUNTER, T. E., HEDAYAT, A. Cooperative communication in wireless networks. *IEEE Communications Magazine,* 2004, vol. 42, no. 10, p. 74–80. DOI: 10.1109/mcom.2004.1341264

[4] VAN DER MEULEN, E. C. Three-terminal communication channels. *Advances in Applied Probability,* 1971, vol. 3, no. 1, p. 120–154. DOI: 10.2307/1426331

[5] LANEMAN, J. N., WORNELL, G. W., TSE, D. N. An efficient protocol for realizing cooperative diversity in wireless networks. In *Proceedings of IEEE International Symposium on Information Theory,* 2001, p. 294. DOI: 10.1109/ISIT.2001.936157

[6] AVESTIMEHR, A. S., DIGGAVI, S. N., DAVID, N. Wireless network information flow: A deterministic approach. *IEEE Transactions on Information Theory,* 2011, vol. 57, no. 4, p. 1872 to 1905. DOI: 10.1109/TIT.2011. 2110110

[7] COVER, T., GAMAL, A. E. Capacity theorems for the relay channel. *IEEE Transactions on Information Theory,* 1979, vol. 25, no. 5, p. 572–584. DOI: 10.1109/TIT.1979.1056084

[8] SENDONARIS, A., ERKIP, E., AAZHANG, B. User cooperation diversity. Part I. System description. *IEEE Transactions on Communications,* 2003, vol. 51, no. 11, p. 1927–1938. DOI: 10.1109/TCOMM.2003.818096

[9] HUNTER, T. E., NOSRATINIA, A. Cooperation diversity through coding. In *Proceedings of IEEE International Symposium on Information Theory.* Lausanne (Switzerland), 2002, p. 220. DOI: 10.1109/ISIT.2002.1023492

[10] SOLIMAN, T. H., YANG, F., EJAZ, S. Interleaving gains for receive diversity schemes of distributed turbo codes in wireless half-duplex relay channels. *Radioengineering,* 2015, vol. 24, no. 2, p. 481–488. DOI: 10.13164/re.2015. 0481

[11] HU, J., DUMAN, T. M. Low density parity check codes over wireless relay channels. *IEEE Transactions on Wireless Communications,* 2007, vol. 6, no. 9, p. 3384–3394. DOI: 10.1109/TWC.2007.06083

[12] ELFITURI, M., HAMOUDA, W., GHRAYEB, A. A convolutional-based distributed coded cooperation scheme for relay channels. *IEEE Transactions on Vehicular Technology,* 2009, vol. 58, no. 2, p. 655–669. DOI: 10.1109/TVT.2008.927033

[13] EJAZ, S.,YANG, F.-F. Jointly optimized Reed-Muller cCodes for multi-level multi-relay coded-cooperative VANETS. *IEEE Transactions on Vehicular Technology,* 2017, vol. 66, no. 5, p. 4017–4028. DOI: 10.1109/TVT.2016.2604320

[14] SOLIMAN, T. H., YANG, F. Cooperative Punctured Polar Coding (CPPC) scheme based on Plotkin's construction. *Radioengineering,* 2016, vol. 25, no. 3, p. 482–489. DOI: 10.13164/re.2016.0482

[15] BRAVO-SANTOS, A. Polar codes for the Rayleigh fading channel. *IEEE Communications Letters,* 2013, vol. 17, no. 12, p. 2352–2355. DOI: 10.1109/LCOMM.2013.111113. 132103

[16] EJAZ, S., YANG, F.-F. Turbo codes with modified code matched interleaver for coded-cooperation in half-duplex wireless relay networks. *Frequenz,* 2015, vol. 69, p. 171–184. DOI: 10.1515/freq-2014-0072

[17] KIM, H. Coding and modulation techniques for high spectral efficiency transmission in 5G and satcom. In *23rd European Signal Processing Conference (EUSIPCO),* 2015, p. 2746–2750. DOI: 10.1109/EUSIPCO.2015. 7362884

[18] PEREZ, L. C., SEGHERS, J., COSTELLO, D. J. A distance spectrum interpretation of turbo codes. *IEEE Transactions on Information Theory,* 1996, vol. 42, no. 6, p. 1698–1709. DOI: 10.1109/18.556666

[19] HALL, E. K., WILSON, S. G. Convolutional interleavers for stream-oriented parallel concatenated convolutional codes. In *IEEE International Symposium on Information Theory,* 1998, p. 33. DOI: 10.1109/ISIT.1998.708613

[20] RAMSEY, J. Realization of optimum interleavers. *IEEE Transactions on Information Theory,* 1970, vol. 16, no. 3, p. 338 to 345. DOI: 10.1109/TIT.1970.1054443

[21] DIVSALAR, D., POLLARA, F. Turbo codes for PCS applications. In *IEEE International Conference on Communications, ICC'95 Seattle,Gateway to Globalization,* 1995, vol. 1, p. 54–59. DOI: 10.1109/ICC.1995.525138

[22] FENG, W., YUAN, J., VUCETIC, B. S. A code-matched interleaver design for turbo codes. *IEEE Transactions on Communications,* 2002, vol. 50, no. 6, p. 926–937. DOI: 10.1109/TCOMM.2002.1010612

[23] RAMASAMY, K., SIDDIQI, M. U. A new class of asymmetric turbo code for 3G systems. *AEU-International Journal of Electronics and Communications,* 2006, vol. 60, no. 6, p. 447–458. DOI: 10.1016/j.aeue.2005.09.007

[24] JOY, A., CHAKKA, V. Performance comparison of LMS/NLMS based transceiver filters for MIMO two-way relaying scheme. In *International Conference on Communications and Signal Processing (ICCSP).* 2011, p. 105–107. DOI: 10.1109/ICCSP.2011.5739411

[25] CATTIVELLI, F. S., SAYED, A. H. Analysis of spatial and incremental LMS processing for distributed estimation *IEEE Transactions on Signal Processing,* 2011, vol. 59, no. 4, p. 1465 to 1480. DOI: 10.1109/TSP.2010.2100386

[26] GUI, G., XU, L., SHAN, L., et al. Adaptive MIMO channel estimation using sparse variable step-size NLMS algorithms. In *IEEE International Conference on Communication Systems (ICCS).* 2014, p. 605–609. DOI: 10.1109/ICCS.2014.7024874

[27] BERROU C., GLAVIEUX, A. Near optimum error correcting coding and decoding: Turbo-codes. *IEEE Transactions on Communications,* 1996, vol. 44, no. 10, p. 1261–1271. DOI: 10.1109/26.539767

[28] TAKESHITA, O. Y., COLLINS, O. M., MASSEY, P. C., et al. A note on asymmetric turbo-codes. *IEEE Communications Letters,* 1999, vol. 3, no. 3, p. 69–71. DOI: 10.1109/4234.752905

[29] ZHAO, B., VALENTI, M. C. Distributed turbo coded diversity for relay channel. *Electronics Letters,* 2003, vol. 39, no. 10, p. 786 to 787. DOI: 10.1049/el:20030526

[30] ZHANG, Z., DUMAN, T. M. Capacity-approaching turbo coding and iterative decoding for relay channels. *IEEE Transactions on Communications,* 2005, vol. 53, no. 11, p. 1895–1905. DOI: 10.1109/TCOMM.2005.858654

[31] VUCETIC, B., YUAN, J. *Turbo Codes: Principles and Applications,* vol. 559: Springer US, 2012. ISBN: 978-1-4615-4469-2

[32] NEKUII, M., KISIALIOU, M., DAVIDSON, T. N., et al. Efficient soft-output demodulation of MIMO QPSK via semidefinite relaxation. *IEEE Journal of Selected Topics in Signal Processing,* 2011, vol. 5, no. 8, p. 1426–1437. DOI: 10.1109/JSTSP.2011.2172772

[33] BENEDETTO, S., DIVSALAR, D., MONTORSI, G., et al. A soft-input soft-output APP module for iterative decoding of concatenated codes. *IEEE Communications Letters,* 1997, vol. 1, no. 1, p. 22–24. DOI: 10.1109/4234.552145

About the Authors ...

Shoaib MUGHAL received his B.S. degree in Electronic Engineering from Sir Syed University of Engineering and Technology, Karachi, Pakistan, in 2008. In 2011, he obtained his M.E. degree in Telecommunication at NED University of Engineering and Technology, Karachi, Pakistan. He worked as a lecturer in the Department of Electronic Engineering at Sir Syed University of Engineering and Technology, Karachi, Pakistan, for three years. He is currently a Ph.D. research scholar at the College of Electronic and Information Engineering, Nanjing University of Aeronautics and Astronautics, China. His major research interests are multiple-input multiple-output systems, channel coding, coded-cooperative diversity and coded-modulation techniques.

Fengfan YANG received the M.Sc. and Ph.D. degrees from the Northwestern Polytechnical University and Southeast University, China in 1993, and 1997, respectively, all in Electronic Engineering. He has been with the College of Information Science and Technology, Nanjing University of Aeronautics and Astronautics since May 1997. From October 1999 to May 2003, he was a research associate at the Centre for Communication Systems Research, University of Surrey, UK, and Dept. of Electrical and Computer Engineering, McGill University, Canada. His major research interests are information theory, channel coding and their applications for mobile and satellite communications.

Saqib EJAZ received his B.Sc. (Hons.) degree in Electrical Engineering from the University of Engineering and Technology, Taxila, Pakistan, in 2006. Later, he did his M.S. Electronic Engineering from Ghulam Ishaq Khan Institute of Engineering Sciences and Technology (GIKI), KPK, Pakistan, in 2008. In 2015, he completed his Ph.D. degree from the College of Electronic and Information Engineering, Nanjing University of Aeronautics and Astronautics, China. He has more than 20 international journal/conference publications. His research interests include information theory, channel coding, multi-input multi-output systems, cooperative diversity, coded-modulation techniques, and spread spectrum systems.

Syed Hassan AHMED completed his B.S. degree in Electronic Engineering from Sir Syed University of Engineering and Technology, Karachi, Pakistan, in 2008. He received his M.S. degree in Wireless Communication at Brunel University, London, England, in 2009. He is currently pursuing his Ph.D. degree in the College of Electronic and Information Engineering, Nanjing University of Aeronautics and Astronautics, Nanjing, China. His research interests are channel coding and coded-cooperative communications.

Rahim UMAR completed his B.Sc. degree in Electrical Engineering at the University of Engineering and Technology, Peshawar, Pakistan, in 2007. He worked as a lecturer in the Electrical Engineering Department at CECOS University Peshawar, Pakistan, for two years. In 2011, he completed his MS degree in Electrical Engineering, specializing in signal processing and wave propagation from Linnaeus University Växjö, Sweden. In late 2011 he joined Ghulam Ishaq Khan Institute of Engineering Science and Technology where he worked as Research Associate in the Department of Engineering Sciences. Presently, he is working as a Ph.D. research scholar in the field of channel coding at Nanjing University of Aeronautics and Astronautics, China. His research interests are signal processing and channel coding.

Permissions

All chapters in this book were first published in RADIOENGINEERING, by Spolecnost pro radioelektronicke inzenyrstvi[CZ]; hereby published with permission under the Creative Commons Attribution License or equivalent. Every chapter published in this book has been scrutinized by our experts. Their significance has been extensively debated. The topics covered herein carry significant findings which will fuel the growth of the discipline. They may even be implemented as practical applications or may be referred to as a beginning point for another development.

The contributors of this book come from diverse backgrounds, making this book a truly international effort. This book will bring forth new frontiers with its revolutionizing research information and detailed analysis of the nascent developments around the world.

We would like to thank all the contributing authors for lending their expertise to make the book truly unique. They have played a crucial role in the development of this book. Without their invaluable contributions this book wouldn't have been possible. They have made vital efforts to compile up to date information on the varied aspects of this subject to make this book a valuable addition to the collection of many professionals and students.

This book was conceptualized with the vision of imparting up-to-date information and advanced data in this field. To ensure the same, a matchless editorial board was set up. Every individual on the board went through rigorous rounds of assessment to prove their worth. After which they invested a large part of their time researching and compiling the most relevant data for our readers.

The editorial board has been involved in producing this book since its inception. They have spent rigorous hours researching and exploring the diverse topics which have resulted in the successful publishing of this book. They have passed on their knowledge of decades through this book. To expedite this challenging task, the publisher supported the team at every step. A small team of assistant editors was also appointed to further simplify the editing procedure and attain best results for the readers.

Apart from the editorial board, the designing team has also invested a significant amount of their time in understanding the subject and creating the most relevant covers. They scrutinized every image to scout for the most suitable representation of the subject and create an appropriate cover for the book.

The publishing team has been an ardent support to the editorial, designing and production team. Their endless efforts to recruit the best for this project, has resulted in the accomplishment of this book. They are a veteran in the field of academics and their pool of knowledge is as vast as their experience in printing. Their expertise and guidance has proved useful at every step. Their uncompromising quality standards have made this book an exceptional effort. Their encouragement from time to time has been an inspiration for everyone.

The publisher and the editorial board hope that this book will prove to be a valuable piece of knowledge for researchers, students, practitioners and scholars across the globe.

List of Contributors

Mahmoud A. Abdalla
Dept. of Electronic Engineering, MTC College Cairo, Egypt

Ahmed A. Ibrahim
Dept. of Electrical Engineering, Faculty of Engineering, Minia University, Egypt
Inst. of Micro- and Sensor Systems, University of Magdeburg, Germany

Hafsa Anam, Ayesha Habib and Syeda Irum Jafri
ACTSENA Research Group, University of Engineering and Technology, 47050- Taxila, Pakistan

Yasar Amin
ACTSENA Research Group, University of Engineering and Technology, 47050- Taxila, Pakistan
iPack VINN Excellence Center, Royal Institute of Technology, SE-16440, Stockholm, Sweden

Hannu Tenhunen
iPack VINN Excellence Center, Royal Institute of Technology, SE-16440, Stockholm, Sweden
TUCS, University of Turku, Turku-20520, Finland

Ting Wang, Yongjun Zhao, Shiwen Chen and Kunfan Zhang
Zhengzhou Institute of Information Science and Technology, Science Road 62, Zhengzhou, People's Republic of China

Ting Wang, Yongjun Zhao and Jiantao Wang
Zhengzhou Institute of Information Science and Technology, Science Road 62, Zhengzhou, People's Republic of China

Hao Wang and Benben Wen
College of Electronic and Information Engineering, Nanjing University of Aeronautics and Astronautics, Nanjing, China

Weiye Xu
School of Communication Engineering, Nanjing Institute of Technology, Nanjing, China

Xiangbin Yu
College of Electronic and Information Engineering, Nanjing University of Aeronautics and Astronautics, Nanjing, China
National Mobile Communications Research Laboratory, Southeast University, Nanjing, China

Supakit Kawdungta, Puwanai Jaibanauem and Rattiya Pongga
Faculty of Engineering, Rajamangala University of Technology Lanna, Chiang Mai 50300, Thailand

Chuwong Phongcharoenpanich
Faculty of Engineering, King Mongkut's Institute of Technology Ladkrabang, Bangkok 10520, Thailand

Ahmed Ibrahim
Dept. of Electrical Engineering, Faculty of Engineering, Minia University, Minia, Egypt
Institute of Micro- and Sensor Systems, University of Magdeburg, Germany

Wael Ali
Dept. of Electronics & Comm. Engineering, College of Engineering, Arab Academy for Science, Technology and Maritime Transport (AASTMT), Alexandria, Egypt

Jan Machac
Dept. of Electromagnetic Field, Faculty of Electrical Engineering, Czech Technical University in Prague, Czech Republic

Chao-Ming Luo, Jing-Song Hong and Muhammad Amin
School of Physical Electronics, University of Electronic Science and Technology of China, No. 4, Section 2, Chengdu 610054, China

Ayesha Habib, Sohaira Ansar, Adeel Akram and Muhammad Awais Azam
ACTSENA Research Group, University of Engineering and Technology, 47050- Taxila, Pakistan

Asha Verma and Manoj Singh Parihar
Dept. of Electronics & Communication Engineering, PDPM-Indian Inst. of Information Technology, Design and Manufacturing Jabalpur (PDPM-IIITDMJ), Jabalpur, India

Cristiano Tomassoni
Department of Engineering, University of Perugia, Perugia, Italy

Maurizio Bozzi
Department of Electrical, Computer and Biomedical Engineering, University of Pavia, Pavia, Italy

Wei-Dong Ma, Guang-Ming Wang, Ya-Wei Wang and Bin-Feng Zong
Air and Missile Defense College, Air Force Engineering University, Shaanxi Province 710051, China

Ming Du, Jun Xu and Xiao Ding
School of Physical Electronics, University of Electronic Science and Technology of China, Chengdu, China

Yuliang Dong
National Institute of Measurement and Testing Technology of China, Chengdu, China

Katherine Siakavara, Sotirios Goudos, Argiris Theopoulos and John Sahalos
Dept. of Physics, Aristotle University of Thessaloniki, 54124 Thessaloniki, Greece

Sonia Sharma and Chandra Charu Tripathi
Dept. of Electronics and Communication Engineering, UIET, Kurukshetra University, Kurukshetra, India

Raafia Irfan and Haroon Ur Rasheed
Dept. of Electrical Engineering, Pakistan Inst. of Eng. and Applied Sciences (PIEAS), Nilore, Islamabad-45650, Pakistan

Waqas Ahmed Toor
Dept. of Electrical Engineering, Capital University of Science and Technology (CUST), Express Way, Zone 5, Islamabad-45650, Pakistan

Murli Manohar
Dept. of Electronics and Communications Engineering, IIIT Manipur, 795002 Imphal, India

Rakhesh Singh Kshetrimayum and Anup Kumar Gogoi
Dept. of Electronics and Electrical Engineering, IIT Guwahati, 781039 Guwahati, India

Kun Wei, Jian-Ying Li and Ling Wang Rui Xu
School of Electronics and Information, Northwestern Polytechnical University, 710072 Xi'an, Shannxi, China

Tomislav Shuminoski and Toni Janevski
Ss. Cyril and Methodius University, Faculty of Electrical Engineering and Information Technologies, Karpos 2 bb, 1000 Skopje, Republic of Macedonia

Mubashir Ul Hassan, Farzana Arshad and Syeda Iffat Naqvi
ACTSENA Research Group, University of Engineering and Technology (UET), Taxila, 47050, Pakistan

Navid Daryasafar and Mohammad Naser-Moghadasi
Faculty of Engineering, Science and Research Branch, Islamic Azad University, Tehran, Iran

Ramazanali Sadeghzadeh
Faculty of Electrical Engineering, K. N. Toosi University of Technology, Tehran, Iran

Pongsathorn Chomtong
Dept. of Teacher Training in Electrical Engineering, Faculty of Technical Education

Prayoot Akkaraekthalin
Dept. of Electrical and Computer Engineering, Faculty of Engineering King Mongkut's University of Technology North Bangkok, 1518 Pracharat 1 Road, Bangsue, Bangkok, Thailand

Soroush Veisee and Shahrooz Asadi
Dept. of Electrical Engineering, University of Shahid Beheshti, Tehran, Iran

Issa Elfergani
Instituto de Telecomunicações, Aveiro, Portugal

Jonathan Rodriguez
Instituto de Telecomunicações, Aveiro, Portugal
University of South Wales, Pontypridd, CF37 1DL , UK

Fathi Abdussalam and Read Abd-Alhameed
Antennas & Applied Electromagnetics Research Group, School of Engineering and Informatics, Univ. of Bradford, Bradford, BD7 1DP, UK

Chan H See
School of Engineering, University of Bolton, Bolton, BL3 5AB, UK

Shoaib Mughal, Fengfan Yang, Saqib Ejaz, Syed Hassan Ahmed and Rahim Umar
College of Electronic and Information Engg., Nanjing Univ. of Aeronautics and Astronautics, Nanjing 210016, China

Index